1996 - 1997
CLASSIC CAR
BUYER'S GUIDE

1996 - 1997
CLASSIC CAR
BUYER'S GUIDE

Rob de la Rive Box & Ton Lohman
Edited by Chris Rees

Published 1996 by
Bay View Books Ltd
The Red House
25-26 Bridgeland Street
Bideford, Devon EX39 2PZ

© Copyright Uitgeverij Elmar B.V. 1996
English edition © Copyright Bay View Books Ltd 1996

All rights reserved.
No part of this publication may be reproduced
in any form or by any means
without the prior written permission
of the publisher.

ISBN 1 870979 70 2

Printed in England by
Butler & Tanner Ltd, Frome, Somerset

CONTENTS

Introduction	6	Denzel	67	Lanchester	122	Rochdale	185
Abarth	9	De Soto	67	Lancia	123	Rolls-Royce	185
AC	10	De Tomaso	67	Lea-Francis	129	Rover	188
Aero	14	DKW	68	Leyland	129	Saab	190
Alfa Romeo	14	Dodge	70	Ligier	130	Salmson	191
Allard	20	Edsel	70	Lincoln	130	Seat	192
Alpine	21	Elva	70	Lloyd	131	Siata	192
Alvis	22	Enzmann	70	LMX	131	Simca	193
AMC	23	Excalibur	71	Lotus	132	Singer	196
Amphicar	24	Facel Vega	71	Marcos	134	Skoda	198
Armstrong-Siddeley	24	Fairthorpe	72	Maserati	135	Spatz	199
Arnolt	25	Falcon	73	Matra	137	Standard	199
ASA	25	Ferrari	73	Matra-Simca	138	Stanguellini	200
Aston Martin	26	Fiat	79	Mazda	138	Steyr-Puch	201
Audi	28	Ford (D)	87	Mercedes-Benz	139	Studebaker	201
Austin	29	Ford (F)	88	Mercury	144	Stutz	201
Austin-Healey	35	Ford (GB)	89	Messerschmitt	144	Subaru	201
Autobianchi	37	Ford (USA)	94	Metropolitan	145	Sunbeam	202
Bedford	38	Ford (Aus)	96	MG	145	Sunbeam-Talbot	202
Bentley	38	Frazer Nash	96	Mini	149	Swallow	205
Berkeley	40	Fuldamobil	97	Mitsubishi	151	Talbot	205
Bitter	41	Ghia	97	Monica	152	Tatra	205
Bizzarrini	41	Gilbern	98	Monteverdi	152	Thurner	206
BMW	41	Ginetta	98	Moretti	153	Toyota	207
Bond	46	Glas	99	Morgan	153	Trabant	208
Bonnet	47	Goliath	100	Morris	155	Trident	208
Borgward	48	Gordon-Keeble	101	Moskvich	160	Triumph	209
Bristol	49	GSM	101	Muntz	161	Tucker	214
Bugatti	51	Gutbrod	101	Nash	161	Turner	215
Buick	51	Healey	102	Nissan	161	TVR	215
Cadillac	51	Heinkel	102	NSU	162	Unipower	217
Caterham	52	Hillman	103	Ogle	163	Vanden Plas	217
Chaika	53	Hino	105	Oldsmobile	163	Vauxhall	218
Champion	53	Honda	105	Opel	164	Veritas	222
Checker	53	Hotchkiss	106	OSCA	168	Vespa	222
Chevrolet	53	HRG	106	Packard	168	Vignale	222
Chrysler	55	Hudson	107	Panhard	168	Volga	223
Cisitalia	57	Humber	107	Panther	169	Volkswagen	224
Citroën	57	Innocenti	109	Peerless/Warwick	170	Volvo	228
Clan	61	Intermeccanica	109	Pegaso	170	Warszawa	231
Connaught	62	Iso	110	Peugeot	171	Wartburg	231
Cord	62	Isuzu	111	Plymouth	173	Willys	232
DAF	62	Jaguar	111	Pontiac	174	Wolseley	232
Daihatsu	63	Jensen	117	Porsche	175	Zagato	235
Daimler	63	Jowett	119	Princess	178	ZAZ	235
DB	66	Kaiser	119	Puma	178	ZIL	236
Delage	66	Lada	120	Reliant	178	ZIM	236
Delahaye	66	Lagonda	120	Renault	180	ZIS	236
Dellow	67	Lamborghini	121	Riley	184	Zündapp	237

INTRODUCTION

This buyer's guide is intended to provide an overview of what is available in the world of classic cars, what to expect from particular models, and how much to pay. We have tried to mingle historical perspective with practical advice, and to combine the passion and pleasure of classic cars with the hard facts which cannot be ignored.

Like most objects of value, the classic car is subject to the whims of supply and demand and is prey to fluctuating values. However, the classic car world is not the chaotic world it used to be. The speculative side of the market has well and truly settled down after the notorious price bubble of the late 1980s, but the level of interest remains strong. Fixing values has never been more realistic than at present.

Naturally there are boundaries. We have tried to include every major British and European car built from 1945 up until 20 years ago (ie. 1976). Sadly, this cannot include every one of the hundreds of eligible specialists, although the most historically significant and interesting to the collector are listed.

Our most severe limits apply to American and Japanese cars. Generally these are confined to the most interesting examples to (admittedly distorted) European eyes. When you realise that Toyota made over 60 variants of the Corolla in just one model year in the mid-1970s, you appreciate the enormity of the problem, especially as no-one really cares for '70s Corollas.

Japanese inclusions are almost invariably the most interesting of those imported to Britain, plus others which are of historic or contextual significance. Naturally there will be controversy over our selections in some cases, but we have aimed for the widest possible spread given the limitations of space. Likewise, the spread of American cars is tremendously diverse and far too wide for this publication. But the availability of American cars has become very good if you are prepared to look abroad and we have attempted to include some of the more interesting choices.

Further exclusions are commercial vehicles, cars converted by specialists, invalid carriages and models which are known to be effectively extinct.

Trends

The classic car market is in a permanent state of flux and, despite its more settled nature recently, there are trends to be discerned.

Not every car over 20 years old is a valuable one; exceeding that age does not make every car a classic. There is at least one good reason why the 20-year gap has been selected. Enthusiasts who grew up dreaming of cars in their school years often reach the age where their own 'dream car' becomes a realistic ownership proposition after about 20 years.

Therefore a lasting trend is the rolling acceptance of newly eligible models into the classic car fold. Cars from the 1970s are gaining interest as the rapidly falling number of bangers clears the way for an appreciation of the few that are left. The most obvious icons of 20 years ago which are now undoubtedly on the verge of classic status include the VW Golf GTI Mk1, BMW 6 Series, Fiat X1/9, Ford Capri, Datsun 240/260Z, Rover SD1, Toyota Celica and Triumph TR7. There are now even owners' clubs for such unexotic models as the Austin Allegro, Princess and Morris Marina.

Classics as investments continue to be frowned upon, although fine examples of historic thoroughbreds will always be in demand. Excellent examples of recognised classics have consistently been achieving good prices, though ropier ones are still available at surprisingly low prices. Thus the best E-Types can change hands at well over £30,000 but it is still possible to buy a poor, yet running one as low as £5000. Likewise, a top Silver Shadow with impeccable history can fetch almost £15,000, whereas neglected specimens plunge well below £5000.

Essentially the market is stable. The announcement in the autumn 1995 budget that the licence disc is to be scrapped for cars aged 25 years and over makes owning a classic just a little bit easier, so there may well be some knock-on effect on values. Certainly classic car enthusiasts are feeling heartened.

Explanation of tables

Engine Capacity in cubic centimetres (cc), followed by bore and stroke dimensions in millimetres, number of cylinders and valve type, and finally engine position and driven wheels.

IOE	Inlet over exhaust valve
OHV	Overhead valve
OHC	Overhead cam
DOHC	Double overhead cam
QOHC	Quad overhead cam
SV	Sidevalve
TS	Two-stroke
F/F	Front engine/front-wheel drive
F/R	Front engine/rear-wheel drive
M/R	Mid-mounted engine/rear-wheel drive
R/R	Rear engine/rear-wheel drive
F/4x4	Front engine/four-wheel drive

Max power Figures quoted are DIN where possible, though some continental figures are in the near-identical German PS rating. American outputs are usually SAE (gross) and Japanese outputs are often JIS.

Max speed Claimed manufacturer's top speed in mph, or test figures where available.

Body styles All passenger bodywork variations are listed.

Saloon	Usually has separate boot, two or foor doors
Estate	Full extended rear end with tail-gate
Coupé	Invariably two-door, having low roof-line
Hatchback	Saloon/estate cross-over with opening tail-gate
Limousine	Cars intended for chauffeur drivers, ie with a division between front and rear seats and usually with an extended wheelbase
Touring limousine	Saloon converted with a division
Landaulette	Fixed-roof limousine with convertible rear section
Convertible/drophead coupé	Permanently attached folding hood
Tourer	Four-seater open-top saloon-type
Targa	Solid removable roof panel
Sports	Two-seater sporting, usually with removable hood
Rolltop	Extended-length fabric sunroof
Hardtop	Either removable roof section or pillarless coupé
Utility	Basic jeep-type transport, usually open

Prod years Refers to years model was actually in production, although in some cases the date may be year of debut. This includes American cars, although in accordance with accepted custom, years given in the titles for American cars are *model* years (eg the car generally known as the '59 Cadillac was actually launched in 1958).

Prod numbers Total production within given life-span, usually excluding commercial variants. Model splits are provided where possible.

PRICES On the whole, these are realistic UK mainland market values. They are compiled, where possible, from genuine private and dealer sales, with additional input from auctions. They are not compiled from prices asked, but not achieved, within the trade.

The increasing international trade in classic cars means that foreign markets cannot be ignored. For cars which are particularly rare in the UK, quoted prices often reflect the accepted values of more plentiful markets such as the USA and Europe.

Prices are given as a guide only: individual cars rarely conform to an exact price category and there should always be room for negotiation. They assume that no warranty is given (which will put up the price). Other factors governing values include whether the car is RHD or LHD, whether it has an automatic, manual or overdrive gearbox, whether it has service history, whether it has a good ownership history, whether it is original or restored, whether it has its original registration plate and whether it has been imported.

Three conditions are quoted wherever possible, as follows:

Condition A Excellent general condition, ie no obvious areas where expenditure is required. Should be in original condition, or restored to authentic specification. Concours, very low mileage and examples of special historic interest will fetch more.

Condition B Sound, usable examples which should be presentable. Some cosmetic attention will probably be required, as well as perhaps a few small mechanical jobs. Any rust should not compromise the integrity of the structure, although there may well be areas of corrosion requiring minor attention.

Condition C Complete, running cars, perhaps with an MoT, which need some sort of fairly major restoration work (body panels, engine overhaul, interior refurbishment etc) but certainly not basket cases. Unlikely to be 100% original, but should still have no major items missing. 'Restoration projects' will fetch less.

Other abbreviations

As well as the tabular abbreviations listed above, the following terms are commonly abbreviated:

4WD	Four-wheel drive
DHC	Drophead Coupé
FHC	Fixed-Head Coupé
FWD	Front-wheel drive
IFS	Independent front suspension
IRS	Independent rear suspension
LHD	Left-hand drive
LSD	Limited slip differential
PAS	Power-assisted steering
RHD	Right-hand drive
RWD	Rear-wheel drive

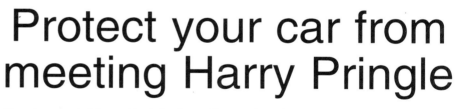

Protect your car from meeting Harry Pringle

Harry loves rust, it keeps him in business. Your car is rusting every day and Harry can't wait to get his hands on it.

Now you could ignore that small rust patch under and around your car or you could stop rust in its tracks with the ultimate Waxoyl* combination.

Waxoyl, the original, thick sticky liquid and tough flexible Underbody Seal don't just protect against rust - they actively stop any rust that's already there from developing. Both products use powerful rust inhibitors.

Spray Waxoyl into your car's hidden nooks and crannies. Instantly, it chases out moisture to form a flexible weatherproof skin. Brush Underbody Seal onto your car's underside and it creates a barrier against the blasting of stones, gravel, water and salt. It also remains flexible and won't crack like ordinary underseals.

Use them together, spend a couple of hours once a year, and for as little as £25 you won't be meeting Harry for years. Now isn't that better than £50 Harry would 'generously' offer for scrap?

WAXOYL® *The Original*

Available at:
Halfords, Charlie Brown, Motor World and all good motorists shops.

Waxoyl is a Registered Trademark. *Application may invalidate some car warranties, check before applying.

QUALITY PARTS FOR BRITISH CLASSICS

ALL THE PARTS YOU NEED... *by Mail Order*

WE HAVE THE PARTS
WE GUARANTEE THE QUALITY
WE'RE KEEN ON PRICE
WE PROVIDE THE SERVICE
WE DELIVER FAST
WE KNOW YOUR CAR

Founded in 1981, Rimmer Bros are the UK's largest independant specialist supplier of replacement parts for Triumph and Rover SD1.

Our degree of specialisation means two things.

Firstly, our people know the cars we're dealing with inside out, and can give you any advice you might need.

Secondly, we are able to stock in depth just about any part you might need all under one roof. Our warehouse has over 40,000 square feet of parts storage space alone, and plans are already in hand to increase the capacity. It currently holds over 30,000 recorded part numbers with over 15,000 stock lines, and we hold as many of each as necessary to make it almost impossible to run out of stock.

All other aspects of our operation receive equal emphasis
- Fully computerised 'Live' stock control system
- The latest racking and storage facilities for fast and accurate 'order picking'
- Careful packaging and despatch procedures

All parts are offered off the shelf for either immediate despatch by fast mail order or collection at our sales counter where you can view our display of Classic Triumph and Rover cars.

Our Customer Card scheme ensure more customers can get more parts for more cars, even more quickly and efficiently.

RIMMERS COMMITMENT TO CUSTOMER SERVICE
Helpful - Accurate - Fast - Efficient

The overall objective at Rimmer Bros is to achieve customer satisfaction at all times.

Our people know your car well and every member of our staff is dedicated in helping preserve Classic Cars.

Our computer system gives up-to-the-minute information regarding availability and the price of any item. We have an enviable reputation for despatching goods when the customer needs it, quickly and efficiently.

We try to make all our systems and documentation easy to follow and use, and make sure that our catalogues and price guide really do help our customers choose the right parts easily, first time.

IF YOU'VE GOT ONE OF THESE CARS...

YOU NEED ONE OF OUR CATALOGUES

ONE FREE CATALOGUE TO EACH CAR OWNER.
ADDITIONAL CATALOGUES AVAILABLE AT £2 EACH, OR FULL PACK (7 CATALOGUES) AT £10, REFUNDABLE ON YOUR NEXT PARTS PURCHASE.

NEW PRICE GUIDE

Our up to date price guide is out now, mailed automatically to all Rimmer Bros Customer Card holders who own a Stag, TR6, TR7/8, Spitfire, GT6, Herald or Vitesse. If you have not received your Price Guide, please call.
For all Dolomite, 2000/2500 and SD1 owners, our parts/price listing catalogues will also be mailed automatically as updated versions are produced.

SHOP BY PHONE FOR FAST MAIL ORDER

WE EXPORT TO THE EC & WORLDWIDE

ORDER BY PHONE
USE YOUR CARD.......

UK SALES: 01522 568000
EXPORT SALES: (+44) 1522 526200

HELPING TO PRESERVE BRITISH CLASSIC CARS

Rimmer Brothers Ltd, Triumph House, Sleaford Road, Bracebridge Heath, Lincoln LN4 2NA
Tel: UK Sales: (01522) 568000. Export Sales: (+44) 1522 526200.
Customer Service: (01522) 567432. Fax: (01522) 567600.

OPENING HOURS:
MON – FRI 8.30am – 5.30pm,
SAT 8.30am – 1.30pm.

1996-1997 CLASSIC CAR BUYER'S GUIDE

Abarth-Fiat 750

Abarth caught the tail end of the 1950s Italian coachbuilt era and its superb engines in bespoke bodies by the likes of Allemano (photo) were an intoxicating combination. Abarth bored the Fiat 600 engine out to 750cc which, in light bodywork, provided peppy performance for those days. A vast range of bodies was offered by other coachbuilders like Bertone, Pininfarina, Vignale, Viotti and Zagato, the latter being particularly celebrated with its so-called 'double bubble' roof.

Engine 747cc 61x64mm, 4 cyl OHV, R/R
Max power 33bhp at 4400rpm to 47bhp at 6000rpm
Max speed 90-100mph
Body styles coupé, convertible
Prod years 1957-60
Prod numbers n/a

PRICES
A 18,000 **B** 12,000 **C** 8500

Abarth-Fiat 2200

Launched at the 1959 Turin Show, Abarth's grown-up sports car was based on the Fiat 2100. Various coupé and convertible bodies were offered, the most popular being by Allemano. Abarth bored the engine out and boosted power by more than 50%, so not surprisingly it was one of the fastest smaller sports cars of its day. Much rarer than the Fiat 600 based Abarths, so prices today tend to be out of most enthusiasts' reach. Eminently raceable but bank on big bills for repairs to that fragile engine.

Engine 2161cc 79x73.5mm, 6 cyl OHV, F/R
Max power 135bhp at 6000rpm
Max speed 125mph
Body styles coupé, convertible
Prod years 1959-61
Prod numbers n/a

PRICES
A 30,000 **B** 22,000 **C** 15,000

Abarth-Fiat 850/1000

Abarth's conversions of the Fiat 600 became legendary for their buzz-box performance and tail-happy manners after winning the European Group 1 Touring title several times. Abarth bored and stroked the engine out to extremes. Mild versions looked pretty much like standard 600s but the wild Corsa and screaming TCR 1-litre semi-race versions with extended front valances and their engine lids on stilts are the ones to have. Popular with historic racers but darned expensive for what they are.

Engine 847cc 62.5x69mm/982cc 65x74mm, 4 cyl OHV, R/R
Max power 47bhp at 4800rpm to 112bhp at 8200rpm
Max speed 81/125mph
Body styles saloon
Prod years 1960-70
Prod numbers n/a

PRICES
850 **A** 5000 **B** 4000 **C** 3000
1000 **A** 14,000 **B** 7500 **C** 5000

Abarth-Fiat 595/595SS/695/695SS

In appearance only mildly altered from Fiat's baby 500 but tuned engines provided performance which matched many sports cars of the day. Certainly almost 90mph from less than 700cc is good in anyone's book. Handling was also improved by Abarth tweaks. An attractive package but rather expensive in its day. Zagato-bodied 500 extremely rare. Very eligible for historic racing. Parts situation is extremely good these days. Make sure your 'genuine' Abarth is not a dressed-up imitation with painted flanks and scorpion badges.

Engine 594cc 73.5x70mm/690cc 76x76mm, 2 cyl OHV, R/R
Max power 27bhp at 5000rpm/ 32bhp at 4900rpm/30bhp at 4900rpm/38bhp at 5200rpm
Max speed 75-88mph
Body styles saloon
Prod years 1963-71
Prod numbers n/a

PRICES
595/595SS **A** 4000 **B** 3000 **C** 2000
695/695SS **A** 5000 **B** 3500 **C** 2250

Abarth-Simca 1150/Corsa

In 1961, Abarth made his first foray away from Fiat with a variety of models based on Simcas. The 1150 was merely a hotted-up Simca 1000 saloon with an Abarth grille but the Corsa used the rear-engined Simca floorpan and added a beautiful sports coupé body. Versions were also built with 1.3, 1.6 and 2.0-litre engines. In the UK, Radbourne built a dozen Simca-Abarths with Fiat 124 engines bored out to 1280cc between 1968 and 1971. Other Simca-Abarths used engines up to 2.0 litres and were capable of speeds in excess of 150mph.

Engine 1137cc 69x76mm, 4 cyl OHV, R/R
Max power 55bhp at 5600rpm to 85bhp at 6500rpm
Max speed 94-110mph
Body styles saloon, coupé
Prod years 1963-65
Prod numbers n/a

PRICES
1150 **A** 5000 **B** 3000 **C** 2000
Corsa **A** 20,000 **B** 12,000 **C** 9000

Abarth-Fiat OT850/OT1000/OT1300/OT1600/OT2000

Abarth pounced on the Fiat 850 almost as soon as it was launched and produced the Omologato Turismo series with an eye on production car racing. Engines ranged from a simple tuned 850 unit through a stroked 1-litre up to scorching 1.3, 1.6 and 2.0-litre monsters with modified Fiat engines and wide alloy wheels. OTS had front-mounted radiators, while OTR versions were semi-racers with extra power. All varieties of 850 bodyshell were available with Abarth tune. Interesting Mini-Cooper rivals. Coupé, cabriolet and OTR/OTS attract premiums, while 1300/1600/2000 are even more expensive.

Engine 843cc 65x63.5mm/982cc 65x74mm/1280cc 75.5x71.5mm/1592cc 86x68.5 mm/1946cc 88x80mm, 4 cyl OHV/OHV/OHV/DOHC/DOHC, R/R **Max power** 44bhp at 5400rpm/62bhp at 6150rpm to 74bhp at 6500rpm/75bhp at 6000rpm/ 154bhp at 7600rpm/185bhp at 7200rpm
Max speed 85/99/102/105/149/151mph **Body styles** saloon, coupé, convertible **Prod years** 1964-71
PRICES
OT850/1000 **A** 7500 **B** 5000 **C** 3000

Abarth-Fiat Scorpione

Francis Lombardi's quirky coupé body on a Fiat 850 platform was borrowed by Abarth. You might also find this basic shape with Giannini badges (and a 994cc DOHC engine) or with OTAS badges (with linered-down Giannini engines). The Abarth version was the hairiest – the bored-out 1.3-litre Fiat 124 engine provided excellent performance – and had a few minor styling changes over the Lombardi (about 10 of which were imported to the UK). As such the Scorpione is worth about twice as much as a Lombardi Grand Prix, and more than a Giannini.

Engine 1280cc 75.5x71.5mm, 4 cyl OHV, R/R
Max power 75bhp at 6000rpm
Max speed 116mph
Body styles coupé
Prod years 1968-71
Prod numbers n/a

PRICES
A 10,000 **B** 6000 **C** 4500

Abarth-Fiat 124 Rally Spider

In 1971, three years after Fiat's official return to racing and rallying, it took over Abarth, which was a natural to develop competition Fiats. Its homologation version of the 124 Spider had the 1.8-litre engine with 128bhp, light alloy and glassfibre panels, independent rear suspension, roll cage (so no rear seats) and a permanently fixed hard top. A 16-valve head was optional in 1975 only. Since only just over 1000 were made, genuine examples are rare and mightily sought-after, so be careful of painstaking fakes.

Engine 1756cc 84x79.2mm, 4 cyl DOHC, F/R
Max power 128bhp at 6200rpm
Max speed 120mph
Body styles coupé
Prod years 1972-76
Prod numbers 1013

PRICES
A 15,000 **B** 10,000 **C** 6000

AC 2-Litre/Buckland Tourer

At introduction in 1947 the AC 2-Litre looked fairly up-to-date but it could not hide the archaic mechanicals. Engine basically dated from 1920, both axles were rigid and the two-door body was aluminium-on-ash. Rather upper-crust inside, large glass area makes it airy for a 1940s saloon, but hard ride. No rear spats from 1948, rare drophead coupé from '49 (only 20 built), hydraulic brakes and the option of four doors from '52. Buckland Tourer of 1949 had all-new drophead coachbuilt body.

Engine 1991cc 65x100mm, 6 cyl OHV, F/R
Max power 74bhp at 4500rpm
Max speed 84mph
Body styles saloon, drophead coupé, tourer
Prod years 1947-56/1949/1949-54
Prod numbers c 1290

PRICES
Saloon **A** 5000 **B** 3500 **C** 2000
DHC/Buckland Tourer **A** 9000 **B** 6000 **C** 4000

AC Ace/Ace-Bristol/Ace 2.6

AC's first true sports car had a chassis by John Tojeiro and power from AC's rather asthmatic prewar 2-litre six. This was Britain's first post-war sports car to have four-wheel independent suspension (by transverse leafs and wishbones). Overdrive arrived in '56 and front discs in '57. Most important advance was the availability of Bristol's 2-litre engine from 1956, greatly enhancing performance. Highly praised for its roadholding. Ruddspeed Ford Zephyr 2.6 engine was factory available from 1961. An encouraging new start for AC and much prized these days.

Engine 1991cc 65x100mm/1971cc 66x96mm/2553cc 82.5x79.5mm, 6 cyl OHC/OHV/OHV, F/R **Max power** 75bhp at 4500rpm/125bhp at 5750rpm/170bhp at 5500rpm **Max speed** 97/125/130mph **Body styles** sports **Prod years** 1953-63/1956-63/1961-63 **Prod numbers** 223/463/37 **PRICES** Ace/Ace 2.6 **A** 30,000 **B** 25,000 **C** 18,000. Ace-Bristol **A** 40,000 **B** 30,000 **C** 22,000

Calorex

Just because your Classic Car is stored under cover doesn't mean it's safe.

Damp is a killer! Its favourite victim is the rare and often expensive classic car. Moisture causes not only rust but also damage to timber and upholstery. It can best be eliminated, not by expensive heat, as is widely believed, but by **DEHUMIDIFICATION**.

CALOREX is acknowledged as the established specialist in the field and has earned a reputation worldwide, throughout industry, as the expert in moisture removal.

Talk to us **NOW** about the cost effective way of protecting your investment.

Calorex Heat Pumps Limited
The Causeway, Maldon, Essex, CM9 5PU
Tel: 01621 856611 Fax: 01621 850871

UROGLAS
SPECIALIST AUTOMOTIVE GLASS DISTRIBUTORS

WINDSCREENS, REARSCREENS & BODYGLASS

available for **all** types of vehicle, rare or classic.

*

Complete remanufacturing service available subject to pattern. 'One-offs' and low volume production available

*

American autoglass a speciality!

*

Overnight carriage within the UK.
Export service available at competitive rates.

*

For professional and personal service, contact:
Tel: 01527 577477 (3 lines)
Fax: 01527 576577 (24 hour)

TUBE TORQUE EXHAUST SPECIALISTS

MANIFOLDS AND SYSTEMS IN MILD AND STAINLESS STEEL FOR VINTAGE, VETERAN, HISTORIC, CLASSICS, ROAD, RACE AND RALLY

TUBE, BEND AND SILENCER SERVICE TO CUSTOMER SPECIFICATION

TEL/FAX 01625 511153

TTCM

TEL/FAX 01625 503866

UNIT 10, BROOK STREET MILL, MACCLESFIELD, CHESHIRE SK11 7AW

AC Aceca/Aceca-Bristol/Aceca 2.6

Coupé version of the Ace with curved screen, resited instruments, larger timber-framed doors and a sort of hatchback. Technically the cars were virtually identical but you will need to look for rotten woodwork. Although priced the same new, the Aceca today is not worth as much as the Ace, despite its good looks. Like the Ace, it was also available from 1956 with Bristol's derivative of the prewar BMW 328 engine and Ruddspeed-tuned Ford 2.6 (only 8 made). Heavier than Ace so less accelerative.

Engine 1991cc 65x100mm/1971cc 66x96mm/2553cc 82.5x79.5mm, 6 cyl OHC/OHV/OHV, F/R **Max power** 75-90bhp at 4500rpm/105-125bhp at 5750rpm/170bhp at 5500rpm **Max speed** 95-130mph **Body styles** coupé **Prod years** 1954-63 **Prod numbers** 151/169/8

PRICES
Aceca 2.6 **A** 25,000 **B** 18,000 **C** 12,000. Aceca-Bristol **A** 30,000 **B** 22,000 **C** 14,000

AC Greyhound

More than just a four-seater Ace, the Greyhound had all-coil springing, a longer wheelbase and was substantially restyled, although certainly with less pleasing lines. Rather expensive in its day (25% more than an E-Type) so always a rarity. Almost all engines came from Bristol (2.0 or 2.2 litres) but a tiny handful were fitted with 2.6-litre Ford Zephyr units and some used the old AC unit. The heaviest and least nimble of the sporting ACs, with poorer handling too, so although it's rarer than the Ace, it's worth a lot less.

Engine 1971cc 65x100mm/2216cc 68.6x99.6mm/2553cc 82.5x79.5mm, 6 cyl OHV, F/R **Max power** 105-126bhp at 6000rpm/107bhp at 4700rpm/170bhp at 5500rpm **Max speed** 107 to 125mph **Body styles** saloon **Prod years** 1959-63 **Prod numbers** 83

PRICES
A 16,000 **B** 11,000 **C** 8000

AC Petite

AC's return to the days of the Sociable were induced by the times: austerity brought a rash of 1950s microcars. This Villiers two-stroke powered micro could hardly have been more different from the Ace but it was one step above other microcars of the time, boasting a full sunroof and seating for three abreast (just). MkII of 1955 had expanded 353cc engine and equal diameter wheels all round (the MkI had a smaller front wheel). Alloy body panels don't rust but the steel chassis/frame does. AC was more successful with its invalid carriages.

Engine 346cc 70x90m/353cc 75x88mm, 1 cyl TS, R/R **Max power** 8bhp at 3500rpm **Max speed** 50mph **Body styles** coupé **Prod years** 1952-57 **Prod numbers** c 4000

PRICES
A 2000 **B** 1200 **C** 600

AC Cobra 260/Cobra 289/289 Sports

Take an Ace and put in an American V8 and you have an instant legend. Texan racing driver and chicken farmer Carroll Shelby's cocktail included Ford power, a tougher but still inadequate chassis and disc brakes. Initially for export only but sold in UK in RHD from 1964 with rack-and-pinion steering in place of cam-and-peg. 260 and 289 refer to engine sizes in cubic inches. Run-out AC 289 Sports (the Cobra was dropped) had the flared-wing 427 body with the 4.7-litre engine. Weighs less than a ton, so hairy road manners. Very sought-after.

Engine 4261cc 96.5x72.9mm/4727cc 101.6x72.9mm, V8 cyl OHV, F/R **Max power** 264bhp at 5800rpm/300bhp at 5750rpm **Max speed** 144/155mph **Body styles** roadster **Prod years** 1962-63/1963-65/1965-68 **Prod numbers** 75/571/27

PRICES
A 90,000 **B** 75,000 **C** 60,000

AC Cobra 427/427SC/428

As if the 4.7-litre engine wasn't enough, Shelby's shoehorn job with a Ford 7-litre engine produced a car with phenomenal acceleration. The chassis got larger diameter tubes and the suspension was upgraded to coil springs, so roadholding improved. Bodywork went on a course of steroids and only the bonnet and doors remained the same. The 427 was more than just a sledgehammer job: it won the world GT title in its debut year. More powerful 427SC was touted as 'the fastest production car in the world'. Milder 428 (6997cc) unit came from the Thunderbird.

Engine 6984cc 107.7x96mm/6997cc 104.9x101.1mm, V8 cyl OHV, F/R **Max power** 425bhp at 6500rpm/480bhp at 6500rpm/390bhp at 5200rpm **Max speed** 165/180/155mph **Body styles** roadster **Prod years** 1965-68 **Prod numbers** 306

PRICES
427/428 **A** 140,000 **B** 100,000 **C** 85,000. 427SC **A** 200,000 **B** 170,000 **C** 150,000

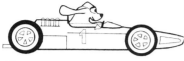

CHRIS ALFORD RACING & SPORTSCARS

NEWLAND COTTAGE, HASSOCKS, WEST SUSSEX BN6 8NU
PHONE: 01273 845966   FAX: 01273 846736

1st FOR HISTORIC RACING CARS!

PHONE: 01273 845966
FAX: 01273 846736
MOBILE: 0585 722962

CHAPTER & SCOTT Ltd. Estd. 1956
trading as
CHATER'S

MOTORING BOOKSELLERS

SPECIALISTS IN NEW AND OUT OF PRINT MOTORING LITERATURE

THE BOOKSHOP FOR MOTORING ENTHUSIASTS

8 SOUTH STREET, ISLEWORTH, MIDDX. TW7 7BG
TEL: (44) 0181 568 9750 • FAX: (44) 0181 569 8273
FREE CATALOGUE AVAILABLE
OPEN MON-SAT 9-5.30

Nearest tube station: Hounslow East – Piccadilly Line. Then take H37 bus to South Street from Hounslow Bus Garage

OPEN ON RACE DAYS AT BRANDS HATCH & SILVERSTONE

Main Street, Great Longstone, Bakewell, Derbyshire.
Telephone 01629 640227. Fax: 01629 640533.

LONGSTONE GARAGE
MIKE HIRST

TYRES AND WHEELS FOR VETERAN, VINTAGE AND CLASSIC VEHICLES

Whatever your tyre and wheel requirements, here at Longstone Garage we have the solution for your pre-war vehicles:

★ Unbeatable stocks of tyres including Michelin, Firestone, Goodrich, Dunlop, Avon etc.
★ Wheels made to your specification
★ Wheel balancing service
★ Worldwide Mail Order service
★ Tyre fitting & balancing available at some VSCC meetings

Whatever your requirements, please call in and see us, we shall be pleased to help you.

NEW CATALOGUE
NOW OUT FROM HOLDENS

Many new items not previously available

Expanded range of Electrical, Lighting, Switches, Ignition, Wiper Equipment, control & Fuse Boxes, Starters & Dynamos, Gauges, Accessories, Race & Rally Equipment, Period clothing.

Specialists In Electrical Equipment and Historic Motoring accessories for most 1930's-1970's vehicles.

For the complete Mail Order Motoring catalogue
Send £3.00 U.K. & Europe £4.00 outside Europe inc. p+p

CCG/14
PLEASE QUOTE THIS REF. NO WHEN PLACING YOUR ORDER

MAIL ORDER OR DIRECT EXPORT ORDERS WELCOMED. POSTAGE PACKING AND VAT NOT INCLUDED

Holden Vintage & Classic Limited
Linton Trading Estate, Bromyard, Herefordshire HR7 4QT Great Britain
Tel: 01885 488 000   Fax: 01885 488 889
International Tel: 0044 (0)1885 488 000   International Fax: 0044 (0)1885 488 889

AC 428/428 Convertible

This was a more sophisticated 'Cobra for the businessman' based on a Cobra chassis lengthened by 6in. The elegant Carrozzeria Frua styled body revived an old favourite concept: Italian bodywork, American engine and British construction. Always less powerful engines than Cobra 427, and manual and automatic gearboxes both available (the latter being more common). Many 428s were rebodied as Cobras so scarcity has pushed prices up, although a 428 will never have the same allure as a Cobra. Italian steel means corrosion on an epic scale.

Engine 6997cc 104.9x101.1mm/ 7017cc 104.9x101.1mm, V8 cyl OHV, F/R **Max power** 345bhp at 6000rpm
Max speed 144mph
Body styles coupé, convertible
Prod years 1965-73
Prod numbers 29/51

PRICES
428 **A** 22,500 **B** 18,000 **C** 12,000
428 Convertible **A** 29,000 **B** 22,000 **C** 15,000

AC ME3000

With the 428, the glory years ended for AC. To replace the 428, the company went about-face and bought a prototype from Bohanna Stables called the Diablo, which they proceeded to modify. Shown as early as 1973, the ME3000 was a transverse mid-engined coupé with Ford V6 power. A strict two-seater, it looked good for 1973 but by the time production started in 1979 it was all a bit tired. Rare turbo version (only 11 built) made wayward handling on the limit even more tricky. A very British curiosity.

Engine 2994cc 93.7x72.4mm, V6 cyl OHV, M/R
Max power 138bhp at 5000rpm
Max speed 125mph
Body styles coupé
Prod years 1979-84
Prod numbers 82

PRICES
A 14,000 **B** 10,000 **C** 8500

Aero Minor

The Aero Minor was a short-lived postwar attempt by a Czechoslovakian firm, known for its motor cars and aeroplanes before the war, to launch a popular car in difficult times. Based on a prototype which had been built in secret during the war, the Aero Minor was developed by Jawa engineers and featured a 615cc two-stroke twin engine. The saloon or estate bodies were too bulky for the power. It was available for export as early as 1945 but the project lasted only a few years. Virtually unknown in Britain and hardly more so on the continent.

Engine 615cc 70x80mm, 2 cyl TS, F/F
Max power 20bhp at 3000rpm
Max speed 56mph
Body styles saloon, estate
Prod years 1945-52
Prod numbers 14,100

PRICES
A 2800 **B** 1500 **C** 700

Alfa Romeo 6C2500

Some enthusiasts regard the 6C2500 as the last 'real' Alfa Romeo as this was the final bespoke built Alfa. Basically a prewar design, the 6C chassis was clothed in a wide variety of bodies, from sporty four-door saloons (like the Ghia version pictured), through desirable cabriolets and coupés. Most desirable are the short wheelbase 110bhp Super Sport (SS) cabriolet and coupé and the ultimate Villa d'Este coupé. All are glorious to look at and fabulous in engineering terms. Only ever made in right-hand drive, the final legacy of prewar Alfa attitudes. Very expensive then, as now.

Engine 2443cc 72x100mm, 6 cyl DOHC, F/R **Max power** 87/90bhp at 4600rpm/110bhp at 4800rpm
Max speed 97/112mph
Body styles saloon, coupé, convertible **Prod years** 1939-52
Prod numbers 1591

PRICES
Saloon **A** 32,000 **B** 22,000 **C** 12,000
SS Coupé/Convertible **A** 65,000 **B** 55,000 **C** 45,000

Alfa Romeo 1900/1900 Super

A landmark model in Alfa's history, with its unitary body and beautiful twin cam engine designed by Dr Puliga. Four-door saloon styling was an in-house job and handsome for the time. Excellent handling and there was improved performance from the larger Super engine, standard from 1953, but unpleasant column shift for four-speed 'box. Sadly rust takes its toll in spectacular fashion but if you find a good one and keep it that way, the 1900 could be a practical and good value classic. Rarest version was the 1955-57 Primavera two-door saloon.

Engine 1884cc 82.4x88mm/1975cc 84.5x88mm, 4 cyl DOHC, F/R
Max power 90bhp at 5200rpm/115bhp at 5500rpm
Max speed 93/100mph
Body styles saloon
Prod years 1950-58
Prod numbers 17,423

PRICES
A 10,000 **B** 7000 **C** 3500

Alfa Romeo 1900 Sprint/Super Sprint

The sporting derivatives of the 1900 range, available in two distinct body types: the Coupé with bodywork by Touring and the Cabriolet made by Pininfarina. As well as these attractive offerings, many one-offs were also built by various coachbuilders, nearly always with aluminium bodies (notably the fabulous Bertone BAT series). Every example came with centre-lock wire wheels and nearly always LHD. Sprint had four-speed 'box and Super Sprint had five speeds, a choice of column or floor shifts, twin carbs and 115bhp.

Engine 1884cc 82.4x88mm/1975cc 84.5x88mm, 4 cyl DOHC, F/R
Max power 100/115bhp at 5500rpm
Max speed 105/112mph
Body styles coupé, convertible
Prod years 1954-58
Prod numbers 1796

PRICES
A 35,000 **B** 30,000 **C** 25,000

Alfa Romeo Giulietta Sprint/Giulia Sprint

The coupé version of the Giulietta made it out before the Berlina. Seductive 2+2 bodywork was designed and built by Bertone. Sprint Veloce version had 90bhp, or 100bhp in twin carb Speciale guise, which was exceptional for 1955, part-alloy bodywork and sliding perspex windows. Both versions initially had four-speed transmission and column change; from 1958 this changed to five speeds and a floor change, coinciding with a longer wheelbase. Giulia Sprint from 1962 had the 1.6-litre engine (the Giulietta was renamed 1300 Sprint from this time).

Engine 1290cc 74x75mm/1570cc 78x82mm, 4 cyl DOHC, F/R
Max power 65bhp at 6100rpm to 100bhp at 6500rpm/92bhp at 6200rpm to 112bhp at 6500rpm
Max speed 96-112mph
Body styles coupé
Prod years 1954-65/1962-65
Prod numbers 27,147/11,171

PRICES
Sprint **A** 10,000 **B** 7000 **C** 4500
Sprint Veloce **A** 12,000 **B** 8500 **C** 5500

Alfa Romeo Giulietta Berlina/Berlina TI

Alfa Romeo perpetrated its reputation as a purveyor of sports cars with saloon bodies through cars like the Giulietta Berlina. Fabulous handling, sparkling performance (particularly in TI guise from 1957) and fairly roomy too. Column shift the norm but less likely on TI. Estate version was virtually non-existent. RHD availability from 1961 on. Expensive in its day (three times the price of a Beetle) but surprisingly good value today. You will need a good magnet to check all the bodywork though.

Engine 1290cc 74x75mm, 4 cyl DOHC, F/R
Max power 53bhp at 5200rpm/74bhp at 6200rpm
Max speed 87/102mph
Body styles saloon, estate
Prod years 1955-64
Prod numbers 54,052/138,865

PRICES
A 6000 **B** 4000 **C** 2500

Alfa Romeo Giulietta Spider/Giulia Spider

Delightful open-top bodywork by Pininfarina on the base of the Giulietta initiated Alfa's noble line of Spiders. Veloce version was significantly faster and much rarer (only 2796 built). All Spiders benefited from a gear lever mounted on the floor but the Giulietta was exclusively LHD until 1961. 1.6-litre engined Giulia replaced the Giulietta in 1962, offering more power and five speeds. Again the twin carb Veloce version was rare (1091 built) and, with 112bhp on tap, quite a flyer. Expect rampaging rust problems.

Engine 1290cc 74x75mm/1570cc 78x82mm, 4 cyl DOHC, F/R
Max power 65bhp at 6500rpm to 100bhp at 6500rpm/92bhp at 6200rpm to 112bhp at 6500rpm **Max speed** 93-112mph **Body styles** convertible
Prod years 1955-62/1962-65
Prod numbers 17,096/10,341

PRICES
Spider **A** 12,000 **B** 8000 **C** 5000
Spider Veloce **A** 14,000 **B** 10,000 **C** 6500

Alfa Romeo Giulietta/Giulia SS

A distinctive and different coupé treatment on the Giulietta base again designed and built by Bertone, the SS (or Sprint Speciale) was a virtually bespoke machine with bodywork partly in aluminium. 1.6-litre engined Giulia SS replaced it in 1962. Always fitted with an engine in the Veloce tune and always with the five-speed 'box. In its day, popular in hillclimbs and circuit racing (but overshadowed by SZ) and not as expensive as you might think. As ever with Alfas, watch out for rust and fragile ancillaries.

Engine 1290cc 74x75mm/1570cc 78x82mm, 4 cyl DOHC, F/R
Max power 100bhp at 6500rpm/112bhp at 6500rpm
Max speed 120/124mph
Body styles coupé
Prod years 1959-62/1962-66
Prod numbers 1366/1400

PRICES
A 19,000 **B** 14,000 **C** 9000

Alfa Romeo Giulietta SZ/SZ2

This was Zagato's effort on the Giulietta Sprint floorpan, SZ standing for Sprint Zagato. Bodywork was all-alloy and very lightweight, and so more suitable for competition work than Bertone's SS, a fact proven by the string of racing successes for the SZ. Stubby styling was changed for the SZ2 from 1961 with a longer Kamm-style tail. Most SZ2s have four-wheel disc brakes. All SZs make superbly evocative historic racers and are avidly hunted down for just such purposes. Very rare in any form but the SZ2 does carry a premium.

Engine 1290cc 74x75mm, 4 cyl DOHC, F/R
Max power 100bhp at 6500rpm
Max speed 126mph
Body styles coupé
Prod years 1959-60/1961-62
Prod numbers 210/30

PRICES
SZ **A** 35,000 **B** 28,000 **C** 24,000
SZ2 **A** 40,000 **B** 32,000 **C** 28,000

Alfa Romeo 2000 Spider

Alfa's 2000 range was nothing but a reworked 1900, sharing the same chassis and engine but with new bodies. The convertible version of the 2000 range had bodywork made by Touring of Milan and very elegant it looked. You even got two occasional seats in the rear. The twin cam engine was tweaked to give an extra 10bhp over the 2000 saloon but brakes remained drums all round. Five-speed gearbox with a floor shift was standard and a detachable hardtop was available as an option. Problems? Left-hand drive only and dreadful rust record.

Engine 1975cc 84.5x88mm, 4 cyl DOHC, F/R
Max power 115bhp at 5500rpm
Max speed 112mph
Body styles convertible
Prod years 1958-61
Prod numbers 3443

PRICES
A 10,000 **B** 7000 **C** 4500

Alfa Romeo 2600 Spider

1962's 2000 Spider replacement had almost identical bodywork to the 2000 but was distinguishable by its revised front end, full-width front bumper and lack of air ducts behind the front wheelarches. The big news was under the bonnet where a new twin cam six-cylinder engine was installed. 145bhp spelt a high top speed but the car was too heavy to be much of a serious cornerer. You got standard front discs and, from 1964, rear discs and a factory fitted RHD option as well. An undoubted classic and not too expensive either.

Engine 2584cc 83x79.8mm, 6 cyl DOHC, F/R
Max power 145bhp at 5900rpm
Max speed 125mph
Body styles convertible
Prod years 1962-65
Prod numbers 2255

PRICES
A 12,500 **B** 8500 **C** 5500

Alfa Romeo 2000/2600 Berlina

The replacement for the 1900, the 2000 was styled in-house by Alfa with sharp lines which were not to everyone's taste. A lovely engine, five-speed column-shift gearbox and room for five people sound good on paper but the new Berlina cost twice as much as the already-expensive Giulietta. As a result, the Berlina sold fewer examples than its Spider and Sprint brethren, so it is very rare today. 2600 six-cylinder engine from 1962, plus floor gearchange and front discs. UK available from 1964. Some coachbuilt saloons, such as the OSI (54 built). Also made in Brazil until 1974.

Engine 1975cc 84.5x88mm/2584cc 83x79.8mm, 4/6 cyl DOHC, F/R
Max power 105bhp at 5300rpm/130bhp at 5900rpm
Max speed 102/109mph
Body styles saloon
Prod years 1958-62/1962-68
Prod numbers 2927/2092

PRICES
A 7000 **B** 4000 **C** 2500

Alfa Romeo 2000/2600 Sprint

Third of the 2000 family was the Sprint. Its 2+2 bodywork was one of the first designs to come from a youthful Giugiaro while at Bertone, which also made the bodies. An attractive car from most angles, helped by the up-to-the-minute use of double headlamps. Because of its short production life the 2000 Sprint is very rare but the 2600 version (from 1962) was the most popular of all the 2000/2600 variants. It also had front disc brakes (all-wheel discs and RHD from 1964). Excellent value today although plagued by constant rust headaches.

Engine 1975cc 84.5x88mm/2584cc 83x79.8mm, 4/6 cyl DOHC, F/R
Max power 115bhp at 5500rpm/145bhp at 5900rpm
Max speed 112/124mph
Body styles coupé
Prod years 1960-62/1962-66
Prod numbers 700/6999

PRICES
A 8500 **B** 5500 **C** 3500

Alfa Romeo 2600 Sprint Zagato

Zagato's highly distinctive body on the 2600 was an all-aluminium and hence very lightweight creation. Unmistakable with its scalloped headlamps, large Alfa grille and quarter bumpers. Unlike the Giulietta SZ, the 2600 SZ's main purpose in life was not as a racing machine but as a fast touring car, and it had a high compression engine and triple-barrel carbs. Expensive in its day and so very few were built. That means you'll have to do a lot of hunting to find one these days and be prepared to fork out a pretty penny.

Engine 2584cc 83x79.8mm, 6 cyl DOHC, F/R
Max power 165bhp at 5900rpm
Max speed 135mph
Body styles coupé
Prod years 1965-67
Prod numbers 105

PRICES
A 30,000 **B** 20,000 **C** 12,000

Alfa Romeo Giulia 1300/1600

An upright but boxy saloon, the Giulia 105 series was the replacement for the 1950s Giulietta. Initially only the 1600 engine was offered, with five speeds and a column shift, rigid axle, drum brakes and front bench seat. After 1964 came the 1300 with single headlamps and four speeds. Spec gradually uprated on both to four-wheel discs. Giulia Nuova (1974) had evolved body style. All are 100mph cars but TI versions are most highly regarded, especially the rare 1600 TI (501 built). A very successful car for Alfa and cheap to buy today.

Engine 1290cc 74x75mm/1570cc 78x82mm, 4 cyl DOHC, F/R
Max power 78bhp at 6200rpm to 112bhp at 6500rpm
Max speed 100-109mph
Body styles saloon
Prod years 1964-78/1962-78
Prod numbers 258,550/228,251

PRICES
A 5000 **B** 2800 **C** 1400

Alfa Romeo Giulia TZ/TZ2

The successor to the Giulia SZ, Zagato's fabulous TZ was really only ever intended for competition but cars were road-equipped so we're including it here. The initial 'T' was for *Tubolare*, indicating its special tubular chassis with a shorter wheelbase than the Berlina or the Sprint (86.5in). Independent suspension all round. Base engine was the 112bhp 1.6-litre unit but in practice most were competition tuned and could exceed 150mph. Ten TZs had glassfibre bodies, a revolution for Alfa. The TZ2 had lower and quite different bodywork.

Engine 1570cc 78x82mm, 4 cyl DOHC, F/R
Max power 112/129bhp at 6500rpm
Max speed 118/125mph
Body styles coupé
Prod years 1963-64/1965-66
Prod numbers 120/50

PRICES
TZ **A** 80,000 **B** 60,000 **C** 50,000
TZ2 Almost never offered for sale, but likely to be around the same price as the TZ.

Alfa Romeo Giulia Sprint GT/GT Junior/GTV

Giulia nomenclature gets very confusing in the early 1960s but, while the 101 series Giulia Sprint continued, a new Giulia Sprint GT was launched in 1963 based on the 105 series with a short wheelbase. Bodywork was by Giugiaro out of Bertone and a masterpiece of subtlety, as proven by its 14-year run with hardly a line altered. Launched with 106bhp 1600 engine but also eventually sold as GT Veloce (GTV) with an extra 3bhp, plus 1300 Junior, 1600 Junior, 1750 GTV and 2000 GTV, the later versions with slatted grille. Rust has decimated most.

Engine 1290cc 74x75mm/1570cc 78x82mm/1779cc 80x88.5mm/1962cc 84x88.5mm, 4 cyl DOHC, F/R
Max power 87bhp at 6000rpm to 131bhp at 5500rpm **Max speed** 105-125mph **Body styles** coupé
Prod years 1963-68/1966-77/1967-77
Prod numbers 21,850/106,352/94,236

PRICES
1300 **A** 6000 **B** 3000 **C** 1000
1600/1750/2000 **A** 7500 **B** 4000 **C** 2000

Alfa Romeo Giulia Sprint GTA 1600/1300

The ultimate Giulia Sprint was the light-alloy bodied GTA. This was really a stripped-out racer which could also be used on the road and it did without bumpers and had light alloy wheels to keep weight down. All GTAs boasted the first incarnation of what is now called 'twin spark' – an engine with two plugs per cylinder. Initially that engine was the 1600 unit, superseded from 1968 by a twin-plug 1300cc engine with twin electric fuel pumps. Even though they're not that much faster, their rarity in either form means bumped-up prices.

Engine 1570cc 78x82mm/1290cc 78x67.5mm, 4 cyl DOHC, F/R
Max power 133bhp at 6000rpm/110bhp at 6000rpm
Max speed 116/109mph
Body styles coupé
Prod years 1965-68/1968-72
Prod numbers 493/447

PRICES
GTA 1600 **A** 30,000 **B** 23,000 **C** 20,000
GTA 1300 **A** 20,000 **B** 15,000 **C** 12,000

Alfa Romeo Giulia Sprint GTC

Years before the BMW 1600 Cabriolet or the Triumph Stag, Alfa Romeo was making a four-seater sporting convertible based on the Giulia Sprint. Called the GTC, it was a particularly neat conversion: the hood folded away out of sight behind the rear seats, enhancing Bertone's lines even more. Only offered for two seasons, and always with the 105bhp 1600 engine, the GTC is extremely rare and therefore much more highly valued than the fixed head Sprint or Junior. Rust situation is critical to the GTC's structural integrity.

Engine 1570cc 78x82mm, 4 cyl DOHC, F/R
Max power 105bhp at 6000rpm
Max speed 112mph
Body styles convertible
Prod years 1966-67
Prod numbers 1000

PRICES
A 12,500 **B** 11,000 **C** 9000

Alfa Romeo Duetto

Replacing the earlier Spider, the Duetto was the last major member of the Giulia 105 family. Styling was by Pininfarina but opinion was divided about it. With its 109bhp 1600 twin cam engine, the Duetto was a good performer, a delicate handler and a safe braker thanks to all-round discs. All Duettos had the 'coda lunga' boat-tail treatment and the model was renamed after 18 months as the Spider and fitted with a 1779cc engine. As the genesis of the Spider dynasty, the Duetto carries some kudos – just watch that rust!

Engine 1570cc 78x82mm, 4 cyl DOHC, F/R
Max power 109bhp at 6000rpm
Max speed 117mph
Body styles convertible
Prod years 1966-67
Prod numbers 6325

PRICES
A 11,500 **B** 8500 **C** 5000

Alfa Romeo 1750 Spider/1300 Spider/2000 Spider

The Duetto reverted to the Spider badge with the introduction of the 1779cc engine, officially called 1750 Spider Veloce. This was joined in 1968 by the 1300 Spider Junior, identifiable by its lack of headlamp covers and scarce in UK. Boat-tail rear end was truncated in 1970 and the 1750 replaced by the 2000 in 1971. RHD imports ceased in 1977, just before US laws brought in rubber bumpers. '70s 1600 Spider was Italy-only. Alloy engines need care and bodywork is as rust-prone as ever, but the Spider is a rewarding drive and distinctively Alfa.

Engine 1779cc 80x88.5mm/1290cc 74x75mm/1962cc 84x88.5mm, 4 cyl DOHC, F/R **Max power** 113bhp at 5000rpm/87bhp at 6000rpm/131bhp at 5500rpm **Max speed** 118/106/124mph **Body styles** convertible
Prod years 1967-71/1968-72/1971-93
Prod numbers 8721/7680/82,500

PRICES
1750 **A** 10,000 **B** 6500 **C** 4000
2000 **A** 9000 **B** 6000 **C** 4000
1300 **A** 7500 **B** 5000 **C** 3000

Alfa Romeo Junior Zagato 1300/1600

In 1969 Zagato styled a modern coupé body on the Spider wheelbase. The lines were not what you would call harmonious but you did have the advantage of guaranteed weatherproofing and some space for kids in the back. 1600 engine and slightly longer chassis arrived in 1972, but neither ever came to UK. The trouble was, the Junior Zagato cost more than either the Spider or the 1750 GTV, so it was never going to be a best-seller. Its very scarcity today makes it safe to say that the value of this unloved Alfa will rise in years to come.

Engine 1290cc 74x75mm/1570cc 78x82mm, 4 cyl DOHC, F/R
Max power 87bhp at 6000rpm/113bhp at 5000rpm
Max speed 105/118mph
Body styles coupé
Prod years 1969-72/1972-75
Prod numbers 1108/402

PRICES
1300 **A** 7500 **B** 5500 **C** 4500
1600 **A** 9000 **B** 7000 **C** 5500

Alfa Romeo 1750/2000 Berlina

The successor for the Giulia followed the old formula very closely with mildly updated and enlarged bodywork which, however, lost all the character features of its predecessor. The twin-choke twin carb 1779cc engine provided excellent performance while power-assisted all-round disc brakes pulled the car up effectively. 2-litre version from 1970 had new grille and revised dash, although production overlapped. Galloping rust and unhappy image have wiped out most of these good-selling saloons but prices are still very cheap.

Engine 1779cc 80x88.5mm/1962cc 84x88.5mm, 4 cyl DOHC, F/R
Max power 113bhp at 5500rpm/131bhp at 5500rpm
Max speed 112/116mph
Body styles saloon
Prod years 1967-72/1970-77
Prod numbers 101,880/89,840

PRICES
A 4000 **B** 2500 **C** 1500

Alfa Romeo Montreal

As a brave attempt by Alfa to muscle into Maserati territory, the Montreal was only partially successful. The bodywork derived from Bertone's show car of 1967 but, when it arrived in 1970, the front-engined Montreal looked quite different and perhaps a little too fussy. Engine was a detuned version of the T33 quad cam, dry sump, fuel-injected alloy racing V8 and is enough to make any DIY man wince. Four seats and real grand touring ability are strong suits but delicate nature and high running costs have kept prices down.

Engine 2593cc 80x64.5mm, V8 cyl DOHC, F/R
Max power 200bhp at 6500rpm
Max speed 136mph
Body styles coupé
Prod years 1970-77
Prod numbers 3925

PRICES
A 10,000 **B** 8000 **C** 6000

Alfa Romeo Alfasud

Drastically new Alfa with fresh range of engines, smart styling and (shock horror!) front-wheel drive. Intended to be an 'Alfa for the common man' it was built at a brand new factory in the deprived south of Italy (hence Alfasud). A real gem to drive, with pin-sharp handling and a wonderfully sweet flat-four 1186cc engine. Two-door TI from 1974 had more power, boot spoiler and twin headlamps. Five speeds from 1976, 1286cc from '77, 1350cc/1490cc from '78 in Super and TI tunes. Two/four-door only until '81 hatchbacks. But all its brilliance was ruined by monkey metal which rusted within months.

Engine 1186cc 80x59mm/1286cc 80x64mm/1350cc 80x67.2mm/1490cc 84x67.2mm, 4 cyl OHC, F/F
Max power 63-73bhp at 6000rpm/76bhp at 6000rpm/71-86bhp at 5800rpm/85-95bhp at 5800rpm
Max speed 93-109mph
Body styles saloon
Prod years 1972-83
Prod numbers 567,093

PRICES
A 2000 **B** 1000 **C** 400

Alfa Romeo Alfasud Sprint

Giugiaro's coupé body on the Alfasud floorpan was sharp and clean, with two doors plus a hatchback and four headlamps. Arriving in 1976, it had a perky 1.3-litre engine from the outset, uprated in 1978 to 1.5 litres which was also available in Veloce tune (95bhp). Still spacious interior, lighter weight and all the Alfasud's dynamic qualities heightened. A very attractive car which had a long production run. Inevitably, the Sprint suffered from all the usual Alfa rust problems. Find a nice one and look after it.

Engine 1286cc 80x64mm/1490cc 84x67.2mm, 4 cyl OHC, F/F
Max power 76bhp at 6000rpm/85bhp to 95bhp at 5800 rpm
Max speed 103-108mph
Body styles coupé
Prod years 1976-90
Prod numbers 96,450

PRICES
A 3200 **B** 2000 **C** 1000

Alfa Romeo Alfetta

All change for Alfa Romeo's new medium-sized saloon with a brand new floorpan and modern styling. Engines were the same as ever, though: initially 1800, then 1600 from 1975 and 2000 from 1977, all running concurrently. A VM turbodiesel version (Alfa's first ever diesel!) did not come to the UK. The name Alfetta derived from the glory years of Alfa's grand prix victories; the racing allusion was not totally lost as the Alfetta had a De Dion transaxle and integral clutch. Not a great Alfa but it did sell well. Should be cheap but buy your Waxoyl now.

Engine 1570cc 78x82mm/1779cc 80x88.5mm/1962cc 84x88.5mm 1995cc 88x82mm 4 cyl DOHC/DOHC/DOHC/diesel F/R
Max power 108bhp at 5600rpm/118bhp at 5300rpm/130bhp at 5400rpm/82bhp at 4300rpm
Max speed 109/112/115/96mph
Body styles saloon
Prod years 1972-84
Prod numbers c 450,000

PRICES
A 1800 **B** 1000 **C** 400

Alfa Romeo Alfetta GT/GTV

Beautiful coupé body on shortened Alfetta floorpan was designed by Giugiaro, one of his last jobs while still at Bertone. A lot more spacious inside than any previous Alfa coupé and initially featured a bizarre dash layout with all gauges except rev counter grouped in centre console. Like Alfetta on which it was based, available successively with 1.8, 1.6 and 2.0-litre engines and, from 1981, with the fine alloy V6 from the Alfa 6. Rust-prone, naturally, but there are still enough around to keep prices down – for the moment.

Engine 1570cc 78x82mm/1779cc 80x88.5mm/1962cc 84x88.5mm/2492cc 88x68.3mm, 4/V6 cyl DOHC, F/R **Max power** 108bhp at 5600rpm/118bhp at 5300rpm/130bhp at 5400rpm/160bhp at 5800rpm
Max speed 112-127mph
Body styles coupé
Prod years 1974-87
Prod numbers c 120,000

PRICES
A 4500 **B** 2500 **C** 1200

Allard

Allard J1/K1/L/M1/K2

One of the first manufacturers to get going after the war, Allard took full advantage of demand with the 1946 J1, K1 and L. The J was a short-chassis two-seater with cutaway sides, the K slightly longer with doors, the L longer still with four seats. Standard engine and gearbox were Ford Pilot V8, although the J could also be had with a 3.9 and the later K2 (photo) also with a 4.4-litre unit. All have divided front axles and a similar appearance but the K2 sported fuller bodywork and opening boot. Sturdy and fast. J1 is extremely rare as many were rebodied for trials use. M1 is the drophead coupé.

Engine 3622cc 77.8x95.2mm/3917cc 81x95.2mm/4375cc 84.1x98.4mm, V8 cyl SV/OHV/SV F/R **Max power** 85bhp at 3600rpm/140bhp at 4000rpm/ 110bhp at 3800rpm **Max speed** 86/102/95mph **Body styles** sports/roadster/tourer/drophead coupé/roadster **Prod years** 1946-47/1946-50/1946-50/1947-50/1950-53 **Prod numbers** 12/151/191/500/119
PRICES
K1/L/M1 **A** 22,000 **B** 17,000 **C** 11,000
K2 **A** 25,000 **B** 22,000 **C** 14,000

Allard P1/M2X

Front-end styling was similar to the first post-war Allards but the P1 was a 4/5 seater two-door saloon with full-width rear bodywork. Hardly what you'd call pretty but aluminium panelling kept weight down and performance up. Nasty column gearchange, rectified on drop-top M2X version. M2X also had forward radius arms and curious 'A' shaped grille. Coil springs from 1950 on. Again standard engines were 3.6-litre Ford or modified Mercury (4.4-litre). Easily the most popular Allards and the one in which Sydney Allard won the Monte in 1952.

Engine 3622cc 77.8x95.2mm/4375cc 84.1x98.4mm, V8 cyl SV, F/R
Max power 85bhp at 3600rpm/110bhp at 3800rpm
Max speed 90/95mph
Body styles saloon/drophead coupé
Prod years 1949-52/1951-53
Prod numbers 559/25

PRICES
P1 **A** 20,000 **B** 15,000 **C** 10,000
M2X **A** 24,000 **B** 18,000 **C** 12,000

Allard J2/J2X

Of all the Allards, the J2 is the most evocative. Stark in the extreme, it had a split front axle, De Dion rear and a removable alloy body. There wasn't much of it – cycle wings and aero screens were usual fitments – but at under a ton dry weight it went like stink. With the 'standard' 4.4-litre Mercury V8 in place, 0-60mph took 5.9 sec – sensational stuff for 1949. Could be even quicker with the Ardun OHV head for the Ford V8, or American options of Chrysler/Cadillac V8s. J2X of 1952 had radius arms ahead of the front axle and a longer nose. Uncomplicated, occasionally frightening, highly successful on the track.

Engine 3917cc 81x95.2mm/4375cc 84.1x98.4mm/5??0cc 96.8x95.25mm, V8 cyl OHV/SV, F/R
Max power 140bhp at 4000rpm/110bhp at 3800rpm/160-180bhp at 4000rpm
Max speed 100-130mph
Body styles sports
Prod years 1949-54
Prod numbers 173

PRICES
A 60,000 **B** 40,000 **C** 25,000

Allard P2 Monte Carlo/Safari

The 1952 replacement for the P1 had a new full-width body with much better proportions, aluminium-on-wood construction and a new space frame. Grille mimicked the M2X's 'A' pattern and the enormous one-piece bonnet folded forward with the wings. The saloon was called the Monte Carlo (celebrating Allard's 1952 victory) while the imposing (16½ft) woody estate was called the Safari and boasted three rows of seats, enough for eight passengers. Huge and heavy, so one of the optional American V8 powertrains might be advisable (Lincoln/Chrysler/Cadillac).

Engine 3622cc 77.8x95.2mm/4375cc 84.1x98.4mm, V8 cyl SV, F/R
Max power 85bhp at 3600rpm/110bhp at 3800rpm
Max speed 85/90mph
Body styles saloon/estate
Prod years 1952-55
Prod numbers 11/10

PRICES
Very rarely turn up for sale, but should be in the £10-25,000 range

Allard K3

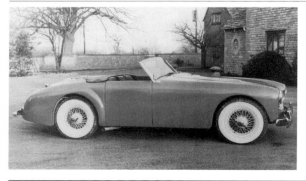

The full-width K3 was launched at the same time as the smaller Palm Beach and looked rather like it. Based on a shortened P2 chassis, like the P2 it had a gear lever sited on the floor on the right-hand side. Winding windows were a luxury, while the three-passenger wide bench seat betrayed the K3's role as a sporting car for the man with a small family. Most were exported to the USA where, like most Allards, they were fitted on entry with Lincoln, Cadillac or Chrysler V8 engines. Not really a match for the XK120, which was a cheaper car.

Engine 3622cc 77.8x95.2mm/4375cc 84.1x98.4mm, V8 cyl SV, F/R
Max power 85bhp at 3600rpm/110bhp at 3800rpm
Max speed 87/96mph
Body styles roadster
Prod years 1952-54
Prod numbers 62

PRICES
A 26,000 **B** 20,000 **C** 13,000

Allard Palm Beach 21C/21Z

A radical departure for Allard: a sports car with less than eight cylinders! Ford Zephyr or Consul engines powered Allard's budget roadster. Chassis was similar to the K3, with split axle front and rigid axle rear ends but a central gearchange and detachable side screens. Ford steel wheels were standard and knock-on wires optional. Sadly the Palm Beach rather had the carpet pulled from under it by the Austin-Healey, launched at the same time for the same price. The Palm Beach's dodgy handling consigned it to the fringes.

Engine 1508cc/2262cc 79.4x76.2mm, 4/6cyl OHV, F/R
Max power 47bhp at 4400rpm/68bhp at 4000rpm
Max speed 75/87mph
Body styles roadster
Prod years 1952-56
Prod numbers 8/65

PRICES
A 20,000 **B** 14,000 **C** 9000

Allard JR/J2R

The JR series was built for competition work and was based on the mechanical specification of the Palm Beach. However there was a De Dion rear end and a wide choice of engines, the factory preferred unit being the 5.4-litre Cadillac V8 with which American drivers scored many notable successes (and Allard himself led briefly in the 1953 Le Mans). Bodywork was even more compact than the Palm Beach and windscreens were eschewed for cut-down deflectors. The fastest of all the Allards and certainly the most collectable.

Engine 5420cc 96.8x92.25mm, V8 cyl OHV, F/R
Max power 250-270bhp at 4800rpm
Max speed 140mph
Body styles sports
Prod years 1953-56/1956-57
Prod numbers 17

PRICES
Likely to be in excess of £50,000

Allard Palm Beach MkII

By 1956, Allard was something of a spent force. Demand had dried up in the face of more sophisticated competition and only one model was listed after 1956, a restyled version of the Palm Beach with an entirely new body and a new X-braced chassis. This was the first Allard to abandon the divided front axle system (for struts and torsion bars). Engines were either Ford Zephyr or Jaguar XK but only a tiny handful of MkIIs were sold, two of those as GT coupés with neat fastback rear ends.

Engine 2553cc 82.5x79.5mm/3442cc 83x106mm, 4/6cyl OHV/DOHC, F/R
Max power 90bhp at 4400rpm/190-210bhp at 5500rpm
Max speed 97/120mph
Body styles convertible, coupé
Prod years 1956-50
Prod numbers 8

PRICES
Too rare to value accurately, but likely to be slightly more than the earlier Palm Beach

Alpine A106

Rally driver and Renault dealer Jean Rédélé's first production car was the A106, also known as the Coupé Mille Miles after he won his class in the 1956 Mille Miglia in one. The formula was simple: a Renault 4CV floorpan with a glassfibre body fitted over tubular reinforcers. The result was highly successful. Even with the standard 4CV engine fitted, the lightweight A106 could out-perform many bespoke machines with tuned engines. Renault was impressed too, and sold them through its dealers. Rare A106 Cabriolet of 1957 was also styled by Michelotti.

Engine 747cc 54.5x80mm, 4 cyl OHV, R/R
Max power 21bhp at 5000rpm to 43bhp at 6200rpm
Max speed 80-93mph
Body styles coupé, convertible
Prod years 1955-61
Prod numbers c 150

PRICES
A 15,000 **B** 12,000 **C** 8000

Alpine A108

This was Rédélé's development model between the A106 and the definitive A110 and, in 1960 Berlinette Tour de France guise, it established the pretty coupé style which Alpine would use for some 16 years. All A108s had a separate chassis. As before, 4CV engines were used in the tail, later uprated to the Dauphine (845cc) in various stages of tune from 31 to 68bhp, with swing axle rear suspension. The A108 design was licensed simultaneously to Willys of Brazil (where it was made as the Interlagos), Diesel Nacional in Mexico (and called Dinalpine) and Spain. There was also a rare 2+2 fixed head model.

Engine 747cc 54.5x80mm/845cc 58x80mm, 4 cyl, R/R
Max power 21bhp at 5000rpm to 43bhp at 6200rpm/31bhp at 5000rpm to 68bhp at 6200rpm
Max speed 90-105mph
Body styles coupé
Prod years 1958-65
Prod numbers c 500

PRICES
A 15,000 **B** 12,000 **C** 8000

Alpine · Alvis

Alpine A110

The seminal Alpine. Now using Renault 8 and 8 Gordini bits (and later Renault 16), the A110 stormed its way to dozens of rally victories and the 1973 world rally championship, establishing Alpine as France's leading sports car maker. Very pretty, the shape hardly changed in 14 years. Few concessions to road comfort but very quick; handling can be tricky. Four speeds standard but most were built with optional five-speed 'box. Spares support is adequate and you don't have to worry about rust. LHD only and never marketed in UK. Also made in Spain, Mexico, Brazil and Bulgaria (!).

Engine 956cc 65x72mm/1108cc 70x72mm/1255cc 74.5x72mm/1289cc 73x77mm/1565cc 77x84mm/1647cc 79x84mm, 4 cyl OHV, R/R **Max power** 48bhp at 5200rpm to 138bhp at 6000rpm **Max speed** 108-133mph **Body styles** coupé, convertible **Prod years** 1962-77 **Prod numbers** 8139
PRICES
1100/1300/1500 **A** 12,000 **B** 10,000 **C** 7000. 1600S/Cabriolet **A** 20,000 **B** 16,000 **C** 12,000

Alpine A110 GT4

Rédélé was not as interested in road cars as racers, and the GT4 was a market-induced attempt to make a practical family version of the A110. It shared the same basic underpinnings as the A110 but had a longer wheelbase to accommodate four passengers. Using a reworked A108 2+2 front, it had a frankly rather frumpy body but it served its purpose. Although it did not last long in production, the GT4 presaged much which was later found in the A310 chassis. Rarity has not so far pushed up prices and the GT4 remains the unloved runt of the Alpine litter.

Engine 1108cc 70x72mm, 4 cyl OHV, R/R
Max power 44bhp at 4900rpm to 95bhp at 6500rpm
Max speed 90-115mph
Body styles coupé
Prod years 1963-65
Prod numbers 112

PRICES
A 7500 **B** 5500 **C** 4000

Alpine A310/A310 V6

The A310 was intended to take Alpine firmly into Porsche's territory but in reality it fell short as a road car and as a racer. The layout was the same as the A110 – backbone chassis, glassfibre body, rear-mounted Renault engine – but it was bigger, a genuine 2+2 and with a very '70s appearance. First cars had the Renault 17TS engine but post-1976 ones got the PRV 2664cc V6 unit with much-improved torque. Much softer and quieter than the A110, and a better performer, with five gears from 1978. Like all Alpines up to this time, the A310 was never sold in the UK.

Engine 1605cc 78x84mm/2664cc x mm, 4/V6 cyl OHV/OHC, R/R
Max power 140bhp at 6450rpm/150bhp at 6000rpm
Max speed 130/137mph
Body styles coupé
Prod years 1971-76/1976-85
Prod numbers 2340/9276

PRICES
A310 **A** 7000 **B** 5000 **C** 3000
A310 V6 **A** 9000 **B** 6000 **C** 3000

Alvis TA14

Essentially a modified version of the prewar 12/70, the TA14 was a fairly conventional sporting saloon: mechanical brakes, rigid axles, leaf springs, four-speed 'box. Biggest visual change was the switch from 17in wires to 16in disc wheels. Although a heavy car powered by a mere 1.9-litre engine, the Alvis returned reasonable performance and good economy. Range expanded in 1948 to encompass a pair of two-door drophead coupés by Carbodies and Tickford. Also popular with specialist coachbuilders, and estate car versions were relatively common.

Engine 1892cc 74x110mm, 4 cyl OHV, F/R
Max power 65bhp at 4000rpm
Max speed 75mph
Body styles saloon, drophead coupé
Prod years 1946-50
Prod numbers 3213

PRICES
Saloon **A** 10,000 **B** 7500 **C** 4000
DHC **A** 15,000 **B** 11,000 **C** 8000

Alvis TB14/TB21

Ghastly sports version of the TA14 aroused passions among traditional Alvis clientele in 1948. Gone was the sober grille in favour of an odd cloverleaf shape with concealed headlamps, while the proportions of this imposing machine were awkward. Mechanically identical to the TA14 except for twin SU carbs and an extra 3bhp. Fold-flat windscreen and folding hood were standard. TB21 of 1950 (photo) updated the spec with a six-cylinder engine and coil spring/wishbone front suspension, plus traditional Alvis grille and wing-mounted headlamps.

Engine 1892cc 74x110mm/2993cc 84x90mm, 4/6 cyl OHV, F/R
Max power 68bhp at 4000rpm/95bhp at 4000rpm
Max speed 84/94mph
Body styles sports
Prod years 1948-50/1950-52
Prod numbers 100/31

PRICES
A 22,000 **B** 18,000 **C** 14,000

Alvis TA21/TC21/Grey Lady

1950's replacement for the TA14 was larger and had a more elegant, sweeping body style. Biggest news was the introduction of a new six-cylinder 3-litre seven-bearing pushrod engine, hydraulic brakes and coil sprung independent front suspension. Initially a single Solex carburettor, uprated in 1952 to twin SUs. TC21 introduced in 1953 had minor mechanical tweaks but more signifcant was the TC21/100 Grey Lady, good for 100mph thanks to its 100bhp engine. Grey Lady also had slim pillars, concealed hinges, Rudge wires and bonnet scoops.

Engine 2993cc 84x90mm, 6 cyl OHV, F/R **Max power** 83/93/100bhp at 4000rpm **Max speed** 84/92/100mph **Body styles** saloon, drophead coupé **Prod years** 1950-53/1953-54/1953-56 **Prod numbers** 1314/757/48

PRICES
TA21/TC21 **A** 10,000 **B** 8000 **C** 5000
TC21/100 Grey Lady **A** 13,000 **B** 10,000 **C** 6500
TA21/TC21 DHC **A** 20,000 **B** 14,000 **C** 9500

Alvis TC108G/TD21

After Tickford was acquired by Aston Martin, Alvis turned to Graber of Switzerland for its coachbuilt bodies. Graber was already familiar for its work on TC21 chassis and a modified version of its two-door saloon body was licensed to be built by Willowbrook of Loughborough in Britain. The so-called TC108G was both elegant and fast and came in saloon and drophead coupé styles. Restyled Park Ward TD21 from 1958 had single-piece rear screen and choice of auto 'box, more power and front discs from 1959, all-wheel discs and five speeds from Series II (1962).

Engine 2993cc 84x90mm, 6 cyl OHV, F/R **Max power** 104bhp at 4000rpm/115bhp at 4500rpm **Max speed** 95/110mph **Body styles** saloon, drophead coupé **Prod years** 1955-58/1958-63 **Prod numbers** 16/1070

PRICES
Saloon **A** 16,000 **B** 10,000 **C** 7500
DHC **A** 28,000 **B** 20,000 **C** 13,000

Alvis TE21/TF21

Subtly improved TD went into production in 1963 as the TE21. Easily identifiable by its stacked twin headlamps and reshaped rear wings, it also boasted more power and improved steering and front suspension. Saloon or drophead bodies, manual or automatic 'boxes, power steering option. TF21 from 1966 had triple carbs (in RHD form) and 150bhp, redesigned facia, improved springing and was offered alongside the TE21. Steel wheels often fitted, or traditional wires. These were the last Alvises as new master Rover killed off the marque.

Engine 2993cc 84x90mm, 6 cyl OHV, F/R **Max power** 130bhp at 5000rpm/150bhp at 4750rpm **Max speed** 108/118mph **Body styles** saloon, drophead coupé **Prod years** 1963-67/1966-67 **Prod numbers** 352/106

PRICES
Saloon **A** 17,500 **B** 13,000 **C** 8000
DHC **A** 30,000 **B** 22,000 **C** 15,000

AMC Javelin

American Motors' Mustang competitor never really got it together. It had neither the charisma nor the ability to make much of an impact, despite some competition success. Shape changed year by year, gradually getting bigger and more contrived and selling fewer and fewer examples. Major restyle in 1971 added curious wing lines. Large variety of engines used in seven-year production life, split into straight sixes and larger V8s. Imported to UK but never in great quantity.

Engine 3799cc 95.2x88.8 mm to 6572cc 105.6x93.4mm, 6/V8 cyl OHV, F/R **Max power** 100bhp at 3600rpm to 255bhp at 4600rpm **Max speed** 92-122mph **Body styles** coupé **Prod years** 1967-74 **Prod numbers** 235,497

PRICES
A 5000 **B** 3000 **C** 1500

AMC AMX

A short-wheelbase version of the Javelin, the AMX was a strict two-seater. The styling was similar and frankly far too dumpy to be considered a true rival for American exotics like the Corvette. Yet this was how the AMX was marketed. Only ever available with V8 engines, it was a rather feeble competitor among American muscle cars. Buyers steered clear of it in large numbers and the AMX lasted only three years. Any kudos the AMX might have had was lost when the AMX badge was subsequently used on a version of the longer Javelin. Almost never heard of outside the USA.

Engine 4749cc 95.2x83.3/5622cc 103.6x83.3mm/6383cc 105.8x90.8mm, V8 cyl OHV, F/R **Max power** 228bhp at 4700rpm/284bhp at 4800rpm/319bhp at 4600rpm **Max speed** 102/118/127mph **Body styles** coupé **Prod years** 1968-71 **Prod numbers** 19,134

PRICES
A 7000 **B** 4000 **C** 2500

AMC Gremlin

Launched on 1 April 1970, the Gremlin was no joke. A landmark model in a way, since the Gremlin was America's first sub-compact car. In reality, it was an AMC Hornet cut off behind the front doors and grafted on to a short-wheelbase platform and a hatchback rear end. The result was a car only 2½in longer than a VW Beetle but the interior was desperately cramped. And only in America could such a small car be ordered with a 5-litre V8, although sixes were more usual, and an Audi four-cylinder unit was even offered from 1976 to 1978.

Engine 1984cc to 4979cc, 4/6/V8cyl OHC/OHV, F/R
Max power 100bhp at 5000rpm to 152bhp at 4200rpm
Max speed 90-108mph
Body styles hatchback
Prod years 1970-78
Prod numbers c 650,000

PRICES
A 2500 **B** 1500 **C** 800

AMC Pacer

Perhaps the motoring world's greatest lemon was the AMC Pacer. Described by AMC as a 'wide but small car', this was its attempt to crack the valuable Ford Pinto market. Bizarre styling with a longer door on the passenger's side than the driver's, huge glass area and disastrous packaging left the Pacer looking like an oddball. If this 'economy' car hadn't been launched in the oil crisis it might have bombed even more spectacularly but AMC managed to shift 64,939 in the first six months. Thereafter it was downhill all the way.

Engine 3799cc 95.2x88mm/4235cc 95.2x99mm/4979cc 95.2x87.3mm, 6/V8 cyl OHV, F/R
Max power 100bhp at 3600rpm/110bhp at 3200rpm/125bhp at 3200rpm
Max speed 95/96/99mph
Body styles hatchback, estate
Prod years 1975-80
Prod numbers 280,858

PRICES
A 2500 **B** 1500 **C** 1000

Amphicar

Hans Trippel spent the war years developing and building amphibious vehicles for the German army. He returned to water-borne cars with the 1961 Amphicar, the only example of an amphibious car sold in large numbers. In the water it could do 6 knots (7mph) and on land only 65mph and it was inadequate in either role. The engine was a rear-mounted Triumph Herald unit driving the rear wheels or, in water, two screws. Novelty value may be high but it soon wears off; rust and sealing problems are a constant headache. Most went to America.

Engine 1147cc 69.3x76mm, 4 cyl OHV, R/R
Max power 38bhp at 4750rpm
Max speed 65mph
Body styles convertible
Prod years 1961-68
Prod numbers c 3000

PRICES
A 9000 **B** 5500 **C** 3500

Armstrong-Siddeley Lancaster/Hurricane/Typhoon

The Lancaster and Hurricane were Britain's first all-new postwar cars, launched in May 1945 in the very week that war ended. Named after wartime aircraft, each model was solidly built, durable and somewhat staid, though styling exciting at the time. The Lancaster was a four-door saloon, the Hurricane a drophead coupé and the 1946 Typhoon a two-door hardtop coupé with fabric roof covering. Torsion bar independent front suspension, hydro-mechanical brakes, four-speed 'box or Wilson preselector. Initially a 2-litre engine, bored out to 2.3 litres in 1949, along with lower bonnet and new facia.

Engine 1991cc 65x100mm/2309cc 70x100mm, 6 cyl OHV, F/R
Max power 70bhp at 4200rpm/ 75bhp at 4400rpm **Max speed** 70/75mph **Body styles** saloon/drophead coupé/coupé
Prod years 1945-52/1945-53/1946-50
Prod numbers 12,470

PRICES
Lancaster/Typhoon **A** 9000 **B** 5000 **C** 3000
Hurricane **A** 11,500 **B** 7000 **C** 4500

Armstrong-Siddeley Whitley

The Whitley was a slightly more dashing version of the Lancaster with a restyled, more angular rear roof section giving greater rear headroom. It was launched at the same time as the 18HP (2.3-litre) models, so only ever offered with this engine. Four-light saloon was later also made as a six-light and was built alongside a very rare long-wheelbase seven-passenger Limousine (photo) with extra luxury touches. Rarest versions of all were the Station Coupé and Utility pick-ups, hardly ever seen in Britain.

Engine 2309cc 70x100mm, 6 cyl OHV, F/R
Max power 75bhp at 4400rpm
Max speed 75mph
Body styles saloon, limousine
Prod years 1949-54 (Limousine 1950-51)
Prod numbers n/a

PRICES
A 8000 **B** 4000 **C** 2500

Armstrong-Siddeley Sapphire 346/Star Sapphire

More elegant style for the all-new Armstrong-Siddeley with new chassis and independent coil sprung front suspension. Brand new 3.4-litre hemi-head engine was good for 100mph with optional twin carbs. Gearbox choice was four-speed synchro, electronic preselector or (from 1954) Rolls-Royce Hydramatic auto. Power steering option was the first ever to appear on a British car. Four-light and six-light bodies available. Lwb limousine offered from 1955. Star Sapphire had 4-litre engine, front discs, auto 'box and power steering, and was also offered as a Limousine.

Engine 3435cc 90x90mm/3990cc 97x90mm, 6 cyl OHV, F/R
Max power 125bhp at 4400rpm/150bhp at 4700rpm/165bhp at 4250rpm
Max speed 95/100/102mph
Body styles saloon, limousine
Prod years 1952-58/1958-60
Prod numbers 7697/980

PRICES
A 10,000 **B** 6000 **C** 3500

Armstrong Siddeley Sapphire 234/236

Rather gormless looking sporting saloons which never really caught on. 236 revived the old Whitley 18HP engine, while the 234 used the Sapphire 346 engine minus two cylinders. The four was rather unrefined but had much more power than the 236 and would do 100mph. Inside, the 234/236 were well-equipped and roomy five-seaters. Recommended gearboxes were Manumatic automatic for the 236 and four-speed manual with Laycock overdrive for the 234. Both models had floor-mounted gear levers and the option of wire wheels. Well-built, long-lasting saloons.

Engine 2290cc 90x90mm/2309cc 70x100mm, 4/6 cyl OHV, F/R
Max power 120bhp at 5000rpm/85bhp at 4400rpm
Max speed 100/82mph
Body styles saloon
Prod years 1955-58/1955-57
Prod numbers 803/603

PRICES
A 7500 **B** 5500 **C** 3000

Arnolt MG

When Stanley 'Wacky' Arnolt, the MG importer for the USA, came across the Bertone stand at the 1952 Turin Show, his eye was caught by two special bodied MG TDs. The TDs had been bought with Bertone's last money and he must have been delighted when Arnolt asked the carrozzeria to build 200 of them. Arnolt became a director of Bertone to provide the finance. The pretty, enveloping bodywork of the Arnolt MG was, however, rather heavy and only half of the 200 planned cars were ever built. A prized rarity for the determined MG collector.

Engine 1250cc 66.5x90mm, 4 cyl OHV, F/R
Max power 55bhp at 5200rpm
Max speed 75mph
Body styles coupé, convertible
Prod years 1952-56
Prod numbers c 100

PRICES
A 25,000 **B** 18,000 **C** 10,000

Arnolt Bristol

When the MG TD left production, Arnolt took the opportunity of trying to find a more powerful engine/chassis combination to clothe. As he also happened to be the US importer for Bristol, he secured their co-operation to supply him with new 404 chassis. Bertone built a handsome sports body on the chassis and the result was a race winner and a commercial success. When a new Bristol 404 cost more than $10,000, Arnolt offered his Bristol for just $3995. Production only ended when Bristol stopped playing ball. Highly desirable.

Engine 1971cc 66x96mm, 6 cyl OHV, F/R
Max power 132bhp at 5500rpm
Max speed 112mph
Body styles sports, coupé
Prod years 1956-60
Prod numbers 142

PRICES
A 40,000 **B** 25,000 **C** 18,000

ASA 1000 GT/Spyder

It was always Enzo Ferrari's ambition to build a small sports car. With this in mind he began work on a small coupé with a four-cylinder engine, a chassis by Bizzarrini and bodywork by Bertone. It had almost been finished when Ferrari changed his mind and sold the project to the De Nora family, which formed ASA to build the car in series. Powerful 1032cc OHC engine delivered sparkling performance, gearbox was a four-speeder with double overdrive and there were four disc brakes. Last examples had GRP bodies and six-cylinder 1.3-litre engines.

Engine 1032cc 69x69mm, 4 cyl OHC, F/R
Max power 84bhp at 6800rpm
Max speed 115mph
Body styles coupé/convertible
Prod years 1962-68
Prod numbers 70/30

PRICES
A 25,000 **B** 15,000 **C** 10,000

Aston Martin 2-Litre (DB1)

Retrospectively known as the DB1, Aston Martin's first true post-war production model only entered production in 1948. It was a rather elegant aluminium drophead coupé with sweeping wings, a split windscreen and seating for 2+2. The 2-litre pushrod engine developed a meagre 90bhp but the model gained a reputation for fine handling and robust build. Independent coil spring front suspension, coil-spring live rear axle, Girling brakes. Sports version with fold-flat screen from 1949. Rarity lends value.

Engine 1970cc 82.5x92mm, 4 cyl OHV, F/R
Max power 90bhp at 4750rpm
Max speed 96mph
Body styles drophead coupé, sports
Prod years 1948-50
Prod numbers 14

PRICES
A 50,000 **B** 40,000 **C** 30,000

Aston Martin DB2

The DB2's 105bhp six-cylinder engine was a twin cam unit designed by W.O. Bentley for Lagonda. Based on the DB1 chassis, the smart new full-width body was initially sold in coupé form then, from 1949, as a rare drophead coupé (49 built), both strict two-seaters unless optional column change fitted, when three-abreast was possible. 120bhp Vantage engine from 1950 (125bhp from '51), one-piece windscreen and horizontal-slatted grille from 1952. All DB2 fixed heads are correctly referred to as saloons. Some say this is the finest car Aston Martin ever made.

Engine 2580cc 78x90mm, 6 cyl DOHC, F/R
Max power 105-125bhp at 5000rpm
Max speed 109-121mph
Body styles saloon, drophead coupé
Prod years 1949-53
Prod numbers 411

PRICES
Saloon **A** 35,000 **B** 25,000 **C** 17,000
DHC **A** 50,000 **B** 35,000 **C** 22,000

Aston Martin DB2/4 MkI/II/ DB MkIII

The DB2/4 from 1953 broadened the DB's appeal by offered 2+2 seating, with a higher rear roof section and an opening rear hatch. 125bhp Vantage engine was standard, but the extra weight blunted performance, cured by uprating to 3 litres and 140bhp from 1954. MkII of 1955 had more pronounced rear wings and stronger rear axle, plus a new model – a notchback hardtop (34 made). This and DHC were to special order only from 1956. MkIII got a 'droop snoot', standard 162bhp engine, hydraulic clutch, front discs and optional overdrive and automatic.

Engine 2580cc 78x90mm/2922cc 83x90mm, 6 cyl DOHC, F/R
Max power 125bhp at 5000rpm/ 140bhp at 5000rpm/162bhp at 5500rpm
Max speed 116/120/125mph **Body styles** saloon, dhc, hardtop
Prod years 1953-55/1955-57/1957-59
Prod numbers 565/199/551
PRICES MkI/II Saloon **A** 30,000 **B** 22,000 **C** 15,000
MkI/II/III DHC **A** 45,000 **B** 30,000 **C** 25,000. MkII/hardtop/MkIII Saloon **A** 37,000 **B** 28,000 **C** 22,000

Aston Martin DB4/DB4 DHC

Although it looked like an elegant evolution of the MkIII shape, the DB4 was in fact designed by Touring of Milan to the *superleggera* principle (light alloy panels over steel tube frame) and built in Britain. All-alloy DOHC engine was all-new and took Aston into a different performance league, especially in 266bhp Vantage guise (1962 on). Servo'd discs all round. An instant success despite its high price. Most coveted model is the Drophead Coupé (1961-63), of which 32 were built with the Vantage engine and 38 with the standard power unit.

Engine 3670cc 92x92mm, 6 cyl DOHC, F/R
Max power 240bhp at 5500rpm/266bhp at 5750rpm
Max speed 140/148mph
Body styles saloon, drophead coupé
Prod years 1958-63
Prod numbers 1040/70

PRICES
Saloon **A** 45,000 **B** 37,000 **C** 30,000
DHC **A** 70,000 **B** 55,000 **C** 40,000

Aston Martin DB4 GT/DB4 GT Zagato

The GT version of the DB4 had a wheelbase shortened by 5in and was therefore a strict two-seater. Intended mainly for competition work, it had a twin-plug engine capable of 302bhp, limited-slip diff and cowled headlamps. As it weighed considerably less than the DB4 saloon, it was a fiery machine. Zagato's muscle-bound body on the same chassis (photo) was even lighter (under 2800lb all-up) and its 314bhp meant shattering (6.1secs to 60mph) performance, barely contained by the chassis. Without doubt these are the most coveted of all classic Astons.

Engine 3670cc 92x92mm, 6 cyl DOHC, F/R **Max power** 302/314bhp at 6000rpm **Max speed** 149/155mph
Body styles coupé
Prod years 1959-63/1960-63
Prod numbers 75/19

PRICES
DB4 GT **A** 120,000 **B** 100,000 **C** 90,000
DB4 GT Zagato **A** 250,000 **B** 200,000 **C** 150,000

Aston Martin DB5/DB5 DHC

James Bond made the DB5 famous and some regard it as the most beautiful of all the Astons with its standard cowled headlamps. Standard engine grew to 4 litres and had three SU carbs and, from 1964, came in Vantage form with twin-choke Webers and 314bhp. Earliest cars had four speeds and optional overdrive but five gears became standard within three months, plus an automatic option. Some 37 cross-over ('short-chassis') DB5 Volantes made 1965-66. Vantage models carry a 10% premium and automatics worth correspondingly less.

Engine 3995cc 96x92mm, 6 cyl DOHC, F/R
Max power 282bhp at 5500rpm/314bhp at 5750rpm
Max speed 149/155mph
Body styles saloon, drophead coupé
Prod years 1963-65
Prod numbers 898/123

PRICES
Saloon **A** 50,000 **B** 38,000 **C** 30,000
DHC **A** 70,000 **B** 55,000 **C** 40,000

Aston Martin DB6/Volante

A longer wheelbase, more steeply raked screens and cut-off tail helped make the DB6 more aerodynamic and quicker, despite the extra weight. Recognisable by split front and rear bumpers and higher rear roofline to improve rear seat headroom. Interior appointments grew steadily more comfortable (air conditioning was popular). Extra power for Vantage engine, and options include power steering, LSD and automatic. Drophead renamed Volante and has power hood. 1969 MkII has flared arches, power steering, optional fuel injection. Add 10% for Vantage, deduct 10% for automatic.

Engine 3995cc 96x92mm, 6 cyl DOHC, F/R
Max power 282bhp at 5500rpm/325bhp at 5750rpm
Max speed 145/155mph
Body styles saloon, drophead coupé
Prod years 1965-70
Prod numbers 1567/215

PRICES
Saloon **A** 38,000 **B** 30,000 **C** 22,000
Volante **A** 70,000 **B** 55,000 **C** 40,000

The Aston Workshop

- Total body and chassis rebuilds to the highest standards.
- Complete or partial restoration.
- Interior trimming by ex-factory trimmer.
- Full engine rebuilds/ uprated specification.
- Suspension rebuilds and upgrades.
- Accident repair.
- Repainting – full bare metal or individual panels.
- Routine maintenance or concours restoration.

From feeler gauges for your tool kit, to a complete front end for your DB5. The Aston Workshop can supply any currently available part that your Aston Martin may require. In addition we offer a bespoke manufacturing service for normally unobtainable items. Limited supplies of rare used parts. The Aston Workshop DB6 illustrated parts catalogue and price guide: £15.00 exc. postage. Free

GOSLING HILL, RED ROW, BEAMISH, COUNTY DURHAM
TEL: 01207 233525 FAX 01207 232202

Aston Martin DBS/DBS V8

William Towns' attractive American-influenced shape was a radical departure for Aston in 1967. The DBS was a full four-seater, now with a De Dion rear end. Its engine was the same six-cylinder unit as the DB6, including the Vantage option, but increased weight really knocked the oomph out of it. Quad cam fuel-injected V8 power from 1969 was necessary, in which case power steering and adjustable dampers were standard. Manual and Chrysler Torqueflite 'boxes available. Expensive to maintain but then they're cheap to buy. Manuals worth 25% more.

Engine 3995cc 96x92mm/5340cc 100x85mm, 6/V8 cyl DOHC/QOHC, F/R **Max power** 282bhp at 5500rpm to 325bhp at 5750rpm/375bhp at 5000rpm **Max speed** 140/150/159mph **Body styles** saloon **Prod years** 1967-73/1969-72 **Prod numbers** 857/405

PRICES
DBS **A** 15,000 **B** 11,000 **C** 7500
DBS V8 **A** 18,000 **B** 12,500 **C** 8500

Aston Martin V8

The DBS V8 was renamed just V8 in 1972. For 1973 Aston eschewed injection for four Weber carbs and increased power (though never specified). Always with single headlamps instead of double ones and a lower front wing line. Many detail changes inside and increasing levels of opulence as production progressed. Automatic is the norm but a (very heavy) manual clutch was always available. 438bhp Vantage available from 1977 (320 built) and convertible Volante from 1978 (915 made). Huge, heavy, unsubtle things, but uniquely Aston Martin.

Engine 5340cc 100x85mm, V8 cyl QOHC, F/R **Max power** est 340-438bhp **Max speed** 160mph **Body styles** saloon, convertible **Prod years** 1972-90 **Prod numbers** c 1600

PRICES
V8 **A** 28,000 **B** 20,000 **C** 15,000
V8 Vantage **A** 35,000 **B** 25,000 **C** 18,000. V8 Volante **A** 50,000 **B** 35,000 **C** 26,000

Aston Martin Lagonda

The revival of the Lagonda name after an absence of ten years proved to be something of an anomaly. A V8 had its wheelbase stretched by 12in and was clothed in a four-door bodyshell to a design penned by William Towns, closely based on the V8 shape. It all looked rather odd, however, especially with that unhappy Lagonda-style grille and long cabin. Its purpose was a luxury express but such machines were not in vogue in the mid-1970s oil crisis. Production faltered around Aston Martin's 1974 liquidation and the car was replaced by the startling new Lagonda in 1976. Rare, so expensive.

Engine 5340cc 100x85mm, V8 cyl QOHC, F/R **Max power** est 340bhp **Max speed** 150mph **Body styles** saloon **Prod years** 1974-76 **Prod numbers** 7

PRICES
Hardly ever sold, but a fine example should fetch around £40,000

Aston Martin Lagonda Series 1

The world was stunned by William Towns' drastic restyle of the Lagonda in 1976. Razor-edge lines, the narrowest grille you ever did see, and a stunning digital electronic dash. The latter was beset by gremlins and, when deliveries finally began in 1979, had been replaced by more conventional instruments. Once *de rigueur* in Arab circles, Lagonda sales were respectable but the car is today regarded as somewhat vulgar. Always automatic. Ghastly Tickford bodykit version and stretch limo were marginal; superseded by Series 2 restyle in 1987.

Engine 5340cc 100x85mm, V8 cyl QOHC, F/R **Max power** est 340bhp **Max speed** 143mph **Body styles** saloon, limousine **Prod years** 1976-87 **Prod numbers** 610 (incl Series 2)

PRICES
A 22,000 **B** 18,000 **C** 12,000

Audi 60/70/75/80/90

Famous German name revived under VW in 1965 with the F103 series, basically a facelifted DKW F102 with a front-drive Mercedes-designed four-stroke engine. Four-speed all-synchromesh gearbox and inboard discs standard. Two- and four-door saloon and three-door Variant estate bodyshells with a variety of engines: 1.5 not seen in UK, 1.7 and 1.8, the latter known as the Super 90. Very good sellers in Germany but hardly the stuff dreams are made of.

Engine 1496cc 80x74.4mm/1696cc 80x84.4mm/1770cc 81.5x84.4mm, 4 cyl OHV, F/F **Max power** 55bhp at 4750rpm to 90bhp at 5200rpm **Max speed** 87-102mph **Body styles** saloon, estate **Prod years** 1965-72 **Prod numbers** 416,853

PRICES
A 1800 **B** 900 **C** 400

Audi 100

Handsome, larger new generation Audi which was targeted at BMW followed the conventional layout of the previous models, including front-wheel drive. Three states of tune on the standard 1.8-litre engine were offered: base, S and LS. GL with 1.9-litre engine and double headlamps followed. Manual gearbox only to start with but automatic option from 1970. Two and four-door versions available. Built Audi's reputation - especially in UK – for sound engineering, strong performance and an air of luxury.

Engine 1760cc 81.5x84.4mm/1871cc 84x84.4mm, 4 cyl OHV, F/F
Max power 80bhp at 5000rpm to 112bhp at 5800rpm
Max speed 97-106mph
Body styles saloon
Prod years 1968-76
Prod numbers 796,787

PRICES
A 1600 **B** 800 **C** 300

Audi 100 Coupé S

The only collectable Audi (at least until the Quattro) was the 100 Coupé S. Based on the floorpan and mechanicals of the 100 saloon, it had its own distinctive and rather attractive coupé body. Only ever offered with the 1.9-litre engine which equated to a top speed of 112mph. Although you could not describe this rebodied saloon as particularly sporting it does have a strong following, with dedicated owners' clubs across Europe. Almost no body parts shared with 100 saloon, so some parts headaches.

Engine 1871cc 84x84.4mm, 4 cyl OHV, F/F
Max power 115bhp at 5500rpm
Max speed 112mph
Body styles coupé
Prod years 1970-76
Prod numbers 30,687

PRICES
A 2500 **B** 1200 **C** 700

Audi 80

Replacement for the DKW derived medium-sized Audis, the 80 was a smart all-new model. Engines were all Audi's own work and all overhead cam units: L had 1.3, LS and GL were 1.5 till 1975, 1.6 engine was introduced in 1974 and powered the manual-only GT and high-spec versions after 1975. Two- and four-door saloons, five-door estate, manual or auto 'boxes, MacPherson strut front suspension. Entirely competent cars which founded a generation of Audis – the 'new' 80, Coupé and Quattro were all derived from this floorpan.

Engine 1296cc 75x73.4mm/1470cc 76.5x80mm/1588cc 79.5x80mm, 4 cyl OHC, F/F
Max power 55bhp at 5500rpm to 110bhp at 6100rpm
Max speed 90-112mph
Body styles saloon, estate
Prod years 1972-78
Prod numbers 939,931

PRICES
A 1500 **B** 700 **C** 300

Austin 8HP/10HP

Austin's 1939 saloon was reintroduced in 1945 in four-door form only. A cumbersomely styled six-light saloon with a hint of American lines, it was sturdily built and followed prewar practice with one exception: a floor which was integral with the chassis, providing a semi-platform. Beam axles, cart springs and rod-operated brakes. 10HP was slightly larger and had raised edges to the wheelarches. 8HP engine derived from 1937 Big Seven, while the 10HP's went back to 1932. Both were deathly slow and iffy handlers but were useful in kick-starting Austin after the war.

Engine 900cc 56.7x89mm/1125cc 63.5x89mm, 4 cyl SV, F/R
Max power 27bhp at 4400rpm/32bhp at 4000rpm
Max speed 61/62mph
Body styles saloon
Prod years 1939-47
Prod numbers 56,103/55,521

PRICES
A 3500 **B** 2200 **C** 1200

Austin 12HP/16HP

Stylistically enlarged 8HPs but with a separate chassis. Very roomy inside and with a standard sliding sunroof. 12HP's 1.5-litre four-cylinder engine was inadequate for a car of this size, and the 16HP's 2.2 litres did a much better job. It was the very first OHV engine used in an Austin passenger car. From 1947, there was a Countryman 'woody' estate which is charming in a way. Only badges distinguish the 12HP from the 16HP. Not dynamic machines but worthy after their own fashion.

Engine 1535cc 69.3x101.6mm/2199cc 79.3x111mm, 4 cyl SV/OHV, F/R
Max power 42bhp at 4000rpm/67bhp at 3800rpm
Max speed 65/75mph
Body styles saloon, estate
Prod years 1939-47/1945-49
Prod numbers 8698/35,434

PRICES
A 4000 **B** 2500 **C** 1600

Austin

Austin A110/A125 Sheerline

A massive saloon in the traditional style aimed more at the owner-driver than the carriage trade. Built on a huge X-braced box-section chassis, the semi-razor edge bodywork was by Austin style guru Dick Burzi. Wood and leather filled the interior, although instrument styling was not in keeping. Prewar truck six engine with single Stromberg carb (bored out to 4 litres in first year), coil-and-wishbone independent front suspension and hydraulic brakes but woeful column-shift four-speed 'box. Long wheelbase six-light limousine available from 1949, some supplied as chassis for ambulances and hearses.

Engine 3460cc 85x101.6mm/3993cc 87x111mm, 6 cyl OHV, F/R
Max power 110/125bhp at 3700rpm
Max speed 82/83mph
Body styles saloon, limousine
Prod years 1947/1947-54
Prod numbers c 9000

PRICES
A 6000 **B** 3500 **C** 1800

Austin A120/A135 Princess

Sister model for the Sheerline with same chassis and coachwork by newly acquired Vanden Plas. Despite aluminium construction, heavier and more bulky than Sheerline but better looking, with faired-in headlamps, more subtle wing line and rear spats. More power thanks to triple SU carbs (but single carb after '53). 1950 MkII has reshaped rear door, 1953 MkIII new front end. Automatic 'box optional from 1955. Touring Limousine version had a glass division. Long wheelbase saloon and limousine models sold from 1952, but renamed Vanden Plas from 1957 and better known as a VP (see separate entry).

Engine 3460cc 85x101.6mm/3993cc 87x111mm, 6 cyl OHV, F/R
Max power 120/135bhp at 3700rpm
Max speed 89mph
Body styles saloon, touring limousine, lwb saloon, limousine
Prod years 1947-57
Prod numbers 1910

PRICES
A 7000 **B** 4000 **C** 2200

Austin A40 Dorset/Devon/Countryman

The 8HP replacement had a separate chassis and more modern lines reminiscent of American Chrysler, unhappily squeezed to fit European size. Hydromechanical brakes now standard. The new OHV engine was a shrunken version of the 16HP unit and would evolve into the B-series engine – which lasted until 1980. Dorset had two doors and was mainly for export, dropped in 1949. Devon was four-door, six-light model, discontinued 1952. Countryman estate was merely an A40 van with windows which lasted until 1956. Likeable, but usual dull Austin dynamics.

Engine 1200cc 65.5x89mm, 4 cyl OHV, F/R
Max power 40bhp at 4300rpm
Max speed 71mph
Body styles saloon, estate
Prod years 1947-56
Prod numbers 15,939/273,958/26,587

PRICES
A 2500 **B** 1500 **C** 700

Austin A70 Hampshire

A step up from the A40, the A70 Hampshire had six-light styling in the A40's vein, albeit with rear spats. A40-style chassis housed OHV engine from the 16HP. In theory the Hampshire could seat six but short wheelbase (only 4in more than A40) left rear passengers cramped. Column change and split bench front seat. At last, reasonable performance in an Austin saloon but the same barge-like handling, helped only a little by independent front suspension. Partially-wooded Countryman is cute and rare (901 built). More than 20,000 pick-ups were built (including Herefords), most going for export.

Engine 2199cc 79.3x111mm, 4 cyl OHV, F/R
Max power 67bhp at 3800rpm
Max speed 83mph
Body styles saloon, estate
Prod years 1948-50
Prod numbers 35,461

PRICES
A 3000 **B** 1700 **C** 800

Austin A70 Hereford

Updated Hampshire retained its chassis and mechanicals virtually unmodified but the bodywork pioneered Austin's move to four-light styling. Slightly longer wheelbase freed up space for a genuine six passengers. Handling vagaries remain but braking is better thanks to standard fully-hydraulic brakes. A70 Countryman estate (as ever with wood panelling) is rare. Even less common Drophead Coupé model (266 built), with a body by Carbodies and the option of a power hood, has '50s character and sunshine for six passengers.

Engine 2199cc 79.3x111, 4 cyl OHV, F/R
Max power 67bhp at 3800rpm
Max speed 80mph
Body styles saloon, estate, convertible
Prod years 1950-54
Prod numbers 50,421

PRICES
Saloon **A** 3200 **B** 2000 **C** 1000
DHC **A** 6000 **B** 3200 **C** 1600

Austin A90 Atlantic Convertible/Saloon

For conservative Austin, the extravagant A90 Atlantic was a moment of pure bravado. A flop commercially, as the American market took only 350. Controversial Dick Burzi styling with triple headlamps, gaudy chrome, bold uninterrupted wing line, rear wheels covered by impractical spats. Gold-faced gauges, power hood and window options (a first for a British mass-produced car). Saloon version 1949 on. Mechanically, the Atlantic was merely an A70 but its bored-out engine and twin SU carbs were its best feature, going on to power the Austin-Healey 100.

Engine 2660cc 87.3x111.1mm, 4 cyl OHV, F/R
Max power 88bhp at 4000rpm
Max speed 96mph
Body styles convertible/saloon
Prod years 1948-50/1949-52
Prod numbers 7981

PRICES
Convertible **A** 11,000 **B** 7500 **C** 5000
Saloon **A** 9000 **B** 5000 **C** 3000

Austin A30/A35

Belated successor to the prewar Seven; indeed, initially called the Seven. Austin's first unitary car, with Anglo-American style. Tall, narrow body could seat four and squeezed in four doors (two-door followed in 1953). 803cc engine would become the basis for the A-series unit, still in production. Not really a match for the faster, roomier Morris Minor but a strong seller. A35 of 1956 had 948cc engine, remote gearchange, smaller wheels, larger rear window and body-colour grille. Saloon died in 1959 but Countryman estate survived until 1962 and van until 1968. Pick-up very rare.

Engine 803cc 58x76mm/948cc 62.9x76.2mm, 4 cyl OHV, F/R
Max power 28bhp at 4800rpm/34bhp at 4750rpm
Max speed 62/73mph
Body styles saloon, estate
Prod years 1951-56/1956-62
Prod numbers 222,264/294,892

PRICES
A 2500 **B** 1700 **C** 1000

Austin A40 Somerset

The Somerset attempted to modernise the antiquated Devon, with bulbous lines in the same family style as the A30 and A70 Hereford. The characteristic wingline was retained but there was new frontal treatment. However, the A40 was still hardly in the vanguard of British car design, even for 1952. Chassis and mechanicals not significantly revised from the previous A40 range. Interesting A70-style two-door drophead version with a body made for Austin by Carbodies of Coventry is rarest Somerset (7243 made). Plentiful and cheap.

Engine 1200cc 65.5x89mm, 4 cyl OHV, F/R
Max power 40bhp at 4300rpm
Max speed 68mph
Body styles saloon, convertible
Prod years 1952-54
Prod numbers 173,306

PRICES
Saloon **A** 2700 **B** 1400 **C** 700
DHC **A** 5000 **B** 3500 **C** 2400

Austin A40 Sports

Austin had for some years supplied engines to Jensen of West Bromwich for use in its luxury grand tourers, before Jensen returned the compliment by designing and building an alloy sports body for Austin in 1950 on an A40 Devon chassis. The A40 Sports looked somewhat like a scaled-down Jensen Interceptor, had a folding hood and boasted four seats. Only slightly modified engine (twin SU carbs were fitted) and unmodified Devon chassis wasn't the best recipe for a sports car. Most of the meagre production was exported, the greatest number to the USA.

Engine 1200cc 65.5x89mm, 4 cyl OHV, F/R
Max power 46bhp at 5000rpm
Max speed 79mph
Body styles convertible
Prod years 1950-53
Prod numbers 4011

PRICES
A 6000 **B** 4000 **C** 2800

Austin A40/A50/A55 Cambridge

At last unitary construction for a family-size Austin and, with it, a modern-ish integral-wing body. Reworked Devon/Somerset engine for the A40 and new B-series 1489cc engine for the A50, both with Zenith carbs and mechanical fuel pumps. Wishbone ifs, hydraulic brakes and clutch, four-speed column-change 'box. Two-door version at launch never made it for public consumption. De Luxe trim adds chrome and leather. A55 superseded both in 1957: longer boot, larger rear window, small fins, most with floor shift. Vans and pick-ups lasted until 1971.

Engine 1200cc 65.5x89mm/1489cc 73x89mm, 4 cyl OHV, F/R
Max power 42bhp at 4500rpm/50bhp at 4400rpm/51bhp at 4500rpm
Max speed 69/74/77mph
Body styles saloon
Prod years 1954-56/1954-57/1957-58
Prod numbers 30,666/114,867/ c 154,000

PRICES
A 2000 **B** 1200 **C** 600

Austin

Austin A90/A95/A105 Westminster

The new 85bhp A90 Six Westminster bore an obvious family resemblance to the A40/A50 Cambridge, launched at nearly the same time, but only the doors were interchangeable. The Westminster was a bigger car all round. Austin finally offered a six-cylinder engine in a family-sized car with the C-series, basically a bored-out B-series. 1956 front and tail re-style produced A95 (92bhp) and the 102bhp twin-SU A105 with snazzy wheel trims. All available with overdrive (standard on '56-on A105), auto option on A95 and A105. Rare leather-and-walnut Vanden Plas A105 from 1958 (500 made).

Engine 2639cc 79.4x89mm, 6 cyl OHV, F/R
Max power 85bhp at 4000rpm/92bhp at 4500rpm/102bhp at 4600rpm
Max speed 90/91/96mph
Body styles saloon, estate (A95)
Prod years 1954-59
Prod numbers 60,367

PRICES
A 3200 **B** 2000 **C** 1000

Austin Taxi FX3/FL1

Austin dominated the London taxi market from the 1930s and its postwar effort, the FX3, had a virtual monopoly. Unique chassis with leaf springs and mechanical brakes, and an upright body which resembled the 16HP. Engine came from the 16HP, joined in 1954 by BMC's new 2.2-litre diesel. London spec taxis had an open luggage platform where the front passenger's seat should be but there was a Hire Car (FL1) version with an extra front door and passenger seat. Austin even tried to export to New York, with predictable results.

Engine 2199cc 79.3x111/2178cc 82.6x101.6mm, 4 cyl OHV/diesel, F/R
Max power 52/67bhp at 3800rpm/55bhp (diesel) at 3500rpm
Max speed 60mph
Body styles taxi, saloon
Prod years 1948-58
Prod numbers 13,500

PRICES
A 3500 **B** 2200 **C** 1300

Austin Taxi FX4/FL2

Modern bodywork for 1958 was the FX4's biggest advance but cabbies also got fully automatic transmission as standard, independent front suspension and hydraulic brakes. Manual gearbox was offered from 1961 and the good old 2.2-litre petrol engine from 1962, but this was marginal and, like the FL2 hire car version, sold only with the manual gearbox. BMC 1100/1300 rear lights from 1970 and a new 2.5-litre diesel from 1971. Carbodies gradually took over complete manufacture and still makes it today as the Fairway, powered by Nissan engines.

Engine 2199cc 79.3x111/2178cc 82.6x101.6mm/2520cc 88.9x101.6mm, 4 cyl OHV/diesel/diesel, F/R
Max power 56bhp at 3750rpm/55-60bhp (diesel) at 3500rpm
Max speed 75mph
Body styles taxi, saloon
Prod years 1958-date
Prod numbers 43,000 (to end 1981)

PRICES
A 3000 **B** 1800 **C** 900

Austin A40 'Farina'

The first post-war Austin saloon without an input from Dick Burzi was the crisp Pininfarina-styled A40 of 1958 – also the first ever 'two-box' design so typical of modern hatchbacks. The split tailgate Countryman (1959) predated the hatchback. In layout, very similar to the A35: the same 948cc engine, similar suspension, steering and the unhappy hydro-mechanical brakes, even the same instrument panel. MkII of 1961 has 4in longer wheelbase, new grille, hydraulic brakes. 1098cc engine from 1962 injected much-needed extra pep. Not the best-loved BMC product ever seen.

Engine 948cc 62.9x76.2 mm/1098cc 64.6x83.7mm, 4 cyl OHV, F/R
Max power 34bhp at 4670rpm/37bhp at 5000rpm/48bhp at 5100rpm
Max speed 72/75/82mph
Body styles saloon, hatchback
Prod years 1958-67
Prod numbers 342,280

PRICES
A 2000 **B** 1200 **C** 700

Austin A55/A60 Cambridge

Farina's hand created Austin's new Cambridge in 1958, hence the similarity to Peugeot's new 404. Sharp style with rear fins, but looked rather obese and was very heavy. 1489cc engine carried over from old Cambridge, now with SU carb. IFS and semi-elliptic rear, floor-mounted gear lever or column shift option. Popular Countryman estates but slow and distinctly unpopular diesel. 1961 A60 lost the fins (except in estate form), gained 1in in wheelbase, got a new grille, 1622cc engine, auto option. Honest, tough, reliable, but ultimately dull.

Engine 1489cc 73x89mm/1622cc 76.2x89mm, 4 cyl OHV/diesel/OHV, F/R
Max power 52bhp at 4350rpm/40bhp at 4000rpm (diesel)/61bhp at 4500rpm
Max speed 81/66/84mph
Body styles saloon, estate
Prod years 1959-69
Prod numbers 426,528

PRICES
A 2000 **B** 1400 **C** 700

Austin

Austin A99/A110 Westminster

The same Farina lines seen on the '59 Cambridge were quite pleasingly applied to the bigger '59 Westminster. BMC C-series was now bored out to 3 litres and fitted with twin SUs. All-synchro three-speed 'box (as ever controlled from the steering column) had double overdrive; auto optional. One innovation: servo-assisted disc front brakes (a first for an Austin saloon). Longer-wheelbase A110 of 1961 gained power and luxury, plus power steering and air con options from 1962. Four-speed MkII of 1964 is the best of the lot.

Engine 2912cc 83.3x88.9mm, 6 cyl OHV, F/R
Max power 103bhp at 4500rpm/120bhp at 4750rpm
Max speed 98/102mph
Body styles saloon
Prod years 1959-61/1961-68
Prod numbers 15,162/c 26,100

PRICES
A 3000 **B** 2000 **C** 1000

Austin 1100/1300

Austin's version of Issigonis's front-drive family car appeared in 1963. The 1100 was the first car to use the new Hydrolastic interconnected fluid suspension, and offered disc front brakes and a sealed cooling system. Mini-like packaging, including bored-out A-series engine driving the front wheels and a spacious, if austere, interior. 1300, two-door saloon and Countryman estate versions followed. Was Britain's best-selling car for years. Apart from the 70bhp 1969 1300GT, the 1100/1300 range ultimately lacked character despite its technical brilliance.

Engine 1098cc 64.6x83.7mm/1275cc 70.6x81.3mm, 4 cyl OHV, F/F
Max power 48bhp at 5100rpm/58bhp at 5250rpm/70bhp at 6000rpm (1300 GT)
Max speed 78/92/96mph
Body styles saloon, estate
Prod years 1963-75/1967-75
Prod numbers c 1,060,800

PRICES
A 1500 **B** 900 **C** 500

Austin 1800/2200

The last of Issigonis's trio of front-wheel drive cars, resembling an inflated 1100 and nicknamed 'Landcrab'. Mechanically advanced for its day: front-wheel drive and Hydrolastic suspension. Sturdy, extremely spacious and dependable, it was even Car of the Year. Minus: dreadful gearchange, plain interior, poor driving position and heavy steering (optional PAS arrived 1967). De-tuned MGB engine or, as 1800S, with twin carbs. 2200 of 1972 had 'one-and-a-half' Maxi 1500 engines – the only in-line six-cylinder engine ever fitted transversely in a front-wheel drive car.

Engine 1798cc 80.3x88.9mm/2227cc 76.2x81.3mm, 4/6cyl OHV/OHC, F/F
Max power 80bhp at 5000rpm/96bhp at 5700rpm (1800S)/110bhp at 5250rpm
Max speed 90/99/108mph
Body styles saloon
Prod years 1964-75
Prod numbers c 210,000/c 11,000

PRICES
A 1500 **B** 800 **C** 400

Austin 3-Litre

Expanding such an unprestigious car as the 1800 to compete in the luxury car sector never worked in practice. The replacement of the A110 Westminster with a car which used the central hull and much of the tooling from the 1800 may have saved money, but it produced an ungainly car with little presence or appeal. Entirely new 2912cc seven-bearing six drove the rear wheels and was unique to this and MGC. Self-levelling Hydrolastic rear dampers, standard PAS plus auto and overdrive options. Too little room, too sparse an interior, no image and poor performance to boot.

Engine 2912cc 83.3x88.9mm, 6 cyl OHC, F/R
Max power 124bhp at 4500rpm
Max speed 101mph
Body styles saloon
Prod years 1967-71
Prod numbers 9992

PRICES
A 2000 **B** 1000 **C** 500

Austin Maxi

The story of the Maxi is one of missed opportunities. Basically a good car but poor build quality and design faults like the gearchange left it in the cold. Britain's first hatchback and you could also fold the seats to make a double bed. New E-series 1485cc unit mated to a five-speed gearbox did not make a happy combination. Hydrolastically suspended, of course, but later Hydragas. 1750 version was more popular but the Maxi's sales were well below target. Even more lowly-regarded than the Allegro, which is saying something. Twin-carb 1750HL and HLS had 93bhp.

Engine 1485cc 76.2x81.2mm/1748cc 76.2 x 95.75mm, 4 cyl OHC, F/F
Max power 74bhp at 5500rpm/84bhp at 5000rpm/93bhp at 5350rpm (HL & HLS)
Max speed 88/92/101mph
Body styles hatchback
Prod years 1969-81
Prod numbers 472,098

PRICES
A 1400 **B** 800 **C** 400

Austin

Austin Allegro

BMC's most infamous creation. Successor to the 1100/1300 range, the Allegro followed BMC practice mechanically, with the A-series engines carried over, plus 1500 and 1750 Maxi E-series and automatic option. Hydragas suspension was new. Shape was the first work of Harris Mann to reach production and its dumpy, pregnant look and lack of a hatchback were serious flaws. 1975's estate was bizarre, as was infamous Quartic steering wheel which lasted only two years. Keen club in UK blows Allegro trumpet – who knows, cult status may follow?

Engine 998cc 64.6x76.2mm/1098cc 64.6x83.7mm/1275cc 70.6x81.3mm 4cyl OHV. 1485cc 76.2x81.2mm/ 1748cc 76.2 x 95.75mm, 4 cyl OHC, F/F
Max power 44bhp at 5250rpm/49bhp at 5250rpm / 59bhp at 5300rpm/72bhp at 5500rpm/ 80bhp at 5000rpm/91bhp at 5250rpm (1750 twin-carb)
Max speed 82/84/85/88/94/104mph
Body styles saloon, estate **Prod years** 1973-82 **Prod numbers** c 716,250
PRICES A 1200 **B** 750 **C** 300

Austin 18-22 Series

Following Harris Mann's dumpling Allegro came his wedge of cheese in the form of the 18-22 series – 'not a car for Mr Average', claimed ads. These direct replacements for the 1800/2200 used their engines unchanged. Allegro-style Hydragas suspension, plush and spacious interiors. Austin differed from Morris and Wolseley by unique grille and trapezoidal headlamps. This lasted just six months, when all models regrouped under Princess name (see separate entry). Lack of tailgate took seven years to rectify when the Ambassador bowed in.

Engine 1798cc 80.3x88.9mm/2227cc 76.2x81.3mm, 4/6cyl OHV/OHC, F/F
Max power 82bhp at 5250rpm/110bhp at 5250rpm
Max speed 96/105mph
Body styles saloon
Prod years 1975
Prod numbers c 8550

PRICES A 1200 **B** 700 **C** 250

CENTRAL ENGLAND SPORTSCARS

The Frogeye Specialists

CONTACT US FOR ANYTHING AND EVERYTHING FROGEYE!

CARS: Usually 2 or 3 drive away cars for sale and several restoration projects.

SPARES: We have masses of new and secondhand spares: Bonnets, hardtops, restored body shells, engines, gearboxes, etc... everything!

OUR WONDERFUL NEW CATALOGUE IS NOW AVAILABLE

RESTORATION: Our workshop undertakes beautiful restorations at competitive prices.

VISITORS: You are welcome to visit and browse but please phone first.

Unit 4B, Enstone Airfield, Enstone, Chipping Norton, Oxfordshire, OX7 4NP
Tel: 01608 677141 (Evenings: 01608 642018)
Fax: 01608 645411
Hours: 9.00-5.00 Mon-Fri., 9.00-12.00 Sat.

Unit 7/8, Southam Industrial Estate, Southam, Warwickshire CV33 0JH
Telephone: 01926 817181 Fax: 01926 817868

We carry the most comprehensive stock of parts for Austin-Healey and Jensen-Healey models.
Our staff have been involved solely with these vehicles for many years and can assist with all your spares requirements.

WORLDWIDE MAIL ORDER SERVICE

Austin-Healey

Austin-Healey 100/100M

Donald Healey's Austin parts bin sports car was a landmark in all respects. Seductive, rapid, sharp-handling, practical and above all cheap, it took America by storm. Engine came from the A90 Atlantic. Three-speed gearbox replaced by four-speeder in 1955, both available with overdrive, and bigger brakes at the same time. 100M of 1955 is high-compression 110bhp version with front anti-roll bar and bonnet straps, but beware fakes. Corrosion is any Healey's worst enemy, but striking resilience of post-boom values shows enduring appeal.

Engine 2660cc 87.3x111.1mm, 4 cyl OHV, F/R
Max power 90bhp at 4000rpm/110bhp at 4500rpm
Max speed 102/110mph
Body styles sports
Prod years 1954-56
Prod numbers 14,612

PRICES
100 **A** 20,000 **B** 14,000 **C** 7500
100M **A** 25,000 **B** 18,000 **C** 9000

Austin-Healey 100S

The rare and ultra-desirable competition version of the Healey had a 132bhp light alloy cylinder head engine modified by Weslake, mostly aluminium bodywork, and was the first production sports car to have four-wheel disc brakes. Each example was high-speed tested by the factory. Identified by shallow perspex wind deflector, oval air intake, knock-on wires. Close-ratio four-speed 'box allowed 0-60mph in 7.8 secs. Extremely rare and very expensive. Ensure yours is genuine and not a crafty fake.

Engine 2660cc 87.3x111.1mm, 4 cyl OHV, F/R
Max power 132bhp at 4700rpm
Max speed 121mph
Body styles sports
Prod years 1954-56
Prod numbers 55

PRICES
A 45,000 **B** 38,000 **C** 32,000

Austin-Healey 100/6

In 1956 the Austin-Healey was heavily revised. The 100/6 tag referred to the six-cylinder Westminster engine, fitted with twin carbs. Actually slower than the old four-cylinder car until 1957 when six-port head boosted power, although always smoother. Four gears plus optional electric overdrive, steel wheels standard (wires optional), oval grille. Two-inch longer wheelbase allowed space for occasional rear seats, although a two-seater version was offered from 1958. Less of a sports car and therefore the least desirable of all the Healeys.

Engine 2639cc 79.3x88.9mm, 6 cyl OHV, F/R
Max power 102bhp at 4600rpm/117bhp at 4750rpm
Max speed 105/109mph
Body styles sports
Prod years 1956-59
Prod numbers 14,436

PRICES
A 16,000 **B** 10,000 **C** 5500

Austin-Healey 3000 MkI/II/III

A capacity increase to 2.9 litres brought the new 3000 the nickname 'Big Healey'. Standard front disc brakes featured on this more powerful version, though overdrive was optional. MkII from 1961 had triple SU carbs, a new gearbox and vertical bar grille. MkIIa has twin carbs, wind-up windows and curved screen. MkIII of 1963 (photo) was most powerful, had servo brakes, better equipment and is best Healey; plentiful too. MkIIa and MkIII available only as 2+2 seaters. Premature death in 1968 occurred because of creeping US safety laws. MkII worth 20% more, MkIII worth 50% more.

Engine 2912cc 83.4x88.9mm, 6 cyl OHC, F/R
Max power 124bhp at 4600rpm to 148bhp at 5250rpm
Max speed 112-121mph
Body styles sports
Prod years 1959-61/1961-63/1963-68
Prod numbers 42,926 (13,650/11,564/17,712)

PRICES
MkI **A** 18,000 **B** 13,000 **C** 8000
MkII **A** 22,000 **B** 16,000 **C** 9500
MkIII **A** 27,000 **B** 19,000 **C** 10,000

Austin-Healey Sprite MkI

The Sprite goes down in history as the car which brought sporting motoring to the people. In 1958, it was easily the cheapest mass-produced sports car and close on 50,000 sales in three years says it all. Low price came from using A35 bits, though the 948cc engine had twin carbs and an extra 9bhp. Morris Minor steering and MG clutch and master cylinders were rogue non-A35 parts. Simple unitary construction with large fold-forward bonnet whose high headlamps engendered 'Frogeye' nickname. Quite crude (no bootlid) and not very fast, but bags of character and very 'chuckable'.

Engine 948cc 62.9x76.2mm, 4 cyl OHV, F/R
Max power 43bhp at 5000rpm
Max speed 83mph
Body styles sports
Prod years 1958-61
Prod numbers 48,987

PRICES
A 6000 **B** 4500 **C** 3000

Austin-Healey

Austin-Healey Sprite MkII/III/IV

Conventional bonnet and front end restyled by Donald Healey. Longer rear end and proper opening boot made it much more practical as an everyday sports car. Front discs and optional wire wheels, plus bigger engine from 1962. No door handles until MkIII, when winding windows and extra 3bhp arrived. MkIV changed up to tuned 1275cc engine and at last a hood which attached to the body. Black sills from '69 and last 1022 badged just as Austin Sprites. Eternal rust problems but absolutely no worries about parts.

Engine 948cc 62.9x76.2mm/1098cc 64.6x83.7mm/1275cc 70.6x81.3mm, 4 cyl OHV, F/R **Max power** 47bhp at 5500rpm/55bhp at 5400rpm/65bhp at 6000rpm **Max speed** 87/90/96mph **Body styles** sports, convertible **Prod years** 1961-64/1964-66/1966-71 **Prod numbers** 80,360 (31,665/25,905/22,790)

PRICES
A 4500 **B** 3000 **C** 2300

Stevensons Garage

Buyers and Sellers of Classic Sports Cars, Valuations, Professional Restoration Services, M.O.T.s, Servicing

SPECIALISTS IN AUSTIN-HEALEY

✶ SALES ✶ SPARES ✶
✶ SECONDHAND SPARES ✶
✶ CLASSIC REPAIRS ✶

39-43 SOUTH STREET, BARROW-ON-SOAR, LEICESTERSHIRE LE12 8LY

Telephone: Quorn (01509) 412469 or 880026

Fax: Quorn (01509) 880026

CAMBRIDGE MOTORSPORT

SPECIALISTS IN 1950's AND 1960's BRITISH SPORTS CARS

We offer a complete service for the classic sports car owner.

HISTORIC RACE AND RALLY PREPARATION
Winning cars prepared for all rally and circuit events to F.I.A. specification. Precision engine building by *Chris Conoley*. Our customers are consistent winners at home and abroad, combining radical technology with reliability.

We will be pleased to discuss the possibilities for your car.

- **FULL ENGINEERING FACILITIES INCLUDING LEAD FREE CONVERSION**
- **CARS FOR SALE. INSPECTION & VALUATION**
- **SERVICING & RESTORATION TO THE HIGHEST STANDARDS**

PARTS FOR YOUR CLASSIC CAR
Standard and exclusive competition parts for your classic sports car. Check our prices.

FAST & FRIENDLY MAIL ORDER SERVICE WORLDWIDE
• Se habla Español •

START & FINISH WITH CAMBRIDGE MOTORSPORT!

Specialists in Austin Healey, Triumph, MG and Jaguar.

CAMBRIDGE MOTORSPORT
Caxton Road, Great Gransden, Beds., SG19 3AH
☎ **01767 677 969**
Fax: **01767 677 026**

JME – The Austin Healey Specialists

4a Wise Terrace, Leamington Spa, Warwickshire CV31 3AS

Workshop: 01926 425038
Office/Fax: 01926 640031

- Recognised world-wide as the leaders in Austin Healey restoration and maintenance.
- A small team of highly skilled craftsmen led by Jon Everard, an ex-works employee with over 30 years experience, offer a comprehensive restoration service working only to the highest standards.

Autobianchi Bianchina Trasformabile/Cabriolet

Part of Fiat since 1955, Autobianchi's first project was a rebody of the new Fiat 500 in 1957. The result was a pleasing little rolltop coupé with room for 2+2 passengers. Mechanical spec followed the little Fiat. Even cuter Cabriolet model arrived in 1960 and was known in France as the Eden Roc. More cramped than a 500 but at least you can drape over the edges. Not sold in UK but there's a strong following on the continent (especially Italy, where most examples are to be found) and prices have risen with increased interest in Fiat 500 derivatives.

Engine 479cc 66x70mm/499cc 67.4x70mm, 2 cyl OHV, R/R
Max power 13bhp at 4000rpm to 21bhp at 4000rpm
Max speed 62/68mph
Body styles coupé/convertible
Prod years 1957-62/1960-68
Prod numbers c 35,500/c 9300

PRICES
Trasformabile **A** 4000 **B** 2800 **C** 1500
Cabriolet **A** 5000 **B** 4000 **C** 2500

Autobianchi Bianchina Berlina/Panoramica

Third derivative of the Bianchina range was the Panoramica estate car (photo) based on the longer wheelbase Fiat 500 Giardiniera estate platform. The Autobianchi family resemblance remained in a pretty little all-purpose car for the small family. The final 500-based model was the Berlina Quattroposti ('four-seat'), with an extended fixed roof and space for four adults – just about. The Panoramica remained in production until 1970, after which Autobianchi made the Fiat 500 estate as the Autobianchi Giardiniera until 1977 (see page 72).

Engine 499cc 67.4x70mm, 2 cyl OHV, R/R
Max power 18bhp at 4000rpm/21bhp at 4000rpm
Max speed 60/68mph
Body styles saloon/estate
Prod years 1960-70/1962-68
Prod numbers c 69,000/c 160,000

PRICES
A 2800 **B** 1700 **C** 1000

Autobianchi Stellina

First shown at Turin in 1963, the Stellina was an attempt to make a sports car on a Fiat 6000 basis. The result was not overly pretty, the plexiglass cowled headlamps and pointy tail lamps looking particularly unhappy. Two seats and a folding roof do not make a sports car, however, as customers discovered when they struggled to get past 60mph. One positive aspect of the Stellina was that it was, at least, affordable. Glassfibre body is rust-free, too. Rarity is the only reason for highish values.

Engine 767cc 62x63.5mm, 4 cyl OHV, R/R
Max power 32bhp at 4800rpm
Max speed 65mph
Body styles sports
Prod years 1963-65
Prod numbers n/a

PRICES
A 6000 **B** 4000 **C** 3000

Autobianchi Primula

This was Fiat's answer to the BMC 1100. It had all the British car's qualities – front-wheel drive, transverse engines, big interior space – but was also fairly stylish, and it had a hatchback option. Based mechanically on the Fiat 1100 (later 124 engines) with a choice of two, three, four or five doors. The Primula Coupé with cut-down glasshouse was styled and built by Touring and was highly attractive, especially with optional wire wheels, and is the pick of the Primula range. Never sold in UK and parts supply is dubious, though plenty can still be seen on Italian roads.

Engine 1197cc 73x71.5mm/1221cc 72x75mm/1438cc 80x71.5mm, 4 cyl OHV, F/F
Max power 59bhp at 5400rpm to 75bhp at 5500rpm
Max speed 84-96mph
Body styles saloon, hatchback, coupé
Prod years 1964-70
Prod numbers n/a
PRICES
Saloon/hatchback **A** 1900 **B** 1000 **C** 600
Coupé **A** 3000 **B** 1800 **C** 1200

Autobianchi A112

If the Primula was a BMC 1100 competitor, the A112 answered the Mini. A formula which the Italians do so well: a small car with stylish lines and a certain dash. Engines derived from Fiat 850 Sport and later the 127, but were always mounted in front and drove the front wheels. Hatchback body was only 10ft 7in long. If there is a collectable A112, it is the Abarth version with its tuned engine delivering up to 70bhp and a top speed of nearly 100mph, identified by matt black bonnet, special grille and Abarth badges. Never sold in UK.

Engine 903cc 65x68mm/1050cc 76x57.8mm, 4 cyl OHV/OHC, F/F
Max power 44bhp at 5600rpm to 70bhp at 6600rpm
Max speed 84-97mph
Body styles hatchback
Prod years 1969-86
Prod numbers 1,254,178

PRICES
A112 **A** 1800 **B** 1200 **C** 700
A112 Abarth **A** 3200 **B** 2000 **C** 1200

Bedford Beagle

Unromantic Martin Walter conversion of the Bedford HA van derived from the Vauxhall Viva HA. Engines uprated in 1967 (1159cc) and 1972 (1256cc) but spongy drum brakes remained throughout. Stripped out interior for the most utilitarian motoring but very practical thanks to large side-opening rear doors. Sliding windows for rear passengers. Van version lasted until 1982 so body panel spares are no problem (everything aft of the A-pillar was different to the saloon). Possibly useful for carting autojumble stuff around.

Engine 1057cc 74.3x61mm/1159cc 77.7x61mm/1256cc 81x61mm, 4 cyl OHV/OHC, F/R
Max power 40bhp at 5200rpm/41bhp at 5000rpm/53bhp at 5200rpm
Max speed 73-80mph
Body styles estate
Prod years 1964-77
Prod numbers n/a

PRICES
A 1000 B 600 C 250

Bentley MkVI

Bentley's postwar effort was the MkVI. Traditional construction with sturdy chassis, IFS, servo brakes, centralised chassis lubrication, four-speed gearbox. The first Bentley offered with standard all-steel bodywork (shown), but there were innumerable special bodies on the MkVI chassis by coachbuilders like Park Ward, HJ Mulliner, James Young, Graber and even Pininfarina. These tended to be in aluminium. Larger engine from 1951. Unburstable mechanicals but Standard Steel bodywork prone to rust. Individual coachbuilt cars vary in price according to style and constructor.

Engine 4257cc 88.9x114.3mm/ 4566cc 92x114.3mm, 6 cyl IOE, F/R
Max power 137/150bhp
Max speed 94/102mph
Body styles saloon, various specials
Prod years 1946-52
Prod numbers 5201

PRICES
Standard Steel A 20,000 B 12,000 C 7500
Park Ward Coupé/DHC A 40,000 B 22,000 C 14,000

Bentley R-Type

Replacement for the MkVI with more elegant lengthened tail. Again the vast majority of bodies were the Standard Steel four-door saloon but an equally wide variety of coachbuilt bodies was on offer, although proportionately fewer such specials were built. At over two tons in weight, the R-Type is hardly economical (expect less than 15mpg). Four-speed automatic option was available from 1952 for export, 1953 for UK. New facia in '53 too. 1950s Bentleys look indestructible but just sit back and watch the rust come through...

Engine 4566cc 92x114.3mm, 6 cyl IOE, F/R
Max power est 150bhp
Max speed 106mph
Body styles saloon, various specials
Prod years 1952-55
Prod numbers 2320

PRICES
Standard Steel A 20,000 B 12,000 C 8000
Park Ward DHC A 50,000 B 35,000 C 18,000

Bentley Continental R

Some fabulous coachwork was practised on the R-Type chassis during the 1950s. The most celebrated is the light alloy bodied Continental R produced by HJ Mulliner from 1952. Glorious fastback shape is both proportioned and exotic and probably the pinnacle of Bentley's post-war production. Aerodynamic advantage proven by its top speed of nearly 120mph on standard R-Type engine. Later cars had 4.9-litre unit. Manual or auto 'boxes, LHD cars usually with column shift. Vastly expensive then – and now.

Engine 4566cc 92x114.3mm/4887cc 95.25x114.3mm, 6 cyl IOE, F/R
Max power est 150bhp/est 175bhp
Max speed 117mph
Body styles coupé
Prod years 1952-55
Prod numbers 207

PRICES
A 100,000 B 80,000 C 50,000

Bentley S1

This was the first badge-engineered Bentley, a straight duplication of the Rolls-Royce Silver Cloud announced concurrently. Therefore the same landmark full-width bodywork, now in aluminium, new front suspension and better brakes. Uses the last of the Rolls/Bentley 'sixes', with inlet-over-exhaust valves. Standard four-speed auto, PAS option from '56, long wheelbase Park Ward limousine from '57 (35 built). The S1 outsold the Roller by 50%, so prices today are cheaper for a Bentley, but you'll need to check for chassis rust.

Engine 4887cc 95.25x114.3mm, 6 cyl IOE, F/R
Max power est 175bhp
Max speed 105mph
Body styles saloon, limousine
Prod years 1955-59
Prod numbers 3107

PRICES
A 23,000 B 17,000 C 10,000

Bentley S1 Continental

New Continental based on the S1 could be had with a variety of coachbuilt bodies, generally in aluminium. Standard automatic 'box, like S1, but there was also a (rare) option of all-synchromesh manual until 1957. Twin carbs from '56, power steering option for export from '56, UK from '57. Air conditioning and electric windows optional from '58. Mulliner Flying Spur four-door is pretty, Park Ward two-door saloons easiest to find, James Young and Hooper bodies are rare, all dropheads desirable and not many around.

Engine 4887cc 95.25x114.3mm, 6 cyl IOE, F/R
Max power est 175bhp
Max speed 107mph
Body styles saloon, drophead coupé
Prod years 1955-59
Prod numbers 431

PRICES
Saloons/Flying Spur **A** 75,000 **B** 50,000 **C** 32,500
DHC **A** 110,000 **B** 90,000 **C** 65,000

Bentley S2

A new V8 6230cc engine for the 1959 S2 was its major distinguishing feature, the first of a long line of Rolls V8s. Poorer fuel consumption and not significantly improved performance have relegated the S2 to a less favoured position than the S1 or S3. Power steering and automatic transmission were standard on all S2s, minor interior improvements from 1962. Long wheelbase limousine rare (57 built) and Mulliner drophead coupé is most desirable of the standard-type S2s. Apart from T Series, the cheap route into Bentley ownership.

Engine 6230cc 104.1x91.4mm, V8 cyl OHV, F/R
Max power est 200bhp
Max speed 110mph
Body styles saloon, drophead coupé
Prod years 1959-62
Prod numbers 1922

PRICES
Saloon **A** 17,000 **B** 11,000 **C** 7000
DHC **A** 40,000 **B** 27,000 **C** 20,000

Bentley S2 Continental

Mechanical changes as per the S2 saloon, except that you also got four-shoe front brakes and a higher axle ratio than standard cars. Large variety of coachbuilt bodies, including two-door saloons by Mulliner and James Young, four-door James Young saloon, four-door Mulliner Flying Spur and Park Ward Drophead Coupé (pictured), which had an electrically-operated roof. The latter is the 'connoisseur' choice among S2s, although generally speaking the S1 Continental is preferred. Bespoke body parts are problematic if they need to be replaced.

Engine 6230cc 104.1x91.4mm, V8 cyl OHV, F/R
Max power est 200bhp
Max speed 120mph
Body styles saloon, drophead coupé
Prod years 1959-62
Prod numbers 388

PRICES
Saloon **A** 50,000 **B** 35,000 **C** 30,000
DHC **A** 52,000 **B** 38,000 **C** 30,000

Bentley S3

Easily identified by its controversial double headlamps, the S3 of 1962 benefited from a high compression engine, better power steering. Other changes were a lower bonnet line, reshaped front wings, recessed indicator/side light units and modified seating. Once again, a long wheelbase version was offered but only taken up by 32 customers. Mulliner Drophead Coupé version is also rare. Wider rear wheels from 1964. Better performance and so higher priced than S2, but still cheaper than the equivalent Rolls-Royce Silver Cloud.

Engine 6230cc 104.1x91.4mm, V8 cyl OHV, F/R
Max power est 210bhp
Max speed 116mph
Body styles saloon, drophead coupé
Prod years 1962-65
Prod numbers 1318

PRICES
Saloon **A** 25,000 **B** 17,000 **C** 10,000
DHC **A** 50,000 **B** 35,000 **C** 25,000

Bentley S3 Continental

Quad headlamps were transferred to all Continental S3s, arranged in a distinctive 'Chinese Eye' pattern on Park Ward's Drophead Coupé and new two-door saloon derivative (photo). James Young still offered its saloon and there was Mulliner's Flying Spur four-door, as ever. Park Ward and Mulliner joined forces in 1963, so all bodies known under MPW name after that. Some Continental bodies made with Rolls-Royce grilles and badges (premium prices). All mechanical changes as S3 saloon. The last of the great coachbuilt Bentleys.

Engine 6230cc 104.1x91.4mm, V8 cyl OHV, F/R
Max power est 210bhp
Max speed 120mph
Body styles saloon, drophead coupé
Prod years 1962-66
Prod numbers 312

PRICES
Saloon **A** 40,000 **B** 35,000 **C** 28,000
DHC/Flying Spur **A** 65,000 **B** 45,000 **C** 32,000

Bentley T1/T2

First-ever unitary Bentley, now with four disc brakes and independent suspension all round, with self-levelling courtesy of Citroën. This is basically a Rolls-Royce Silver Shadow clone, though far rarer and therefore commanding a slight premium. Post-1970 cars with 6750cc engines are preferable. Refrigeration standard from 1969, cruise control from '73, facelifted T2 from 1977. Coachbuilt James Young two-door (98 built) and MPW drophead with power hood (41 built), made 1967-71, are very collectable. Rarest is T2 LWB (10 made).

Engine 6230cc 104.1x91.4mm/6750cc 104.1x99mm, V8 cyl OHV, F/R
Max power n/a
Max speed 118/120mph
Body styles saloon, convertible
Prod years 1965-77/1977-80
Prod numbers 2280 (1712/568)

PRICES
Saloon **A** 13,000 **B** 9000 **C** 6000
Convertible **A** 30,000 **B** 22,000 **C** 15,000

Bentley Corniche

The Mulliner Park Ward two-door convertible evolved into the Bentley/Rolls-Royce Corniche in 1971. Only minor changes (new facia, deeper grille), but a brand new saloon version with the same styling below the waistline was launched at the same time. Ventilated front discs from '72, split level air conditioning from '76, T2-style spoiler and rubber-faced bumpers from '77. Saloon dropped in 1981 and Corniche revised and renamed Continental in 1984. Very rare compared to identical Rolls-Royce Corniche so expect to pay a slight premium.

Engine 6750cc 104.1x99mm, V8 cyl OHV, F/R
Max power n/a
Max speed 120mph
Body styles saloon, convertible
Prod years 1971-84
Prod numbers 149

PRICES
Saloon **A** 28,000 **B** 18,000 **C** 13,000
Convertible **A** 38,000 **B** 27,000 **C** 20,000

Berkeley Sports 322/328

Intriguing little sports car designed by Lawrie Bond and made by a caravan manufacturer in Biggleswade from 1956. Pretty glassfibre bodywork came in three pieces. Power was initially by 322cc Anzani two-stroke twin motorcycle engine, from 1957 with a 328cc Excelsior twin. Drive was by chains to the front wheels, usually with three speeds but last ones have four. Surprisingly fast. Avoid examples converted to Mini power, which have usually been modified to oblivion. Spares situation is very healthy.

Engine 322cc 60x57mm/328cc 58x62mm, 2 cyl TS, F/F
Max power 15/18bhp at 5000rpm
Max speed 60/65mph
Body styles sports
Prod years 1956-58
Prod numbers 146/1272

PRICES
A 2500 **B** 1700 **C** 1200

Berkeley Sports 492/Foursome

In 1958, Berkeley uprated its engine to the triple-carb three-cylinder air-cooled Excelsior 492cc unit. With 30bhp pulling under 800lb of car, performance was unnerving: a top speed of 90mph was advertised. Column shift now swapped for floor-mounted lever. Body was strengthened in all versions and, if you had a family, you could go for the Foursome with its wider body and full four-seat accommodation (just about). Optional integral hard top available for both models. Not much of a match for the Frogeye Sprite, similarly priced when new.

Engine 492cc 58x62mm, 3 cyl TS, F/F
Max power 30bhp at 5000rpm
Max speed 90mph
Body styles sports, coupé
Prod years 1958-59
Prod numbers 666/16

PRICES
A 2700 **B** 1800 **C** 1200

Berkeley B95/B105

Numerical system referred to the claimed top speed of Berkeley's 1959 range. Thanks to the new parallel twin Royal Enfield Constellation motorbike engine (in high compression tune on B105), top speed could exceed 100mph. These models are easily distinguished by their upright grilles, needed to clear the taller engine, but you'll have trouble finding one today since production was limited and most were exported, mainly to America. Permanent hard top with sliding perspex windows was optional.

Engine 692cc 70x90mm, 2 cyl OHV, F/F
Max power 40bhp at 5500rpm/50bhp at 6250rpm
Max speed 95/105mph
Body styles sports, coupé
Prod years 1959-60
Prod numbers c 200

PRICES
A 3000 **B** 2200 **C** 1750

Berkeley · Bitter · Bizzarrini · BMW

Berkeley T60

The trike version of Berkeley's little roadster was its most successful. Single rear wheel on a trailing arm instead of swing axles and coils, therefore even lighter weight. Only ever offered with the 328cc Excelsior engine combined with four-speed 'box. All T60s have some room behind the front seats but a few of the very last ones had a larger rear seat (called T60-4). T60 died with collapse of Berkeley's caravan business. Nippy, attractively styled and a lot of fun for very little outlay. Many have sadly been fitted with Mini subframes.

Engine 328cc 58x62mm, 2 cyl TS, F/F
Max power 18bhp at 5000rpm
Max speed 60mph
Body styles sports, coupé
Prod years 1959-60
Prod numbers c 1830

PRICES
A 2000 **B** 1500 **C** 1000

Bitter CD

In 1969, Opel presented a concept car on its stand at Frankfurt. It would have remained a one-off had not former racing driver and Intermeccanica importer Erich Bitter seen it. He persuaded Opel to support his reinterpretation of this show car and entered production in 1973. Built by Baur, the Bitter CD used a shortened Opel Diplomat floorpan complete with its Chevy sourced V8 and GM auto 'box. A full four-seater with luxury trim and stunning looks, it was expensive in its day. Reliable and well-made but no RHD.

Engine 5354cc 101.6x82.6mm, V8 cyl OHV, F/R
Max power 230bhp at 4700rpm
Max speed 129mph
Body styles coupé
Prod years 1973-79
Prod numbers 395

PRICES
A 20,000 **B** 14,000 **C** 7000

Bizzarrini GT Strada 5300

After Giotto Bizzarrini had designed the Ferrari 250 GTO he had an argument with the factory and left to set up on his own. The Strada was a modified lightweight Iso Grifo which Bizzarrini ended up building himself. An extremely low-slung (43in) body styled by Giugiaro was offered in aluminium or glassfibre and proved very lightweight. De Dion rear, four discs, Muncie four-speed 'box. The road car version was a hairy machine with its 365bhp Chevy V8 (later 350bhp) and stripped-out interior. A no-frills supercar.

Engine 5343cc 101.6x82.6mm, V8 cyl OHV, F/R
Max power 365bhp at 6000rpm/350bhp at 5800rpm
Max speed 160mph
Body styles coupé, convertible
Prod years 1965-69
Prod numbers 149

PRICES
A 70,000 **B** 50,000 **C** 35,000

Bizzarrini 1900 Europa

Italy's Europa was launched in the same year as the eponymous Lotus version. Effectively a shrunken GT Strada body fitted with an Opel 1.9-litre engine available in standard or tuned forms. Steel platform chassis with glassfibre body. Independent suspension and disc brakes all round, so a safe handler. Not too strong on the performance front yet it was almost as expensive as its V8 powered larger brother. A marketing flop which hastened the demise of Bizzarrini in 1969. Interesting marginal sports car but almost impossible to find.

Engine 1897cc 93x69.8mm, 4 cyl OHC, F/R
Max power 110/135bhp at 6000rpm
Max speed 122/128mph
Body styles coupé
Prod years 1967-69
Prod numbers 15

PRICES
A 30,000 **B** 24,000 **C** 18,000

BMW Isetta

BMW launched its bubble car at exactly the right time: the market for cheap, basic, weatherproof transport was huge. BMW acquired a licence to build the Isetta from its Italian inventor, Iso, fitted its own motorbike engine and proceeded to churn them out. Novel hinged front door pulled up jointed steering column to ease entry and a rolltop sunroof was obligatory. Three or narrow-track four wheeled set-up. British produced Isettas (1957-64) added full convertible and pick-up styles. Unlikeliest Mille Miglia entrant (1954/55).

Engine 247cc 68x68mm/298cc 72x73mm, 1 cyl OHV, R/R
Max power 12bhp at 5800rpm/13bhp at 5200rpm
Max speed 52/58mph
Body styles saloon, convertible
Prod years 1955-62
Prod numbers 161,728 (plus c 41,000 in other countries)

PRICES
A 4500 **B** 2800 **C** 1800

BMW

BMW 600

Isetta customers ready to move up a step were catered for by the BMW 600. Front-opening door was retained but you also got a conventional door on one side for entry to the rear seats. Expanded BMW flat twin 'bike engine sat in the tail, which was suspended by the first example of BMW's trademark semi-trailing arm system. More comfortable and quicker than the Isetta but still pretty crude for a car which only just undercut the Beetle on price. Only ever available in UK to special order and exclusively with LHD. Plenty still around.

Engine 585cc 74x68mm, 2 cyl OHV, R/R
Max power 20bhp at 4500rpm
Max speed 63mph
Body styles saloon
Prod years 1957-59
Prod numbers 34,813

PRICES
A 3000 **B** 1800 **C** 1000

BMW 700 Coupé/Cabriolet

The first of the 700 series to break cover was in fact the Coupé. Its bodywork was styled by Michelotti (certainly not one of his best) and was BMW's first ever unitary effort. Engine initially developed only 30bhp but high compression twin-carb Sports unit from 1961 extracted an extra 10bhp, leading to it being called the 'poor man's Porsche' – and that's the one to go for. Sport engine was also fitted to the new Cabriolet in '61, which had bodywork by Baur. Final LS models of 1963-65 had longer wheelbase and more space for rear passengers.

Engine 697cc 78x73mm, 2 cyl OHV, R/R
Max power 30/32bhp at 5000rpm/40bhp at 5700rpm
Max speed 73/75/86mph
Body styles coupé/convertible
Prod years 1959-65/1961-65
Prod numbers 13,758 (11,166/2592)

PRICES
Coupé **A** 2700 **B** 1700 **C** 1000
Cabriolet **A** 3700 **B** 2500 **C** 1500

BMW 700

Larger and more conventional microcar built by BMW in saloon form from September 1959. Considerably uglier than the 700 Coupé but at least you got more rear headroom. Rear engine, naturally, but you now had a trailing arm rear end and independent front by wishbones and Dubonnet coils. 30bhp means perky performance for what was an economy car. Four speeds and floor shift. LS (1962 on) had a wheelbase extended by 5.5in and luxury trim. Success with the 700 turned BMW's fortunes around and it's certainly more interesting than a Beetle.

Engine 697cc 78x73mm, 2 cyl OHV, R/R
Max power 30/32bhp at 5000rpm
Max speed 73/75mph
Body styles saloon
Prod years 1959-65
Prod numbers 174,253

PRICES
A 1800 **B** 1000 **C** 600

BMW 501

At the time it was launched, the BMW 501 was a sensation. BMW had got itself up on to its feet rapidly after the war and everyone was amazed to see it come back with a massive six-cylinder saloon stuffed with luxury goodies. Rather grand styling inspired its 'Baroque Angel' nickname. New chassis with torsion bar suspension all round, four-speed column-change 'box, prewar 326 engine, bored out to 2.1 litres from 1955. Desirable two-door coupé and cabriolet (pictured) were glorious white elephants in austere postwar Germany.

Engine 1971cc 66x96mm/2077cc 68x96mm, 6 cyl OHV, F/R
Max power 65bhp at 4400rpm/72bhp at 4500rpm **Max speed** 84/88mph
Body styles saloon, coupé, convertible **Prod years** 1952-58
Prod numbers 8936

PRICES
Saloon **A** 12,000 **B** 8000 **C** 6000
Coupé **A** 18,000 **B** 12,000 **C** 9,000
Cabriolet Too rare to price

BMW 501 V8/502/2600/3200

The fitment of a 95bhp 2.6-litre V8 engine transformed the 501 into a real Mercedes-beater and a genuine 100mph autobahnstormer. Bored-out 3.2-litre engine joined the 2.6 in 1955 and went up as far as 160bhp in the final two years of production to make the V8 the fastest production saloon in Germany. Models renamed 2600 and 3200 in 1960. Once again, coupé and cabriolet conversions by coachbuilders Baur and Authenreith were available. RHD imports were very limited and they all had floor-mounted gear levers.

Engine 2580cc 74x75mm/3168cc 82x75mm, V8 cyl OHV, F/R
Max power 95-110bhp at 4800rpm/120bhp at 4800rpm to 160bhp at 5600rpm **Max speed** 100-118mph
Body styles saloon, coupé, convertible **Prod years** 1954-63
Prod numbers 13,044

PRICES
Saloon **A** 16,000 **B** 12,000 **C** 9000
Coupé **A** 22,000 **B** 16,000 **C** 12,000
Cabriolet Too rare to price

BMW 503

Built on the same wheelbase as the 501, the sporting 503 had a modern full-width body styled by Count Albrecht Goertz, not entirely happily it must be said. Offered in coupé and cabriolet forms and only with the 3.2-litre V8 engine. First cars had column shift. Targeted at the American market, although with little success – too expensive and no image at that time. This was one of the cars which lost BMW so much money in the 1950s, bringing it to near-bankruptcy in 1959. BMW needed a drastic rethink of its polarised bubble car and bank manager range.

Engine 3168cc 82x75mm, V8 cyl OHV, F/R
Max power 140bhp at 4800rpm
Max speed 118mph
Body styles coupé, convertible
Prod years 1955-59
Prod numbers 412

PRICES
Coupé **A** 32,000 **B** 28,000 **C** 20,000
Cabriolet **A** 40,000 **B** 35,000 **C** 28,000

BMW 507

First seen at the same time as the 503, BMW's 507 model was another attempt to muscle in on Mercedes' business, targeting the 300SL. Light alloy bodywork was again by Goertz but this time sensational in its proportions and lines. Based on the 502 chassis shortened by a massive 14in, the 507 was distinctly sporting with its high compression 150bhp V8 but never had much of a competition career. Servo brakes, front discs on later cars. Gorgeous, and cheaper than a 300SL, but very rare because BMW lost fistfuls on each one sold.

Engine 3168cc 82x75mm, V8 cyl OHV, F/R
Max power 150bhp at 5000rpm
Max speed 137mph
Body styles coupé, roadster
Prod years 1956-59
Prod numbers 252

PRICES
A 80,000 **B** 60,000 **C** 45,000

BMW 3200 CS

Intended as a belated successor to the 503, the 3200 CS (for Coupé Sport) was designed by Giugiaro while employed by Bertone. Based on the venerable 502 saloon chassis, the pillarless shape hid its bulk well thanks to a low waist line. The kicked-up shape of the rear pillars was later duplicated in the CS series of 1965-75. As a very expensive, hand-built machine, the 3200 CS sold in tiny numbers and was never offered in the UK, so you may have trouble finding one. The last of the loss-making BMW V8 dinosaurs.

Engine 3168cc 82x75mm, V8 cyl OHV, F/R
Max power 160bhp at 5600rpm
Max speed 125mph
Body styles coupé
Prod years 1961-65
Prod numbers 538

PRICES
A 20,000 **B** 14,000 **C** 7000

BMW 1500/1600

'Neue Klasse' BMW was an absolutely pivotal model in its history. The first medium-sized BMW struck a chord with Germany's middle classes and sold well enough to return the company to profit. State-of-the-art spec included unitary construction, shapely styling by Michelotti, front disc brakes, MacPherson struts at the front, semi-trailing rear and a gem of an engine. Bored-out 1600 version from 1964 has more power, 13in wheels. Alas, historical significance counts for nought in the marketplace and 1500s are worth very little.

Engine 1499cc 82x71mm/1573cc 84x71mm, 4 cyl OHC, F/R
Max power 80bhp at 5700rpm/95bhp at 5800rpm
Max speed 93/100mph
Body styles saloon
Prod years 1961-64/1964-66
Prod numbers 23,554/10,278

PRICES
A 2500 **B** 1500 **C** 700

BMW 1800/1800TI/1800TI/SA

BMW's forward-looking approach was confirmed in 1964 when it launched a 1.8-litre sports version of the 1500 just after a standard 1800 version. The so-called 1800TI (for Touring Internazionale) boasted 110bhp from its twin-carb engine. The TI/SA was a 130bhp special which could only be bought if you had a valid racing licence (and never in UK) – only 200 made, so bumped-up prices. Auto option from '66. New 102bhp 1766cc overbored version of 1500/1600 engine replaced 1773cc unit in 1968, dual-circuit brakes at the same time.

Engine 1773cc 84x80mm/1766cc 89x71mm, 4 cyl OHC, F/R **Max power** 90bhp at 5250rpm/110bhp at 5800rpm/130bhp at 6100rpm/102bhp at 5800rpm **Max speed** 100/108/120/101mph **Body styles** saloon **Prod years** 1963-68/1964-66/1964-65/1968-72 **Prod numbers** 164,989

PRICES
1800/1800TI **A** 3000 **B** 2000 **C** 1000
1800TI/SA **A** 7500 **B** 5500 **C** 3000

BMW 2000/2000TI/2000Tii

Largest engine for the BMW 1500 family was the 1.8-litre bored out to 1990cc. Plain 2000 has single carb, TI has twin carbs (120bhp), Tilux has luxury trim, Tii has 135bhp fuel-injected engine and stripped-out interior. Latter never came to UK but we did have the unique TI Frazer-Nash with vinyl top, wood rim wheel and torsion bars. Dual-circuit brakes and vertical bar grille from '68 and always optional auto. Rectangular headlamps the norm. Popular in its day and still quite plentiful today, so few prospects for increased value.

Engine 1990cc 89x80mm, 4 cyl OHC, F/R
Max power 100bhp at 5500rpm/120bhp at 5500rpm/135bhp at 5800rpm
Max speed 105/109/115mph
Body styles saloon
Prod years 1966-72/1966-70/1969-72
Prod numbers 151,655

PRICES
A 3500 **B** 2300 **C** 1000

BMW 2000C/CS

BMW's first mass-produced luxury coupé set a tradition which would last decades. Slim-pillar in-house styling by Wilhelm Hofmeister looked great from the rear and sides but curious wrap-around headlamps looked awful (UK versions had four separate lamps, which were even worse). Bodies built by Karmann. 2000C had a single carb engine, 2000CS twin carbs, which gave it a good turn of speed – although nothing to match the later six-cylinder CS models, which have left the 2000 coupés rather in the wilderness. CA is single carb auto version.

Engine 1990cc 89x80mm, 4 cyl OHC, F/R
Max power 100bhp at 5500rpm/120bhp at 5500rpm
Max speed 105/115mph
Body styles coupé
Prod years 1965-70
Prod numbers 11,720 (2837/8883)

PRICES
A 5000 **B** 3500 **C** 2000

BMW 1502/1600/1600TI/1802/2002/2002TI/2002tii

The '02 was BMW's crucial entry level range. Compact body style and big range of engines. 1600 was first on scene, then 105bhp 1600TI (1967-68). 2002 from 1968 was most common. Twin carb TI not sold in UK but we did get the fiery 130bhp fuel-injected tii. Automatic optional, as was five-speed gearbox. All models suffer from lots of oversteer but are wonderful drivers' cars with strong following, well built and returning reasonable economy for their power. 1502 was run-out base model, 1802 not imported to UK. Hatchback Touring, launched in 1971, worth 10% more.

Engine 1573cc 84x71mm/1766cc 89x71mm/1990cc 88.9x 80m, 4 cyl OHC, F/R **Max power** 1.6: 75bhp/85bhp/96bhp/105bhp 1.8: 90bhp 2.0: 100bhp/113bhp/120bhp/130bhp
Max speed 96-120mph
Body styles saloon, hatchback
Prod years 1966-77
Prod numbers 698,943

PRICES
1602/2002 **A** 3000 **B** 1900 **C** 1000
2002tii **A** 6000 **B** 3500 **C** 1600

BMW 1600/2002 Cabriolet

One of the few true four-seater convertibles of the 1960s, the 1600 Cabriolet (pictured) was an attractive fully open-topped version of the 1600. Bodies were built by Baur. No roll-over bar, so rust compromises rigidity, therefore few 1600s remain. More common is the 2002 Cabriolet, which replaced it in 1971. This had quite different roof treatment, with a fixed roll-over bar fitted with side glass, targa panels and a folding section at the rear. It was a better performer, too, with the 100bhp version of the 2.0-litre hemi-head engine.

Engine 1573cc 84x71mm/1990cc 88.9x 80m, 4 cyl OHC, F/R
Max power 85bhp at 5700rpm/100bhp at 5500rpm
Max speed 100/105mph
Body styles convertible
Prod years 1967-71/1971-75
Prod numbers 1682/2272

PRICES
1600 **A** 7500 **B** 5000 **C** 3000
2002 **A** 8500 **B** 6000 **C** 3000

BMW 2002 Turbo

The very first turbo road car made in Europe, the 2002 Turbo was startlingly quick – at least if you kept the revs high. Immediately identifiable by its wider wheelarches, rear spoiler and deep chin spoiler, and initially supplied with mirror-image '2002 Turbo' script (soon removed to appease scare-mongers). Basically a 2002tii under the skin but LHD only. Turbocharging was still underdeveloped at this time and the fuel-hungry system did not please oil crisis buyers. Very short production life means low production numbers and high prices.

Engine 1990cc 88.9x 80m, 4 cyl OHC, F/R
Max power 170bhp at 5800rpm
Max speed 130mph
Body styles saloon
Prod years 1973-74
Prod numbers 1670

PRICES
A 12,500 **B** 9000 **C** 7000

BMW 2500/2800/3.0/3.3

The 2500 may be regarded as the car which established BMW's reputation across the world. An elegant saloon body, powerful straight six-cylinder engine and luxury appointments made it a credible Mercedes competitor. All-independent suspension, all-wheel discs, manual/auto option, power steering on most. 3.0-litre engine was also available with fuel injection as the 3.0Si, also fitted to the later (3210cc) 3.3Li, which had a 4in longer wheelbase. The non-injection 3295cc engine was unique to this model. All models have an abominable rust record.

Engine 2494cc 86x71.6mm/2788cc 86x80mm/2985cc 89x80mm/3295cc 89x88.4mm/3210cc 89x86mm, 6 cyl OHC, F/R **Max power** 2.5: 150bhp at 6000rpm. 2.8: 170bhp at 6000rpm. 3.0: 180/200bhp at 6000rpm. 3.3: 200bhp at 5500rpm **Max speed** 118-131mph **Body styles** saloon **Prod years** 1968-77 **Prod numbers** 222,001

PRICES
2500/2800 **A** 2800 **B** 1400 **C** 900
3.0/3.3 **A** 3500 **B** 2000 **C** 1200

BMW 2.5CS/2800CS/3.0CS/3.0CSi

With a redesigned front end and BMW's new six-cylinder engine, the 2000CS at last got the looks and performance to make it a winner on an international scale. Wheelbase was 3in longer, front suspension uprated and power steering made standard. 2800CS replaced by 3.0CS in '71, with saloon's big engine, four-wheel discs and reworked chassis. Injected 3.0CSi has 200bhp. Fabulous handling, good looks, cruising ability and broad-shouldered performance make this perhaps *the* classic BMW. Budget 2.5CS (1974-75) not imported to UK.

Engine 2494cc 86x71.6mm/2788cc 86x80mm/2985cc 89x80mm, 6 cyl OHC, F/R
Max power 150/170/180-200bhp at 6000rpm
Max speed 120-133mph
Body styles coupé
Prod years 1974-75/1968-71/1971-75
Prod numbers 844/9399/11,063/8199

PRICES
A 6000 **B** 3500 **C** 1500

Jaymic — The Classic BMW Specialists

Jaymic Offer The Classic BMW Owner The Complete Range Of Services In House

- World Wide Mail Order Parts
- Free 02 & CS Parts Catalogues
- List Of 220 Classic BMWs For Sale
- Full Workshop Facilities
- AA/VBRA Approved Bodyshop
- Restoration & Repairs
- Mechanical Repairs
- MOT Work
- Engine Tuning
- Engine Rebuilds
- Transmission Overhauls
- Complete Interior Retrims
- Seat Repairs

Jaymic Ltd
Cromer
Norfolk
NR27 0HF
England
Tel : (44) 01263 511710
Parts Line (44) 01263 512883
Fax: (44) 01263 514133

MUNICH LEGENDS

SALES Exciting BMW's for those who know more about understeer and oversteer than hands-free dialling and on-board computers...and all cars are inspected by BMW-trained engineers to ensure no nasty post-purchase surprises. Buy with confidence...whatever the model.

SPARES Everything you need to care for your Coupe from the leading 3.0 CS specialists... now offering the same enthusiastic support to owners of all classic and performance BMW's. Call 01825 740456 - the only part number you will ever need to know.

SERVICE Full workshop support from a service to a complete mechanical rebuild including performance modifications for road or track, all carried out by BMW-trained technicians who never forget...only your satisfaction guarantees our success.

SUPPORT Restoring a classic 3.0 Coupe or searching for a late model M6? Whatever your interest in the Munich Marque, we're here to help...with experience and enthusiasm that combine to provide a standard of service you can trust.

THE CLASSIC AND PERFORMANCE BMW SPECIALISTS
Lewes Road, Chelwood Gate, Sussex RH17 7DE
TEL:01825 740456 FAX:01825 740094

BMW 3.0CSL

The homologation CS 'Lightweight' saved kilos by adopting aluminium for the bonnet, boot and door skins and substituting Plexiglass side windows, stripped-out interior and no front bumper. Slightly bored-out 3.0 engine gave the same output but later 3153cc unit had an extra 6bhp. About half CSL production came to the UK, where cars were returned to CS luxury spec. Power steering was now optional and suspension was by stiffer Bilsteins. Very rare 'Batmobile' version (39 built) had glassfibre rear spoiler, air dam, front wing ribs and roof spoiler. Prices have been known to approach £40,000.

Engine 3003cc 89.25x80mm/3153cc 89.25x84mm, 6 cyl OHC, F/R
Max power 200bhp at 5500rpm/206bhp at 5600rpm
Max speed 135/137mph
Body styles coupé
Prod years 1972-75
Prod numbers 1039

PRICES
CSL **A** 10,000 **B** 7000 **C** 4000

BMW 3 Series

As a replacement for the 2002, the 3 Series was incredibly successful, selling almost 1.5 million units in seven years. Smarter body style on a slightly longer wheelbase, but the same recipe of all-independent suspension, powerful engines and high build quality. Four-cylinder cars were 316, 318 (single headlamps), 320 and 320i (double headlamps); six-cylinder cars from '77 were 320 and scorching 323i, which earned a reputation for tail-happy manners. Entertaining but not yet appreciating in value – could be a real 'sleeper'.

Engine: 1573cc 84x71mm/1766cc 89x71mm/1990cc 88.9x 80m 4 cyl OHC, 1990cc 80x66mm/2315cc 80x76.8mm, 6 cyl OHC, F/R
Max power 90bhp at 6000rpm/98bhp at 5800rpm/109bhp at 5800rpm/125bhp at 5700rpm/122bhp at 6400rpm/143bhp at 6000rpm **Max speed** 101-118mph **Body styles** saloon, convertible **Prod years** 1975-82
Prod numbers 1,364,039

PRICES
A 2500 **B** 1400 **C** 750

BMW 5 Series

Archetypal 1970s business express, which gained BMW a strong following in the UK. New mid-sized BMW which slotted in as a slightly larger replacement for the 1800/2000. Updated styling but underneath everything remained very much as before, including the engines. Six-cylinder engines (525 and 528i) arrived in '73, in which case four-wheel discs were standard, plus a new six-cylinder 2.0-litre engine from '77. An attractive car in its day but now disappearing fast as rust claims its dues, and worth very little as a 'classic'.

Engine 1766cc 89x71mm/1990cc 88.9x 80m 4 cyl OHC, 1990cc 80x66mm/2494cc 86x71.6mm/2788cc 86x80mm 6 cyl OHC, F/R
Max power 90bhp at 5500rpm/115bhp at 5800rpm/122bhp at 6000rpm/145bhp at 6000rpm/165bhp at 5800rpm/175bhp at 5500rpm
Max speed 99-124mph **Body styles** saloon **Prod years** 1972-81
Prod numbers 699,094

PRICES
A 1900 **B** 1200 **C** 600

BMW 6 Series

BMW CS replacement borrowed its mechanicals from the forthcoming 7 Series, but the recipe was as familiar as ever: independent suspension, disc brakes, power steering, large six-cylinder engines. New bodywork (initially built by Karmann) not as pretty as outgoing CS, but slightly less rust-prone. Familiar models in UK are 628CSi and 635CSi, and we also got the early 633CSi, but not the 630CS. Post-1982 635CSi engine's dimensions changed completely. Increasing luxury and always good performers, solid handlers and beautifully engineered.

Engine 2788cc 86x80mm/2985cc 89x80mm/3210cc 89x86mm/3453cc 93.4x84mm/3430cc 92x86mm, 6 cyl OHC, F/R **Max power** 184bhp at 5800rpm/185bhp at 5800rpm/200bhp at 5500rpm/218bhp at 5200rpm/218bhp at 5200rpm **Max speed** 132-142mph **Body styles** coupé
Prod years 1976-89
Prod numbers 80,361

PRICES
A 4000 **B** 2500 **C** 1500

Bond Minicar

Lawrie Bond's answer to post-war shortages was the 1949 Minicar three-wheeler, which had open doorless bodywork, rear brakes only and no rear suspension. A Villiers two-stroke engine drove the single front wheel by chain. 1951 Mark B had rear suspension, Mark C (1952) expanded bodywork and one door, Mark D (1956) 12 volt electrics, Mark E (1957) a proper chassis and restyled body, Mark F (1958) a 246cc engine, Mark G hydraulic brakes and wind-up windows. All now have a strong following.

Engine 122cc 50x62mm/197cc 59x72mm/246cc 66x72mm, 1 cyl TS, F/F
Max power 5/8/12bhp at 4800rpm
Max speed 30-50mph
Body styles tourer, saloon
Prod years 1949-66
Prod numbers 24,484

PRICES
A 2000 **B** 1500 **C** 1000

Bond 875

The 875 was a radically different three-wheeler using the glassfibre expertise Bond had gained from the Equipe. Its four-seater closed saloon body was much larger than previous Bonds. Using a detuned Hillman Imp in the tail gave it unexpectedly good (some said frightening) performance and an aptitude for oversteer on a grand scale. Van and estate versions were also sold. MkII model of 1967 had a restyled nose with rectangular headlamps. Reliant bought Bond in 1969 and axed the 875, which sat uncomfortably next to its own Regal.

Engine 875cc 68x60.4mm, 4 cyl OHC, R/R
Max power 34bhp at 4800rpm
Max speed 80mph
Body styles saloon, estate
Prod years 1965-70
Prod numbers 3431

PRICES
A 1100 **B** 600 **C** 300

Bond Equipe GT4/GT4S

Bond took the simple course of buying in Herald chassis from Triumph and putting its own glassfibre body on top to create a sports coupé. Surely no-one took Bond seriously when they suggested that the Equipe was "the most beautiful car in the world", especially as it was compromised by its use of Herald doors. Engines from the Spitfire (1300cc from '67), plus front disc brakes from the same source. GT4S from '64 has four headlamps, proper opening boot and better headroom. Herald chassis and doors rust but the plastic body doesn't.

Engine 1147cc 69.3x76mm/1296cc 73.7x76mm, 4 cyl OHV, F/R
Max power 63bhp at 5750rpm/67bhp at 6000rpm
Max speed 83/92mph
Body styles coupé
Prod years 1963-70
Prod numbers 2956 (451/2505)

PRICES
A 2500 **B** 1300 **C** 900

Bond Equipe 2-Litre

The GT4 was Herald based, the 2-Litre was Vitesse based. Naturally, that meant the compact straight six engine with its fine spread of power, but also its dodgy rear suspension (improved on MkII version from 1968). All-new heavier body style with quad headlamps, steel doors and scuttle plus glassfibre bodywork. MkII gained increase in power, recessed instruments, Magnum wheel trims and the option of a convertible top. It may not have been elegant but the convertible was then just about the only cheap four-seater open car around.

Engine 1998cc 74.7x76mm, 6 cyl OHV, F/R
Max power 95bhp at 5000rpm/104bhp at 5300rpm
Max speed 108mph
Body styles coupé, convertible
Prod years 1967-70
Prod numbers 1432

PRICES
Coupé **A** 2500 **B** 1500 **C** 1000
Convertible **A** 3500 **B** 2300 **C** 1500

Bond Bug

When Reliant inherited Bond, it took the opportunity of exploring an idea from Tom Karen of Ogle design to create a three-wheeler that was fun. The Bond Bug was certainly that – a triangular shaped pod with a flop-forward canopy, detachable side screens, bug-eye headlamps and sawn-off tail. Strictly a two-seater with fixed seating. Aircraft style decals adorned the paintwork, which was any colour you liked, as long as it was bright tangerine. 700E had 29bhp Reliant engine, 700ES had 31bhp and more equipment. Larger 748cc engine from '73.

Engine 701cc 60.5x61mm/748cc 62.5x61, 4 cyl OHV, F/R
Max power 29/31bhp at 5000rpm/32bhp at 5500rpm
Max speed 73/77/78mph
Body styles coupé
Prod years 1970-74
Prod numbers 2270

PRICES
A 1700 **B** 1200 **C** 750

Bonnet Djet

In 1961, René Bonnet left his long-standing partner, Charles Deutsch, with whom he had been building sports and racing cars under the name CD. His new project was the design, production and marketing of the Djet, the world's first mid-engined road car. Using Renault parts, it had a backbone chassis, all-independent suspension and four-wheel discs. Its unusual and extremely narrow glassfibre body kept weight down to just 1350lb (612kg). Production was taken over by Matra in 1964.

Engine 956cc 65x72mm/1108cc 70x72, 4 cyl OHV, M/R
Max power 60bhp at 6000rpm to 80bhp at 6500rpm
Max speed 107-117mph
Body styles coupé
Prod years 1961-64
Prod numbers n/a

PRICES
A 10,000 **B** 6000 **C** 4000

Borgward Hansa 1500/1800/1800D

The Borgward was the first all-new German post-war car and, in 1949, made all other German cars appear antique. Full-width bodywork was inspired by American Kaisers and fitted over a backbone chassis. Suspension was all-independent, gearbox had synchromesh on all four gears, OHV engines were powerful and reliable. Diesel option from '53. Various body styles were offered, including two- and four-door saloons, an estate and a two-door convertible. In UK from '53 onwards in 1800 guise only.

Engine 1498cc 72x92mm/1758cc 78x92mm, 4 cyl OHV/OHV/diesel, F/R
Max power 48bhp at 4000rpm/60bhp at 4200rpm/42bhp at 3700rpm
Max speed 65-78mph **Body styles** saloon, estate, coupé, convertible
Prod years 1949-52/1952-54/1953-54
Prod numbers 22,504/8111/3226

PRICES
Saloon **A** 4000 **B** 2500 **C** 1500
Convertible **A** 6000 **B** 4500 **C** 3000

Borgward Hansa 2400

The 2337cc six-cylinder version of the Hansa was marketed in two versions: as a Sport saloon with the same wheelbase as the smaller-engined Hansas, and in Pullman form with a longer wheelbase. These both had one-piece curved windscreens as opposed to 'V' screens and were both six-light four-door saloons with a sloping back on the Sport and a notchback on the Pullman. The first European car to be offered with its own automatic transmission and, unusually, it had standard electric windows. More power and optional 2240cc from 1955. Not a great marketing success.

Engine 2337cc 78x81.5mm/2240cc 75x84.5mm, 6 cyl OHV, F/R
Max power 82-100bhp at 4500rpm/100bhp at 5000rpm
Max speed 90/96mph
Body styles saloon
Prod years 1952-58
Prod numbers 1388

PRICES
A 7500 **B** 5000 **C** 3000

Borgward Isabella

Undoubtedly the best-known Borgward was the Isabella, named after the wife of Dr Karl Borgward, the man who was responsible for most of the car's development and its unitary construction. Initially launched with a 60bhp 1.5-litre alloy-head engine, a 75bhp version was added in 1955 with the name TS (Touring Sport). An estate car followed in 1956 and a de luxe version was also offered from 1957. Isabellas gained the reputation of being indestructible and seem to attract fervent loyalty among owners. They're also relatively cheap to buy.

Engine 1493cc 75x84.5mm, 4 cyl OHV, F/R
Max power 60bhp at 4700rpm/75bhp at 5200rpm
Max speed 85/93mph
Body styles saloon, estate
Prod years 1954-61
Prod numbers 202,862 (all types)

PRICES
A 4000 **B** 2500 **C** 1000

Borgward Isabella Coupé/Cabriolet

More than one classic car publication has called the Isabella Coupé one of the best-looking German cars of all time. In style, it was totally different from the staid saloon, with flowing wing features, good proportions and a certain glamour. It also boasted lots of luxury fittings, confirming its role as a comfortable sporting coupé. But it was highly priced for a 1.5-litre car. Coachbuilder Deutsch of Cologne was responsible for the Cabriolet, which was so costly that only 29 customers ever bought one – today, it's the Borgward collector's dream piece.

Engine 1493cc 75x84.5mm, 4 cyl OHV, F/R
Max power 75bhp at 5200rpm
Max speed 93mph
Body styles coupé, cabriolet
Prod years 1957-61
Prod numbers n/a/29

PRICES
Coupé **A** 8000 **B** 5000 **C** 2000
Cabriolet Rarely arises for sale, but expected to be twice the price of the coupé

Borgward 2.3-Litre P100

An ill-fated last-ditch attempt to make a car to rival Opel and Mercedes which ended in Borgward's bankruptcy. Unitary body with American lines (even down to the little tailfins). With only 100bhp, the 2.3-litre engine was somewhat underpowered in a car weighing 2720lb (1235kg). Interesting suspension set-up of independent front and swing axles at the back, plus optional air suspension. An intriguing car but very rare these days, even if production continued in Mexico until the late 1960s.

Engine 2240cc 75x84.5mm, 6 cyl OHV, F/R
Max power 100bhp at 5000rpm
Max speed 100mph
Body styles saloon
Prod years 1959-61
Prod numbers 2587

PRICES
A 8000 **B** 5000 **C** 2000

Bristol 400

Following the war, aeroplane maker Bristol turned its attention to cars and acquired licences for BMW products. Hence the 400 looked very like a prewar BMW 327, and was based on a BMW 326 chassis with a BMW 328 pushrod six-cylinder engine. Beautifully built (to aircraft standards), with independent front suspension, rigid rear axle and first gear freewheel on the four-speed 'box. Rather underpowered on single carb engine, not significantly helped by later fitment of triple carbs. Convertibles extremely rare and worth approximately 50% more.

Engine 1971cc 66x96mm, 6 cyl OHV, F/R
Max power 80bhp at 4200rpm/85bhp at 4500rpm
Max speed 92mph
Body styles saloon, convertible
Prod years 1947-51
Prod numbers 700

PRICES
A 22,000 **B** 14,000 **C** 9000

Bristol 401/402/403

A definitive change for Bristol: handsome wind tunnel tested bodies after the Touring *superleggera* construction method of light alloy panels over a tubular frame. Chassis and mechanicals as 400. 401 was the saloon, 402 the handsome but very rare convertible with a slightly lengthened chassis. Close-ratio gearbox from '51. 403 sported an uprated 100bhp version of the 2.0-litre engine (105bhp from '54), Alfin drum brakes, front anti-roll bar and better steering. All models were excellent handlers and the 403 was Bristol's first 100mph car.

Engine 1971cc 66x96mm, 6 cyl OHV, F/R
Max power 85bhp at 4500rpm/100-105bhp at 5000rpm
Max speed 94/102mph
Body styles saloon, convertible
Prod years 1948-53/1948-49/1953-55
Prod numbers 650/20/281

PRICES
401/403 **A** 20,000 **B** 12,000 **C** 7000
402 **A** 35,000 **B** 25,000 **C** 20,000

Bristol 404

Bristol coined the phrase 'The Businessman's Express' to characterise the 404. Based on a shortened version of the 403 chassis which also sat under the Arnolt-Bristol (see page 25). Considerably shorter, lower and lighter than the 403, it was also the prettiest Bristol, with 'hole in the wall' grille, forward-hinging bonnet and tail fins. Construction was light alloy panels on a pine wood framework (prone to rotting out). More powerful engines were delightful and powered many other sports and racing cars. But at twice the price of an XK140, it was for purists only.

Engine 1971cc 66x96mm, 6 cyl OHV, F/R
Max power 105bhp at 5000rpm/125bhp at 5500rpm
Max speed 105/110mph
Body styles coupé
Prod years 1953-55
Prod numbers 52

PRICES
A 35,000 **B** 25,000 **C** 20,000

Bristol 405 Saloon/Drophead Coupé

Lengthened version of the 404 (114in wheelbase) and the only four-door Bristol ever made. The first Bristol with a proper opening boot and a wrap-around rear window. Much heavier than 404 and without its power options, so a bit of a slowcoach. Standard features included Laycock overdrive on top gear and Michelin radial tyres. Very rare two-door Drophead Coupé was built by Abbott of Farnham until 1956. Front discs from '58. Like the 404, timber framing can deteriorate, but is mostly in the upper structure and therefore less serious.

Engine 1971cc 66x96mm, 6 cyl OHV, F/R
Max power 105bhp at 5000rpm
Max speed 100mph
Body styles saloon, drophead coupé
Prod years 1954-58/1954-56
Prod numbers 265/43

PRICES
Saloon **A** 13,000 **B** 8500 **C** 5000
DHC **A** 32,000 **B** 22,000 **C** 17,500

Bristol 406

The 406 of 1958 got a completely new notchback body which looked more staid but more modern at the same time. Two doors only and steel-framed construction. Mechanically almost identical to the earlier Bristols, but an enlarged 2.2-litre version of the BMW based six gave better torque but the same 105bhp output. All cars came with standard disc brakes all round, Watts linkage for the rear axle and overdrive but their weight (3009lb/1365kg) counted against them. Bristol really needed the forthcoming V8 engines.

Engine 2216cc 69x100mm, 6 cyl OHV, F/R
Max power 105bhp at 4700rpm
Max speed 104mph
Body styles saloon
Prod years 1958-61
Prod numbers 174

PRICES
A 15,000 **B** 10,000 **C** 7500

Bristol 406 Zagato

Zagato used a shortened version of the 406 chassis and clothed it with bodywork of its own design. Still recognisably a Bristol, it had cowled headlamps, a shorter cabin and a strong kick in the rear wing line. The result was some 450lb (204kg) lighter than Bristol's saloon version. Triple Solex carbs and Abarth manifolding increased the 2.2-litre six's power output to 130bhp, to make this Bristol's most rapid six-cylinder car. Some standard wheelbase 406 chassis were also supplied to the Swiss coachbuilder Beutler for saloon bodies.

Engine 2216cc 69x100mm, 6 cyl OHV, F/R
Max power 130bhp at 5750rpm
Max speed 120mph
Body styles coupé
Prod years 1960-61
Prod numbers 7

PRICES
A 40,000 **B** 35,000 **C** 30,000

Bristol 407

A sea-change for Bristol: now its engines were sourced from Chrysler in North America and were big V8s, and it never looked back. At last sufficient power to take these substantial cars into Jaguar's performance class. Always with three-speed Torqueflite automatic transmission and now coils up front instead of transverse leaf springs, plus some minor chassis mods. Performance gained at the expense of fuel consumption. Most 407s went to the 'States. The 407 is the least well regarded of all the 1960s V8 Bristols.

Engine 5130cc 98x84mm, V8 cyl OHV, F/R
Max power 250bhp at 4400rpm
Max speed 125mph
Body styles saloon
Prod years 1961-63
Prod numbers 88

PRICES
A 15,000 **B** 10,000 **C** 7500

Bristol 408/409

Major restyle for the 1963 408: lower overall, squarer roof line and a wider grille incorporating a second pair of headlamps. Also a second chrome side strip identifies the 408/409. Armstrong Selectaride dampers at the rear. 409 introduced in 1965 had larger bore Chrysler V8, better steering (power assisted option from '66), Girling brakes, an alternator and a mechanical parking lock in the transmission. Also boasts an improved ride. V8 Bristols offer high levels of performance and refinement, plus a certain air of dignity, and reasonable value.

Engine 5130cc 98x84mm/5211cc 99.3x84mm, V8 cyl OHV, F/R
Max power 250bhp at 4400rpm/251bhp at 4400rpm
Max speed 125/129mph
Body styles saloon
Prod years 1963-65/1965-67
Prod numbers 83/74

PRICES
A 15,000 **B** 10,000 **C** 7500

Bristol 410/411

New front end styling for the 410 included part recessed headlamps, also two full-length chrome strips on side. Smaller (15in) wheels made car sit 1½in lower on the road. Mechanically much the same as before, except for standard power steering, dual-circuit brakes and a floor-mounted gear lever instead of push button. 411 of 1969 has 6.3-litre engine, revised suspension, limited-slip differential, wider grille. Self-levelling suspension on SII (1970), full-width grille on SIII (1973), 6.6-litre engine on SIV (1974), matt black grille on SV (1975). Air conditioning option.

Engine 5211cc 99.3x84mm/6277cc 107.9x85.7mm/6556cc 110.3x85.7mm, V8 cyl OHV, F/R
Max power 251bhp at 4400rpm/335bhp at 5200rpm/264bhp at 4800rpm
Max speed 130/140mph
Body styles saloon
Prod years 1967-69/1969-76
Prod numbers 79/287

PRICES
A 17,000 **B** 12,000 **C** 9000

Bristol 412

After almost 20 years of basically the same product, the 412 was about as radical as Bristol could get. The chassis and mechanical package remained the same as ever but the bodywork was all-new and very 1970s. Zagato did the 'breeze block' styling and this became Bristol's first ever factory convertible. From 1976, the roof arrangement altered to become a 'convertible saloon' with a fixed roll-over bar and detachable targa roof and rear panels. Smaller engine from 1977. All have leather, wood, PAS and speed hold. Gave way to Beaufighter in 1982.

Engine 6556cc 110.3x85.7mm/ 5900cc 101.6x90.9mm, V8 cyl OHV, F/R
Max power 264bhp at 4800rpm/n/a
Max speed 140mph
Body styles convertible, saloon
Prod years 1975-82
Prod numbers n/a

PRICES
A 20,000 **B** 12,000 **C** 6000

Bristol 603

Another drastic Bristol restyle: some say ghastly, others characterful. Underneath it all lay basically the same chassis as Bristol was making in the 1940s. Alloy body panels, as ever. Launched in two forms: 603E with 5.2-litre V8 and 603S with four-barrel carb 5.9-litre V8. 'E' supposedly stood for economy and this version lasted just one year. S2 in 1977 had standard air conditioning. Chrome bumpers from 1980 and revised air con and new wheels from '81. Updated in 1982 to become the Brigand and Britannia. A true English eccentric.

Engine 5211cc 99.3x84.1mm/5900cc 101.6x90.9mm, V8 cyl OHV, F/R
Max power n/a
Max speed 132mph
Body styles saloon
Prod years 1976-82
Prod numbers n/a

PRICES
A 14,000 **B** 9000 **C** 5000

Bugatti 101

The post-war attempt to restart Bugatti's car manufacturing career was always going to be an uphill struggle. When it appeared at the 1951 Paris Salon, the 'modern' Bugatti had a full-width body by Gangloff, but underneath lay a Type 57 chassis of 1934 vintage and its rigid front and rear axles were way behind the times for a top-class sports car. The engine was a 3.3-litre eight-cylinder with or without supercharging. Four or five-speed manual 'boxes, or Cotal electric transmission. RHD only, in race-bred Bugatti tradition. The price was outrageous and only seven were built.

Engine 3257cc 72x100mm, 8 cyl DOHC, F/R
Max power 135bhp at 5500rpm/188bhp at 5200rpm
Max speed 105/116mph
Body styles saloon, coupé, convertible
Prod years 1951-56
Prod numbers 7

PRICES
A 100,000 **B** 80,000 **C** 70,000

Buick Riviera '63-'65/'65-'70

The Riviera started out life as GM chief designer Bill Mitchell's conception of a Ford Thunderbird beater. The original hardtop coupe (photo) looked convincing and sold very well, although it never matched the T-bird. Replacement of 1965 was less characterful, sharing its structure with the Cadillac Eldorado but adding rather bland panelling. Still had a separate chassis, engines expanding as years go by up to 7.5 litres in 1970. Drum brakes, iffy handling, but up to 130mph!

Engine 6572cc 106.4x92.5mm/ 6970cc 109.5x92.5mm/7046cc 106.4x99.1mm/7468cc 109.5x99mm, V8 cyl OHV, F/R
Max power 225bhp at 4400rpm to 365bhp at 5000rpm
Max speed 110-130mph
Body styles coupe
Prod years 1962-65/1965-70
Prod numbers 112,244/227,639
PRICES
'63-'65 **A** 10,000 **B** 6500 **C** 3000
'65-'70 **A** 7500 **B** 5000 **C** 2500

Buick Riviera '71-'73

Brief return to character styling, with lines penned by Donald Lasky. Sculpted wing lines and a dramatic boat-tail rear end reminiscent of 1963 Corvette. Based on a shortened Electra chassis with coil sprung rear end. Big 7.5-litre V8 good for 330bhp in Grand Sport guise but emissions clipped that back to 250bhp by 1973. Facelifted slightly in '72. There's space inside for six people and a lot of goodies came as standard, but then this car cost almost twice as much new as a Skylark Coupe. Fourth generation Riviera ('73-'76) was utterly dull.

Engine 7468cc 109.5x99mm, V8 cyl OHV, F/R
Max power 250bhp at 4000rpm to 330bhp at 4400rpm
Max speed 115-130mph
Body styles coupe
Prod years 1970-73
Prod numbers 101,618

PRICES
A 9000 **B** 6500 **C** 3000

Cadillac Eldorado '53

The car that started the Eldorado dynasty is one of the most sought-after mainstream post-war American cars. Built for just one year, only 532 examples left the Cadillac works, making it extremely rare today. The reason so few sold was the price tag: at $7750, it was almost twice the price asked for a Cadillac 62 Convertible. Cadillac justified the cost of what was essentially a 62 by stuffing in just about every extra imaginable as standard gear, plus cut-down 'Panorama' screen and metal hood cover.

Engine 5424cc 96.8x92.1mm, V8 cyl OHV, F/R
Max power 210bhp at 4000rpm
Max speed 105mph
Body styles convertible
Prod years 1953
Prod numbers 532

PRICES
A 45,000 **B** 35,000 **C** 27,000

Cadillac Eldorado Brougham '57-'58/'59-'60

The '57 Eldorado had it all: hardtop style, a choice of powerful V8 engines and glitz by the pound. Four headlamps were an industry first for 1956 and a Harvey Earl touch was the stainless steel roof. One novelty was air suspension, but the system leaked and was soon abandoned. The extras were piled on to produce the most expensive post-war Cadillac yet made, with a retail price of $13,074 (when an 'ordinary' Caddy Sedan de Ville cost just $4713). Even more valuable than the '57 Eldorado Biarritz convertible. Replaced by restyled Pininfarina-built car for '59.

Engine 5972cc 101.6x92.1mm/ 6384cc 101.6x98.4mm, V8 cyl OHV, F/R
Max power 300/330bhp at 4800rpm/345bhp at 4800rpm
Max speed 118/120mph
Body styles hardtop saloon
Prod years 1956-58/1958-60
Prod numbers 704/200

PRICES
A 30,000 B 22,000 C 15,000

Cadillac Coupé De Ville/Eldorado '59

If there is one car which comes to mind as the quintessential American car, it is the '59 Caddy. This was the zenith of fin mania, when Harvey Earl slapped on those 'plane shaped fins and rocket ship tail-lights as a parting gesture. The '59 De Ville was offered in sedan and coupe forms, while the Eldorado was the classic convertible (must it *always* be pink?). Power seats and windows were standard and, on the convertible, you got a power hood. Dynamically challenged, but who cares? There will always be a ready market for the '59 Cadillac.

Engine 6384cc 101.6x98.4mm, V8 cyl OHV, F/R
Max power 325bhp at 4800rpm
Max speed 112mph
Body styles coupe, convertible
Prod years 1958-59
Prod numbers 21,924/11,130

PRICES
Coupe A 18,000 B 13,000 C 10,000
Convertible A 30,000 B 20,000 C 15,000

Cadillac Eldorado '67-'70

One of the most interesting post-war American cars was the '67 Eldorado. Yes, this was the first front-wheel drive Cadillac, its system derived from the Oldsmobile Toronado, whose basic body under-structure it shared. The notion of hooking up a 400bhp 8.2-litre V8 to the front wheels could only come from America and it was accomplished by chain drive via an auto 'box. Massive separate chassis, torsion bar front suspension, disc front brakes. Huge dimensions but cramped inside. A big seller for Cadillac, so plenty still around to choose from.

Engine 7025cc 104.9x101.6mm/ 7729cc 109.2x103.1mm/8194cc 109.2x109.3mm, V8 cyl OHV, F/F
Max power 345bhp at 4600rpm/ 375bhp at 4400rpm/400bhp at 4400rpm **Max speed** 115-122mph
Body styles coupe
Prod years 1966-70
Prod numbers 89,633

PRICES
A 9000 B 5500 C 3000

Cadillac Eldorado '71-'78

Still front-wheel drive for this enormous top-flight Cadillac, which was now even bigger than before. Styling was typically glitzy 1970s glam: enormous overhangs, mock sedanca hardtops, two-tone paint schemes yet amazingly cramped inside. Annual sheet metal changes kept the fashion cycle going but power output was slowly but surely strangled as US emissions laws took effect. Smaller engine and all-wheel disc brakes from '77. Most desirable version is the convertible, withdrawn in 1976, and the last of GM's big open-topped cars.

Engine 8194cc 109.2x109.3mm/ 6966cc 104x103mm, V8 cyl OHV, F/F
Max power 235bhp at 4000rpm/180bhp at 4000rpm
Max speed 110-120mph
Body styles coupe, convertible
Prod years 1970-78
Prod numbers 347,401

PRICES
Coupe A 8500 B 5000 C 2500
Convertible A 15,000 B 10,000 C 8000

Caterham Super Seven

Caterham Cars bought the rights to the Lotus 7 in 1973 and proceeded to build the same car, only better. Production began with the unloved S4 but switched to the S3 from 1974 after only 38 examples. The S3 used an uprated Lotus Twin Cam SS chassis. Early cars nearly always with Lotus Twin Cam engines, plus four-speed Ford 'box, Escort rear axle (Morris Ital axle from '81). Uprated Ford engines typical in the 1980s, De Dion rear end from '85, five speeds from '86. Stonkingly quick, benchmark handling, zero practicality. An enduring legend.

Engine 1558cc 82.5x72.7mm/1599cc 81x77.6mm/1691cc 83.3x77.6mm/ 1699cc 83.5x77.6mm, 4 cyl DOHC/ OHV/ OHV/OHC, F/R **Max power** 126bhp at 6500rpm/84 to 150bhp at 6500rpm/135bhp at 6000rpm/170bhp at 6500rpm **Max speed** 110-120mph
Body styles sports
Prod years 1973 to date
Prod numbers 6200 (to 1995)

PRICES
A 12,000 B 9000 C 7000

Chaika GAZ-13

The Chaika was built at the Molotov factory in the Russian city of Gorky, as a replacement for the ZIM (see page 236). In the hierarchy of Russian cars, it slotted in as one of the choices of communist party heads, but always took second place to the ZIS/ZIL. Its styling was very American inspired and it was offered as an eight-passenger limousine or as a huge convertible. Almost always V8 powered (some early cars kept the old six-cylinder GAZ-12 engine), and always push-button auto. Rare outside Russia. Convertibles worth perhaps double.

Engine 3480cc 82x110mm/4890cc 92x92mm, 6 V8 cyl SV/OHV, F/R
Max power 94bhp at 3600rpm/180bhp at 4600rpm
Max speed 75/100mph
Body styles saloon, convertible
Prod years 1958-65
Prod numbers n/a

PRICES
A 2500 **B** 1300 **C** 600

Champion 250/400/500

Directly after the war, the German Hermann Holbein designed a prototype of a microcar which would be built by the famous ZF gearbox factory. They were tiny (78in/198cm) motorbike powered open cars which grew gradually in size and engine capacity until the definitive 400 model of 1951. This had near-symmetrical front and rear wing pressings and semi-circular windows which rotated out of sight. In 1955, the Maico motorcycle factory took the project over and made over 6000 examples, mostly with 452cc Heinkel engines and enlarged bodywork. There were even 10 Beutler convertibles (photo).

Engine 248cc/398cc/452cc, 1 cyl/2 cyl/ 2 cyl TS, R/R
Max power 10bhp/14bhp/18bhp at 4000rpm
Max speed 45/52/65mph
Body styles saloon, convertible
Prod years 1948-51/1951-56/1954-58
Prod numbers 400/5050/6342

PRICES
A 3000 **B** 1800 **C** 800

Checker Superba/Marathon

Until the 1980s, the name Checker was synonymous with the New York taxi. It had been a manufacturer of taxis since 1923 but, from 1959, began supplying private buyers. Engines were initially six-cylinder Continental units, uprated to Chevy sixes and, briefly, V8s. Styling did not change one iota and Checkers gained a reputation for lasting forever. Long wheelbase saloons, estates and enormous eight or ten-door Aerobus models were all available. Usually auto transmission. A handful are now in the UK.

Engine 3704cc 84.1x111.25mm to 5739cc 101.6x82.6mm, 6/V8 cyl OHV, F/R
Max power 90bhp at 3600rpm to 300bhp at 4800rpm
Max speed 80-114mph
Body styles saloon, estate
Prod years 1959-82
Prod numbers ca 110,000

PRICES
A 6500 **B** 4000 **C** 2000

Chevrolet Bel Air '57

The cars which Chevrolet made in the years 1955-57 have a certain magical charm to American collectors, who refer to them as 'Tri-Chevys'. Alongside the short-lived two-door Nomad station wagon, the most favoured is the '57 Bel Air Convertible. It set the scene for mid-fifties Americana: chrome, fins and duo-tone paint. Coupe and convertible styles (plus estates and saloons). Lots of power output choices in six and V8 engines up to 283bhp Ramjet (sub-10 sec 0-60mph time).

Engine 3849cc 90.5x100mm/4637cc 98.4x76.2mm, 6/V8 cyl OHV, F/R
Max power 138bhp at 4200rpm to 283bhp at 6000rpm
Max speed 87-110mph
Body styles coupe, convertible
Prod years 1956-57
Prod numbers 166,426/47,562

PRICES
Coupe **A** 12,000 **B** 8000 **C** 6000
Convertible **A** 20,000 **B** 13,000 **C** 8000

Chevrolet Corvette '53-'55

In the early 1950s, Detroit was facing a flood of European sports cars. Chevrolet's response was the Corvette, a Harvey Earl cross between Italian and American dream car styles. The panoramic windscreen and jukebox dash were just what Uncle Sam ordered but the flimsy soft top, glassfibre bodywork – the world's first mass-produced plastic body – and detachable side windows were distinctly un-American. Combined with rather feeble power outputs, the Corvette emerged as a sales flop. But it is a design classic which inspired a generation of American sports cars and is today very much sought-after.

Engine 3849cc 90.5x100mm/4342cc 95.3x76.2mm, 6/V8 cyl OHV, F/R
Max power 150bhp at 4200rpm/195bhp at 5000rpm
Max speed 102/108mph
Body styles sports
Prod years 1953-55
Prod numbers 4640

PRICES
A 22,500 **B** 18,000 **C** 10,000

Chevrolet

Chevrolet Corvette '56-'62

Chevy might have dropped the Corvette had not Ford launched the Thunderbird in 1955. But the T-bird was never a true sports car and the Corvette was becoming so: thanks to Zora Arkus Duntov, who gave the Chevy V8 the power it needed, the 'Vette became a match for imported exotica on performance. Roadholding was much improved and the styling totally changed, with scalloped sides, an attractive rounded rear end and duo-tone paint schemes. Sales responded dramatically. As America's only true sports car, Corvettes of all descriptions have a big following.

Engine 4342cc 95.3x76.2mm/5359cc 101.6x82.5mm, V8 cyl OHV, F/R
Max power 225bhp at 5200rpm/360bhp at 6000rpm
Max speed 115/138mph
Body styles roadster
Prod years 1955-62
Prod numbers 64,375

PRICES
A 15,000 **B** 10,000 **C** 6000

Chevrolet Corvette Sting Ray '63-'67

Earl's successor Bill Mitchell styled the new Sting Ray which, in 1963, looked like a dream car which you could actually buy. Its dramatic boat-tail rear end and humped wings inspired the name. For the first year only, there was a 'split' rear window, so these cars are very rare. However, many owners of '63 cars cut out the window bar to make them look up-to-date, only to put them back in – along with many owners of younger cars – when rarity pushed prices up, so beware. Huge power from big 7-litre V8 option. The Sting Ray is much fancied these days.

Engine 5359cc 101.6x82.5mm/6997cc 108x95.5mm, V8 cyl OHV, F/R
Max power 250bhp at 4400rpm/435bhp at 5800rpm
Max speed 118/150mph
Body styles coupe, convertible
Prod years 1962-67
Prod numbers 45,546/72,418

PRICES
A 18,000 **B** 12,500 **C** 8000

---

claremont CORVETTE

THE ORIGINAL CORVETTE SPECIALIST

Recognised as the Corvette specialist, our inventory of genuine, new and used GM parts is second to none.

Europe's largest Corvette parts specialist.

12 mint Corvettes always in our showroom.
Established 1977.

EXIT 4 ON M20, SNODLAND,
KENT ME6 5NA, ENGLAND.
TELEPHONE: 01634 244444 FAX: 01634 244534

FROST AUTO RESTORATION TECHNIQUES

Shaping metal made easy

Use the Shrinker/Stretcher to fabricate repair panels anywhere on your car. Works up to 1.2mm steel or 2mm aluminium. Set comes with body and interchangeable Shrinker and Stretcher jaws. **F400 Set £179.50**

Learn from the best in the Trade.
This video series will take you into the best restoration workshops in Europe and North America. See repair work to a concours winning Ferrari Monza; a Wheeling Course, the restoration of a walnut veneered interior etc. **L600 Classic Restoration Profiles: Intro: VHS: 60 minutes. £14.95**

Quote this Advert to receive a free copy of our catalogue.

Kills rust
...and stop more rust forming. Neutra Rust dries to a gloss black, can be top coated with most paints.
S400 250ml **£7.14**
S405 1 litre **£14.22**

Scratch Remover
New M.O.T. regulations could mean your car will fail its inspection because of a scratched windscreen. Now you can restore your glass with our glass polishing kit and your electric drill! **S130 Windscreen Scratch Repair Kit £14.50**

All prices include V.A.T.

Frost Auto Dept., Crawford Street,
Rochdale
OL16 5NU
England
Tel: 01706 58619
Fax: 01706 860338

Chevrolet Corvette Stingray '67–'84

The '67 Corvette metamorphosed again, to a shape which would become known as the 'Coke bottle'. Same wheelbase as before, and there were convertible and coupe versions, but the latter was now a T-Top targa coupe with removable roof panels. Power output reached its zenith with the 7.4-litre 465bhp V8 of 1970, but thereafter the 'Vette was strangled, dropping to just 180bhp by 1976. Reasonable handling. Major revisions for '73 and '77 altered body styling details and added creature comforts. Convertibles ended in 1975.

Engine 5359cc 101.6x82.5mm/5735cc 101.6x88.4mm/6997cc 108x95.5mm/7440cc 107.9x101.6mm, V8 cyl OHV, F/R **Max power** 180bhp at 4000rpm to 465bhp at 5200rpm **Max speed** 118-155mph **Body styles** coupe, convertible **Prod years** 1966-84 **Prod numbers** 480,536/86,147
PRICES
Coupe '67-'72 **A** 12,500 **B** 8000 **C** 5000
Convertible **A** 15,000 **B** 10,000 **C** 7500
Coupe '72-'84 **A** 8000 **B** 5000 **C** 2500

Chevrolet Camaro '67–'70

GM's belated reply to Ford's super-successful Mustang. By the time it arrived in 1966, the heyday of the 'personal coupe' had already passed. Nevertheless GM sold 220,000 Camaros in the first year. Based on the Chevy Nova body/chassis and mechanical package with styling by Bill Mitchell. Various states of tune were available, the most extreme being the SS (Super Sport) – 96,275 built – and the RS (Rallye Sport) – 143,592 built. These are now the most prized of all Camaros and attract small premiums. Convertibles are also more desirable (SS pictured), but harder to find.

Engine 4097cc 98.4x89.7mm/5367cc 108x82.7mm/5385cc 101.6x88.5mm/5735cc 101.6x88.4mm/6489cc 104x95.5mm, 6/V8cyl OHV, F/R **Max power** 155bhp at 4200rpm to 375bhp at 4800rpm **Max speed** 104-130mph **Body styles** coupe, convertible **Prod years** 1966-70 **Prod numbers** 350,411/63,154
PRICES
Coupe **A** 10,000 **B** 6000 **C** 3000
Convertible **A** 14,000 **B** 9000 **C** 6000

Chevrolet Camaro '70–'81

Its own purpose-designed monocoque body distinguished the Camaro of the 1970s, a big dollar earner for GM, selling almost two million. Less subtle styling (especially from the mid-1970s when the nose was reshaped to satisfy safety laws). Nowhere near as powerful as the original series Camaro, even in Z28 form, the most desirable of 1970s Camaros; around 250,000 of these were built, so they are not that rare. V6 engines from 1979. Cloned as a Pontiac Firebird with different engines. Squashed accommodation for four and no convertible option.

Engine 3751cc 94.9x88.4mm to 6573cc 104.8x95.4mm, 6/V6/V8 cyl OHV, F/R **Max power** 90bhp at 3600rpm to 360bhp at 6000rpm **Max speed** 96-125mph **Body styles** coupe **Prod years** 1970-81 **Prod numbers** 1,811,973

PRICES
Camaro **A** 5000 **B** 3000 **C** 1000
Camaro Z28 **A** 8000 **B** 4500 **C** 2000

Chevrolet Corvair

Detroit finally woke up to the fact that some 600,000 cars were imported into America in 1959. The General's reaction was the quirky Corvair, a car which combined the popular Beetle's rear-engined layout in a body the size of a Mercedes. No other American car looked like it, or drove like it: wayward handling got Ralph Nader going. Also unusual was the flat-six engine. Four-door saloon generates little enthusiasm, but the 1960 two-door coupé and 1962 convertible are more interesting. If there is a collectible Corvair, it is the 1962-64 Monza with its 150bhp turbocharged engine.

Engine 2288cc 85.9x66m/2377cc 87.3x66m/2684cc 88.9x72.4mm, 6 cyl OHV, R/R **Max power** 80bhp to 150bhp at 4400rpm **Max speed** 87-105mph **Body styles** saloon, coupe, convertible **Prod years** 1959-64/1960-64/1962-64 **Prod numbers** 1,271,089
PRICES
Saloon **A** 4000 **B** 2500 **C** 1200
Monza Spyder **A** 7500 **B** 4000 **C** 2200

Chrysler C300 Hardtop

The responsibility for starting the great American horsepower race can be laid at Chrysler's door. Its celebrated V8 Hemi engine (short for hemispherical combustion chambers) achieved greatness in the Chrysler 300 in 1955. As the name suggested, the engine boasted 300bhp, the biggest output of any American road car at that time. The clean, aggressive bodywork was by Virgil Exner and the whole car cost a reputed $100 million to develop. The '55 Hardtop sold few, but boosted Chrysler's yearly sales.

Engine 5426cc 96.8x92.2mm, V8 cyl OHV, F/R **Max power** 300bhp at 5200rpm **Max speed** 125mph **Body styles** coupe **Prod years** 1955 **Prod numbers** 1725

PRICES
A 22,000 **B** 15,000 **C** 10,000

Chrysler

Chrysler 300F

Chrysler made sure it stayed ahead of the industry-wide rush for the greatest number of horses under the bonnet. The 300F broke the 400bhp barrier, keeping it America's most rapid production car. It was unusual in another respect for a big American car: it had unitary bodywork. Automatic was the norm but a four-speed manual was made in France by Pont-à-Mousson. The 300F was a popular club racing car, all the more so after a modified turbo version did nearly 190mph on the Bonneville salt flats. Very rare, very fast and very sought after.

Engine 6746cc 106.4x95.25mm, V8 cyl OHV, F/R
Max power 380-400bhp at 5000rpm
Max speed 135mph
Body styles coupe, convertible
Prod years 1960
Prod numbers 964/248

PRICES
A 28,000 **B** 18,000 **C** 9000

Chrysler Imperial '57-'63

Imperial was actually a separate marque under which Chrysler sold its most opulent models, which were supposed to rival Cadillac but never could. The most ostentatious of all Imperials were those made between 1956 and 1963. Virgil Exner's tail fin experiments reached their height in this period and other fancies included optional false bootlid spare wheels and stainless steel roof panels. Le Baron was premium-priced hardtop model. Non-Hemi 6.8 engine from '59. Ghia converted just 112 to Crown Imperial Limousine spec on a 149.5in wheelbase. Almost sci-fi in their outlandishness.

Engine 6423cc 101.6x99.1mm/6769cc 106x95mm, V8 cyl OHV, F/R
Max power 340/350bhp at 4400rpm
Max speed 112mph
Body styles saloon, limousine, coupe, convertible
Prod years 1956-63
Prod numbers 129,430
PRICES
Saloon **A** 7000 **B** 4000 **C** 2000
Coupe/Le Baron **A** 14,000 **B** 8000 **C** 5000
Convertible **A** 18,000 **B** 12,000 **C** 7500

Chrysler 180/2-Litre

In Europe, Chrysler took over Simca in France and the Rootes Group in Britain, in theory creating a force to be reckoned with. In reality it stumbled from one crisis to the next with dreadful cars like the French-built 180. High waist-line styling was utterly dull, chassis uninspired and dynamics well behind the competition. Saving graces were decent engines (though 1.6-litre and diesel not sold in UK) and reasonable ride. Usually auto transmission. Unloved by anyone in the 1970s and even less so now, so absolutely no classic status.

Engine 1812cc 87.7x75mm/1981cc 91.7x75mm, 4 cyl OHC, F/R
Max power 100/110bhp at 5800rpm
Max speed 106/109mph
Body styles saloon
Prod years 1970-80
Prod numbers n/a

PRICES
A 1000 **B** 600 **C** 200

Chrysler Alpine

The Alpine caused quite a stir when it arrived in 1975. Here was a modern-looking front-wheel drive hatchback with lots of space and a class-leading ride. It even won the Car of the Year award. Looking back, that decision seems baffling because the Alpine was, in reality, a rather mediocre affair. Engines and gearboxes came from the old Simca 1100 and were the car's worst features, along with soggy handling. Most have rusted into oblivion by now. Rebadged as a Talbot in 1980 and joined by a three-box Solara version in 1982. Almost no merit.

Engine 1294cc 76.7x70mm/1442cc 76.7x78mm/1592cc 80.6x78mm, 4 cyl OHV, F/F
Max power 68bhp at 5600rpm/85bhp at 5600rpm/88bhp at 5400rpm
Max speed 94-102mph
Body styles hatchback
Prod years 1975-86
Prod numbers n/a

PRICES
A 1000 **B** 500 **C** 200

Chrysler Charger

After the big Humbers died at Chrysler's hand in the late 1960s, it attempted to fill the gap in the UK with its Australian production. Aussie cars were basically similar to American designed Plymouths and Dodges. The most interesting was undoubtedly the Valiant Charger coupé, described in the Antipodean press as 'the best-looking car ever produced in Australia'. Big range, all with Hemi V8s, up to the very powerful R/T of 1971-72 – the fastest Aussie car then made. Today these cars have a semi-cult following, despite their rarity in Britain.

Engine 5210cc 99.3x84.1mm/5572cc 102.6x84.1mm, V8 cyl OHV, F/R
Max power 200-300bhp
Max speed 100-120mph
Body styles coupé
Prod years 1971-73
Prod numbers n/a

PRICES
A 5000 **B** 3000 **C** 1800

Cisitalia 202

Army uniform manufacturer Piero Dusio made many sports and racing cars after the war, most of them one-offs. However, the 202 was the exception, some 170 being built with different bodies. The Pinin Farina full-width coachwork (pictured) is widely regarded as a lynchpin in post-war styling. Engines were tuned Fiat 1100 units, mounted in a tubular space frame along with largely standard 1100 mechanicals. However brilliant its design, the 202 was simply not quick enough for its price.

Engine 1089cc 68x75mm, 4 cyl OHV, F/R
Max power 66bhp at 5500rpm
Max speed 93mph
Body styles coupé, convertible
Prod years 1947-52
Prod numbers 153/17

PRICES
A 80,000 **B** 50,000 **C** 25,000

Cisitalia 750/850

Dusio emigrated to Argentina in 1949 and Cisitalia's new owners continued the marque with a variety of rebodied Fiats through the 1950s. Production ceased in 1958, only to restart in 1961 with a 'Tourism Special' body on the Fiat 600 chassis. This was typical of many Italian coachbuilt sports cars of the period and could be had in coupé and convertible styles. Enlarged 750 or 850 engines gave the car a little more pep than standard rebodied 600s. Cisitalia bowed out in 1964, its cars unable to live up to the standards set by the 202.

Engine 735cc/847cc, 4 cyl OHV, R/R
Max power 40/54bhp at 6200rpm
Max speed 100mph
Body styles coupé, convertible
Prod years 1961-64
Prod numbers n/a

PRICES
Too rare for accurate valuation, but expect to pay between £5000 and £15,000

Citroën 11CV

The fabulous Traction Avant (front-drive) Citroën arrived as early as 1934 and lasted 23 years. It set standards for roadholding and ride which would not be bettered for years to come. Short boot version supplemented by long boot model. Called Light 15 and Big 15 in Britain (latter with extra 7in in wheelbase), where cars were also built, with wood and leather trim (about 25,000 were Slough-built and are more sought after). Commerciale had 'hatchback' of sorts, Familiale had eight seats. Coupés and Cabriolets prewar only.

Engine 1911cc 78x100mm, 4 cyl OHV, F/F
Max power 59bhp at 4000rpm
Max speed 75mph
Body styles saloon
Prod years 1934-57
Prod numbers 759,123

PRICES
A 9000 **B** 6000 **C** 3000

Citroën 15CV/Big 6

A bigger wheelbase and an in-line six-cylinder engine distinguished the big brother of the Traction Avant range. Even heavier and tougher on the steering biceps, the 15CV was a large, comfortable and rapid saloon well suited to long-distance travel. There was a Familiale version and, before the war, some Cabriolets. Slough-built cars are rare (only about 1300 were made), while the ultimate Traction is the very rare 6H (only 3079 were built from 1954). The 'H' stood for hydropneumatic, Citroën's pioneering self-levelling suspension (though on the rear end only).

Engine 2866cc 78x100mm, 6 cyl OHV, F/F
Max power 77bhp at 3800rpm/80bhp at 4000rpm
Max speed 85mph
Body styles saloon
Prod years 1938-55
Prod numbers 47,670

PRICES
A 14,000 **B** 10,000 **C** 6500

Citroën 2CV '48-'60

Although its development began before the war, the Deux Chevaux was not launched until 1948 – and then lasted for over 40 years. Absolute basic motoring aimed at getting the population moving. Front-drive air-cooled flat twin engine was gutless but unburstable. Interconnected suspension, roll-top canvas roof, hammock seats, washboard bonnet. AZ version from 1954 had 425cc engine, second tail-light in '55 and bigger rear window from '56, but otherwise no changes until 1960. 'Luxury' AZLP from '57 had steel boot lid. Amazing in all respects.

Engine 375cc 62x62mm/425cc 66x62mm, 2 cyl OHV, F/F
Max power 9/12bhp at 3500rpm
Max speed 40/43mph
Body styles saloon
Prod years 1948-60
Prod numbers 128,685

PRICES
A 3000 **B** 1800 **C** 750

Citroën

Citroën 2CV '60-'90

Major facelift in 1960 banished corrugated steel for smooth bonnet, plus a new grille and better trim. Folding rear seat option from '62, AZAM with 602cc joins range in '63 and boasts six-light treatment. Uprated 435cc engine ousts 425cc unit in '68, known as 2CV4 from '70 (and 602cc version as 2CV6). UK sales, abandoned in 1960, restart from 1974. Special editions flowed: Spot, Beachcomber, Charleston, Dolly, Bamboo. Disc front brakes from 1982 a big advantage. Production transferred to Portugal in 1988, only to be halted two years later. Inimitable.

Engine 425cc 66x62mm/435cc 68.5x59mm/602cc 74x70mm, 2 cyl OHV, F/F
Max power 12bhp at 3500rpm/18bhp at 5000rpm/23bhp at 6750rpm/31bhp at 7000rpm
Max speed 44/59/63-71mph
Body styles saloon
Prod years 1960-90
Prod numbers 3,743,915

PRICES
A 3000 **B** 1200 **C** 500

Citroën 2CV Sahara

Citroën's exploits in North Africa were legendary from the 1920s and, for Citroën, the 2CV Sahara was just a logical extension of the half-tracks it used to make. For everyone else, it was demented. Four-wheel drive was achieved by using two engines and two transmissions, one at either end. One engine could be turned off for two-wheel drive. Saharas are easily spotted by their cutaway rear wheel arches and filler cap in the passenger's door. Although in production for many years, it sold poorly and surviving examples are extremely rare.

Engine 2 x 425cc 66x62mm, 2 x 2 cyl OHV, F&R/4x4
Max power 12bhp at 3500rpm (16.5bhp at the wheels with both engines)
Max speed 65mph
Body styles saloon
Prod years 1958-66
Prod numbers 694

PRICES
Very rarely offered for sale, but likely to fetch around £10,000

Citroën DS

An absolutely unique car in automotive history. For 1955, it was astonishingly bold in virtually every area. A pressurised self-levelling gas-and-oil system replaced suspension springs and also powered the brakes, steering, clutch and even gearchange (a semi-automatic). 1.9-litre engine was the only non-futuristic item, deriving from the Traction. Rare Prestige from '58, conventional manual 'box available from '63, Pallas luxury version from '64. DS21 (2.2-litre) choice from '65, DS20 (2-litre) from '68, DS23 (2.3-litre) from '72. Fabulous to drive but a maintenance/restoration challenge.

Engine 1911cc 78x100mm/1985cc 86x85.5mm/2175cc 90x85.5mm/2347cc 93.5x85.5mm, 4 cyl OHV, F/F
Max power 63bhp at 4500rpm to 141bhp at 5250rpm
Max speed 84-117mph
Body styles saloon
Prod years 1955-75
Prod numbers 1,455,746 (all D models)

PRICES
DS19/DS21 **A** 7000 **B** 5000 **C** 3000
DS20/DS23 **A** 8000 **B** 6000 **C** 4000

Citroën ID/D

If DS conjured up 'Déesse' (Goddess), the ID moniker stood for 'Idée'. This was always the entry-level D, launched in 1957 with no hydraulic assistance and a detuned engine. Pneumatics gradually introduced for brakes and steering (Slough-built IDs always had this) and optional power steering from '69. ID20 from '68, DW is DS spec car with ID manual gearbox. D Special and D Super replace IDs in '69, the latter with five-speed option from '70. What's the point of having a DS without those pneumatic aids? At least there is less to worry about.

Engine 1911cc 78x100mm/1985cc 86x85.5mm, 4 cyl OHV, F/F
Max power 62bhp at 4000rpm to 84bhp at 5250rpm
Max speed 82-105mph
Body styles saloon
Prod years 1957-75
Prod numbers see DS

PRICES
A 6000 **B** 4000 **C** 2500

Citroën DS/ID Décapotable

Henri Chapron was one of the many French coachbuilders who pounced on the DS. His convertible was probably the most elegant of the lot and Citroën recognised this when it offered the conversion on its official list from 1961 (in UK from '62). Only two doors and full fold-away hood make this the most dramatic of all Ds. Was twice the price of a saloon in the 1960s, so only 1365 official Décapotables (plus quite a few more unofficially by Chapron) means high prices. Mechanically identical to the DS/ID but structural problems are more common.

Engine 1911cc 78x100mm/1985cc 86x85.5mm/2175cc 90x85.5mm, 4 cyl OHV, F/F
Max power 83bhp at 4000rpm to 109bhp at 5250rpm
Max speed 85-108mph
Body styles convertible
Prod years 1961-71
Prod numbers 1365

PRICES
A 20,000 **B** 14,000 **C** 9000

Citroën D Safari/Familiale

The estate version of the D was first shown in 1958 but UK sales did not begin until '59. Rear styling inspired by American station wagons looked somewhat gawky, and spatless narrow-track rear wheels looked lost in wheel arches. But you got masses of luggage space and – bliss! – self-levelling suspension for constant ride height. Safari was usually a seven seater with inward-facing rear seats, but the Familiale could seat eight in three rows. Generally followed ID spec, with all its upgrades, but all engine sizes were eventually offered.

Engine 1911cc 78x100mm/1985cc 86x85.5mm/2175cc 90x85.5mm/2347cc 93.5x85.5mm, 4 cyl OHV, F/F
Max power 62bhp at 4500rpm to 115bhp at 5500rpm
Max speed 82-111mph
Body styles estate
Prod years 1958-75
Prod numbers 93,919

PRICES
A 6000 **B** 3500 **C** 2000

Citroën Bijou

Citroën built 2CVs at Slough from 1953 to 1960, during which time the economy car failed to make an impression this side of the channel. The company was left with quite a few rolling chassis lying around and so conceived the Bijou ('jewel'). This was a supposedly stylish little coupé based on standard 2CV parts. The all-glassfibre bodywork was more compact but actually heavier than the standard saloon body, so performance suffered. It was strictly a 2+2 inside, so it wasn't nearly as practical, either. No more than a curiosity for Citroën fanatics.

Engine 425cc 66x62mm, 2 cyl OHV, F/F
Max power 12bhp at 3500rpm/18bhp at 5000rpm
Max speed 60/62mph
Body styles coupé
Prod years 1959-64
Prod numbers 207

PRICES
A 3000 **B** 1600 **C** 900

Citroën Ami 6/8

This extraordinary attempt by Citroën to launch an up-market 2CV shared the 2CV chassis but was only ever offered with the 602cc engine (except for Ami Super, below). Utterly bizarre styling quintessentially French and, until the 1969 Ami 8, had reverse angle rear window. Some practicality sacrificed for comfort. Ami 8 had fastback rear end treatment, a more conventional grille and front disc brakes. Was France's best-selling car for many years. Dreadful rust problems sent most to the scrappie but an Ami 6 is about as cult 1960s as you can get.

Engine 602cc 74x70mm, 2 cyl OHV, F/F
Max power 20bhp at 4500rpm to 32bhp at 5750rpm
Max speed 65-75mph
Body styles saloon, estate
Prod years 1961-69/1969-79
Prod numbers 1,039,384/755,955

PRICES
Ami 6 **A** 3000 **B** 2200 **C** 1500
Ami 8 **A** 1600 **B** 900 **C** 600

Citroën Ami Super

Most people are agreed that the Ami Super's GS flat-four engine was too powerful for the 2CV chassis to handle. The Super didn't weigh very much so the 61bhp of the 1015cc engine gave it eyebrow-raising Mini-Cooper style performance: over 90mph and 0-60mph in 16.4 seconds – and still that roly-poly cornering style. Comfortable, interesting interior architecture but still too noisy. Short production run means low numbers and most have been butchered to make go-faster 2CVs and specials, so a genuine Ami Super is a rare find today.

Engine 1015cc 74x59mm, 4 cyl OHC, F/F
Max power 61bhp at 6750rpm
Max speed 90mph
Body styles saloon, estate
Prod years 1972-76
Prod numbers 44,820

PRICES
A 3000 **B** 2000 **C** 1200

Citroën M35

As masters of the unconventional, it is unsurprising that Citroën should have wanted to dabble in rotary engine technology. However, even Citroën was wary of the Wankel engine and, rather than leap in with a production model, it built a pilot batch of cars for favoured customers to test, each agreeing to notch up at least 20,000 miles a year. That car was the M35, a 2+2 coupé based on the Ami chassis and sharing a family look. The suspension was DS hydropneumatic. In the event, only 267 individually numbered prototypes were built, most of which returned to Citroën.

Engine single-rotor Wankel, F/F
Max power n/a
Max speed 90mph
Body styles coupé
Prod years 1970
Prod numbers 267

PRICES
Too rare to price accurately, but likely to be around the same as the GS Birotor (page 61)

Citroën

Citroën Dyane

The car that tried to replace the 2CV but could not. The third member of the 2CV family shared exactly the same mechanical basis as the 2CV but had a more modern body shape, extra equipment and a hatchback, which improved its practicality considerably. Full-length canvas sunroof remained. Front discs from '68. The Dyane cannot be considered a failure (it sold nearly 1.5 million) but 2CV drivers preferred the original shape and down-to-earth nature of the old tin snail. Will never be in the same boat as the 2CV and most have now rusted away.

Engine 425cc 66x62mm/435cc 68.5x59mm/602cc 74x70mm, 2 cyl OHV, F/F
Max power 18bhp at 5000rpm/23bhp at 6750rpm/31bhp at 7000rpm
Max speed 65/70/78mph
Body styles hatchback
Prod years 1967-85
Prod numbers 1,443,583

PRICES
A 1000 **B** 600 **C** 300

Citroën Méhari

Yet more fun and games on the 2CV chassis. This time the intention really was just fun and games. Launched at a time when carefree fun machines were at their zenith, the Méhari was a jeep as only the French could conceive it. The open bodywork was glassfibre and the doors and soft top were designed to be removed in minutes. Popular in the south of France but never sold in UK (so LHD only) and not in Germany after it was discovered that the body was flammable. Rarities are the panel van, blue-and-white Côte d'Azur and desirable 4 x 4.

Engine 602cc 74x70mm, 2 cyl OHV, F/F & F/4x4
Max power 31bhp at 7000rpm
Max speed 67mph
Body styles utility
Prod years 1968-86
Prod numbers 143,747

PRICES
A 2800 **B** 1700 **C** 1000
4 x 4 **A** 5000 **B** 3500 **C** 2500

ANDREW BRODIE ENGINEERING LTD.

50 Sapcote Trading Centre ● 374 High Road ● Willesden
London NW10 2DJ ● United Kingdom
Telephone: 0181 459 3725   Facsimile: 0181 451 4379

WORLDWIDE SUPPORT FOR CITROËN SM
EXTENSIVE PARTS LIST FOR SM,
DS & CX, MASERATI
SERVICING & RESTORATIONS CARRIED OUT
STAINLESS EXHAUSTS FOR ALL CITROENS
HYDRAULIC SPECIALISTS FOR MASERATI
OUR FLYING MECHANIC WILL COME TO YOU

A & R Sheldon THE MAIL ORDER PEOPLE
SUPERIOR QUALITY HAND TOOLS
33 Bramhall Park Rd, Bramhall, Stockport, Cheshire, SK7-3JN
TEL & FAX (24hrs):- 0161-440-0821 for FREE Catalogue and Ordering
Trade, Export & Club enquires Welcome

EXAMPLES FROM OUR EXTENSIVE RANGE

A/F 1/4" - 6"
Whitworth 1/8" - 2"
Metric 6 - 235mm

SETS OF O/E & COMBINATION SPANNERS (6, 8 & 10's) SOCKET RAILS AND SETS. ALSO SOLD SEPARATELY
LUBRICATION - GREASE GUNS, OIL SYRINGES, ALSO HUGE RANGE OF GREASE NIPPLES, FUNNELS IN ALL SHAPES & SIZES. OIL CANS + MUCH, MUCH MORE.

A/F 1/4" - 4.5/8"
Whit 1/8" - 3.1/2"
Metric 4 - 115mm

FULLY POLISHED O/E, RING & COMBINATION SPANNERS & SOCKETS UP TO 1" SQ. DRV. INCLUDING IMPACT & DEEP IMPACT SOCKETS
ALSO B.A. SIZES IN O/E SPANNERS & 1/4" SQ. DRV. SOCKETS

HIGHEST ENGINEERING QUALITY
DRILLS, TAPS & DIES. ALL CURRENT SIZES AND MOST IMPERIAL SIZES AVAILABLE.

Beta SUPPLIERS TO INDUSTRY & FORMULA 1
HUGE RANGE OF QUALITY TOOLS FOR ANY JOB, FROM TOOL BOXES, WALL CABINETS, ROLL CABS, TOP BOXES ETC. TO PLIERS, SCREWDRIVERS, DEAD BLOW HAMMERS, STUD EXTRACTORS, PULLERS, FLAT RING SPANNERS, ETC.

HAVE A PROBLEM WITH SOME THING? WE WILL DO OUR BEST TO HELP. JUST RING WITH REQUIREMENTS AND WE WILL POINT YOU IN THE RIGHT DIRECTION.

Citroën GS

The first all-new Citroën since the DS, the GS occupied a neglected market area between the Ami and the DS. And what a car! The then-most aerodynamic small car ever (Cd 0.34), all-new air-cooled flat four engine and gearbox, self-levelling suspension, futuristic interior – a real Citroën, and Car of the Year. 1.2-litre engine and estate from '72, GS X was 'sports trim' model (X3 from '79 had 1.3 engine and more power). Hatchback GSA from '79. In the end, too complex for small car buyers. Weak camshafts and rust mean rock-bottom prices.

Engine 1015cc 74x58mm/1222cc 77x65.5mm/1129cc 74x65.6mm/1299cc 79.4x65.6mm, 4 cyl OHC, F/R
Max power 55-61bhp at 6500rpm/59-65bhp at 5750rpm/56bhp at 5750rpm/65bhp at 5500rpm
Max speed 92-98mph
Body styles saloon, estate
Prod years 1970-79
Prod numbers 2,473,150 incl GSA

PRICES
A 900 **B** 500 **C** 200

Citroën GS Birotor

Following the experimental M35, Citroën went into production with a Wankel-engined GS in 1973. Originally a special body had been planned. The twin-rotor engine was built by Comotor and was admirably silent and powerful but hopelessly thirsty (about 20mpg was normal), which was the main reason why sales bombed. Wider wheels and flared arches identify. After a short production run, Citroën canned the GS Birotor and proceeded to buy most of them back off customers to prevent having to support them with spares – which makes them extremely rare today.

Engine 2 x 497cc, 2-rotor Wankel, F/F
Max power 107bhp at 6500rpm
Max speed 109mph
Body styles saloon
Prod years 1973-75
Prod numbers 847

PRICES
A 10,000 **B** 6000 **C** 3500

Citroën SM

In 1968 Citroën assumed a controlling interest in Maserati and the result of this marriage arrived two years later. A technical tour de force, the SM used a quad cam Maserati V8 with two cylinders lopped off to make it a V6. Otherwise the SM was pure Citroën: self-levelling suspension, self-centring speed-sensitive power steering, hydropneumatic brakes. Five speed 'box joined by 3.0-litre auto from '73, fuel injection from '72. LHD only. Cramped interior and timing gear/gasket problems. Avoid US imports with hideous front end treatment.

Engine 2670cc 87x77mm/2965cc 91.6x75, V6 cyl QOHC, F/F
Max power 178-188bhp at 5500rpm/180bhp at 5750rpm
Max speed 127-142mph
Body styles coupé
Prod years 1970-75
Prod numbers 12,920

PRICES
A 13,000 **B** 8500 **C** 5000

Citroën CX

Expanded GS-style replacement for the DS, gloriously eccentric and the last true Citroën. Naturally, self-levelling suspension, power brakes and steering (also Vari-Power self-centring on all but the earliest cars). Star Trek interior, fabulous ride and assured handling made it Car of the Year. Safari estate, lwb Prestige, diesels and optional C-Matic semi-auto from '76, 5 speeds and GTi from '77. Pre-'82 cars rusted terribly and got a bad reputation. Series 2 cars from 1985 had body-coloured bumpers and conventional instruments. The CX's charms are now finally dawning on a small band of enthusiasts.

Engine 1985cc 86x85.5mm/1995cc 88x82mm/2175cc 90x85.5mm/2165cc 88x89mm/2347cc 93.5x85.5mm/2500cc 93x92mm, 4 cyl OHC/OHC/OHC/OHC & diesel/OHC/OHV/OHV8 diesel F/F
Max power 66bhp at 4500rpm to 168bhp at 5000rpm
Max speed 108-138mph
Body styles saloon, estate
Prod years 1974-91
Prod numbers 1,042,300
PRICES
A 3500 **B** 1200 **C** 400

Clan Crusader

Two ex-Lotus employees, Paul Haussauer and John Frayling, founded Clan to build their own car, the Crusader. This was a quirky glassfibre monocoque design of very compact dimensions. Engine came from the Hillman Imp Sport, as did most of the running gear. Expensive when sold complete (£1400), so a kit version was also available. Economic crisis and the imposition of VAT killed the project. Many attempts were made to revive the Crusader in the 1980s but none was as attractive as the original.

Engine 875cc 68x60.4mm, 4 cyl OHC, R/R
Max power 50bhp at 6100rpm
Max speed 100mph
Body styles coupé
Prod years 1971-74
Prod numbers 315

PRICES
A 3000 **B** 2200 **C** 1200

Connaught L/L3

Two Bugatti specialists created the Connaught, which was always intended as a dual-purpose road/race car. On a shortened Lea-Francis 14HP Sports chassis, two body styles were offered: an attractive enveloping body whose front end hinged forward and, in the later L3/SR, an Abbott bodied cycle-wing affair. Semi-elliptic springs all round, later IFS. Tuned LeaF twin cam engine and optional alcohol fuel system. Fantastic roadholding and performance for its day, but Connaught moved back into pure racing in the 1950s.

Engine 1484cc 75x84mm/1767cc 75x100mm, 4 cyl DOHC, F/R
Max power 98bhp at 5500rpm to 140bhp at 6000rpm
Max speed 104-125mph
Body styles sports
Prod years 1948-54
Prod numbers 6/10

PRICES
Very seldom seen but likely to be over £30,000

Cord 8/10

American firms started producing replicas of early classics in the 1950s. One of the most celebrated was the Cord 8/10 which, as the name hinted, was an 8/10 scale replica of the great 1936 Cord 810. Produced by the Cord Automobile Company of Tulsa from 1964, it did not share the original's exotic specification, having a mere Corvair engine under the bonnet, although it was front-wheel drive like the '36 car. Ownership changed in 1968 and Ford/Chrysler V8s and rear drive were substituted.

Engine 2377cc 87.3x66m/2684cc 88.9x72.4mm/4948cc 101.6x76.2mm/7211cc 109.7x95.2mm, 6/V8 cyl OHV, F/F or F/R
Max power 103bhp at 4400rpm to 245bhp at 4400rpm
Max speed 105-120mph
Body styles convertible
Prod years 1964-72
Prod numbers 91

PRICES
A 15,000 **B** 10,000 **C** 7500

DAF 600/750/Daffodil/33

The success of the Van Doorne brothers' truck business persuaded them to try car manufacture. The DAF 600 was first shown in 1958 and it looked a rather unexceptional little saloon car with an air-cooled twin engine. But the gearbox certainly was unusual – or rather the lack of one. An infinitely variable transmission by twin V-belt rubber bands to the rear wheels meant merely selecting forward or reverse. 750 engine from '61, square roof treatment from '63, restyled 33 and more power from '67.

Engine 590cc 76x65mm/746cc 85.5x65mm, 2 cyl OHV, F/R
Max power 19bhp at 4000rpm/26-30bhp at 4000rpm
Max speed 59/70mph
Body styles saloon
Prod years 1959-61/1961-63/1963-67/1967-75
Prod numbers 312,367

PRICES
650/750 **A** 2000 **B** 1200 **C** 700
Daffodil/33 **A** 1000 **B** 700 **C** 400

DAF 44/46/55/66

The DAF's bigger sister arrived in 1966 with an 8in longer wheelbase and styling by Michelotti. Still powered by the two-pot engine but now 844cc. The 55 of 1967 was the first four-cylinder DAF, using the engine from the Renault 10, and it had front disc brakes. Marathon version was named after the 1968 London-Sydney Marathon (a DAF was placed 17th) and had a tuned engine, alloy wheels and better front suspension. 66 of 1972 had new nose, De Dion rear and 1.3-litre option from '73. Always Variomatic. DAF was taken over by Volvo and rebadged.

Engine 844cc 85.5x73.5mm, 2 cyl OHV, 1108cc 70x72mm/1289cc 73x77mm, 4cyl OHV, F/R **Max power** 34bhp at 4500rpm/45-63bhp at 5600rpm/47-57bhp at 5200rpm
Max speed 78/83/90/89mph
Body styles saloon, estate
Prod years 1966-75/1974-76/1967-72/1972-75 **Prod numbers** 510,786 (167,905/32,353/164,231/146,297)

PRICES
A 1000 **B** 600 **C** 300

DAF 55/66 Coupé

The only vaguely collectable DAFs are the coupé models built between 1967 and 1975. With these cars, DAF hoped to shake off its image as a supplier of transport to old aunts and nurses, and it almost succeeded. Michelotti did the restyling, in which it was fairly successful. Offered in standard and Marathon forms and 55 and 66 versions, with bodywork and mechanical changes as per the saloons. The idea of a small 90mph automatic coupé may not sound very exciting but the point-it-and-go driving character was actually a lot of fun.

Engine 1108cc 70x72mm/1289cc 73x77mm, 4cyl OHV, F/R
Max power 55-63bhp at 5600rpm/47-57bhp at 5200rpm
Max speed 83-90mph
Body styles coupé
Prod years 1967-72/1972-75
Prod numbers see above

PRICES
A 1500 **B** 1000 **C** 700

Daihatsu Compagno/800/Spider

Daihatsu has the distinction of being the first Japanese car firm to sell its products in Europe: the Compagno arrived in the UK in 1966. This was Daihatsu's first four-wheeled car, launched in 1963. It was very conventional, with a separate chassis, drum brakes, torsion bar front end and semi-elliptic rear. Britain only ever got the two-door saloon (and in tiny numbers) but an estate, four-door saloon and an attractive 65bhp Spider four-seater convertible were marketed in Japan.

Engine 797cc 62x66mm/958cc 68x66, 4 cyl OHV, F/R
Max power 45bhp at 5000rpm/58bhp at 5500rpm to 65bhp at 6500rpm
Max speed 72/81/90mph
Body styles saloon, estate, convertible
Prod years 1963-70
Prod numbers ca 120,000

PRICES
Saloon/Estate **A** 1500 **B** 1200 **C** 900
Spider **A** 4000 **B** 2200 **C** 1500

Daimler DB18/DB18 Sports Special/DB18 Consort

Britain's oldest car maker returned after the war with its 1939 DB18 2½-litre in six-light saloon and Barker or Tickford drophead forms. Four-speed fluid flywheel and pre-selector gearbox, IFS and live rear axle. 1949 Consort was an update, with faired-in headlamps, curved grille and bumpers, hydromechanical brakes and hypoid final drive. Sister model was the 85bhp Sports Special with a Barker 2/3 seater DHC alloy-on-ash body (or Hooper specials). Too sedate for 'sports' tag.

Engine 2522cc 69.6x110.5mm, 6 cyl OHV, F/R
Max power 70bhp at 4200rpm/85bhp at 4200rpm
Max speed 78/85mph
Body styles saloon, drophead coupé
Prod years 1939-50/1948-53/1949-53
Prod numbers 3355/608/4250
PRICES
DB18/Consort **A** 6000 **B** 3500 **C** 2000
DB18 DHC **A** 12,000 **B** 8000 **C** 5000
Sports Special **A** 20,000 **B** 14,000 **C** 9000

Daimler DE27/DH27/DE36

In addition to the DB18, Daimler returned in 1946 with its enormous DE27 and Straight Eight (DE36) models. These were supplied in chassis form only to various coachbuilders, notably Hooper, Freestone & Webb and Windover. A variety of body styles was available, from saloons to dropheads, ambulances to limousines, which were the most popular. DE27 had 138in wheelbase, DH27 and DE36 considerably longer. 27s have six cylinders, 36s the straight eight, the last such engine made in Britain. Huge, ponderous and favoured by royalty.

Engine 4095cc 85.1x120mm/5460cc 85.1/120mm, 6/8 cyl OHV, F/R
Max power 110bhp at 3600rpm/150bhp at 3600rpm
Max speed 81/85mph
Body styles saloon, limousine, drophead coupé
Prod years 1946-51/1946-53
Prod numbers 205/50/205

PRICES
DE27/DH27 **A** 12,000 **B** 8000 **C** 6000
DE36 **A** 25,000 **B** 17,000 **C** 10,000

Daimler Regency/Empress

First shown in 1951, the Regency was a significant model in that it anticipated the style of the Conquest, albeit on a larger scale. It was based on the Consort chassis but had an all-new 3-litre engine, a more substantial body and a one-piece curved windscreen. Hooper offered a scaled-down MkII Empress body on this chassis and Barker advanced a two-door drophead coupé in 1952 only. But the Regency never actually entered proper production and only just over 50 chassis were built. The style metamorphosed to become the Regency II in 1954 (see next entry).

Engine 2952cc 76.2x108mm, 6 cyl OHV, F/R
Max power 90/100bhp at 4100rpm
Max speed 81mph
Body styles saloon, drophead coupé
Prod years 1951-52
Prod numbers 52

PRICES
A 9000 **B** 7000 **C** 4500

Daimler Regency MkII/Sportsman/Empress

It was 1954 by the time the Regency actually got into proper production and it was with a bigger 3.5-litre engine and longer, lower body incorporating a more squared-up rear end. The Sportsman (photo), with its 140bhp alloy-head engine, had more curved rear bodywork, wrap-around rear window, four-light styling, hydraulic brakes and overdrive top. Both models were available with an optional 4.6-litre engine, in which case the Sportsman's mechanical improvements were standard. Once again, Hooper offered its stately Empress body on this chassis.

Engine 3468cc 82.5x108mm/4617cc 95x108mm, 6 cyl OHV, F/R
Max power 107bhp at 4000rpm to 140bhp at 4200rpm/127bhp at 3600rpm to 167bhp at 3800rpm
Max speed 82-95/90-98mph
Body styles saloon
Prod years 1954-57
Prod numbers 400/69/39

PRICES
Regency **A** 6500 **B** 4500 **C** 2500
Sportsman **A** 12,000 **B** 8500 **C** 4000

Daimler

Daimler Conquest/Conquest Century/Drophead Coupé

The successor to the Consort with a curvaceous body all but identical to the Lanchester 14. Much more sporting performance thanks to reduced size and weight and far better handling than earlier Daimlers. Conquest saloon soon joined by Century Drophead Coupé, with 100bhp alloy-head twin carb engine and a Carbodies four-seater drop-top body (and half-power roof – you had to pull it forward from the sedanca position). Century spec also available on saloon from '54, auto option from '56 and standard from '57. MkII with small body changes in '56.

Engine 2433cc 76.2x88.9mm, 6 cyl OHV, F/R
Max power 75bhp at 4000rpm/100bhp at 4400rpm
Max speed 81/90mph
Body styles saloon, drophead coupé
Prod years 1953-57
Prod numbers 4568/4818/234

PRICES
Conquest/Century **A** 5000 **B** 3500 **C** 1800
DHC **A** 12,000 **B** 9000 **C** 6000

Daimler Conquest Roadster/New Drophead Coupé

Concurrent with the Century DHC, Daimler launched a two-seater sports version of the Conquest with a much lower open body. However, it all looked rather odd since too much effort was directed towards a family look, making the Roadster (photo) resemble a squashed saloon. You got tail fins, though. Discontinued in late 1955 and replaced the following year by a modified version called the New Drophead Coupé, whose minor changes included wind-up windows and three seats. Aluminium bodywork does not rust and this was the first 100mph Daimler.

Engine 2433cc 76.2x88.9mm, 6 cyl OHV, F/R
Max power 100bhp at 4400rpm
Max speed 101mph
Body styles sports, drophead coupé
Prod years 1953-55/1956-57
Prod numbers 65/54

PRICES
A 15,000 **B** 11,000 **C** 7500

Daimler Regina/DK400

The Straight Eight DE36 had been available to special order only since 1951, so the new Regina of 1954 really marked Daimler's re-entry into the limousine market. It was a huge (18ft), slab-sided thing with seven-seater bodywork made by Carbodies. Redesignated DK400 in 1955 and given an extra occasional seat. Largely redesigned to become DK400B in 1956, with a much improved 167bhp engine, and straight-through wing feature line (with fins!). Very rare luxury Hooper limousine version (photo) also listed from '56. Too cumbersome for casual collectors.

Engine 4617cc 95.25x108mm, 6 cyl OHV, F/R
Max power 130bhp at 3600rpm/167bhp at 3800rpm
Max speed 88/94mph
Body styles limousine
Prod years 1954-55/1955-60
Prod numbers 132

PRICES
A 6000 **B** 4000 **C** 2500

Daimler One-O-Four/Four-Light Saloon

Replacement for the Regency MkII with a stiffer structure, 137bhp engine and servo twin trailing-shoe hydraulic brakes. At launch, both 3.5 and 4.5-litre models were listed but no 4.5s were ever made. The so-called Lady's Model (of which only 50 were built) had a special interior with a walnut facia and satin chrome instrument panel, gold propelling pencil, umbrella, shooting stick, electric windows and picnic gear. The old Sportsman body was kept going, too, now renamed Four-Light Saloon. No premium over the earlier Regency.

Engine 3468cc 82.5x108mm, 6 cyl OHV, F/R
Max power 137bhp at 4400rpm
Max speed 98mph
Body styles saloon
Prod years 1955-59
Prod numbers 561

PRICES
A 6500 **B** 4500 **C** 2500

Daimler Majestic

Another reworking of the Regency/One-O-Four theme. Bodywork was widened and made more bulbous, easily recognised by featureless rear wing line and lack of front air intake grilles. The engine was bored out to 3.8 litres (the same as Jaguar's rival MkIX), allowing a top speed of over 100mph, and disc brakes became standard. Automatic transmission was obligatory. In 1960, axle ratio was changed, facia revised and power steering joined the options list. After Jaguar's 1960 take-over, it clashed with the MkX and the plug was pulled in 1962.

Engine 3794cc 86.4x108mm, 6 cyl OHV, F/R
Max power 147bhp at 4400rpm
Max speed 100mph
Body styles saloon
Prod years 1958-62
Prod numbers 1490

PRICES
A 6000 **B** 4000 **C** 2500

Daimler

Daimler Majestic Major/DR450

Announced by Daimler in 1959, the Majestic Major went into production after the 1960 take-over by Jaguar. The bodywork was the same as the Majestic except for a longer boot and, from 1960, horn grilles with 'V' motifs at the front end. The 'Major' name was added to denote the fabulous alloy-head 4.6-litre V8 with its hemispherical combustion chambers. This made it a very fast car for its bulk. Power steering optional from 1960, standard from '64. DR450 lwb limousine version from '61 weighed over two tons and was nearly 19ft long.

Engine 4561cc 95.25x80mm, V8 cyl OHV, F/R
Max power 220bhp at 5500rpm
Max speed 120/110mph
Body styles saloon/limousine
Prod years 1960-68/1961-68
Prod numbers 1180/864

PRICES
A 8000 **B** 5500 **C** 3000

Daimler SP250

Daimler's all-new sports car was announced as the Dart in 1959, but had to be changed to SP250 after Chrysler objected. Nothing else looked like it and, with a beautifully flexible Edward Turner designed 2.5-litre V8 under the bonnet, nothing drove like it either. Glassfibre bodywork tended to crack on early cars because near-bankrupt Daimler cut development corners. Post-1961 'B' cars had stronger chassis and bodywork, 'C' versions (1963-64) had standard heater. Disc brakes all round, optional automatic from '61. Unique, durable, quirky and delightful.

Engine 2548cc 76.2x69.9mm, V8 cyl OHV, F/R
Max power 140bhp at 5800rpm
Max speed 120mph
Body styles roadster
Prod years 1959-64
Prod numbers 2648

PRICES
A 15,000 **B** 11,000 **C** 8000

Daimler 2½-Litre/V8 250

Putting the Turner V8 from the SP250 into a Jaguar Mk2 hull created the first 'Daimlerised' Jag in 1962. The V8 engine was a natural for the Mk2: flexible, powerful and refined. Traditional Daimler fluted grille and 'D' insignia were the external distinguishing marks and automatic was compulsory until 1967, when manual/overdrive was offered as an option. V8 250 of 1967 had slimmer bumpers but was kept superior to late 1960s Jags by having leather upholstery. Not rated highly by Jaguar enthusiasts, so it's possible to find a bargain.

Engine 2548cc 76.2x69.8mm, V8 cyl OHV, F/R
Max power 140bhp at 5800rpm
Max speed 115mph
Body styles saloon
Prod years 1962-69
Prod numbers 17,620

PRICES
A 12,000 **B** 6500 **C** 4000

Daimler DS420

With Jaguar/Daimler part of BMC, a new corporate limo was required to replace the archaic Vanden Plas Princess and Daimler DR450. Vanden Plas was asked to design it and Motor Panels of Coventry extended a Jaguar 420G floorpan by 21in to form the base. The result was the DS420, a rather portly limo with prewar styling themes, weighing over two tons and measuring 18ft 5in long. Automatic only and a detuned 4.2-litre XK engine, so spirited driving a no-no. It was easily the cheapest purpose-built limo around and mayors lapped them up.

Engine 4235cc 92x106mm, 6 cyl DOHC, F/R
Max power 177bhp at 4250rpm
Max speed 110mph
Body styles limousine, landaulette
Prod years 1968-92
Prod numbers 5043

PRICES
A 12,000 **B** 6000 **C** 1500

Daimler Sovereign/Double Six

The first Sovereign was simply a badge-engineered clone of the Jaguar 420 (see Jaguar) and its 1969 replacement was little more than a rebadged XJ6, with standard manual/overdrive gearbox and minor trim adjustments. Same engines as Jaguars, the V12 version known as the Double Six. Vanden Plas version was unique to Daimler: from 1972, the London coachbuilder would add bespoke seating, better trim, vinyl roof and fog lamps. Daimler Series 2 Coupés are very rare (1677 4.2s and 407 5.3s). The very last XJ off the line was a Daimler, a 1992 Double Six.

Engine 2791cc 83x86mm/4235cc 92.1x106mm, 6 cyl DOHC, 5343cc 90x70mm, V12 cyl OHC, F/R
Max power 180bhp at 6000rpm/ 245bhp at 5500rpm/253bhp at 6000rpm
Max speed 117/124/140mph
Body styles saloon, coupé
Prod years 1969-87/1972-92
Prod numbers ca 57,000/15,655
PRICES
Saloon **A** 7000 **B** 4500 **C** 2000
4.2 Coupé **A** 9000 **B** 6000 **C** 4000
5.3 Coupé **A** 11,000 **B** 7500 **C** 5000

DB HBR

'DB' stood for Charles Deutsch and René Bonnet, two friends who had collaborated on their first sports car before the war. After the war they built cars in series on Panhard parts, the most popular of which was the HBR 5 first shown in 1954. As it weighed only 1411lb (640kg), its 850cc Panhard two-cylinder engine was able to return excellent performance. Previous cars had aluminium bodies, while the DBR used glassfibre. 1-litre and even supercharged 1.3-litre engines were offered as options.

Engine 851cc 85x75mm/954cc 90x75mm, 2 cyl OHC, F/R
Max power 55bhp at 5700rpm/70bhp at 6000rpm
Max speed 96/115mph
Body styles coupé
Prod years 1955-61
Prod numbers 660

PRICES
A 12,000 **B** 8000 **C** 4000

Delage D-6

Taken over by Delahaye in 1935, Delage returned post-war with a model based on its master's prewar products. The D-6 used the Delahaye 148 (124in) chassis with the Cotal electric gearbox fitted as standard, along with hydraulic brakes. The six-cylinder engine had a shortened stroke so was more free-revving. Most D-6s were sold as rolling chassis, the customer being responsible for asking his favourite coachbuilder to knock something up; otherwise the factory offered a saloon body by Guilloré.

Engine 2988cc 83.7x90.5mm, 6 cyl OHV, F/R
Max power 82bhp at 4000rpm/100bhp at 4500rpm
Max speed 80/90mph
Body styles saloon, drophead coupé
Prod years 1946-54
Prod numbers ca 250

PRICES
Saloon **A** 18,000 **B** 12,000 **C** 9000
DHC Depends on coachbuilder – up to double saloon's value

Delahaye 134/135M/135MS

Delahaye had a brave stab at being the French Bentley in its post-war years but never stood much chance against French luxury car taxes. If there was a 'volume' Delahaye, it was the 135. Supplied as a rolling chassis (weighing one ton!). Triple carb MS version had 135bhp, rather too much for the mechanical brakes. Usually Cotal four-speed 'box, but manuals offered as well. A Type 134 (2.3-litre four) was listed for 1949 only. Longer wheelbase 148 also available post-war. All versions inevitably RHD and inevitably very expensive.

Engine 2336cc 84x107mm/3557cc 84x107mm, 4/6cyl OHV, F/R
Max power 55bhp at 3200rpm/95bhp at 3200rpm to 135bhp at 4200rpm
Max speed 80/86/105mph
Body styles saloon, coupé, drophead coupé
Prod years 1936-52
Prod numbers ca 2000

PRICES
134/135M **A** 40,000 **B** 25,000 **C** 15,000
135 MS DHC **A** 80,000 **B** 65,000 **C** 50,000

Delahaye 175/178/180

Whereas the 135 was a revival of a prewar model, the 175 was new in 1948. It had a brand new and modern chassis with Dubonnet independent front suspension, De Dion rear and hydraulic brakes. Equally important was the expansion of the six-cylinder engine to 4.5 litres. According to carburation and compression ratio, outputs varied from 130 to 185bhp. The 178 and 180 models had a longer wheelbase suitable for saloon and limousine bodywork. The main custom for these big Delahayes seemed to be French colonial heads of state.

Engine 4453cc 94x107mm, 6 cyl OHV, F/R
Max power 130-185bhp at 4000rpm
Max speed 100mph
Body styles saloon, coupé, convertible
Prod years 1948-52
Prod numbers n/a

PRICES
A 80,000 **B** 60,000 **C** 45,000

Delahaye 235

The 235 was the successor to the 135MS Competition model and, like its forebear, had triple carbs but with a higher compression ratio and modified head, so that power rose to 152bhp. Performance was on a par with the 4.5-litre 175 series. Delahaye sold chassis only and a wide variety of bodies was offered by firms such as Chapron, Le Tourneur & Marchand and Vanden Plas. A new 'family' grille was oval-shaped and most bodies were full-width sports types. The 235 was the last Delahaye built, as the marque was subsumed by Hotchkiss in 1954.

Engine 3557cc 84x107mm, 6 cyl OHV, F/R
Max power 152bhp at 4200rpm
Max speed 110mph
Body styles coupé, convertible
Prod years 1951-54
Prod numbers ca 90

PRICES
A 60,000 **B** 40,000 **C** 25,000

Dellow

For trials drivers, post-war motoring meant dual-purpose cars and the Dellow was the ultimate in such machines. Mostly Ford parts, beam front axle, leaf springs. MkII (1951) gains coil sprung rear, MkIII has four seats. MkV (1954) has inlet-over-exhaust valve conversion, coils front and rear, four speeds and oval grille. Final MkVI (1957) was all-new enveloping bodied model with new ladder chassis, IFS, Ford 100E engine. Some supercharged engines. Spasmodic production throughout.

Engine 1172cc 63.5x92.5mm, 4 cyl SV/IOE, F/R
Max power 31-42bhp at 4500rpm
Max speed 65-74mph
Body styles sports
Prod years 1949-57
Prod numbers ca 500

PRICES
A 9000 B 7000 C 5000

Denzel

The Austrian Wolfgang Denzel started his manufacturing career after the war by fitting wooden bodies on VW Kübelwagens. Then he turned to making sports car bodies on either the VW Beetle floorpan or his own box-section chassis. Engines were usually Beetle and often modified with twin carbs and other tweaks, such that he could extract as much as 86bhp from 1.5 litres. For a while, the Denzel was a credible Porsche rival but never really shook off its Beetle origins, so ducked out in 1960.

Engine 1281cc 78x67mm/1488cc 80x74mm, 4 cyl OHV, R/R
Max power 61bhp at 5400rpm/86bhp at 5400rpm
Max speed 100-110mph
Body styles roadster
Prod years 1953-60
Prod numbers n/a

PRICES
A 18,000 B 12,500 C 7500

De Soto '57-'61

A sub-Chrysler species, De Soto attempted to offer big style for a cheap price but, in the post-war period at least, never really succeeded. The marque's swansong years of the late 1950s saw a range of models with Chrysler engineering and their own striking Exner bodywork. Firesweeps were the entry-level, Firedomes slightly higher, Fireflites higher still and the Adventurer convertible was top-of-the-range. Shopper and Explorer were estates. De Soto was a casualty of the need for compact cars, and these were the last to wear De Soto badges.

Engine 5326cc 93.7x96.5mm/5588cc 96x96.5mm/5907cc 104.6x84.8mm/6286cc 108x84.8mm, V8 cyl OHV, F/R
Max power 225bhp at 4400rpm to 350bhp at 4600rpm
Max speed 103-115mph
Body styles saloon, coupe, convertible, estate
Prod years 1956-61
Prod numbers 241,798

PRICES
Coupe A 9000 B 6000 C 4000
Convertible A 15,000 B 10,000 C 5000

De Tomaso Vallelunga

Argentinian Alejandro de Tomaso rented a small corner of the 1963 Turin Motor Show to present his first road car, the Vallelunga. The open and coupé bodies hid one of the first examples of road-going mid-engined chassis, a backbone affair fitted with a tuned Ford Cortina engine. Production took a while to get going and the new coupé bodywork – styled by Giugiaro – was in glassfibre, not aluminium as on the show cars. Chassis flex and vibration consigned it to an early grave. Something of an oddball.

Engine 1498cc 81x72.7mm, 4 cyl OHV, M/R
Max power 102bhp at 6000rpm
Max speed 120mph
Body styles coupé
Prod years 1965-67
Prod numbers ca 50

PRICES
A 20,000 B 14,000 C 9000

De Tomaso Mangusta

It is said that the mangusta (mongoose) is the only animal which has no fear of the cobra and this fact was not lost on de Tomaso. His Mangusta was designed to beat Shelby's Cobra. It too had a Ford V8, but had it mounted amidships in basically the same backbone chassis as the Vallelunga. What 306bhp did in such a wobbly frame was give it phenomenal straight-line performance but drastic handling. The bodywork was another Giugiaro design but was too cramped inside. Basically not a very pleasant car but a reasonable seller.

Engine 4728cc 101.6x72.9mm, V8 cyl OHV, M/R
Max power 306bhp at 6000rpm
Max speed 155mph
Body styles coupé
Prod years 1966-72
Prod numbers 401

PRICES
A 35,000 B 25,000 C 15,000

De Tomaso Pantera

De Tomaso got serious with his follow-up to the Mangusta. The Pantera had a unitary chassis designed by Giampaolo Dallara, independent wishbone suspension front and rear, an even more powerful Ford V8 mid-mounted engine and smart Italianate bodywork by Tom Tjaarda. De Tomaso pulled off a coup by selling part of his business to Ford, so that the car had ready distributorship in the US, where it sold by the bucketload. But rust, overheating, poor ventilation and dire build quality made Ford chuck it in 1974. 350bhp GTS version from '73, smooth restyle in 1990.

Engine 5763cc 101.6x88.9mm, V8 cyl OHV, M/R
Max power 310bhp at 5400rpm/350bhp at 6000rpm
Max speed 155/160mph
Body styles coupé
Prod years 1970-date
Prod numbers ca 10,000

PRICES
A 25,000 **B** 15,000 **C** 8000

De Tomaso Deauville

The Tom Tjaarda-designed prototype of De Tomaso's first four-door car was shown in 1970 with a front-mounted Ford V8 fitted with overhead cams. When it entered production in 1971, a standard Ford V8 engine was installed, plus automatic transmission, power steering, disc brakes, coil spring independent suspension and air conditioning. The styling was a plain XJ6 rip-off and bodies were built by Ghia. Lots of goodies but no match for other 1970s luxury expresses. Bad rust problems and parts difficulties are major headaches.

Engine 5763cc 101.6x88.9mm, V8 cyl OHV, F/R
Max power 270bhp at 5600rpm to 330bhp at 5400rpm
Max speed 140-145mph
Body styles saloon
Prod years 1971-88
Prod numbers 244

PRICES
A 12,000 **B** 6000 **C** 3500

De Tomaso Longchamp

Not far off a simple two-door version of the Deauville, but as a package it worked much better. Still no match for the Mercedes SL and BMW coupés, though. Mechanically the Longchamp shared its specification with the Deauville, but you had the option of a five-speed manual ZF 'box. Rather cramped interior. GTS version launched in 1980 alongside cabriolet version – both are very rare. Badge-engineered to become Maserati Kyalami with its own engine. Far too expensive to have anything but marginal impact, and cheap these days.

Engine 5763cc 101.6x88.9mm, V8 cyl OHV, F/R
Max power 270bhp at 5600rpm to 330bhp at 5400rpm
Max speed 145-150mph
Body styles coupé, convertible
Prod years 1972-90
Prod numbers 412

PRICES
A 14,000 **B** 8000 **C** 5000

DKW Meisterklasse

Following the war, DKW lost the drawings of a new three-cylinder engine to Russian hands, so it re-entered production of the Meisterklasse in 1950 with the old two-stroke twin. However the bodywork was all-new, a streamlined steel affair, still on a separate chassis with front-wheel drive. This was a more expensive but credible alternative to the VW Beetle. 'Woody' estates, built 1951-53, are rare (6415 made). Go for a later model with the same basic shape but the three-cylinder engine.

Engine 684cc 76x76mm, 2 cyl TS, F/F
Max power 23bhp at 4200rpm
Max speed 60mph
Body styles saloon, estate
Prod years 1950-54
Prod numbers 65,890

PRICES
Saloon **A** 3000 **B** 2000 **C** 1200
Estate **A** 6000 **B** 3500 **C** 2200

DKW Sonderklasse/3=6

In 1953, DKW's three-cylinder engine came on stream. Fitted to the same saloon body, the car was called the Sonderklasse. An alternative body style was the Luxus Coupé, with its hardtop construction and wrap-around rear screen (with ordinary side windows, this became the new saloon in '54). Superseded by the 3=6 in '55: 4in wider body, new grille, and a choice of body styles. The full range consisted of two and four-door saloons (the latter on the 3.5in longer wheelbase of the estate), two-door coupé and two- and four-seater convertibles.

Engine 896cc 71x76mm, 3 cyl TS, F/F
Max power 34bhp at 4000rpm/38bhp at 4200rpm/40bhp at 4250rpm
Max speed 72/75/78mph
Body styles saloon, estate, coupé, convertible
Prod years 1953-57/1955-59
Prod numbers 55,857/157,330

PRICES
Saloon **A** 3000 **B** 1800 **C** 1000
Coupé **A** 3500 **B** 2200 **C** 1500
Convertible **A** 6000 **B** 3500 **C** 2500

DKW 1000

The final development of the post-war DKW body style was the 1958-63 1000, later known as the Auto Union 1000. The biggest change was to bore the three-cylinder engine out to 980cc and the availability of a high compression 'S' model with 50bhp. 1000 models are easily recognised by their wrap-around windscreens. Front disc brakes and wider tyres from 1961. Same large range of body styles as before, except for the convertibles. Like all DKWs, column change gearbox is compulsory (four-speed in the case of the 1000). Has quite a large following.

Engine 980cc 74x76mm, 3 cyl TS, F/F
Max power 44bhp at 4500rpm/ 50bhp at 4500rpm
Max speed 81/85mph
Body styles saloon, coupé, estate
Prod years 1958-63
Prod numbers 171,008

PRICES
A 2500 B 1500 C 1000

DKW Monza

The idea for the Monza sports car came from racing driver Gunther Ahrens. The 3=6 chassis was clothed in a small coupé body made of glassfibre – one of the very first German plastic-bodied cars. In a convoluted production career, coachbuilders Dannenhauer & Stauss built the first 15 examples, then production passed to Massholder (who made a further 90 cars) and finally to Wenk (who built the last 50). Rather too expensive to be popular and superseded by the 1000SP. German DKW collectors have grabbed almost all of them.

Engine 896cc 71x76mm/980cc 74x76mm, 3 cyl TS, F/F
Max power 38bhp at 4200rpm/ 40bhp at 4250rpm/44bhp at 4500rpm/50bhp at 4500rpm
Max speed 87-96mph
Body styles coupé
Prod years 1955-58
Prod numbers 155

PRICES
A 10,000 B 7000 C 5000

DKW 1000SP Coupé/Roadster

When it appeared at the Frankfurt Show in 1957, the 1000SP was inevitably compared with the Ford Thunderbird, announced two years earlier. If it was plagiaristic, it was well received: apparently plenty of people wanted to buy a mini T-bird. Its fashionable looks were enhanced by a wrap-around windscreen and tail fins. Coupé was joined by a roadster in 1961, both with bodies by Baur. Both used the same 3=6 chassis with a tuned version of the 1-litre engine, but neither was a hot shot even so. The 1000SP remains the bee's knees as far as DKW fanatics are concerned.

Engine 980cc 74x76mm, 3 cyl TS, F/F
Max power 55bhp at 4500rpm
Max speed 86mph
Body styles coupé, convertible
Prod years 1958-65
Prod numbers ca 5000/1640

PRICES
Coupé A 6000 B 4500 C 2500
Roadster A 8000 B 6000 C 3500

DKW Junior/F11/F12

The new generation of DKW saloons stuck to three cylinders and the twin-stroke cycle but the engine was, in fact, all-new. So were the body styles – and they weren't a pretty sight. Junior did boast infinitely adjustable IFS by wishbones and an all-synchro gearbox (Saxomat automatic clutch option from '60). 1961 Junior De Luxe had 796cc engine, larger wheels, uncowled headlamps. Replaced in '63 by new style bodies with squared-off roof: F12 had 889cc engine, F11 stuck to the smaller size. The only collectable version is the 1964-65 F12 Roadster (2794 built).

Engine 741cc 68x68mm/796cc 70.5x68mm/889cc 74.5x68mm, 3 cyl TS, F/F
Max power 30bhp at 4300rpm/34-40bhp at 4300rpm/43bhp at 4500rpm
Max speed 68-85mph
Body styles saloon, convertible
Prod years 1959-63/1963-65
Prod numbers 237,605/30,738/82,506

PRICES
Saloon A 2500 B 1500 C 800
Roadster A 7000 B 4000 C 2500

DKW F102

The last of the two-stroke DKWs – and indeed the last DKW ever – was the F102 of 1963. This was the first unitary car produced by the German firm, which was then in Daimler Benz's hands. The three-cylinder engine expanded to 1.2 litres, and you got front disc brakes. Initially two-door only but four-door option from '64. Hardly ever seen in UK (imported for less than 12 months). Volkswagen bought DKW and used the F102 as the basis of the first Audi: it differed in its use of a conventional four-cylinder engine and rectangular headlamps (see page 28).

Engine 1175cc 81x76mm, 3 cyl TS, F/F
Max power 60bhp at 4500rpm
Max speed 85mph
Body styles saloon
Prod years 1963-66
Prod numbers 52,753

PRICES
A 1600 B 1000 C 400

Dodge · Edsel · Elva · Enzmann

Dodge Charger/Charger Daytona

Chrysler's sister marque always produced interesting corporate alternatives, none more so than the late 1960s Charger. For 1968, the Charger had a flying buttress fastback style, surprisingly simple for the time. R/T ('Road/Track') – pictured – had up to 425bhp, firm suspension and bigger brakes. The '69 Charger Daytona was Dodge's star model: aerodynamic 'bullet' nose, hidden headlamps, flush rear window and towering rear spoiler, plus 375 or 425bhp V8 lumps. Only 505 were built to homologate the model for NASCAR racing, in which it was dominant.

Engine 5211cc/6276cc /6974cc / 7206cc, V8 cyl OHV, F/R
Max power 230bhp at 4400rpm to 425bhp at 5000rpm **Max speed** 110-150mph **Body styles** coupe
Prod years 1967-69
Prod numbers 185,307/505

PRICES
Charger **A** 7500 **B** 4500 **C** 3000
Charger R/T **A** 10,000 **B** 7000 **C** 4000
Charger Daytona **A** 20,000 **B** 15,000 **C** 12,500

Edsel

Ford's marketing lemon lost an estimated $300 million in its brief existence. The mistake was in the timing more than the product. The Edsel filled a gap between the more expensive Fords and the budget Mercurys, but it had no identity and people were demanding smaller cars. The full model range consisted of the Ranger, Pacer, Corsair, Citation and Villager station wagon. Engines were V8s but from '59 an entry-level six was offered too. The whole fiasco lasted just three years.

Engine 3655cc, 6 cyl OHV, 4785cc/ 5440cc/ 5768cc/5916cc /6719cc, V8 cyl OHV, F/R **Max power** 147bhp at 4000rpm to 350bhp at 4600rpm
Max speed 100-117mph
Body styles saloon, coupe, estate, convertible **Prod years** 1957-60
Prod numbers 108,532

PRICES
Saloon '58 **A** 7500 **B** 4500 **C** 3000
Convertible '58 **A** 12,000 **B** 7000 **C** 5000 Coupe '60 **A** 5000 **B** 3200 **C** 1800

Elva Courier

Pretty but basic sports cars from a firm already successful on the track. Tubular ladder chassis, Riley 1.5 or MGA engines, Triumph based suspension, glassfibre body. MkI had a split windscreen, MkII a curved screen, MkIII (pictured) was built by Trojan and had a fresh chassis with rack-and-pinion steering, front discs, MGB power. Final MkIV T-Type had IRS and choice of Cortina GT engine. Coupé available with reverse-angle rear screen or fastback. Most production went to America.

Engine 1489cc 73x88.9mm/1588cc 75.4x88.9mm/1622cc 76.2x88.9mm/ 1798cc 80.3x88.9mm/1498cc 81x72.7mm, 4 cyl OHV, F/R
Max power 68bhp at 5500rpm to 95bhp at 5400rpm
Max speed 100-115mph
Body styles sports, coupé
Prod years 1958-65
Prod numbers ca 610

PRICES
A 6500 **B** 4500 **C** 3000

Elva-BMW GT160

This could have been a great car had it not arrived just as Elva's parent company, Trojan, was tiring of car manufacture. The GT160's lines were penned by stylist Trevor Fiore and built by Italian coachbuilders Fissore in aluminium over an Elva Mk7S racing chassis. Only three such bodies were ever made and the car was never offered to the public at its projected price of £4500. The prototype had a tuned BMW 1800 engine fitted amidships, while another was fitted with a 3.5-litre Buick V8. One finished as the fastest British car at Le Mans in 1965.

Engine 1766cc 89x71mm, 4 cyl OHC, M/R
Max power 182bhp at 7200rpm
Max speed 140mph
Body styles coupé
Prod years 1964
Prod numbers 3

PRICES
Too rare to put a value on

Enzmann 506

The Enzmann was probably the most successful Swiss car of the 1950s. The brainchild of a Volkswagen dealer and his son, the Enzmann 506 was a sports car based on VW Beetle chassis. The numeral 506 derived from Enzmann's stand number at the 1957 Frankfurt Show where the car was launched! The doorless glassfibre body could be ordered with a perspex top which folded over for entry. Amazingly, all the tooling still survives and it's possible to buy a brand new Enzmann 506 from Dr Enzmann.

Engine 1192cc 77x64mm, 4 cyl OHV, R/R
Max power 30bhp at 3400rpm to 52bhp at 4250rpm
Max speed 78-93mph
Body styles sports
Prod years 1957-68
Prod numbers ca 60

PRICES
Too rare to value accurately, but a good one should fetch around the £10,000 mark

Excalibur SS S1-S3

The 'neo-classic' phenomenon began with Brooks Stevens' Excalibur. Apart from a short run of Kaiser-engined cars in 1951, his first car was the SS of 1964. It looked like a prewar Mercedes SSK and used the chassis and supercharged V8 of the Studebaker Daytona or, when that left production, Ford or Chevy Corvette parts. Prices were high but quality was excellent and the Excalibur had a knack of endearing itself to Hollywood stars. The S3 lasted until 1979 but subsequent models are still in production, still at exorbitant prices.

Engine 4247cc 90.5x82.5mm/5351cc 101.6x82.5mm, V8 cyl OHV, F/R
Max power 182bhp at 4500rpm/304bhp at 5000rpm
Max speed 110-120mph
Body styles sports, convertible
Prod years 1964-79
Prod numbers 1848

PRICES
A 20,000 **B** 15,000 **C** 10,000

Facel Vega FV

Facel entered the luxury sports car market in 1954 at a time when most competitors had been killed off. The Vega (or just FV) had beautiful bodywork over a tubular chassis, IFS and live rear and a De Soto V8 engine. Quality was excellent and you got four seats and 130mph touring ability. Chrysler V8s followed. Usually two or three-speed auto 'boxes but there was a Pont-à-Mousson four-speed manual option. Front disc brakes optional from '58. No Ferrari but highly desirable nonetheless.

Engine 4524cc 92x84.9mm/4940cc 99.3x79.5mm/5410cc 94.5x96.5mm/5801cc 100x92mm, V8 cyl OHV, F/R
Max power 180bhp at 4400rpm/250bhp at 4600rpm/255bhp at 4400rpm/325bhp at 4600rpm
Max speed 115/121/136mph
Body styles coupé, convertible
Prod years 1954-58
Prod numbers 357
PRICES
Coupé **A** 18,000 **B** 15,000 **C** 12,000
Cabriolet **A** 28,000 **B** 20,000 **C** 16,000

Facel Vega HK500

Definitive Facel Vega grand tourer with same chassis and bodywork as FV, but larger Chrysler V8 engines with a claimed output of 360bhp on twin quadrajet carbs. 6.3-litre engine from '59 and standard power steering and disc brakes from '60 – so the post-1960 model is a far better car than the under-braked early models. Automatic more usual than four-speed manual. American engine means wonderful cruising and shattering acceleration but spirited driving is hampered by chronically unsupportive seats. Quite an experience.

Engine 5907cc 105x85.7mm/6286cc 108x85.7mm, V8 cyl OHV, F/R
Max power 335bhp at 4600rpm/360bhp at 5200rpm
Max speed 142mph
Body styles coupé
Prod years 1958-61
Prod numbers 548

PRICES
A 22,000 **B** 17,000 **C** 12,000

Facel Vega Excellence

When you extend the FV wheelbase by 20in, you get the four-door Excellence. This was a monster of car, over 17ft long and weighing two tons. Interesting 'clap hands' door layout meant pillarless construction. It also means there's a panoramically vast unsupported sill, and sagging chassis is a common problem. Mostly automatic but the usual four-speed manual option existed. Like HK500, bigger engine from '59 and discs from '60 (one wonders what the drum braked Excellence was like to stop). Last ones had less windscreen wrap-around and no tail fins.

Engine 5907cc 105x85.7mm/6286cc 108x85.7, V8 cyl OHV, F/R
Max power 335bhp at 4600rpm/360bhp at 5200rpm
Max speed 130/135mph
Body styles saloon
Prod years 1958-64
Prod numbers 152

PRICES
A 22,000 **B** 17,000 **C** 10,000

Facel Vega Facel II

Restyled HK500 with narrower windscreen, cut-down squarer roof line, larger boot and cowled vertical headlamps. Less headroom but more glass area. More desirable because of its spec: power now up to no less than 390bhp, Hydrosteer on automatic version, standard Borrani wires, servo disc brakes. Chassis barely capable of dealing with the power, so entertaining drives guaranteed. Expensive to keep going. By the early 1960s, the Facel II had a lot more competition from other grand tourers with American V8s (Iso, Gordon-Keeble, Jensen CV8), hence its rarity.

Engine 6286cc 108x85.7, V8 cyl OHV, F/R
Max power 390bhp at 5500rpm
Max speed 150mph
Body styles coupé
Prod years 1961-64
Prod numbers 184

PRICES
A 40,000 **B** 32,000 **C** 25,000

Facel Vega Facellia

The Facellia goes down in history as the car which sank Facel Vega. Former Talbot engineer Carlo Marchetti designed his own four-cylinder alloy-head twin overhead cam engine, which was built at the Pont-à-Mousson factory. Sadly, its unreliability led to crippling warranty claims. It was also not nearly powerful enough for such a heavy car and too peaky for the sporting driver. Improved F2S engine from '61 was more responsive and more reliable and had optional twin Webers. Servo disc brakes, improved interior at the same time. Bodywork much smaller than other Facels.

Engine 1647cc 82x78mm, 4 cyl DOHC, F/R
Max power 115-126bhp at 6400rpm
Max speed 112-118mph
Body styles coupé, convertible
Prod years 1959-63
Prod numbers 1258

PRICES
Coupé **A** 9000 **B** 6000 **C** 4000
Cabriolet **A** 13,000 **B** 9000 **C** 6500

Facel Vega Facel III/6

In an effort to salvage the Facellia and the company's reputation, the smaller car was fitted with the Volvo P1800 twin carb engine. This was not very satisfactory: although more reliable it was less powerful and heavier, and led to poorer handling. Disc brakes all round, IFS, overdrive manual 'box. Recognisable by its headlamps under a single vertical cowl. Extremely rare Facel 6 had a small-bore Austin-Healey 3000 six-cylinder engine fitted, which at least restored the lost performance. But by then it was all too late and Facel Vega went bust in 1964.

Engine 1780cc 84.1x80mm/2860cc 82.5x89mm, 4/6 cyl OHV, F/R
Max power 108bhp at 6000rpm/150bhp at 5250rpm
Max speed 108/120mph
Body styles coupé, convertible
Prod years 1963-64
Prod numbers ca 400/42

PRICES
Coupé **A** 8000 **B** 5000 **C** 3500
Cabriolet **A** 12,000 **B** 8000 **C** 6000

Fairthorpe Atom/Atomota

Air Vice-Marshal Donald Bennett's first attempt at car manufacture was the Atom, a desperately crude microcar. Central backbone chassis, all-independent suspension, hydraulic brakes and glassfibre coupé or convertible bodywork. Motorcycle engines included 250cc BSA (Mk1), 322cc Anzani twin (Mk2), 350cc BSA (Mk2A) and 646cc BSA twin (Mk3). Replacement Atomota (photo) had front-mounted BSA twin engine, car-type synchro four-speed 'box, live hypoid rear axle and big fins at the back.

Engine 249cc 1 cyl OHV, 322cc 2 cyl TS, 348cc 1 cyl OHV, 646cc 2 cyl OHV, R/R and F/R
Max power 11bhp at 5400rpm/ 15bhp at 4800rpm/17bhp at 5500rpm/35bhp at 5700rpm
Max speed 45-75mph
Body styles coupé, convertible
Prod years 1954-57/1957-60
Prod numbers 44/n/a

PRICES
Too few survivors to price

Fairthorpe Electron

Fairthorpe's first sports car and the first Electron was nothing more than a proprietary GRP shell called the Microplas Mistral plonked on a ladder chassis made by Fairthorpe, and powered by a Coventry-Climax engine. Soon Fairthorpe was making its own glassfibre bodyshell but the impression was still rather down-market. However, this was undoubtedly one of the cheapest ways to get into sports car motoring and certainly the cheapest Climax-powered car on the market. Front discs from 1957. Almost impossible to find nowadays.

Engine 1098cc 72.4x66.6mm, 4 cyl OHC, F/R
Max power 84-93bhp at 6900rpm
Max speed 110-120mph
Body styles sports
Prod years 1956-65
Prod numbers 30

PRICES
Accurate prices impossible to formulate, but expect to pay more than an Electron Minor

Fairthorpe Electron Minor/Electrina

The Electron Minor was a great idea: a compact sports car with its own ladder chassis and proprietary parts, spoiled only by the arrival of the Austin-Healey Sprite. Initially Standard 10 suspension and engine were used, later from the Herald; from 1963 that switched to the 1147cc Spitfire engine. EM4 (1965) upped the capacity to 1296cc, EM6 had GT6 chassis instead of Fairthorpe's own. Amazingly, it struggled on in ever decreasing circles until 1973. Electrina was awful, short-lived 2+2 saloon version. Most EMs were sold in kit form.

Engine 948cc 63x76mm/1147cc 69.3x76mm/1296cc 73.7x76mm, 4 cyl OHV, F/R
Max power 35bhp at 4500rpm to 75bhp at 6000rpm
Max speed 85-100mph
Body styles sports, saloon
Prod years 1957-73/1961-63
Prod numbers ca 500/20

PRICES
A 4000 **B** 2700 **C** 1300

Lumenition®

IGNITION SYSTEMS

★ SPOT ON TIMING ALWAYS

★ OVERCOMES WORN DISTRIBUTOR

★ KITS AVAILABLE TO FIT OVER 600 MAKES AND MODELS OF VEHICLES

★ STILL IN PRODUCTION AFTER 25 YEARS

★ EASY TO FIT WITH NO PERMANENT MODIFICATIONS

AUTOCAR ELECTRICAL EQUIPMENT Co Ltd
49/51 TIVERTON STREET
LONDON SE1 6NZ
Tel: 0171-403 4334
Fax: 0171-378 1270

MOUNTINGS & BUSHES — MAIL ORDER

The specialists in

Bushes. Suspension etc.
Mountings. Engine, gearbox, subframes.
Steering. Trackrod ends, Ball joints.

British made parts for: *Austin, Austin Healey, Bentley, BMC/Rover, Ford, Hillman, Chrysler/Talbot, Jaguar, Land Rover, Range Rover, M.G., Rolls Royce, Triumph.*

Example Prices: *Inclusive of P&P and VAT* (Price each)

Ford	Heavy duty track control arm kit, Sierra, Granada, Scorpio.	£10·40
	Fiesta rear radius arm complete	£10·69
Rover	Metro rearsubframe mounting (left or right) 84-88	£12·28
	Metro rearsubframe mounting (left or right) 89-91	£13·04
Jaguar	E Type, V12 engine mounting	£15·50
	E Type, 6 cylinder engine mounting	£ 9·70
MGB	Engine mounting 1976 on	£ 6·70
Triumph	Stag engine mounting	£ 9·87
	Stag gearbox mounting	£ 3·82
	Spitfire engine mounting	£ 7·34
	Spitfire gearbox mounting	£ 3·35

BMS Ltd

92 Co-operative St., Stafford ST16 3DA
Phone: 01785 250850 Fax: 01785 250852

VISA — **24 HOUR ANSWERING SERVICE TRADE ENQUIRIES WELCOME.** — Access

---

IF YOU CARE

about originality in your classic car, 'Bay View Books' acclaimed 'Original' series could be just what you need. Whether you're restoring a car, having one restored, or thinking of buying one and want to be sure it's 'right', our books will provide you with comprehensive detail, in words and colour photographs, on every aspect of original factory specification, equipment, finishes and options.

These beautifully produced hardback books are 96-152 pages A4 format, with 150-250 colour photographs of carefully selected cars, and are priced at £17.95 - £19.95

BAY VIEW BOOKS

For a free colour catalogue of these and other Bay View motoring books, write, fax or phone
Bay View Books, The Red House, 25-26 Bridgeland Street, Bideford, Devon EX39 2PZ
Tel 01237 479225/421285   Fax 01237 421286

Titles so far published

- AC Ace & Cobra
- Aston Martin DB4/5/6
- Austin-Healey (100 & 3000)
- Jaguar XK
- Jaguar E-type
- Jaguar Mark I/II
- MG T Series
- MGA
- MGB
- Mini Cooper
- Morgan
- Morris Minor
- Porsche 356
- Porsche 911
- Sprite & Midget
- Triumph TR
- VW Beetle

coming this year

- Mercedes SL
- Citröen DS
- Land-Rover Series 1

ESSENTIAL SKIN CARE FOR CARS

Crystal-Glo Acrylic Car Polish deep-cleans, restores and reseals in one. Having brought back the shine it wraps it in a protective acrylic shield that is weather-proof and scratch-resistant. The Polish is designed to work in conjunction with Crystal-Glo Acrylic Total Wash to provide a complete and lasting 'skin' treatment on all exterior surfaces. It builds protection with every application and buffs to a clear mirror finish which won't yellow, patch or peel.

CRYSTAL-GLO
ACRYLIC VEHICLE POLISH

● ACRYLIC CAR POLISH •
ACRYLIC TOTAL WASH • &
LEATHER AND UPHOLSTERY
CLEANER

ON TWO WHEELS
DISTRIBUTION LIMITED
Unit 3 • Castell Close
Swansea Enterprise Park
Llansamlet • Swansea SA7 9FH
T 01792 774944
F 01792 781452

LOOK AT WHAT'S RESTING ON OUR REPUTATION.

If you own a classic car, no other manufacturer offers you more choice.

Our *Classic Range* of tyres is made to match authentic specifications, often using the original Dunlop tread patterns.

So when you specify Dunlop quality tyres, you can be sure that your car is safe, reliable and *right*, down to the very last detail.

Send for your copy of 'Keeping The Legend Alive', the essential guide to the *Classic Range* of Dunlop Tyres.

CCBG/96
NAME _____
ADDRESS _____
_____ POSTCODE _____
MAKE OF CAR _____
MODEL _____ CC _____ YEAR _____

D DUNLOP
DRIVING TO THE FUTURE

Post coupon to: 'Keeping The Legend Alive,'
SP Tyres UK Limited, Fort Dunlop, Birmingham B24 9QT.

*S*pecialising in all popular and exotic classic cars, we offer Agreed Value cover for laid up, limited mileage and everyday use. With over **40,000** classic car policy holders we lead the market in price, policies and service.

LANCASTER INSURANCE SERVICES OPERATE SCHEMES FOR THE MG OWNERS CLUB AND OTHER LEADING CLUBS

FOR A FREE QUOTATION OR FURTHER INFORMATION PLEASE TELEPHONE:

01480 484848

(30 lines)

ALL POLICIES TAILORED TO SUIT YOUR REQUIREMENTS

FROM INDIVIDUAL VEHICLES TO GROUP POLICIES

LANCASTER
Classic INSURANCE SERVICES LIMITED

8 STATION ROAD, ST. IVES, HUNTINGDON, CAMBS PE17 4BH   Fax: 01480 464104

MINIATURES LTD
FINE QUALITY MODELS FOR COLLECTORS
Hand Built 1:43 Scale Metal Cars

1959 Jaguar Mk IX Saloon. Also in White and Maroon £52.95

1971 Jaguar E Type V12. Hard Top. Also in Red with wire wheels £54.95

1961 Jaguar Mk X 3.8 Saloon. Also in Maroon £52.95

1951 Studebaker Commander Business Coupé V8. Also in Maroon, Green & Blue £52.95

From your model dealer. Send for your free 1996 catalogue.
25 WEST END, REDRUTH, CORNWALL, TR15 2SA
TELEPHONE 01209 218356   FAX 01209 217983

SEEING IS BELIEVING
One folded Jaguar cushion circa 1962

This side has been liquid leathered, the colour is rich, leather soft, grain distinct and the creases have 'healed'. | This side is dry and full of dirt, hard and heavily creased, anaemic and pale in colour.

THE SMELL IS APPRECIATING
Quick and easy to use, *Liquid Leather* is safe, economical and thorough - it works! It even SMELLS of Jaguars of yesteryear

250ml Cleaner £4.90   250ml Conditioner £5.95
both include p&p (Credit Card sales add 35p, goods dispatched same day)

gliptone *leathercare UK*

Gliptone leathercare, 5 Bridgewater Street, Manchester M3 4NN
Tel: 0161 834 4153/0161 832 8532   Fax: 0161 839 2941
France 1/34864161 Germany 07621/791920 Switzerland 061 303 2303
Holland 0487 513717 USA (603) 622 1050 Australia (07) 5530 2554 Japan 3/3439 4222

Who Covers the World's Finest Motor Cars?..

- Beware of cheap imitations. Only 'Specialised Car Covers' tailor covers to fit all makes of car, each cut to its own individual pattern.
- Extensive colour range available with or without logos.
- Manufactured in the U.K. from the finest breathable sateen cotton.
- Also available Evolution 4 Fabric by Kimberly-Clark water resistance breathable covers or exclusive Nylon Coated Waterproof covers for outdoor use or transportation.
- Prices (including storage bag) from £100 - £180.
- Delivery: 14-21 days
- **IF YOU CHERISH IT – INVEST IN THE BEST**

Step on it

For the true connoisseur of style there has long been a need for a car mat that matches the quality of his car. We have recognised that need and produced a mat of exceptional quality which is specifically aimed at the driver who parallels appearance with durability.

- Unique coloured logos embroidered in silk
- Luxury pile carpet
- Leather cloth edging

WORLDWIDE EXPORT SERVICE

Official Suppliers to: ASTON MARTIN LAGONDA LTD · PORSCHE CARS (GT. BRITAIN) LTD · PORTMAN LAMBORGHINI · JAGUAR CARS · MASERATI LTD · TVR ENGINEERING · ROVER
FOR FURTHER DETAILS CONTACT: CONCOURS HOUSE, MAIN STREET, BURLEY IN WHARFEDALE, YORKSHIRE LS29 7JP
TEL: 01943 864646 (24hrs) FAX: 01943 864365

Fairthorpe Zeta/Rockette

How to inject too much performance into an Electron chassis. Fairthorpe's sledgehammer approach was to squeeze a Ford Zephyr engine into a standard chassis and body, with front disc brakes and rack-and-pinion steering. It then proceeded to offer various states of tune up to 137bhp! Still desperately crude. An alternative engine was the Triumph TR3A. The Rockette was a more practical successor, its Triumph Vitesse six offering 70bhp. Early Rockettes had a third centrally-mounted headlamp which looked horrible. Both models are very rare.

Engine 2138cc 86x92mm, 4 cyl OHV, 2553cc 82.5x79.5mm, 6 cyl OHV, 1596cc 66.75x76mm, 6 cyl OHV, F/R
Max power 100bhp at 5000rpm/90-137bhp at 4400rpm/70bhp at 5000rpm
Max speed 100-120mph
Body styles sports
Prod years 1960-65/1963-67
Prod numbers 3/3

PRICES
Probably worth twice as much as an Electron Minor

Fairthorpe TX-GT/TX-S/TX-SS

Donald Bennett's son Torix joined the fold and immediately made an impression by designing a new independent rear suspension system by transverse rods. The TX-1 prototype appeared in 1965 as an open car but made production as the TX-GT coupé from 1967, although almost always with Triumph GT6 rear suspension (the complete GT6 chassis was used). Strange glassfibre body gained larger rear three-quarter windows from '69. TX-S had tuned GT6 engine, while the TX-SS had fuel injection in either 2-litre or later 2.5-litre forms, all from Triumph.

Engine 1998cc 74.7x76mm/2498cc 74.7x95mm, 6 cyl OHV, F/R
Max power 104bhp at 5300rpm/112bhp at 5300rpm/140bhp at 6000rpm
Max speed 112/115/130mph
Body styles coupé
Prod years 1967-76
Prod numbers ca 30

PRICES
A 4500 **B** 3000 **C** 2000

Falcon Caribbean/Bermuda

Falcon was Peter Pellandine's follow-up to the successful Ashley bodyshell of the 1950s. By far the best known Falcon was the Caribbean. This had a space frame chassis designed by old F1 hand Len Terry, later examples boasting independent front suspension. Ford Ten parts were the norm, but other engine options, including MGA and Coventry-Climax, were available. Bodies always in glassfibre. Bermuda was square-roof four-seater version, uglier and only marginally 2+2.

Engine 1172cc 63.5x92.5mm/1220cc 76.2x66.5mm/1489cc 73x89mm, 4 cyl SV/OHC/OHV, F/R
Max power 30bhp at 4000rpm/75bhp at 6100rpm/68bhp at 5500rpm
Max speed 80-110mph
Body styles coupé, sports
Prod years 1957-63
Prod numbers ca 2000/200

PRICES
A 2500 **B** 1500 **C** 1000

Falcon 515

Perhaps the most mature product of the specials bodyshell boom of the early 1960s was Falcon's 515. A very handsome, almost Ferrari-like glassfibre body styled by Brazilian Tom Rohonyi was bonded to a tubular chassis. Mechanicals were taken from the brand new Ford Cortina; the front end was independent and the rear axle was live. The 515 was intended to be a fully-built car, although many ended up being supplied in kit form. Could have been a great success but sadly the project never progressed beyond small-scale production.

Engine 1498cc 81x72.6mm, 4 cyl OHV, F/R
Max power 64bhp at 4600rpm
Max speed 95mph
Body styles coupé
Prod years 1963-64
Prod numbers 25

PRICES
Verified sales unrecorded, but probably in the £3000-8000 range

Ferrari 166

Ex-racing driver and ex-Alfa team manager Enzo Ferrari founded his own firm in 1940, and post-war set to work promptly on the racing type 125. It arrived in 1947, and soon spawned embryonic production of touring versions, likewise powered by the fabulous V12 engine designed by ex-Alfa man Gioacchino Colombo. Early racing 125s were 1.5 litres, but true road cars started as 166s with 2.0 litres. Double wishbone front end, live rear. Won Le Mans in 1949, and the Mille Miglia. Sport/Inter/MM models with bodies by Allemano, Touring, Vignale, Ghia, Bertone and Pinin Farina. Vastly expensive.

Engine 1995cc 60x58.8mm, V12 cyl DOHC, F/R
Max power 95bhp at 6000rpm to 140bhp at 7000rpm
Max speed 100-125mph
Body styles coupé, sports, convertible
Prod years 1947-53
Prod numbers 71

PRICES
A 150,000 **B** 120,000 **C** 100,000

Ferrari

Ferrari 195

In the early days, Ferrari's numerical designations derived from the size of a single cylinder; hence the V12 195 had a total capacity of 2341cc. This was a bored-out version of the 166 engine on the same chassis. The Inter model was the touring road version, while the Sport was competition-orientated and was offered with a triple carb engine delivering up to 180bhp. All were coachbuilt with aluminium bodies, mostly by Ghia and Vignale, plus some by Pinin Farina, Touring and Ghia-Aigle. Expect to pay three-figure sums for any 195.

Engine 2341cc 65x58.8mm, V12 cyl DOHC, F/R
Max power 130bhp at 6000rpm/160-180bhp at 6000rpm
Max speed 110/125-130mph
Body styles coupé, sports, convertible
Prod years 1950-52
Prod numbers 27

PRICES
A 170,000 **B** 140,000 **C** 120,000

Ferrari 212

The alloy V12 engine was bored out once more in 1951 to create the 212. Although power was not up significantly, the engine was more flexible. The 212 scored many victories on the track and it therefore became much better known than the 195, especially in its best market, the United States. The 212 Export used basically the same chassis as previous models but the 212 Inter had a longer (98.5in) wheelbase. Again a vast range of coachbuilt bodies were created, including Vignale (pictured), Ghia, Touring and Pinin Farina.

Engine 2562cc 68x58.8mm, V12 cyl DOHC, F/R
Max power 130-140bhp at 6000rpm/150-170bhp at 6500rpm
Max speed 120-140mph
Body styles coupé, convertible
Prod years 1951-53
Prod numbers 106

PRICES
A 150,000 **B** 120,000 **C** 100,000

Ferrari 340/342/375 America

The ultimate evolution of the 166 chassis was the 'America' series. Aurelio Lampredi designed a long block version of the V12 engine which was used in Ferrari's F1 effort but also in the America sports cars. The 340 and 342 shared the same 4.1-litre size, the latter with reduced horsepower and all-synchro four-speed 'box to suit its role more as a touring car (it had a longer wheelbase). 375 had an even longer wheelbase and a 4.5-litre engine and was sold in racing or road-going forms. 340 Mexico was an all-out racer. Ultra-desirable and extremely rare, hence out-of-orbit prices.

Engine 4101cc 80x68mm/4523cc 84x68mm, V12 cyl DOHC, F/R
Max power 220bhp at 6600rpm/200bhp at 5000rpm/300bhp at 6300rpm **Max speed** 130/120/149mph
Body styles coupé, sports, convertible
Prod years 1950-55
Prod numbers 22/6/13

PRICES
Individual examples vary in price; perhaps double that of the 195

Ferrari 250 Europa

The first Ferrari tailored specifically for road use, hence four-speed all-synchro 'box and fairly mild state of tune (200bhp). Still the 166 platform, but using a smaller bore version of Lampredi's V12 – the only road-going Ferrari to have this engine. All 250 Europas used the same 110.2in wheelbase chassis as the 375 America but could be ordered with a variety of bodies. By this stage, Pinin Farina was becoming Ferrari's favourite and most Europas were fitted with Farina's coupé body (pictured). Easier to live with. Rare, so high prices.

Engine 2963cc 68x68mm, V12 cyl DOHC, F/R
Max power 200bhp at 6000rpm
Max speed 120mph
Body styles coupé
Prod years 1953-55
Prod numbers 17

PRICES
A 125,000 **B** 100,000 **C** 80,000

Ferrari 250 GT

A volume Ferrari at last. The 102.3in wheelbase chassis has since become known as the 'long wheelbase' to distinguish it from the shorter post-'59 250 GT SWB. The engine was the classic expression of the Colombo V12, a new 2953cc version. Coil springs replaced transverse leafs at the front and there was a rigid axle at the rear. Although the 250 GT was produced in larger numbers than any previous Ferrari, it was still coachbuilt, usually by Pininfarina (one word after '58). From '58, one Farina style was standardised – by far the most common.

Engine 2953cc 73x58.8mm, V12 cyl DOHC, F/R **Max power** 200bhp at 6600rpm to 220bhp at 7000rpm
Max speed 120-125mph
Body styles coupé, convertible
Prod years 1954-62
Prod numbers 905

PRICES
Pininfarina Coupé **A** 45,000 **B** 32,000 **C** 24,000
Cabriolet **A** 100,000 **B** 80,000 **C** 60,000

Ferrari 250 GT Tour de France

To celebrate Ferrari's victory in the Tour de France, a special 250 GT model was created. The handsome coupé bodywork was designed by Pininfarina but built in aluminium by the Modenese coachbuilder Scaglietti. In character, it was very much a sports racer, with just enough comfort to be driven to the race circuit and back. The engine was a tuned version of the classic Colombo V12 (up to 280bhp). The result was the fastest of the original lwb 250 GTs and one of the best Ferraris of all time. Uncommon and therefore exceedingly expensive.

Engine 2953cc 73x58.8mm, V12 cyl DOHC, F/R
Max power 230-280bhp at 7000rpm
Max speed 125-137mph
Body styles coupé
Prod years 1955-59
Prod numbers 84

PRICES
A 400,000 B 350,000 C 300,000

Ferrari 410 Superamerica

Replacement for the 375 America boasted a new chassis, bigger brakes and more power from a bored-out Lampredi V12. Huge range of axle ratios meant top speed could be tailored to suit – up to 165mph! Three distinct series were produced, mostly with coupé bodywork by Farina, though some by Ghia and Scaglietti. Series III from '58 had shorter wheelbase, outboard spark plugs, bigger brakes, higher compression ratio and up to 400bhp. Expensive, yes, but not as much as road/race cross-over Ferraris. Viewed as too 'street' by cognoscenti.

Engine 4962cc 88x68mm, V12 cyl DOHC, F/R
Max power 340bhp at 6000rpm/ 360bhp at 7000rpm/400bhp at 6500rpm
Max speed 135-165mph
Body styles coupé, convertible
Prod years 1956-59
Prod numbers 38

PRICES
A 125,000 B 100,000 C 80,000

Ferrari 250 GT Cabriolet/California

From December 1957, Scaglietti built an open version of the 250 GT for Pininfarina. Originally this was on the 102.3in wheelbase, and from 1960 on the SWB wheelbase of 94.5in. The standard engine came from the GT but the California was fitted with the tuned engine from the Tour de France and could, in theory, be used for racing. As the bodywork was open, extra bracing was added to the chassis. At the time, the most refined Ferrari yet for everyday road use. Beware of convincing conversions of much less valuable coupés.

Engine 2953cc 73x58.8mm, V12 cyl DOHC, F/R
Max power 200-240bhp at 7000rpm
Max speed 125-137mph
Body styles sports
Prod years 1957-63
Prod numbers 241/104

PRICES
LWB A 350,000 B 300,000 C 250,000
SWB A 500,000 B 450,000 C 400,000

Ferrari 250 GT SWB Berlinetta

From 1959, the 250 GT chassis was also made in short wheelbase form (94.5in), which stiffened it considerably and made it handle much more confidently. Another major advance was the fitment of four-wheel disc brakes. Coil springs at the front, various axle ratios, standard Borrani centre-lock wires. Pininfarina styled bodies were again built by Scaglietti, in steel with aluminium doors, boot and bonnet. Again, a black market industry has sprung up to convert standard Ferraris into Berlinettas. Not as smooth as the later Lusso but much more prized.

Engine 2953cc 73x58.8mm, V12 cyl DOHC, F/R
Max power 280bhp at 7000rpm
Max speed 140mph
Body styles coupé
Prod years 1959-63
Prod numbers 175

PRICES
A 400,000 B 360,000 C 330,000

Ferrari 250 GT Berlinetta Lusso

Initially called just 250 GT/L, the Berlinetta Lusso effectively replaced the SWB 250 GT Berlinetta. Its bodywork was again styled by Pininfarina and executed by Scaglietti and featured a fastback ending in a Kamm-style tail. Although it shared the SWB chassis, it was far more of a road-going machine than the old Berlinetta, as the Lusso tag indicated. You got a centre console and plenty of gauges but still no glovebox or provision for a radio. This was the swansong of the long-running 250 GT series, lasting until 1964, and many regard it as the most beautiful of them all.

Engine 2953cc 73x58.8mm, V12 cyl DOHC, F/R
Max power 240bhp at 7500rpm
Max speed 135mph
Body styles coupé
Prod years 1962-64
Prod numbers 350

PRICES
A 120,000 B 90,000 C 70,000

Ferrari

Ferrari 250 GTE

This was Ferrari's first production four-seater (previous coachbuilt 2+2s had been one-offs). As such it tends to get a bad press, especially as the rear seats are all but useless. Engine, gearbox and front seats were moved forward in the standard 102.3in wheelbase to create extra interior space. Pininfarina not only styled but also built the bodywork this time. 1962 revisions saw the model renamed 250 GT 2+2. Many GTEs have been butchered to make SWB and GTO replicas, so spare body parts are plentiful. Regarded as rather common in Ferrari circles.

Engine 2953cc 73x58.8mm, V12 cyl DOHC, F/R
Max power 235bhp at 7000rpm
Max speed 120mph
Body styles coupé
Prod years 1960-63
Prod numbers 950

PRICES
A 45,000 **B** 30,000 **C** 20,000

Ferrari 400 Superamerica

Latest in the fast-disappearing line of custom-bodied Ferraris, the 400 Superamerica was almost always bodied by Pininfarina. It also marked a change in numbering policy, now following total engine capacity in decilitres. Initially based on a 95.3in chassis (Series I), lengthened to 102.3in (Series II) in 1962. The V12 engine was now the Colombo unit, not the Lampredi, and electric overdrive was standard, as were servoed four-wheel discs and Koni shock absorbers. Most had Pininfarina's new 'aerodynamic coupé' body, with choice of open or cowled headlamps. Not as greatly fancied as most coachbuilt Ferraris.

Engine 3967cc 77x71mm, V12 cyl DOHC, F/R
Max power 340bhp at 7000rpm/400bhp at 6750rpm
Max speed 140-160mph
Body styles coupé, sports, convertible
Prod years 1959-64
Prod numbers 54

PRICES
A 130,000 **B** 110,000 **C** 100,000

Ferrari 250 GTO

The letters GTO are all that it takes to get a Ferrari man's pulse racing. Although never built as a road car, this is probably the only true racing Ferrari that is regularly driven on the road. The GTO was essentially a lightened 250 GT SWB with a Testa Rossa engine, five-speed 'box and even sleeker Bizzarrini-shaped lines. Typically, the V12 engine would develop around 300bhp, which made the GTO almost unbeatable on the circuits. Only 39 cars were built to GTO specification, later works racers having notch back styling. The greatest of all Ferraris? Those who buy them certainly think so.

Engine 2953cc 73x58.8mm, V12 cyl DOHC, F/R
Max power 300bhp
Max speed 155mph
Body styles coupé
Prod years 1962-64
Prod numbers 39

PRICES
Around £2.5 million for a first-class example.

Ferrari 500 Superfast

Although it looked like a Superamerica, the 500 Superfast was in fact longer (on a 104.3in wheelbase) and heavier. It also boasted a unique engine: a 5-litre V12 with the same dimensions as the old 410 but a different design, combining the long block Lampredi with Colombo-type removable heads. Pininfarina bodies always had exposed headlamps and Kamm tails. Some christened the Superfast 'Ferrari's Bugatti Royale' because it was large, hand-crafted and ridiculously expensive. But it was never a truly great car and prices reflect this.

Engine 4962cc 88x68mm, V12 cyl DOHC, F/R
Max power 400bhp at 6500rpm
Max speed 160mph
Body styles coupé
Prod years 1964-67
Prod numbers 37

PRICES
A 130,000 **B** 100,000 **C** 90,000

Ferrari 275 GTB/275 GTS/275 GTB4

The 275 GTB was the natural successor to the 250 GT Berlinetta. The larger engine produced a healthy output at 280bhp, but the later GTB4 quad cam/triple carb had 300bhp at a screaming 8000rpm; the softer GTS convertible had only 260bhp. The chassis followed traditional Ferrari practice, but was more modern – all-independent suspension, five-speed gearbox integral with the differential. Series II (1965 on) had longer nose and bonnet bulge. Alloy bodies were available for competition use and today fetch about a 15% premium.

Engine 3286cc 77x58.8mm, V12 cyl DOHC/QOHC, F/R
Max power GTS: 260bhp at 7000rpm/GTB: 280bhp at 7600rpm/GTB4: 300bhp at 8000rpm
Max speed 145-168mph
Body styles coupé, convertible
Prod years 1964-66/1964-66/1966-67
Prod numbers 460/200/350
PRICES
GTB **A** 125,000 **B** 100,000 **C** 80,000
GTS **A** 140,000 **B** 120,000 **C** 100,000
GTB4 **A** 200,000 **B** 175,000 **C** 150,000

Ferrari 330 GT

While the 275 was the two-seater replacement for the 250, the 330 GT was the 2+2 successor to the GTE. Initially launched as the 330 GT America, it had Ferrari's new Tipo 209 4-litre V12, which made it significantly quicker than the old four-seaters. The America is identifiable by its twin headlamps, and it had a four-speed-and-overdrive gearbox. The 1965 GT 2+2 had single headlamps, a stronger five-speed 'box and alloy wheels instead of (now optional) wires. The earlier car is worth slightly less. One of Pininfarina's less successful styling efforts, and dynamics not that great.

Engine 3967cc 77x71mm, V12 cyl DOHC, F/R
Max power 300bhp at 6600rpm
Max speed 130mph
Body styles coupé
Prod years 1963-67
Prod numbers 1080

PRICES
A 30,000 **B** 22,000 **C** 15,000

Ferrari 330 GTC/330 GTS/365 GTC/365 GTS

The two-seater version of the 330 GT may not have looked much prettier than the 'family' car, but it certainly behaved better, thanks to independent rear suspension and a five-speed transaxle: the chassis package was basically the same as the 275. The essentially boulevard role of this Ferrari was confirmed by (gasp) optional air conditioning (standard from '68). Larger 365 engine from '68 had not much extra power but wider torque band. GTS versions were Spyders, as ever. Likeable if soft. Overshadowed by Daytona, therefore fair value.

Engine 3967cc 77x71mm/4390cc 81x71, V12 cyl DOHC/QOHC, F/R
Max power 300bhp at 7000rpm/ 320bhp at 6600rpm
Max speed 125/130mph
Body styles coupé, convertible
Prod years 1966-68/1966-68/1968-70/1968-69
Prod numbers 600/100/200/20

PRICES
GTC **A** 60,000 **B** 45,000 **C** 35,000
GTS **A** 150,000 **B** 135,000 **C** 120,000

LAMBORGHINI   FERRARI
MOTORAPIDE

From a service to a complete restoration to concours.
We can provide the following services:
- FULL MECHANICAL REBUILDS
- PANEL MAKING and ALL FINE LIMIT SHEET METAL FABRICATIONS
- SPECIALISTS in STAINLESS STEEL TRIM SECTIONS
- GENERAL MACHINING: TURNING, MILLING, GRINDING, BORING AND GRIT BLASTING etc
- PAINTING WITH LOW BAKE OVEN
- All types of PLATING and POLISHING
- TRIMMING in all materials to concours

Our rates are competitive
We are only too happy to quote for work or just give friendly advice.
We always have cars for sale, just ring for details

MOTORAPIDE
Hale Oaks Farm, Loxwood Road, Rudgwick,
West Sussex, RH12 2BP
Tel/Fax: 01403 753762 Mobile: 0468 112532

 # VERDI for Ferrari
Performance Car Workshop

Servicing
Mechanical Repairs
Body Repairs
Rebuilds
Spares

Renovations
Retrimming
Sales
Air Conditioning
-Repair & Recharging

Telephone & Fax: 0181 756 0066
9-10 Hayes Metro Centre, Springfield Road, Hayes, Middlesex UB4 0LE

Ferrari 365 California

This was the first Ferrari to receive the quad cam V12 engine. No prizes for guessing the target audience of this one. Pininfarina's elegant open bodywork was based on the long (104in) wheelbase and the level of equipment was high. Five-speed 'box, power steering and wire wheels were standard. Cowled headlamps supplemented by retractable foglamps in the nose. Strictly a two-seater and one of the prettiest convertibles of the 1960s. Very short production run, so ultra-rare: if you can find one, it'll be very expensive.

Engine 4390cc 81x71, V12 cyl QOHC, F/R
Max power 320bhp at 6600rpm
Max speed 130mph
Body styles convertible
Prod years 1966-67
Prod numbers 14

PRICES
A 200,000 **B** 170,000 **C** 145,000

Ferrari 365 GT 2+2

This is getting pretty big by Ferrari standards: well over 16ft long and weighing almost two tons. No surprise that it was nicknamed 'Queen Mary'. Following on from the 330 GT 2+2, the Pininfarina styling borrowed from the Superfast book. All-independent suspension, four-wheel discs, five speeds, power brakes/steering, leather, air conditioning, tape deck, electric windows and – a first for Ferrari – Koni self-levelling rear suspension. Very popular in its day (half of all Ferraris made 1968-71 were 365 GTs), so plenty around to choose from.

Engine 4390cc 81x71, V12 cyl QOHC, F/R
Max power 320bhp at 6600rpm
Max speed 125mph
Body styles coupé
Prod years 1967-71
Prod numbers 801

PRICES
A 35,000 **B** 28,000 **C** 22,000

Ferrari Daytona 365 GTB/4 /GTS/4

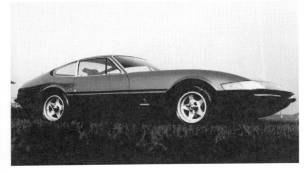

Launched at the Paris Salon in 1968, the 365 GTB/4 was perhaps the last great front-engined Ferrari – until the current 456 GT. Although the 275 chassis was retained, and the quad cam 4.4-litre V12 was by now familiar, it all gelled in the Daytona, a nickname coined by the press, incidentally. Pininfarina's gorgeous and aerodynamic bodywork was what made the Daytona special. Not an easy car to drive. Genuine Spyders are like hen's teeth, so plenty of coupés have been converted (they're worth only 50% of the genuine article). Every serious Ferrari fan must have one.

Engine 4390cc 81x71, V12 cyl QOHC, F/R
Max power 325bhp at 7500rpm
Max speed 150mph
Body styles coupé, convertible
Prod years 1968-74
Prod numbers 1285/123

PRICES
GTB/4 **A** 90,000 **B** 75,000 **C** 60,000
GTS/4 **A** 250,000 **B** 210,000 **C** 180,000

Ferrari Dino 206/246 GT/246 GTS

If you discount the 250 LM competition car, this was Ferrari's first mid-engined road car and the 'baby' of the family. Initially it wasn't even called a Ferrari – just Dino, after Enzo's son. The compact V6 engine was derived from Ferrari's F1/F2 racing unit and was built by Fiat. Driving through a five-speed transaxle, it was capable of powering the slippery Dino to over 140mph. 206 signified 2.0 litres, six cylinders. Iron-block 2.4-litre engine from '69, targa-top GTS from '72. Seminal handling but poor build quality and dreadful rust problems.

Engine 1987cc 77x58.8mm/2418cc 92.5x60mm, V6 cyl QOHC, M/R
Max power 180bhp at 8000rpm/195bhp at 7500rpm
Max speed 142/149mph
Body styles coupé, targa
Prod years 1967-69/1969-74/1972-74
Prod numbers 152/2487/1274

PRICES
206 **A** 50,000 **B** 35,000 **C** 25,000
246 GT **A** 45,000 **B** 30,000 **C** 22,000
246 GTS **A** 50,000 **B** 36,000 **C** 27,000

Ferrari 365 GTC/4

Like the 365 GT 2+2 and GTC which it replaced at one fell swoop, the GTC/4 was by far the best-selling model in Ferrari's range. Chassis was virtually identical to the Daytona except for a conventional five-speed 'box instead of a transaxle. This four-seater fastback Ferrari looked a little gawky and came with the sort of luxuries an owner of ten years previously would have laughed at: power steering and, in America, standard air conditioning. But it was fast and, in its short lifespan, sold extremely well. More highly regarded than later 365 GT4 2+2.

Engine 4390cc 81x71, V12 cyl QOHC, F/R
Max power 320-340bhp at 6200rpm
Max speed 150/155mph
Body styles coupé
Prod years 1971-72
Prod numbers 500

PRICES
A 50,000 **B** 40,000 **C** 28,000

Ferrari 365 GT4 2+2/400 GT

The new four-seater Ferrari had clean, simple lines by Pininfarina (who else?) on a GTC/4 chassis extended by 6in. The result was a rather large and heavy (almost two-ton) car, though it concealed its bulk well and featured pop-up headlamps. Usual quad cam V12 was stroked and injected in '76 to become the 400 GT, which was offered with optional automatic – previously heresy at Ferrari! Very fast touring car but expensive to keep going (single-figure mpg). Replaced by 5-litre 412 in '85, which lasted until '89 – making this Ferrari's most durable model.

Engine 4390cc 81x71/4823cc 81x78mm, V12 cyl QOHC, F/R
Max power 320bhp at 6200rpm/325-340bhp at 6000rpm
Max speed 150/152mph
Body styles coupé
Prod years 1972-76/1976-85
Prod numbers 470/502

PRICES
A 25,000 **B** 18,000 **C** 12,000

Ferrari 365 GT4 BB/512 BB

BB meant Berlinetta Boxer, a name which alluded to the new flat-12 engine which was mounted amidships. Four triple-barrel Webers and toothed-belt cam drive meant 360bhp and initially a claimed top speed of 200mph (in reality 175mph). Chassis was an all-new multi-tube affair. Pininfarina's understated bodywork was stunning and featured matt black glassfibre lower panels. 512BB of '76 had 5-litre engine, small front spoiler and air ducts behind the front wheels. Injection from '81. A total contrast to the Miura/Countach with which it competed.

Engine 4390cc 81x71mm/4942cc 82x78mm, 12 cyl QOHC, M/R
Max power 360bhp at 7000rpm/360bhp at 6200rpm
Max speed 175mph
Body styles coupé
Prod years 1973-76/1976-85
Prod numbers 387/1936

PRICES
A 70,000 **B** 55,000 **C** 40,000

Ferrari Dino 208 GT4/308 GT4

A couple of firsts here: Ferrari's first ever V8 engine and the first mid-engined four-seater. Bigger than the Dino 246 (wheelbase extended by 8.4in) and much more practical. Bertone's angular styling did not come off and that is the main reason why the GT4 is the cheapest Ferrari of any description today – bargains abound if you steer clear of abused ones. 208 was 2-litre tax-break model for Italy alone. Standard 308 3-litre engine gave storming performance. Ferrari badges not officially used until 1975 and 'Dino' name tag dropped in '77.

Engine 1991cc 66.8x71mm/2926cc 81x71mm, V8 cyl QOHC, M/R
Max power 160bhp at 6600rpm/230bhp at 6600rpm
Max speed 125/150mph
Body styles coupé
Prod years 1975-80/1973-80
Prod numbers 840/2826

PRICES
A 20,000 **B** 13,000 **C** 9000

Ferrari 208 GTB/GTS /308 GTB/GTS

Pininfarina's version of the Dino 246 replacement was as balanced and poised as Bertone's GT4 was ungainly, although it took an extra two years to arrive. Strictly a two-seater. Mechanically exactly the same as the GT4, though the chassis was 8.3in shorter. 208 was Italy-only 2-litre version. The first examples had a large proportion of glassfibre body panels but all-steel bodies across the board from 1977 – watch for rust on these. GTS targa roof model launched in '78, the whole range uprated to fuel-injection in '81.

Engine 1991cc 66.8x71mm/2926cc 81x71mm, V8 cyl QOHC, M/R
Max power 170bhp at 6600rpm/230bhp at 6600rpm
Max speed 130/152mph
Body styles coupé, targa
Prod years 1975-81/1978-81/1975-81/1978-81
Prod numbers 160/140/2897/3219

PRICES
308 GTB **A** 30,000 **B** 22,000 **C** 17,000
308 GTS **A** 33,000 **B** 25,000 **C** 20,000

Fiat 500/500B

Dr Dante Giacosa was the man behind the Fiat 500, known affectionately as the 'Topolino' (little mouse). When it arrived in 1936, it offered minimal and, above all, cheap motoring. Perfect for austere post-war times, and the car that got Italy moving. Only two seats and choice of roll-top or fixed-head coupés, plus a woody estate. IFS and quarter-elliptic rear. Tiny gravity-fed side-valve engine replaced by OHV unit in the 500B of 1948. Licence-built in France as the Simca 5. Expect rust problems.

Engine 569cc 52x67mm, 4 cyl SV/OHV, F/R
Max power 13bhp at 4000rpm/16.5bhp at 4400rpm
Max speed 52/55mph
Body styles coupé, convertible, estate
Prod years 1936-48/1948-49
Prod numbers 122,213/21,623

PRICES
A 5000 **B** 2500 **C** 1700

Fiat

Fiat 500C

In 1949, the 500 was restyled with a rather Americanesque nose incorporating faired-in headlamps and an extended rear end with a boot. Still only two seats unless you go for the estate, again a 'woody' until 1951, when it switched to all-steel construction (estate called Giardiniera in Italy and Belvedere in UK). Obligatory roll-top saloon roof allowed transportation of unfeasible loads. Very popular in native Italy but appeal limited by two-seat status elsewhere. Actually a lot of fun to drive, and economical, but difficult to work on.

Engine 569cc 52x67mm, 4 cyl OHV, F/R
Max power 16.5bhp at 4400rpm
Max speed 60mph
Body styles convertible, estate
Prod years 1949-55
Prod numbers 376,371

PRICES
A 5000 **B** 2000 **C** 1500

Fiat 1100/1100B/1100E

Fiat returned to production after the war in 1947 with the 1939 1100, itself a development of the 1937 Balilla. In 1948, this was uprated to become the 1100B, available in two wheelbase lengths, and the following year this metamorphosed again to become the 1100E, with revised steering and the spare relocated to the boot. Each evolution had minor changes to the front grille. There was a Cabriolet version, but this was little more than a roll-top model with the same four-door treatment. Little known in the UK and not especially collectable.

Engine 1089cc 68x75mm, 4 cyl OHV, F/R
Max power 32-35bhp at 4400rpm
Max speed 62mph
Body styles saloon, convertible
Prod years 1939-48/1948-49/1949-53
Prod numbers ca 74,000/22,000/58,000

PRICES
Saloon **A** 4000 **B** 2500 **C** 1500
Cabriolet **A** 6000 **B** 4000 **C** 2200

Fiat 1100S

The 1100S was the post-war relaunch of the trail-blazing 1938 full-width bodied 508C MM (for Mille Miglia, to celebrate its 1938 win), of which only 400 were built prewar. Savio did the prewar bodywork, Farina took up the mantle after 1947. Not what you'd call overpoweringly pretty: lines are rather tank-like, although this was very racy stuff for the time. Pretty anaemic performance, despite the high compression head giving over 50bhp. An almost total lack of rear vision and serious rust problems do not endear it to owner-enthusiasts. Rarity value only.

Engine 1089cc 68x75mm, 4 cyl OHV, F/R
Max power 51bhp at 5200rpm
Max speed 92mph
Body styles coupé
Prod years 1947-50
Prod numbers 401

PRICES
A 12,500 **B** 9000 **C** 6000

Fiat 1400

1950's 1400 was the first unitary Fiat, which boasted up-to-date full-width bodywork. Not overly handsome but very spacious inside (six seats). IFS, 14in wheels, and four-speed 'box geared for economy, not performance. 1400A from 1954 had an improved engine, revised grille, larger rear window, extended rear wings and choice of 1.9-litre diesel power (13,500 diesels built in total). 1400B of '56 changed grilles again, and added duo-tone paint schemes. Two-door Cabriolet (pictured) is most desirable, but was never sold in UK – nor was the diesel.

Engine 1395cc 82x66mm/1901cc 82x90mm, 4 cyl OHV/diesel, F/R
Max power 44bhp at 4400rpm/58bhp at 4600rpm/43bhp at 3200rpm
Max speed 84/62mph
Body styles saloon, coupé, convertible
Prod years 1950-58
Prod numbers 120,356

PRICES
Saloon **A** 3000 **B** 2000 **C** 1000
Cabriolet **A** 5000 **B** 3500 **C** 2500

Fiat 1900

This larger-engined version of the 1400 was introduced two years after its smaller sister. Identifiable by extra body chrome and fog or side lights in the grille. This was Fiat's top model in the 1950s and boasted a 60bhp 1.9-litre engine (later 80bhp), five speeds with fluid coupling, rear coil springs and an average speed calculator. Granluce ('full light') coupé shown had a wrap-around rear window and hardtop two-door coupé style. Cabriolets popular with Italian police. 1900B from '56 with body changes as 1400, plus single central foglamp.

Engine 1901cc 82x90mm, 4 cyl OHV, F/R
Max power 60bhp at 4300rpm/80bhp at 4000rpm
Max speed 87/92mph
Body styles saloon, coupé, convertible
Prod years 1952-58
Prod numbers 15,759

PRICES
Saloon **A** 3500 **B** 2200 **C** 1000
Coupé **A** 4500 **B** 3000 **C** 2000
Cabriolet **A** 6000 **B** 4000 **C** 2500

Fiat 8V

A unique V8 engine and a tubular chassis on to which was welded a striking steel sports coupé body made the 8V an unrepeatable exercise by Fiat in 1952. All-round independent suspension and an all-synchro four-speed 'box were unusual features for the era. With the standard Fiat body it was a strict two-seater, initially with large headlamps built into the grille and, from 1953, with 'Chinese eye' head and sidelamp treatment. Special bodies by Ghia, Pinin Farina and Zagato. The Siata 208, with Vignale-styled bodywork, was a much revised version made 1952-54 (see page 174).

Engine 1996cc 72x61.3mm, V8 cyl OHV, F/R
Max power 105-115bhp at 6000rpm
Max speed 118mph
Body styles coupé, convertible
Prod years 1952-55
Prod numbers 114

PRICES
A 90,000 **B** 70,000 **C** 55,000

Fiat 1100-103/1100D/1100R

Unitary construction for the 1100 arrived in 1953, still with the prewar engine which would last until the bitter end in 1969. Class-leading performance and handling, especially in TV (Turismo Veloce) form – high-compression 48bhp engine and cyclops foglamp. Estate car from '54, larger rear window and little fins from '57, De Luxe with dual-choke carb from '59, squarer lines and front-hinged doors from '61. 1100D of '62 has 1.2-litre engine, 1100R (1966) has standard 48bhp engine, lower bonnet and front discs. Amazingly, it's still made in India as the Premier Padmini.

Engine 1089cc 68x75mm/1221cc 72x75mm, 4 cyl OHV, F/R
Max power 34bhp at 4400rpm/ 48bhp at 4800rpm/52bhp at 5200rpm
Max speed 72-85mph
Body styles saloon, estate
Prod years 1953-62/1962-66/1966-69
Prod numbers 1,768,375 (1,019,378/408,997/340,000)

PRICES
A 2500 **B** 1400 **C** 800

Fiat 1100/1200 Spider

In 1955, Fiat offered a sports convertible version of the 1100-103, which it dubbed the 1100 Trasformabile (Spider in export markets). This was really Fiat's first 'popular' sports car and it was built by Pinin Farina. Fiat also credited Farina with the styling, but the coachbuilder denies it was responsible – just look at the styling and you'll understand why. 1100 TV engine spelt mediocre performance, slightly improved in the 1200 version from 1957. Usual rust headaches of the saloon apply to the Spider, which really has little to recommend it.

Engine 1089cc 68x75mm/1221cc 72x75mm, 4 cyl OHV, F/R
Max power 48bhp at 4800rpm/55bhp at 5300rpm
Max speed 81/88mph
Body styles convertible
Prod years 1955-57/1957-59
Prod numbers 1030/2363

PRICES
A 8000 **B** 5500 **C** 4000

Fiat 1200

This was the 1957 replacement for the 1100 TV. Boring out the 1100 engine created a 1.2-litre unit which gave better performance. Body was distinguished by its wrap-around rear window, larger side windows, restyled nose section and more brightwork. In Italy, the increased glass house was identified by the Gran Luce badge, which in the UK translated as Full Light (available here from 1958). Two-tone paintwork was common, as indeed was the model: in just three years, it notched up over 400,000 sales. Almost no collector value.

Engine 1221cc 72x75mm, 4 cyl OHV, F/R
Max power 55bhp at 5300rpm
Max speed 87mph
Body styles saloon
Prod years 1957-60
Prod numbers 400,066

PRICES
A 2500 **B** 1500 **C** 800

Fiat 600/600D

The 600 caught the tilt towards popular motoring at full swing. The recipe was simple: unitary body, water-cooled rear-mounted engine, all-independent suspension and four seats, all in a package smaller than the old Topolino. Roll-top version from '56 and 24.5bhp from '59. Almost a million were sold by the time the 600D arrived in 1960 with larger engine, winding windows, higher final drive, no transmission brake. Suicide doors abandoned in '64. Production in Italy ended in 1969 but Seat made the 600 in Spain until 1973 and Zastava made theirs in Yugoslavia until 1986!

Engine 633cc 60x56mm/767cc 62x63.5mm, 4 cyl OHV, R/R
Max power 22-24.5bhp at 4600rpm/29bhp at 4800rpm
Max speed 62-65mph
Body styles saloon
Prod years 1955-60/1960-69
Prod numbers 2,452,107 (891,107/1,561,000)

PRICES
A 2200 **B** 1400 **C** 900

Fiat 600 Multipla

A miniature forebear of the current craze for people carriers. The Multipla was based on a 600 floorpan and mechanicals, except for the front suspension which came from the 1100. Forward-control driving position allowed three rows of seats and up to six passengers; four-seater and taxi versions were also offered. The Multipla doubled up as a delivery van for many businesses. 767cc engine from 1960. Bad points: very noisy, a rough ride unless well loaded, rust problems and a soggy gearchange. On the other hand, it has a truly unique character.

Engine 633cc 60x56mm/767cc 62x63.5mm, 4 cyl OHV, R/R
Max power 22-24.5bhp at 4600rpm/29bhp at 4800rpm
Max speed 56-60mph
Body styles saloon
Prod years 1955-66
Prod numbers 160,260

PRICES
A 3000 B 1700 C 1000

Fiat 500/500D/500F/L/500R

Initially this was called the Nuova 500, and was the spiritual successor to the Topolino. Absolutely brilliant packaging got the most out of the least. Air-cooled vertical twin in the rear, all-round independent suspension, initially just two proper seats and crash 'box. Always a roll-top roof except for ultra-rare 1958 500 Sport (fixed roof, 499cc engine). Gutless 479cc engine uprated to 499cc in 1960 to become 500D. The 500F of 1965 got front-hinged doors, bigger windscreen, better interior. 500L was additional 'luxury' model from '68, run-out 500R ('72) had detuned 126 594cc engine.

Engine 479cc 66x70mm/499cc 67.4x70mm/594cc 73.5x70mm, 2 cyl OHV, R/R
Max power 13bhp at 4000rpm/17.5-22bhp at 4400rpm/18bhp at 4800rpm
Max speed 53-62mph
Body styles saloon
Prod years 1957-75
Prod numbers 3,427,648 (181,036/640,520/2,272,092/334,000)

PRICES
500 A 2700 B 1700 C 1000
500 Sport A 6000 B 3500 C 2000

Fiat 500 Giardiniera

In 1960, an estate version of the Nuova 500 was launched. The wheelbase was extended by 4in and overall length went up from 9ft 9in (297cm) to 10ft 5in (318cm). To create the necessary room to make this a useful estate car, the vertical twin engine was turned on its side and mounted under the boot floor. Access to the boot was via a side-hinged rear door. A full-length sunroof was still standard. After 1968, the Giardiniera was built by Autobianchi and badged as such. It outlasted the saloon, finally ducking out in 1977. Watch for rust in the floorpan.

Engine 499cc 67.4x70mm, 2 cyl OHV, R/R
Max power 17.5bhp at 4600rpm
Max speed 59mph
Body styles estate
Prod years 1960-77
Prod numbers ca 327,000

PRICES
A 2500 B 1700 C 1000

Fiat 1800/2100/2300

The first post-war six-cylinder Fiat, except for the short-lived 1500 (1947-50). Lampredi created the four-bearing engines in 1.8 and 2.1-litre forms, and they were admirably powerful. Servo drum brakes, all-synchro 'box, initially coil-sprung rear end, swapped to leaf springs from 1961 (plus front discs at the same time). 2.3-litre engine replaced the 2100 in '61 and, as 1800B, the 1.8 engine got more power. 2100s and 2300s had quad headlamps, 1800s only singles. Interesting lwb variants included 2100 Speciale and 2300 President (6in longer and six-light bodywork).

Engine 1795cc 72x73.5mm/2054cc 77x73.5mm/2279cc 78x79.5mm, 6 cyl OHV, F/R
Max power 75-95bhp at 5000rpm/82bhp at 5000rpm/105-115bhp at 5300rpm
Max speed 85-90/90/105mph
Body styles saloon, limousine, estate
Prod years 1959-68/1959-61/1961-68
Prod numbers ca 185,000

PRICES
A 4000 B 2500 C 1000

Fiat 2300/2300S Coupé

Individual – perhaps even awkward – styling by Ghia made the coupé version of the 2300 saloon stand out. Floorpan was identical to the saloon except you got standard power-assisted four-wheel discs. The engine always came with twin carbs, so power was usefully boosted to 115bhp, or 130bhp on the high-compression twin-choke 2300S. First sold in UK in 1962. Dual-circuit brakes from 1963, gruesome brightwork enhancements in 1965. Somewhat cumbersome to drive, but good performance and high on 1960s style. Your biggest problem is keeping corrosion at bay.

Engine 2279cc 78x79.5mm, 6 cyl OHV, F/R
Max power 115bhp at 5300rpm/130bhp at 5600rpm
Max speed 115-120mph
Body styles coupé
Prod years 1961-68
Prod numbers n/a

PRICES
A 6000 B 4000 C 2000

Fiat 1200/1500 Cabriolet/Coupé

The 1200 Cabriolet replaced the ugly old Spider in 1959. This time Pininfarina certainly was responsible for the styling, and it built the bodies too. The engine and floorpan were shared with those from the 1200 Gran Luce saloon. The 1500 Spider which replaced the 1200 in 1963 had a 72bhp 1481cc pushrod engine (not the twin cam unit fitted to the almost-identical 1500S, for which see below). In contrast to the 1200, the 1500 boasted assisted disc front brakes. More power and five speeds from 1965. Only ever LHD. This body style rusts appallingly.

Engine 1221cc 72x75mm/1481cc 77x79.5mm, 4 cyl OHV, F/R
Max power 58bhp at 5300rpm/72bhp at 5400rpm/80bhp at 5200rpm
Max speed 87/95/102mph
Body styles coupé, convertible
Prod years 1959-63/1963-66
Prod numbers 11,851/22,630 (incl 1500S)

PRICES
A 5000 **B** 3000 **C** 2000

Fiat 1500S/1600S

Introduced at the same time as the 1200 Cabriolet, the 1500S was the performance version, fitted with a fine OSCA twin cam engine. Fiat kept the output down to a manageable 75bhp but still hoped to do battle with Alfa Romeo. Externally, the only way to tell the 1500S apart from the 1200 was its bonnet scoop. Dunlop discs became standard in 1960. When the 1200 was displaced by the 1500 in 1963, the S engine was bored out and the model renamed 1600S. Now there was no bonnet scoop but you could still spot one by its second pair of headlamps in the grille.

Engine 1491cc 78x78mm/1568cc 80x78mm, 4 cyl DOHC, F/R
Max power 75bhp at 5200rpm/90bhp at 6000rpm
Max speed 100/110mph
Body styles coupé, convertible
Prod years 1959-62/1963-66
Prod numbers see above/3089

PRICES
A 6000 **B** 4000 **C** 2500

Fiat 1300/1500

The replacement for the old 1200 was a larger saloon with sharp styling by Pininfarina, basically a shrunken version of the 1800/2100 model. Two brand new engines were offered concurrently: the 1300 and the 1500, whose character was very free-revving. Four speeds on the column, semi-elliptic rear springs, standard front discs (servo from '64). Higher outputs from '64. 1500L of 1964 was basically an 1800 bodyshell with the 1.5-litre engine and all-round discs. The early 1300/1500 range always had quad headlamps. As ever with Fiat, rust will be your sworn enemy.

Engine 1295cc 72x79.5mm/1481cc 77x79.5mm, 4 cyl OHV, F/R
Max power 65-70bhp at 5200rpm/72-80bhp at 5200rpm
Max speed 86-96mph
Body styles saloon, estate
Prod years 1961-67/1961-68
Prod numbers ca 600,000 (excl 1500L)

PRICES
A 2000 **B** 1200 **C** 600

Fiat 850

The old 600 formula, much improved: more room inside (five passengers at a pinch), larger engine with better performance, and sharper handling balance. Basic model with 34bhp and post-'65 Familiare (van-type estate) not imported to UK, but we did get the 1968 Special, with its more powerful engine from the 850 Coupé, front discs and luxury trim. Automatic version (called Idromatic or Idroconvert) was offered from '66 but was dire. Another very strong seller for Fiat, but somehow this little saloon has no mystique, as well as an unfortunate reputation for rusting.

Engine 843cc 65x63.5mm, 4 cyl OHV, R/R
Max power 34bhp at 4800rpm to 47bhp at 6400rpm
Max speed 75-84mph
Body styles saloon, estate
Prod years 1964-71
Prod numbers 1,780,000

PRICES
A 1000 **B** 600 **C** 200

Fiat 850 Coupé

In 1965 Fiat presented its Coupé version of the 850. Unlike the Spider (page 84), the styling was an in-house job. To go with its more sporting image, the engine was given a higher compression ratio and delivered 47bhp, so a top speed of 90mph was feasible. You also got disc brakes at the front. In March 1968, the engine was expanded to 903cc/52bhp, with corresponding gains in performance. This version had double headlamps. Entertaining handling, but still too noisy and not much space for people or luggage. Find a good one cheap and look after it.

Engine 843cc 65x63.5mm/903cc 65x68mm, 4 cyl OHV, R/R
Max power 47bhp at 6400rpm/52bhp at 6400rpm
Max speed 90/94mph
Body styles coupé
Prod years 1965-72
Prod numbers 342,873

PRICES
A 2000 **B** 1200 **C** 800

Fiat

Fiat 850 Spider

In 1964, Fiat gave 850s to all the major Italian coachbuilders and asked them to come up with a convertible body. Bertone's attractive shape won hearts at Fiat and so it got the contract for the production version, which it also built in series. Mechanically followed the same pattern as the 850 Coupé, including the larger engine from 1968. Also in '68, it received larger, recessed headlamps and overriders. Only ever available to special order in the UK, and exclusively LHD. Bertone also made a hardtop version. A lot more fun than you might imagine, and you can still find ones for sensible prices.

Engine 843cc 65x63.5mm/903cc 65x68mm, 4 cyl OHV, R/R
Max power 47bhp at 6400rpm/52bhp at 6400rpm
Max speed 92/96mph
Body styles convertible
Prod years 1965-72
Prod numbers 124,600

PRICES
A 3000 **B** 2000 **C** 1000

Fiat 124

With the 124, launched at the 1966 Geneva Motor Show, Fiat entered a new era. Model numbering policy switched from engine capacity to abstracts and the car itself broke new ground in many respects. The 1200 engine was punchy, you got four-wheel disc brakes and excellent rear coil springing. 1.4-litre engine arrived in the Special in 1968, which also had twin headlamp clusters and uprated interior spec. Twin cam Special T launched 1970, in UK '71, and in long-stroke 1.6-litre form from '72. Any kudos the 124 might have has been totally destroyed by the curse of the Lada.

Engine 1197cc 73x71.5mm/1438cc 80x71.5mm/1592cc 80x79.2mm, 4 cyl OHV/DOHC, F/R
Max power 60bhp at 5600rpm/70bhp at 5400rpm/80bhp at 5800rpm/90bhp at 6600rpm
Max speed 87-99mph
Body styles saloon, estate
Prod years 1966-74
Prod numbers 1,543,000

PRICES
A 800 **B** 300 **C** 100

Fiat 124 Spider 1400/1600/1800

Pininfarina's subtle hand was again evident in the new Fiat Spider of 1966, and the design house also made the bodies for Fiat. Mechanicals were taken from the 124, but the 1.4-litre twin cam engine was offered from the outset, and you always got servo disc brakes and the option of five gears. Suspension and braking improved in '69 along with new 1608cc engine option. Both engines changed in '72: new 1592cc and 1756cc units now standard. 2-litre version replaced the lot in 1979. Almost all went to the USA and that's where most cars for sale in the UK have come from.

Engine 1438cc 80x71.5mm/1592cc 80x79.2mm/1608cc 80x80mm/1756cc 84x79.2mm, 4 cyl DOHC, F/R
Max power 90bhp at 6000rpm to 118bhp at 6000rpm
Max speed 102-115mph
Body styles convertible
Prod years 1966-72/1969-79/1972-79
Prod numbers 129,416

PRICES
A 7000 **B** 4000 **C** 2500

Fiat 124 Coupé

In closed coupé form, the in-house styled 124 lacked grace. Mechanically it duplicated the specification of the Spider, with all its various engine options through the years. First facelift in 1969 saw improved interior and twin headlamps. A less successful second facelift in 1972 included a new grille, wrap-around rear bumper and revised dash. Always good handling, quite quick, but ultimately lacks class. Unlike the Spider, the Coupé was sold in the UK as a regular model but almost all have now rusted into oblivion or been crushed. Cheap.

Engine 1438cc 80x71.5mm/1592cc 80x79.2mm/1608cc 80x80mm/1756cc 84x79.2mm, 4 cyl DOHC, F/R
Max power 90bhp at 6000rpm to 118bhp at 6000rpm
Max speed 100-110mph
Body styles coupé
Prod years 1967-75
Prod numbers 279,672

PRICES
A 2800 **B** 1600 **C** 700

Fiat 125

The bigger brother of the 124 arrived in 1967, its styling very similar to the 124 – indeed its centre hull was identical – but its overall dimensions expanded. The twin cam engine from the sports versions of the 124 gave it a genuine 100mph top speed. Servo disc brakes all round, automatic option. 125 S (Special in UK) arrived in 1968 with 100bhp engine, five-speed gearbox, better trim, extra chrome and twin headlamps. Like the 124, quite a pleasant saloon in its day, but typical Fiat corrosion treatment (none) led 99% to an early grave.

Engine 1608cc 80x80mm, 4 cyl DOHC, F/R
Max power 90-100bhp at 5600rpm
Max speed 100-108mph
Body styles saloon
Prod years 1967-72
Prod numbers 603,870

PRICES
A 1200 **B** 800 **C** 300

Fiat Dino Spider 2000/2400

Ferrari had to have 500 of its Dino engines built to homologate the unit for F2 racing. As well Ferrari's mid-engined Dino, Fiat launched its own Dino, into which most of the quad cam units were eventually fitted. Pininfarina designed and built the beautiful front-engined Spider. You got the advantage of 2+2 seating, that wonderful but fragile triple carb engine, five-speed gearbox and servo discs. 2-litre engine was replaced by 2.4 in 1969, when IRS, ventilated discs, dual-circuit brakes, ZF 'box and revised grille were incorporated.

Engine 1987cc 86x57mm/2418cc 92.5x60mm, V6 cyl QOHC, F/R
Max power 160bhp at 7200rpm/180bhp at 6600rpm
Max speed 125-130mph
Body styles convertible
Prod years 1966-73
Prod numbers 1163/420

PRICES
A 13,000 **B** 10,000 **C** 7000

Fiat Dino Coupé 2000/2400

The Dino Coupé, launched one year later than the Spider in 1967, had a completely different body designed and built by Bertone on a wheelbase a full foot longer than the open car. The style was undoubtedly one of Bertone's better efforts but the Coupé suffers from being the Spider's poor relation. Hence you can buy a Dino Coupé for what seems a bargain price. Mechanical spec and changes exactly the same as the Spider. Both models were sold to special order in the UK, but RHD was never available. Concerns: rust and engine durability.

Engine 1987cc 86x57mm/2418cc 92.5x60mm, V6 cyl QOHC, F/R
Max power 160bhp at 7200rpm/180bhp at 6600rpm
Max speed 122-126mph
Body styles coupé
Prod years 1967-73
Prod numbers 3670/2398

PRICES
A 9000 **B** 6000 **C** 4000

Fiat 130

Fiat has never had a happy record with its larger cars, but the 130 series was perhaps the bravest try. It's a common myth that the 130 shared the Ferrari Dino's engine, but only the bottom end was the same: each head had a single cam, not twin cams, and the drive to the camshafts was by belt, not chain. Torsion bar front, IRS, five-speed manual or (more usually) Borg Warner auto. Larger engine and standard power steering from '72. Bulky four-door bodywork was stuffed with goodies but weight was over 1.5 tons, so not that quick. Nowadays rarer than 130 Coupé.

Engine 2866cc 95x66mm/3235cc 102x66mm, V6 cyl DOHC, F/R
Max power 140bhp at 5800rpm/165bhp at 5600rpm
Max speed 110/114mph
Body styles saloon
Prod years 1969-77
Prod numbers ca 15,000

PRICES
A 2500 **B** 1500 **C** 1000

Fiat 130 Coupé

A true classic of Pininfarina styling, the Coupé version of the 130 arrived in 1971. Although it was actually 3in longer than the saloon, it hid its bulk masterfully. Always sold with the larger 3.2-litre engine, and standard power steering. Automatic was usually fitted but some Coupés were built with the ZF five-speed manual 'box. Plenty of luxury inside, though leather optional. Could have handled much more power to make it a real grand tourer; as it was, the Coupé was even heavier than the saloon and struggled to reach 60mph in 11.4 secs.

Engine 3235cc 102x66mm, V6 cyl DOHC, F/R
Max power 165bhp at 5600rpm
Max speed 114mph
Body styles coupé
Prod years 1971-75
Prod numbers 4491

PRICES
A 7000 **B** 45000 **C** 2500

Fiat 128

It may look like just another brick-like Fiat saloon, but the 128 was possibly the most important model Fiat made in the 1970s. It was their first front-wheel drive, transverse-engined car and had a mechanical spec which set the standards that others had to follow, including all-independent MacPherson suspension and front disc brakes. Wonderful handling, fair pace, good economy. Two- and four-door saloon bodies and an estate were offered, and the same basic floorpan sat under the later Ritmo/Strada. Like all 1970s Fiats, they rusted dreadfully.

Engine 1116cc 80x55.5mm/1290cc 86x55.5mm, 4 cyl OHC, F/F
Max power 55-67bhp at 6000rpm
Max speed 87-93mph
Body styles saloon, estate
Prod years 1969-84
Prod numbers ca 2,776,000

PRICES
A 900 **B** 500 **C** 100

Fiat

Fiat 128 Coupé Sport/3P

How to give your sporting saloon the appearance its pace deserved. 128 Coupé Sport was launched in 1971 on a modified saloon floorpan, sharing its engines/gearboxes and running gear but lopping 9in out of the wheelbase. Quite pretty two-door style, with double headlamps or alternatively single rectangular units. Replaced in 1975 by the 3P ('3 Porte' or three-door): new, cleaner styling with a hatchback but still the same choice of 1.1 and 1.3 engines (only the 1.3 came to the UK). Comparable handling to 128 saloon, slightly better performance.

Engine 1116cc 80x55.5mm/1290cc 86x55.5mm, 4 cyl OHC, F/F
Max power 64bhp at 6000rpm/75bhp at 6600rpm
Max speed 93-100mph
Body styles coupé/hatchback coupé
Prod years 1971-75/1975-78
Prod numbers 330,897

PRICES
A 2500 **B** 1200 **C** 600

Fiat 127

Fiat created a new class of car with the 127, which has since become known as the 'supermini'. Take the engine from the 850, put it in the front to drive the front wheels and add a compact body with enough space to seat four comfortably and you have a car which sold almost 4 million. Two-door saloon and three-door hatchback styles. Bigger 1049cc overhead cam unit also offered from '77, badged 1050. Restyled nose with rubber-faced bumpers from '81. Best model is 127 Sport, with its 70bhp engine and front/rear spoilers. Made until recently in Argentina.

Engine 903cc 65x68mm/1049cc 76x57.8mm, 4 cyl OHV/OHC, F/F
Max power 45bhp at 5600rpm to 70bhp at 6500rpm
Max speed 84-100mph
Body styles saloon, hatchback
Prod years 1971-83
Prod numbers 3,730,000

PRICES
A 1000 **B** 500 **C** 200

Fiat X1/9

Clever package created by Bertone which brought mid-engined handling finesse to the masses. Launched as early as 1972 with the engine/gearbox taken directly from the 1.3-litre 128 Rally and mounted just in front of the rear wheels. Even in later 1.5-litre form (1978 on), there was never enough power to exploit the fine chassis to the full. Targa panels stowed in the front boot, but there was still some luggage space behind the engine. UK RHD imports did not begin until '76. The following year saw impact bumpers, 1500 engine and better interior. Badged Bertone from 1981.

Engine 1290cc 86x55.5mm/1498cc 86.4x63.9mm, 4 cyl OHC, M/R
Max power 66-75bhp at 6000rpm/85bhp at 6000rpm
Max speed 93-112mph
Body styles targa coupé
Prod years 1972-89
Prod numbers 141,108

PRICES
A 3500 **B** 2000 **C** 1000

Fiat 126

The replacement for the 500 did everything right to charm Italian city drivers, but somehow conspired to have a character lobotomy. The recipe was the same as ever: a rear-mounted two-cylinder air-cooled engine, four seats and rock-bottom prices. But it was larger (over 10ft/305cm long), no more spacious inside, just as noisy and didn't even have a standard sun roof. The 594cc engine made it a little faster but even the 652cc version (1977 on) was the slowest car around. Italian production ended in '87, but Poland's hatchback 704cc 126 Bis is still made today.

Engine 594cc 73.5x70mm/652cc 77x70mm, 2 cyl OHV, R/R
Max power 23bhp at 4800rpm/24bhp at 4500rpm
Max speed 65mph
Body styles saloon
Prod years 1972-87
Prod numbers 1,970,000

PRICES
A 800 **B** 400 **C** 100

Fiat 132/Argenta

The 125's replacement could have been a cracker: twin cam engines, standard five-speed 'box on all but the basic 1600, and spacious, luxurious interiors. But the reality was anonymous styling, stodgy road manners and an indifferent ride. However, it sold pretty healthily in its home land, if not anywhere else. Twin cam 2-litre from 1977 had four headlamps and replaced both 1600 and 1800 models. Automatic option on 1800 and 2000 only. Name change to Argenta in 1982 occurred in the middle of the Falklands crisis (don't cry for me, Argenta...)

Engine 1585cc 84x71.5mm 4 cyl OHC, 1592cc 80x79.2mm/1756cc 84x79.2mm/1995cc 88x82mm 4 cyl DOHC, 2445cc 93x90mm 4 cyl diesel, F/R **Max power** 60bhp at 4400rpm to 112bhp at 5600rpm
Max speed 84-105mph
Body styles saloon
Prod years 1972-84
Prod numbers 975,970

PRICES
A 800 **B** 400 **C** 100

Fiat 131 Mirafiori/Abarth Rallye

Mirafiori was the name of the factory where this 124 replacement was built. It never had the same sharp manners and performance as the 124 did in its day – nor its sales success – but a 131 was better than a Cortina (then again, what wasn't?). Twin cam Sport from 1978, all models OHC from 1981. The only really exciting 131 was the 1976 Abarth Rallye, built by Bertone for rally homologation. It had a 16-valve 84×90mm 2.0-litre engine with light-alloy head, Colotti gearbox, 124 Abarth-type IRS, glassfibre body panels, low profile tyres, and lots of spoilers and ducts. Ordinary 131s are banger fodder.

Engine 1297cc 4 cyl OHV, 1367cc 4 cyl OHC, 1585cc OHV/OHC/DOHC, 1995cc 84x90mm 4 cyl DOHC, 1995cc 88x82mm 4 cyl OHC/DOHC, 2445cc 4 cyl diesel **Max power** 60bhp at 4400rpm to 115bhp at 5800rpm/140bhp at 6400rpm **Max speed** 87-112mph/120mph **Body styles** saloon, estate **Prod years** 1974-84/1976 **Prod numbers** 1,513,800/400

PRICES
131 Sport **A** 1800 **B** 1000 **C** 600
Abarth Rallye **A** 10,000 **B** 7500 **C** 5000

Ford (D) Taunus

Henry Ford's German outpost included factories at Berlin and Cologne. The first in a long line of Taunus models appeared in 1939 but, for obvious reasons, did not really enter production until 1948. It had the same side-valve engine as the British Ford Ten but its styling was much more American and up-to-date than Dagenham's immediate post-war output. Cabriolet from '49, Spezial model from '49 had chrome grille and there was a De Luxe for '51 only. Very like a Ford Prefect to drive.

Engine 1172cc 63.5x92.5mm, 4 cyl SV, F/R **Max power** 34bhp at 4250rpm **Max speed** 65mph **Body styles** saloon, estate, convertible **Prod years** 1948-51 **Prod numbers** 74,128

PRICES
Saloon **A** 4000 **B** 2500 **C** 1300
Convertible **A** 6500 **B** 4500 **C** 2500

Ford Taunus 12M/15M

The 1952 replacement for the prewar type Taunus had full-width bodywork which aped the '49 American Ford style. However, underneath what looked vaguely like a Ford Consul sat nothing more remarkable than the good old sidevalve engine. Hence this has only reliability on its side. More power arrived in 1965, when the 15M joined the range: the 1.5-litre OHV engine allowed a top speed of 78mph. More basic 12M model also available. Curious emblem in centre of bonnet was actually a fog lamp. Last 12M Super models had the larger 1.5-litre engine.

Engine 1172cc 63.5x92.5mm/1498cc 82x70.9mm, 4 cyl SV/OHV, F/R **Max power** 38-43bhp at 4250rpm/55-60bhp at 4500rpm **Max speed** 69/78mph **Body styles** saloon, estate, convertible **Prod years** 1952-62/1955-62 **Prod numbers** 430,736

PRICES
Saloon **A** 3000 **B** 2000 **C** 1000
Convertible **A** 5000 **B** 3600 **C** 2200

Ford Taunus 17M (P2)

Even more blatant American pastiche in miniature which tried to do battle with Opel's new Rekord. It didn't quite succeed – this time. Same engine, MacPherson suspension and brakes as the British-built Ford Consul. Larger dimensions took Ford more up-market and the 1.7-litre engine gave it reasonable acceleration (0-60mph in 17 secs). Overdrive and Saxomat automatic clutch options. This was the first German Ford to make it to the UK (from 1959), but you hardly ever see any here. German collectors love them, and that's where the best prices are.

Engine 1698cc 84x76.6mm, 4 cyl OHV, F/R **Max power** 55-67bhp at 4250rpm **Max speed** 80mph **Body styles** saloon, estate, convertible **Prod years** 1957-60 **Prod numbers** 239,978

PRICES
Saloon **A** 4000 **B** 2500 **C** 1200
Convertible **A** 7000 **B** 4000 **C** 2500

Ford Taunus 17M (P3)

Very distinctive styling (ellipsoid wheelarches, cowhorn bumpers and TV screen headlamps) earned the P3 Taunus the nickname 'bathtub'. This was also the first German Ford to be sold in Britain in any quantity, although most preferred Dagenham's Classic and Cortina. Choice of engines, but the best was the 1.8-litre TS, with its 70-75bhp, plus floor shift and sporty interior. Estate car called Turnier (literally 'tournament'). Optional front discs from '62, standard on TS from '62 and on all cars from '63. Some curiosity value over here.

Engine 1698cc 84x76.6mm/1758cc 85x76.6mm, 4 cyl OHV, F/R **Max power** 60bhp at 4250rpm to 75bhp at 4500rpm **Max speed** 85-96mph **Body styles** saloon, estate, convertible **Prod years** 1960-64 **Prod numbers** 669,731

PRICES
Saloon **A** 3000 **B** 1800 **C** 800
Convertible **A** 5500 **B** 3500 **C** 2500

Ford Taunus 12M (P4)/ 12M/15M (P6)

A fascinating against-the-grain Ford. The P4 marked the first-ever front-wheel drive Ford and the first use of the V4 engine. However, the project began in Detroit, where the model would have been built and sold under the name Cardinal. When Lee Iacocca took over Ford's reins, he shunted it across to Cologne as the successor of the long-in-the-tooth '52 12M. There the interest dies, sadly. This was a uniformly dull machine, not helped by a 1966 restyle (P6). TS and RS versions were supposedly sporting machines. Not even the Germans like this one.

Engine 1183cc 80x58.9mm/1288cc 84x58.9mm/1498cc 82x70.9mm/ 1699cc 90x66.8mm, V4 cyl OHV, F/F
Max power 40to 90bhp
Max speed 78-100mph
Body styles saloon, estate, coupé, convertible
Prod years 1962-66/1966-70
Prod numbers 672,695/ 668,187
PRICES
12M/15M Saloon **A** 2000 **B** 1000 **C** 500
12M Convertible **A** 4000 **B** 2500 **C** 1800
15M RS Coupé **A** 3500 **B** 2000 **C** 1500

Ford Taunus 17M/20M (P5)

Although the profile of the new 1964 17M looked rather like the old 17M, the panelwork was all-new, with rounder lines, new grilles and less curvy wheelarches. The range of bodies was expanded by the introduction of a hardtop coupé variant (pictured). Engines were new too: as well as a 1.7-litre version of the V4 unit recently seen in the 12M, there was a new V6 2-litre. There was a TS version of the latter unit, which developed 89bhp. However, rear-wheel drive was retained for the big Cologne Ford, as was a leaf-sprung rear.

Engine 1699cc 90x66.8mm/1998cc 84x60.1mm, V4/V6 cyl OHV, F/R
Max power 65bhp at 4500rpm to 90bhp at 5000rpm
Max speed 85-100mph
Body styles saloon, estate, coupé, convertible
Prod years 1964-67
Prod numbers 710,059

PRICES
Saloon **A** 2500 **B** 1200 **C** 600
Convertible **A** 4500 **B** 3000 **C** 2200

Ford Taunus 17M/20M/26M (P7)

Another new body for the large Taunus which didn't really look like much had happened. That meant the car was restyled after just one season. As a result, the '67 cars are quite rare but certainly not rare enough for collector status. The 2.3-litre RS had 108bhp. Top-of-the-range of the revised P7 range was the 26M with its 2.6-litre V6 engine, available in saloon and hardtop coupé body styles. Imports to the UK ceased in 1971 as Ford moved towards a pan-European model policy for the 1970s. German Fords generally do not have much of a following in the UK.

Engine 1498cc 90x58.9mm/1699cc 90x66.8mm V4 cyl OHV, 1812cc 80x60.1mm/1998cc 84x60.1mm/ 2293cc 90x60.1mm/2550cc 90x66.8mm, V6 cyl OHV, F/R
Max power 60bhp at 4800rpm to 125bhp at 5300rpm **Max speed** 84-112mph **Body styles** saloon, estate, coupé, convertible **Prod years** 1967-71
Prod numbers 723,622
PRICES
Saloon **A** 1800 **B** 1200 **C** 800
Convertible **A** 4500 **B** 3000 **C** 2000

Ford OSI 20M

At the Geneva Motor Show in 1966 was a brand new sports coupé. The stand belonged to the Italian coachbuilder OSI. Underneath the coupé bodywork sat the standard floorpan of a Ford Taunus 20M, whose long wheelbase gave the sports coupé rather odd proportions. However, Ford was impressed enough to agree a contract with OSI whereby the Italian-built cars would be sold through Ford dealerships. The idea of an Italian suit of clothes on German mechanicals was not new (eg VW Karmann-Ghia), but the experiment turned sour for Ford when OSI went bust in 1968.

Engine 1998cc 84x60.1mm/2293cc 90x60.1mm, V6 cyl OHV, F/R
Max power 90bhp at 5000rpm/108bhp at 5100rpm
Max speed 102/110mph
Body styles coupé
Prod years 1967-68
Prod numbers ca 3500

PRICES
A 8000 **B** 6000 **C** 3500

Ford (F) Vedette/Vendôme

Ford had collaborated with Mathis before the war to produce cars under the Matford name. That relationship split up post-war, and Ford France proceeded to build a modern-looking saloon under the name Vedette. Underneath lay a separate chassis and V8 sidevalve engine. At the same time as a facelift in 1952, a more up-market luxury model was launched under the name Vendome, with a 100bhp 3.9-litre engine. Simca bought Ford out in 1954 and continued making the Simca Vedette, a substantially redesigned car (see page 195).

Engine 2158cc 66x78.8mm/2355cc 66x85.7mm/3923cc 80.9x95.2mm, V8 cyl SV, F/R
Max power 60bhp at 4000rpm to 100bhp at 3700rpm
Max speed 75-92mph
Body styles saloon
Prod years 1949-54
Prod numbers ca 99,000

PRICES
A 4500 **B** 2500 **C** 1000

Ford Comète/Monte Carlo

From 1951, Ford France offered a 2+2 coupé body on the basis of the Vedette. The body was made by Facel but the mechanicals were basic, to say the least. A live rear axle and only 60bhp in a car weighing almost 1.5 tons (1370kg) was not a recipe for winning the Monte Carlo. Nevertheless, Ford named its 3.9-litre 100bhp luxury version the Monte Carlo (available from 1954). A convertible body was also offered from 1952. Standard three-speed manual 'box, Cotal four-speed option. Simca kept the Comète on under its own badge until 1955. A real rarity.

Engine 2158cc 66x78.8mm/2355cc 66x85.7mm/3923cc 80.9x95.2mm, V8 cyl SV, F/R
Max power 60bhp at 4000rpm to 100bhp at 3700rpm
Max speed 80-100mph
Body styles coupé, convertible
Prod years 1951-54/1954-55
Prod numbers 4066 (incl Vendôme)
PRICES
Comète **A** 13,000 **B** 8000 **C** 5000
Monte Carlo **A** 16,000 **B** 10,000 **C** 6000

Ford (GB) Anglia/Prefect

'Sit-up-and-beg' prewar style Anglia returned in 1945. Vintage spec included mechanical drum brakes, transverse-leaf springs front and rear, three speeds, no indicators. Sidevalve engine was 933cc, or 1172cc for export and the larger Prefect. Prefect had four doors post-war and a 4in longer wheelbase and, from 1949, a Pilot-style grille and integral headlamps. Post-'49 Anglias have sloping bonnet. Dead basic transportation, massively uncomfortable and tediously slow. Quite a high survival rate.

Engine 933cc 56.6x92.5mm/1172cc 63.5x92.5mm, 4 cyl SV, F/R
Max power 24bhp at 4000rpm/31bhp at 4300rpm
Max speed 58/62mph
Body styles saloon, convertible
Prod years 1939-53
Prod numbers 166,864/379,339

PRICES
A 2500 **B** 1800 **C** 1300

Ford Popular 103E

When the Anglia was replaced in 1953, the body style was continued under the name Popular. Its sole reason for existing was to be the cheapest new car around: at £443, it certainly was. You certainly knew you were in the economy class, too: almost no instruments, just one wiper, headlamps that looked and worked like hand torches, and initially not even indicators, rear lights or reflectors. Non-opening fabric roof. 1172cc engine standard. Beloved by a grateful generation who could at last afford a new car for the first time.

Engine 1172cc 63.5x92.5mm, 4 cyl SV, F/R
Max power 30bhp at 4000rpm
Max speed 60mph
Body styles saloon
Prod years 1953-59
Prod numbers 155,340

PRICES
A 2500 **B** 1800 **C** 1200

Ford Anglia/Prefect/Escort/Squire/Popular 100E/Prefect 107E

Unitary construction for the 1953 sit-up-and-beg replacement. Much improved spec includes MacPherson strut front suspension, semi-elliptic springs at the rear, and hydraulic brakes. Anglia has two doors, Prefect has four. Estates are basic conversions of the 5cwt van: Escort is based on Anglia spec and lasts until '61, Squire (photo) adds Prefect grille and screwed-on wood strips. Popular carries on body style after '59 with even less in the way of trim and adornment, and lasts until '62. Prefect 107E (1959-61) has 39bhp OHV engine. Not much charisma.

Engine 1172cc 63.5x92.5mm/997cc 81x48.4mm, 4 cyl SV/OHV, F/R
Max power 37bhp at 4500rpm/39bhp at 5000rpm
Max speed 68/72mph
Body styles saloon, estate
Prod years 1953-59/1959-62
Prod numbers 345,841/100,554/33,131/17,812/126,115/38,154
PRICES
Saloon **A** 1500 **B** 1000 **C** 600
Escort/Squire **A** 2200 **B** 1300 **C** 800

Ford V8 Pilot

This was really a concoction of prewar parts to kick-start Ford's business after the war. Very much in the line of the big V8 sidevalve Fords of the 1930s, with the torquey old V8 lump doing its reliable service in a big separate chassis. Hydromechanical brakes, column-shift gearbox, imposing vertical chrome grille, free-standing headlamps. Extremely durable and well supported even today. Sedate transport and one to go for if you like dependable old cars built like they used to make them. Not expensive to buy and own, either.

Engine 3622cc 77.8x95.25mm, V8 cyl SV, F/R
Max power 85bhp at 3500rpm
Max speed 83mph
Body styles saloon, estate
Prod years 1947-51
Prod numbers 22,155

PRICES
A 8000 **B** 5000 **C** 3000

Ford (GB)

Ford Consul/Zephyr/Zodiac MkI

The first unitary Ford had dumpy full-width styling, overhead valve engines, MacPherson strut IFS, full hydraulic brakes, but still a three-speed column change. Consul was the entry-level model with four cylinders, Zephyr (pictured) had six cylinders and a different grille with horizontal bars, Zodiac (from '53) had higher compression six, fog lamps and two-tone paint. Consul/Zephyr convertibles by Carbodies from 1951 (7797 built), the latter boasting a power top. Farnham estates were conversions by Abbott of Farnham. Much liked and rare these days.

Engine 1508cc 79.4x76.2mm/2262cc 79.4x76.2mm, 4/6 cyl OHV, F/R
Max power 48bhp at 4400rpm/68-72bhp at 4000rpm **Max speed** 75/85mph **Body styles** saloon, estate, convertible **Prod years** 1950-56
Prod numbers 231,481/152,677/22,643
PRICES
Consul **A** 2700 **B** 1800 **C** 1000
Zephyr/Zodiac **A** 4000 **B** 3000 **C** 1800
Consul/Zephyr Convertible **A** 8000 **B** 5500 **C** 4000

Ford Consul/Zephyr/Zodiac MkII

Enlarged dimensions and more power for the grown-up MkII Fords. Longer wheelbase, more space inside and sleeker styling. Once again, Consul has four cylinders, Zephyr/Zodiac six, both units enlarged over MkI. Automatic transmission option for latter pair from October '56, overdrive always on offer, Farnham estates available as well as manual-roof (Consul) and semi-power (Zephyr/Zodiac) convertibles (16,309 built). In 1959, the roofline was lowered on all three models; front disc option from October '60, standard from May '61, when Consul was rebadged Consul 375.

Engine 1703cc 82.5x79.5mm/2553cc 82.5x79.5mm, 4/6 cyl OHV, F/R
Max power 60bhp at 4200rpm/87bhp at 4400rpm
Max speed 80/86mph
Body styles saloon, estate, convertible
Prod years 1956-62
Prod numbers 682,400
PRICES
Consul **A** 3300 **B** 2300 **C** 1300
Zephyr/Zodiac **A** 4500 **B** 3200 **C** 2000
Convertible **A** 8500 **B** 6000 **C** 4500

Ford Zephyr 4/Zephyr 6/Zodiac MkIII

All-new razor edge styling could not hide the fact that these were bulky creatures, over 10.5in longer than their predecessors. Four-speed all-synchro 'box, auto and overdrive options and servo front discs across the board. Zephyr 4 continues old four-cylinder engine, Zephyr 6 and Zodiac have sixes, identifiable by full-width grille. Zodiac has 107bhp and quad headlamps. Rare Zodiac Executive (1965-66 only) is luxury model. Estate cars available in all three versions but are not common. Like all previous models, these can rust to pieces.

Engine 1703cc 82.5x79.5mm/2553cc 82.5x79.5mm, 4/6 cyl OHV, F/R
Max power 65bhp at 4700rpm/93bhp at 4500rpm/107bhp at 4500rpm
Max speed 81/92/100mph
Body styles saloon, estate
Prod years 1962-66
Prod numbers 292,144 (106,810/107,380/77,709)
PRICES
Zephyr 4 **A** 2200 **B** 1500 **C** 900
Zephyr 6/Zodiac **A** 3300 **B** 2000 **C** 1200

Ford Anglia 105E/123E

1959's Anglia lived up to its name: that breezaway rear window was bold rather than pretty but at least it now gives Ford's economy car something to remember it by. OHV oversquare engine was loved by the tuning brigade and it powered many sports and racing cars. Super (123E) engine was the long-stroke 1.2-litre development from '62 also available on De Luxe and estate car – the latter body style arrived in '61 (129,528 built). Super identifiable by its chrome-edged body flash. The million-selling Anglia quickly became part of the nation's subconscious.

Engine 997cc 81x48.4mm/1198cc 81x58.2mm, 4 cyl OHV, F/R
Max power 40bhp at 5000rpm/53bhp at 5000rpm
Max speed 80/90mph
Body styles saloon, estate
Prod years 1959-67/1962-67
Prod numbers 1,004,737/79,223

PRICES
A 1800 **B** 1000 **C** 500

Ford Consul Classic

Bizarre over-styling in the worst Americanised fashion: fins, chrome, Z-back rear window and hooded double headlamps. Engine was a further development of the Anglia 105E engine, stroked to approach square dimensions. Girling front discs, choice of column or floor gearchange. After July 1962, the 1340cc engine was replaced by a five-bearing 1.5-litre unit and an all-synchromesh gearbox. Both two-door and four-door saloon bodies were available, but neither was attractive. Not a good seller, despite occupying a large market gap between the Anglia and Consul.

Engine 1340cc 81x65.1mm/1498cc 81x72.7mm, 4 cyl OHV, F/R
Max power 55bhp at 4900rpm/58bhp at 4600rpm
Max speed 81/84mph
Body styles saloon
Prod years 1961-63
Prod numbers 111,225

PRICES
A 2500 **B** 1500 **C** 800

Ford (GB)

Ford Consul Capri

The first use of the Capri name occurred in 1961 when Ford launched a two-door coupé version of the Classic. Its transatlantic style did not capture the imagination of the British buying public and it ducked out of production in 1964 – after the Classic, the shortest production run of any Ford to date. Identical to the Classic mechanically and bodily below the waist. Interesting variant was the Capri GT of 1963: its high-compression twin Weber 1498cc engine developed 78bhp for 95mph, but only 2002 were built. A daring design which didn't pay off.

Engine 1340cc 81x65.1mm/1498cc 81x72.7mm, 4 cyl OHV, F/R
Max power 55bhp at 4900rpm/58bhp at 4600rpm/78bhp at 6200rpm
Max speed 82-95mph
Body styles coupé
Prod years 1961-64
Prod numbers 18,716

PRICES
Capri **A** 3700 **B** 2500 **C** 1300
Capri GT **A** 4500 **B** 3000 **C** 1800

Ford Cortina MkI

Four years of development and £12 million of investment yielded Ford a best-seller, but you might have expected more innovation for the outlay. Mechanical package very similar to Consul Classic but slightly more compact and a lot more conventional in appearance. Saloons available with two or four doors. 1.5-litre engine for Super from 1963, which was also available with automatic. Front discs from '64 plus smaller grille incorporating side flashers. GT version had a Weber carb and 84bhp, plus plush trim and auto option. Estates from 1963; Super Estate has fake wood panelling.

Engine 1340cc 81x65.1mm/1498cc 81x72.7mm, 4 cyl OHV, F/R
Max power 46bhp at 4800rpm to 78bhp at 5200rpm
Max speed 78-95mph
Body styles saloon, estate
Prod years 1962-66
Prod numbers 1,010,090

PRICES
Cortina **A** 1500 **B** 1000 **C** 500
Cortina GT **A** 3500 **B** 2800 **C** 1800

Ford Lotus Cortina I

Shrewd Colin Chapman deal led to the supply of his classic twin cam engine to Ford for the first of two Lotus-badged Cortinas. Suspension was lowered, and there were wide wheels, coil rear springing (semi-elliptics from '66) and servo front discs. Easily recognisable by its cream paint scheme with a prominent green flash down the side, black grille and split bumpers. Dangerously competitive on the track in its day and a strong candidate for historic racing today. Very few were made and not many survive, so prices have risen dramatically. Tuning potential also good. Beware of fakes.

Engine 1558cc 82.5x72.75mm, 4 cyl DOHC, F/R
Max power 105bhp at 6000rpm
Max speed 105mph
Body styles saloon
Prod years 1963-66
Prod numbers 3301

PRICES
A 15,000 **B** 10,000 **C** 7000

Ford Corsair

Another distinctively styled Ford, aimed to fill the gap between the Cortina and the Zephyr. Front discs, four speeds, single carb engine, plus a GT with single twin-choke Weber. Superseded in 1965 by a V4 engine with more power than the old straight four but less refinement, plus larger front discs. GT version only lasted 1965-66. Replaced on home market by Corsair 2000 in January 1967; 2000E was luxury version with unique grille, walnut dash and black vinyl roof. Lord Lucan disappeared in one and that's just about the Corsair's only claim to fame.

Engine 1498cc 81x72.7mm/1662cc 93.7x60.3mm/1996cc 93.7x72.4mm, 4/V4/V4 cyl OHV, F/R
Max power 58bhp at 4600rpm to 102bhp at 5500rpm
Max speed 80-102mph
Body styles saloon, estate
Prod years 1963-70
Prod numbers 294,591

PRICES
Corsair **A** 1500 **B** 1000 **C** 500
Corsair GT/2000E **A** 2200 **B** 1500 **C** 900

Ford GT40/MkIII

The legend goes that, when Ferrari jilted Ford at the altar of its marriage in 1963, the corporate bigwigs at Detroit were so incensed that they decided to build a car to beat Ferrari at Le Mans. The GT40 was it – so named because it was only 40in high. Lola's Eric Broadley designed the steel monocoque, with its all-independent suspension, five-speed ZF 'box and Ford 4.7-litre V8. The GT40 did win Le Mans – four times. 31 road cars were built by Ford Advanced Vehicles before the true 'MkIII' road version arrived, with detuned engine, trimmed interior, and silencers.

Engine 4727cc 101.6x72.9mm, V8 cyl OHV, M/R
Max power 335bhp at 6250rpm/306bhp at 6250rpm
Max speed 164/155mph
Body styles coupé
Prod years 1966-72/1967-69
Prod numbers 31/7

PRICES
Depend to a large degree on history, but certainly all command six-figure sums

Ford (GB)

Ford Zephyr/Zodiac/Executive MkIV

Even more cumbersomely large, the Zephyr IV bowled into the late 1960s with a transatlantic feel and exclusively 'V' format engines. Base model was the Zephyr 4 (V4 engine), then the Zephyr 6 (2.5 V6), Zodiac (3-litre V6, different grille and four headlamps) and Executive (lots of goodies including standard sunroof and power steering). Four-speed column-change 'box with auto and floor shift options and optional overdrive on V6 cars. PAS for Zodiac from '67, revised suspension for all in '69. Too barge-like and rust-infected to inspire saviours.

Engine 1996cc 93.7x72.4mm/2495cc 93.7x60.3mm/2994cc 93.7x72.4mm, V4/V6/V6cyl OHV, F/R
Max power 83-94bhp at 4750rpm/104bhp at 4750rpm/128bhp at 4750rpm
Max speed 87/96/103mph
Body styles saloon, estate
Prod years 1966-72
Prod numbers 149,263

PRICES
A 2200 **B** 1500 **C** 700

Ford Cortina MkII

Shoe box styling for new Cortina, but dimensionally almost exactly the same as the Mk1. Launched in 1966 with a choice of 1300, 1500 and twin carb 1500GT engines, two/four-door saloon and estate bodies, Standard/Super/GT trim levels, front disc brakes, automatic option. Within a year, the 1500 engine was dropped in favour of a 1600 unit and both 1300 and 1600 engines become cross-flow. Super gains GT-type remote gearchange from '67. Still-popular 1600E (1967-70) has GT engine, lowered suspension, fog lamps, luxury trim, black grille. Just made the million.

Engine 1297cc 81x63mm/1499cc 81x72.8mm/1599cc 81x77.6mm, 4 cyl OHV, F/R
Max power 53bhp at 5000rpm to 88bhp at 5400rpm
Max speed 83-102mph
Body styles saloon, estate
Prod years 1966-70
Prod numbers 1,023,837

PRICES
1300/1500/1600 **A** 1000 **B** 600 **C** 300
GT **A** 1600 **B** 1100 **C** 700
1600E **A** 3000 **B** 2000 **C** 1000

Ford Cortina Lotus MkII

Same basic idea as the first Lotus Cortina, but more power from Lotus's twin carb, twin cam engine. Servo-assisted brakes standard, as were lowered suspension package, wide wheel rims and tyres. Black radiator grille, and the painted body flash was now an option (but have you ever seen one without it?). Other options included oil cooler and limited-slip diff. The Lotus badge was actually removed after just seven months and replaced by a Twin Cam logo and plain Ford badges. Just as eligible for historics as the first edition, but not as sought-after.

Engine 1558cc 82.5x72.75mm, 4 cyl DOHC, F/R
Max power 109bhp at 6000rpm
Max speed 107mph
Body styles saloon
Prod years 1967-70
Prod numbers 4032

PRICES
A 6500 **B** 4000 **C** 3000

Ford Escort 1100/1300 Mk1

Another crucial best-selling Ford, which took over the reins from the Anglia. Bland Euro-styling was penned in Britain but this was the first post-war Ford to be built simultaneously in Britain and Germany, as well as in other countries around the world. MacPherson struts up front, rack-and-pinion steering, auto option. Engine choice: 1100 or 1300 in standard Escorts. GT version had a Weber carb and 75bhp, lasted until '73. Sport model had same mechanical spec as GT but Mexico shell and suspension (see below), and 1300E ('73 on) was the luxury version of the GT.

Engine 1098cc 81x53.3mm/1298cc 81x62.3mm, 4 cyl OHV, F/R
Max power 53bhp at 5300rpm/61bhp at 5000rpm/75bhp at 5400rpm
Max speed 80-90mph
Body styles saloon, estate
Prod years 1968-74
Prod numbers 2,228,349

PRICES
1100/1300 **A** 800 **B** 500 **C** 200
1300GT **A** 2000 **B** 1200 **C** 600
1300 Sport/1300E **A** 2500 **B** 2000 **C** 1200

Ford Escort Mk1 Twin Cam/RS1600/Mexico/RS2000

Competition-orientated Escorts flowed from the outset. Twin Cam had Lotus Cortina engine and running gear, servo front discs, strengthened two-door bodyshell, split bumpers, flared arches. Joined by RS1600 from '70, with its 16-valve Cosworth BDA twin cam. Massive tuning potential and most went straight into competition. Mexico arrived later in 1970, differed because of OHV Cortina GT engine without an oil cooler, therefore much slower, but it looked the part and was great value. RS2000 ('73-'75) had OHC 2-litre unit, extra trim. All are hairy road cars.

Engine 1558/1601cc 4 cyl DOHC, 1599cc 4 cyl OHV, 1993cc 4 cyl OHC, F/R **Max power** 109/120/86/100bhp
Max speed 115/116/103/110mph
Body styles saloon **Prod years** 1968-71/1970-74/1970-75/1973-75
Prod numbers 1263/947/9382/n/a

PRICES
Twin Cam/RS1600 **A** 7500 **B** 5000 **C** 3000
Mexico **A** 3500 **B** 2200 **C** 1500
RS2000 **A** 3800 **B** 2500 **C** 1700

Ford (GB)

Ford Escort Mk2

Absolutely safe styling for the new Escort in 1975 and underneath everything remained very much as before. However, there was a new standard engine choice: the 1599cc Kent engine. GT renamed Sport, and top-of-the-range version is now the Cologne-built Ghia (sports wheels, vinyl roof, better trim). Stripped-out Popular version had economy engines and was very cheap. Belgian-made 1600L was a surprising Q-car. Harrier was limited edition 1600 Sport with stripes and spoilers. Incredibly tedious by any standard but just what Joe Public wanted.

Engine 1098cc 81x53.3mm/1298cc 81x62.3mm/1599cc 81x77.6mm, 4 cyl OHV, F/R
Max power 41-48bhp at 5300rpm/57-70bhp at 5500rpm/84bhp at 5500rpm
Max speed 76-83/87-94/101mph
Body styles saloon, estate
Prod years 1975-80
Prod numbers ca 2,000,000

PRICES
A 800 **B** 400 **C** 100

Ford Escort Mk2 RS Mexico/RS1800/RS2000

The RS Mexico was the German-made 1.6-litre OHC replacement for the old Mexico. As such, it was really aimed at the boy racer market rather than possible competition work. UK available from '76. It formed the basis of the British-made RS1800: same shell and front/rear spoilers, but 1835cc Cosworth BDA 16-valve engine, special gearbox and harder suspension. Essentially a rally machine and very few ever made it to the road. To satisfy that need, the plastic-droop-snoot RS2000 arrived in '76 (though made in Germany from March '75): 110bhp 2-litre Pinto, choice of Custom trim upgrade from '78.

Engine 1593cc 87.6x66mm/1835cc 86.8x77.6mm/1993cc 90.8x77mm, 4 cyl OHC/DOHC/OHC, F/R
Max power 95bhp at 6000rpm/115bhp at 6000rpm/110bhp at 5500rpm
Max speed 105/114/112mph
Body styles saloon
Prod years 1975-78/1975-77/1976-80
Prod numbers n/a

PRICES
RS Mexico **A** 3500 **B** 1700 **C** 1000
RS1800 **A** 10,000 **B** 8000 **C** 6000
RS2000 **A** 4000 **B** 2500 **C** 1500

Ford Capri I

'The car you always promised yourself' went the ad line of the time. The Capri, a marketing success inspired by the Mustang, was, in mechanical terms, basically a Cortina on a 101.75in wheelbase. MacPherson strut front suspension, rack-and-pinion steering, front discs, beam rear axle. Base 1300 hardly fitted the style, 1600GT better, V4 and V6 2000/3000GT models were 1970s icons. Automatic option on 1600 and above, 'L', 'X' and 'R' trim pack options early on. From 1972, bonnet bulge, revised suspension, better facia, OHC 1593cc engine. Very fashionable when new, gaining a following now.

Engine 1298/1599cc 4 cyl OHV, 1593cc 4 cyl OHC, 1996cc V4 cyl OHV, 2994cc V6 cyl OHV, F/R
Max power 61/72-88/88/113/128-138 bhp, F/R
Max speed 84-122mph
Body styles coupé
Prod years 1969-74
Prod numbers 1,172,900

PRICES
1300/1600 **A** 1500 **B** 900 **C** 300
2000/3000GT **A** 2000 **B** 1000 **C** 300
3000E **A** 3000 **B** 1500 **C** 500

Ford Capri RS2600/RS3100

RS2600 was the first homologation special Capri, built in Germany. It used a long-stroke version of the Cologne-built 2.5-litre V6 with Kugelfischer fuel injection, good for 150bhp. 6in wide cast alloy wheels, four headlamps, quarter bumpers, later ones with ventilated disc brakes. Very rare in the UK. RS3100 was Halewood-built homologation special, even rarer than RS2600. Overbored 3-litre V6 power, ventilated discs and alloy wheels of RS2600, running gear from 3000GT. Big boot spoiler and black bumpers identify. Road-going survivors are very rare and rather expensive.

Engine 2637cc 90x69mm/3093cc 95.2x72.4mm, V6 cyl OHV, F/R
Max power 150bhp at 5800rpm/148bhp at 5200rpm
Max speed 126/122mph
Body styles coupé
Prod years 1970-74/1973-74
Prod numbers 3532/200

PRICES
RS2600 **A** 6000 **B** 4000 **C** 3000
RS3100 **A** 7500 **B** 5000 **C** 4000

Ford Capri II

A smoothed-out body style and a hatchback gave the Capri II a more modern, if blander, profile. Mechanically left almost untouched: same range of engines, except for replacement of V4 2-litre engine by overhead cam Pinto 2-litre. More room inside. Various trim levels: L, XL, GT and a new Ghia version with alloy wheels, vinyl roof, special interior. Ghias were made in Germany, as indeed all Capris were after October 1976. S pack, first seen in '75, was a GT with sporty suspension, alloys and black bumpers (initially called 'Midnight Capri').

Engine 1298cc 4 cyl OHV, 1593/1993cc 4 cyl OHC, 2994cc V6 cyl OHV, F/R
Max power 61/72-88/98/138bhp
Max speed 89/98-106/108/122mph
Body styles coupé
Prod years 1974-77
Prod numbers 403,612

PRICES
1300/1600 **A** 1000 **B** 500 **C** 100
2000/3000 **A** 1300 **B** 700 **C** 200

Ford (GB) · Ford (USA)

Ford Cortina MkIII

All change for the Cortina: new floorpan with wishbone front and coil sprung rear, plus a curvy new body style with American Ford overtones (this differed slightly from the German-built equivalent, the Taunus – which was also offered in coupé form). Huge range of models, from base 1300 up to GT and GXL versions with their twin headlamps, black grilles, better equipment. Suspension improved in 1973, along with new facia, restyled grilles and new OHC 1600 engine in place of 1599cc Kent unit. Best is 1974 luxury 2000E. Not very sophisticated.

Engine 1298/1599cc 4 cyl OHV, 1593/1993cc 4 cyl OHC, F/R
Max power 57/68/72-88/98bhp
Max speed 83-103mph
Body styles saloon, estate
Prod years 1970-76
Prod numbers 1,126,559

PRICES
A 800 **B** 500 **C** 100

Ford Cortina MkIV/V

Even more dreary development of the Cortina theme. Exactly the same under the skin, but a new squarer body style which was instantly anonymous. Same range of engines, plus, from 1977, a new top-model 2.3-litre V6 which had power steering as standard (built in Germany). Ghia became the upper trim level. MkV facelift of 1979 changed grille and dash. Another new grille in '81 could not hide the fact that the Cortina was well past its best, and it died a year later. This model typified the sad but successful wisdom of Ford's policy of playing it safe.

Engine 1298cc 4 cyl OHV, 1593/1993cc 4 cyl OHC, 2293cc V6 cyl OHV, F/R
Max power 50-57/59-88/98/108bhp
Max speed 82-106mph
Body styles saloon, estate
Prod years 1976-79/1979-82
Prod numbers 1,131,850

PRICES
A 700 **B** 400 **C** 100

Ford Consul/Granada/Granada Coupé

Ford's new pan-European big car was yet another solid sales success. Conventional new floorpan featured all-independent suspension and servo front discs. Entry level was the Consul, available in 2000, 2500 and 3000 forms. Granada models were initially top-of-the-range 2.5 and 3 litres only, but after '75 all types were known as Granadas and the Consul tag was dropped. 2-litre V4 engine replaced by OHC Pinto unit in '74. Attractive two-door coupé from '74 (2000 till '76, 3000 '74-'77). Production devolved entirely to Germany for the last 12 months.

Engine 1996cc V4 cyl OHV, 1993cc 4 cyl OHC, 2494/2994cc V6 cyl OHV F/R
Max power 90/98/120/138bhp
Max speed 92-114mph
Body styles saloon, estate, coupé
Prod years 1972-77
Prod numbers 846,609

PRICES
Saloon/Estate **A** 1200 **B** 800 **C** 200
Coupé **A** 2000 **B** 1500 **C** 900

Ford Fiesta Mk1 '76-'83

A completely new avenue for Ford, built to compete with the VW Polo/Fiat 127. Transverse engine and front-wheel drive were novelties, as was the hatchback. New 'Valencia' base engines of 957cc and 1117cc capacity, but 1300 and 1600 Kent versions were also built. Made in Spain, Germany and Dagenham. Four-speed manual gearbox obligatory (no auto option), MacPherson front suspension and front discs fitted across the board. XR2 version of '81-'83 has 1600 engine, alloy wheels, extended wheelarches but not really the power to match its looks. An international success.

Engine 957cc 74x55.7mm/1117cc 74x65mm/1298cc 81x63mm/1599cc 81x77.6mm, 4 cyl OHV, F/F
Max power 40/53/66/84bhp
Max speed 90-106mph
Body styles hatchback
Prod years 1976-83
Prod numbers 1,750,000

PRICES
Fiesta **A** 700 **B** 400 **C** 100
Fiesta XR2 **A** 1700 **B** 1000 **C** 700

Ford (USA) Fairlane 500 Skyliner

The idea of a convertible which shared the benefits of a hardtop was not new (Peugeot's 402 did it in the 1930s), but Ford's '57 Fairlane 500 Skyliner Retractable was the definitive item. Dozens of relays and electric motors wound the hardtop back and under a huge rear deck. The idea was certainly appealing but the reality rather less so: the system was not what you'd call reliable and Ford lost $20 million on warranty claims. Falling demand sealed its fate. Very collectable.

Engine 3654cc 92x91.4mm, 6 cyl OHV, 4458cc 92.1x83.8mm/4785cc 95.25x 83.8mm/5113cc 99.1x87.4mm/5441cc 101.6x87.3mm/5767cc 101.6x89mm, V8 cyl OHV, F/R
Max power 144bhp at 4200rpm to 300bhp at 4600rpm
Max speed 95-110mph
Body styles convertible hardtop
Prod years 1956-59
Prod numbers 48,394

PRICES
A 15,000 **B** 10,000 **C** 6000

Ford (USA)

Ford Mustang '64-'68

The original 'ponycar', or personal coupe, and the greatest marketing coup of all time: Ford sold one million Mustangs in less than two years, then a record. The brainchild of Lee Iacocca, the original 'Stang was launched as a notchback coupe or convertible on a 108in wheelbase (the pretty 2+2 fastback arrived for the '65 model year). Unspectacular sixes powered the first cars, but big block V8s were soon shoehorned in. Infinitely flexible options: handling packages, PAS, front discs, air conditioning, auto, GT suspension. Long nose and full fastback from '67.

Engine 2788cc/3277cc/4097cc/4261cc, 6cyl OHV, 4727cc/4949cc/6390cc/6997cc/7014cc, V8 cyl OHV, F/R **Max power** 101bhp at 4400rpm to 390bhp at 5600rpm **Max speed** 90-130mph **Body styles** hardtop coupe, fastback coupe, convertible **Prod years** 1964-68 **Prod numbers** 2,204,038
PRICES
Hardtop Coupe **A** 7500 **B** 5000 **C** 2500
Fastback Coupe **A** 8000 **B** 6000 **C** 3500
Convertible **A** 13,000 **B** 9000 **C** 6000

Ford Mustang Shelby GT350/GT500

American folk hero Carroll Shelby got into the Mustang in 1965 with the GT350. Near-complete Fastback shells were shipped to Shelby's works in Los Angeles and fitted with 306bhp 289ci engines with high-lift manifolds, four-barrel carbs and free-flow exhausts. Also stronger axles, GRP bonnet, four-speed auto option. Typically painted white with blue stripes. GT500 of '67 had a 428 engine with an advertised 355bhp (probably more like 400bhp). GT500KR (King of the Road) from '69. A few factory convertibles were made – beware fakes. Highly collectable.

Engine 4724cc 101.6x72.9mm/4949cc 101.6x76.2mm/7014cc 104.9x101.1mm, V8 cyl OHV, F/R **Max power** 306bhp at 6000rpm to 425bhp at 5600rpm **Max speed** 100-137mph **Body styles** coupe, convertible **Prod years** 1965-70/1967-70 **Prod numbers** 7403/7000

PRICES
GT350 **A** 20,000 **B** 15,000 **C** 10,000
GT500 **A** 30,000 **B** 20,000 **C** 13,000

Ford Mustang '69-'73

Major restyle for '69 added 4in to the length, two extra headlamps, bulging rear arches and no side sculptures. The revisions also made it lower and wider, and improved interior space and luggage capacity. Notchback, SportsRoof and convertible styles continued and there was a new luxury Grande model. Recessed tail lamps followed for 1970, plus a return to single headlamps. Grew again in '71: heavier, longer, wider. The wind was slowly going out of the Mustang's sales: sales in 1969 were 50% down to 300,000 but rock bottom was '72's 125,000 units.

Engine 3277cc/4097cc/4261cc, 6 cyl OHV, 4727cc/4949cc/5752cc/6390cc/6997cc 107.4x96mm/ 7030cc V8 cyl OHV, F/R **Max power** 95bhp to 370bhp at 5200rpm **Max speed** 90-118mph **Body styles** hardtop coupe, fastback coupe, convertible **Prod years** 1968-73 **Prod numbers** ca 675,000 (excl Mach 1/Boss)
PRICES
Hardtop Coupe **A** 4000 **B** 2500 **C** 1500
SportsRoof **A** 5000 **B** 3000 **C** 2000
Convertible **A** 10,000 **B** 6500 **C** 3500

Ford Mustang Mach 1/Boss 302/Boss 429/Boss 351

Inspired by Shelby's success with the Mustang, Ford launched its own high performance versions for '69. The Mach 1 was a SportsRoof with a standard 250bhp 351cu in V8, sports suspension, special grille and bonnet scoop. Much more exciting was the Boss 302, inspired by Ford's TransAm racer. Ford claimed a conservative 290bhp and you got a chin spoiler, rear wing, lots of stripes and a louvred rear window. On the Boss 429, you got a Cobra Jet big block alloy-head engine (from '71 optional on Mach 1 too) – blindingly quick (0-60mph in 6 secs). The 1971 Boss 351 was less powerful (330bhp).

Engine 4949cc/5752cc/6390cc/7014cc/7030cc V8 cyl OHV, F/R **Max power** 141bhp at 4000rpm to 375bhp at 5200rpm **Max speed** 100-120mph **Body styles** coupe **Prod years** 1968-73/1968-70/1968-70/1970-71 **Prod numbers** 213,042/8253/1356/ca 4000

PRICES
Mach 1 **A** 8500 **B** 5000 **C** 3000
Boss 302 **A** 18,000 **B** 12,000 **C** 9000
Boss 429 **A** 27,000 **B** 22,000 **C** 16,000

Ford Mustang II

By 1973 the Mustang had become a behemoth, way out of step with the oil crisis. The much scaled-down successor (Iacocca called it his 'little jewel') couldn't have been better timed. 20in shorter, 4in narrower and 1in lower, the Mustang II had unitary construction, with a front subframe for the drive train and front suspension. Notchback and hatch-fastback styles, but no convertible. Only a four and the Cologne 2.8-litre V6 (standard on the Mach 1) to start with (V8s from '75). Cobra II and King Cobra were pale add-on packages. No classic status.

Engine 2301cc 95.9x79.5mm/2798cc 92.9x68.5mm/4945cc 101.6x72mm, 4/V6/V8 cyl OHC/OHV/OHV, F/R **Max power** 83-92bhp at 5000rpm/90-105bhp at 4400rpm/122-139bhp at 3600rpm
Max speed 97/103/109mph
Body styles coupe, hatchback coupe
Prod years 1973-78
Prod numbers 1,107,718

PRICES
A 2700 **B** 1500 **C** 800

Ford Thunderbird '55-'57

Arch-rival Chevrolet's Corvette was the gauntlet that Ford could not refuse to pick up. Although the Thunderbird rode the same wheelbase as the Corvette, it ended up as a more comfortable touring car, not a true sports car. You got a power soft top or removable hard top, big V8 engine and a choice of manual or auto transmissions. And it trounced the Corvette in the marketplace. Exterior spare and hard top portholes from '56, new grille, bumpers and bigger fins for '57. Very rare 'F-Bird' was supercharged 300/340bhp semi-racer (208 built).

Engine 4785cc 95.25x83.8mm/ 5113cc 99.1x87.4mm, V8 cyl OHV, F/R
Max power 193-212bhp at 4600rpm/215-340bhp at 4600rpm
Max speed 110-120/125-145mph
Body styles convertible
Prod years 1955-57
Prod numbers 53,166

PRICES
A 20,000 **B** 12,000 **C** 8000

Ford Thunderbird '58-'60

The T-Bird was enlarged and given four seats for 1958. Construction was now unitary and the styling became almost sci-fi (it has since been dubbed the 'squarebird'). There was the convertible, naturally, but also a fixed hard top featuring very deep 'C' pillars. Only one engine, a big block V8, joined for '59 by the huge 430cu in Lincoln V8. Styling changes for 1959 included bright 'bullets' on the side mouldings and a dummy bonnet scoop. Sliding metal sunroof option on the final '60 hard top, which outsold convertibles 8 to 1. Limited edition 'gold roof' model mildly collectable.

Engine 5769cc 101.6x88.8mm/7049cc 109.2x94mm, V8 cyl OHV, F/R
Max power 300bhp at 4600rpm/350bhp at 4800rpm
Max speed 114/124mph
Body styles coupe, convertible
Prod years 1958-60
Prod numbers 198,191

PRICES
Coupe **A** 11,000 **B** 7500 **C** 4500
Convertible **A** 16,000 **B** 10,000 **C** 7000

Ford Thunderbird '61-'63

New 'cigar shape' for the '61 T-Bird on an unchanged wheelbase. Heavy front bumper treatment, big circular tail lamps, miniature fins. Just one engine to start with: a stroked version of the 352 unit developing 300 or 340bhp. Hardtop coupe and convertible styles as before, both with four seats, joined in '62 by the interesting Sports Roadster, with its glassfibre tonneau to cover the rear seats, and wire wheels: expensive, so rare (1882 built). Also in '62 came the Landau, boasting luxury trim and a vinyl roof with fake 'S' bar.

Engine 6392cc 102.9x96mm, V8 cyl OHV, F/R
Max power 300-340bhp at 4600rpm
Max speed 115-125mph
Body styles coupe, convertible
Prod years 1961-63
Prod numbers 214,375

PRICES
Coupe **A** 9500 **B** 6000 **C** 4000
Convertible **A** 12,000 **B** 7500 **C** 5500

Ford (AUS) Falcon Hardtop

Aussie Fords were basically modified versions of American metal and the 1973 Falcon Hardtop (the fastest member of the Falcon family) was no exception. It was the only down-under Ford which made it to the UK with anything like classic credentials. The so-called XB Falcon was the car which took Ford ahead of Holden in the sales stakes for the very first time. Front discs, all-synchro manual 'box, auto option, safety steering column. Lasted three years – much longer than usual for an Australian Ford – so plenty were made.

Engine 3277cc 93.4x79.4mm/4097cc 93.4x99.2mm, 6 cyl OHV, 4945cc 101.6x76.1mm/5752cc 101.6x88.8mm, V8 cyl OHV, F/R
Max power 130bhp at 4600rpm to 300bhp at 5400rpm
Max speed 96-124mph
Body styles coupé
Prod years 1973-76
Prod numbers 211,971 (all Falcons)

PRICES
A 3500 **B** 2000 **C** 1200

Frazer Nash Competition/Le Mans Replica I/II

As the UK concessionaire for BMW pre-war, Frazer Nash relied on these links for its production sports cars, launched in 1948. Hence BMW 328 parts supplied the mechanical basis for the Competition model (the planned High Speed name was never used). Simple cycle-wing bodywork. Renamed Le Mans Replica after 3rd place at '49 Le Mans. Highly successful road/race/sprint machine. Le Mans Replica II of '52 had 132bhp engine, new parallel tube frame and lower bodywork. Highly sought-after, many 'unofficial' cars.

Engine 1971cc 66x96mm, 6 cyl OHV, F/R
Max power 120-132bhp at 5500rpm
Max speed 118mph
Body styles sports
Prod years 1948-49/1949-52/ 1952-53
Prod numbers 34

PRICES
A 80,000 **B** 65,000 **C** 50,000

Frazer Nash Fast Roadster/Cabriolet/Mille Miglia

Marginally 'touring' sisters of the Le Mans Replica: elegant full-width bodywork, featuring one-piece windscreen and front wing-mounted spare, on similar chassis. 85bhp Bristol engine standard, but up to 125bhp to order. 1950 Mille Miglia and four-seater Cabriolet were intended as high-speed tourers, had 'V' screens which slid down into the scuttle. Cabriolet was only FN to have longer wheelbase (9ft). There were several one-off bodies, and specifications varied widely between individual cars. Extremely rare and impossible to price, but will be expensive.

Engine 1971cc 66x96mm, 6 cyl OHV, F/R
Max power 85bhp at 4500rpm to 125bhp at 5500rpm
Max speed 100-125mph
Body styles sports, convertible
Prod years 1949-50/1949-52/1949-53
Prod numbers 12

PRICES
Impossible to quote, but less than Le Mans Replica

Frazer Nash Targa Florio

An all-new alloy body with a straight-through wing line and one-piece front and rear sections for the new touring Frazer Nash of 1952. Named after a win in the tortuous Alpine race, the only British car ever to do so. Available in Turismo guise or, stripped out and fitted with an engine up to 140bhp, as the Gran Sport. Based on the same new chassis as the Le Mans Replica II, plus usual transverse leaf and wishbone IFS, and torsion bars at the rear. Austin Atlantic engined Roadster of 1952 remained a one-off: the cheap new Austin-Healey saw to that.

Engine 1971cc 66x96mm, 6 cyl OHV, F/R
Max power 100bhp at 5000rpm to 140bhp at 5750rpm
Max speed 110-130mph
Body styles sports
Prod years 1952-54
Prod numbers 14

PRICES
Impossible to quote, but less than Le Mans Replica

Frazer Nash Le Mans Coupé/Sebring/Continental

A rare example of a coupé Frazer Nash arrived in 1953: the Fixed Head Coupé (pictured) became known as the Le Mans after its class win in that year's 24-hour race. New style bodywork with diamond-shape grille having 'Venetian blind' slats to regulate engine temperature. De Dion rear end, knock-on wires, various states of tune. 1954 Sebring shared similar front-end styling but a new sports body suited for club racing. BMW V8 powered Continental was intended to run at Le Mans, and never made it as a production model. The mature bodywork was styled by Peter Kirwan-Taylor.

Engine 1971cc 66x96mm/3168cc 82x75mm, 6/V8 cyl OHV, F/R
Max power 100-140bhp at 5750rpm/160bhp at 5600rpm
Max speed 110-135/140mph
Body styles coupé/sports/coupé
Prod years 1953-56/1954/1956-57
Prod numbers 9/3/2

PRICES
Probably in the range £20-50,000

Fuldamobil/Nobel

Production of Fuldamobils began in Germany as early as 1950, and these were pretty primitive affairs (the first ones had aluminium sheets nailed to a wood frame). The most familiar one is the S-7, the first glassfibre Fulda. This is the one that was licensed to York Nobel Industries in Britain, where it was made in three- and four-wheeled forms. Fichtel & Sachs single-cylinder engine (post-'65 Fuldas had 198cc Heinkels), GRP-and-plywood body on steel tube chassis, mechanical brakes, four-speed 'box with electrically-operated reverse. Also sold in kit form.

Engine 191cc 65x58mm/198cc 64x61.5, 1 cyl TS/OHV, R/R
Max power 10bhp at 5200rpm/10bhp at 5500rpm
Max speed 40mph
Body styles coupé
Prod years 1957-69/1959-62
Prod numbers c 700/1000

PRICES
A 1900 **B** 1000 **C** 700

Ghia L.6.4

Ghia's first attempt to sell a car under its own name was the Italo-American Dual-Ghia Firebomb, of which 104 were made 1956-58. The follow-up model, the 1960 L.6.4, was made entirely by Ghia in Italy and was based on the Virgil Exner-styled Dart. A Chrysler V8 engine and Chrysler suspension sat under the hand-built coupé body. America was naturally the target market and Frank Sinatra bought the first one. But the world was woefully devoid of Sinatras prepared to pay the $15,000 tag.

Engine 6279cc 108x85.9mm, V8 cyl OHV, F/R
Max power 335bhp at 4600rpm
Max speed 140mph
Body styles coupé
Prod years 1960-62
Prod numbers 26

PRICES
Too rare to quote, but probably in the region of £20,000

Ghia 450SS

A Los Angeles man called Bert Sugarman saw a Ghia-bodied Fiat 2300S on the cover of a magazine and decided that Ghia was the right company to build his idea of a great sports car. Ghia actually said yes (Mr Sugarman was either very persuasive or very rich) and proceeded to build the 450SS from 1965. Essentially this was nothing more than a rebodied Plymouth Barracuda. However, the body was stunningly beautiful, having exceptionally clean lines with the soft top folded (there was also a coupé version). But at four times the price of the 'Cuda, it's not surprising that demand was limited.

Engine 4490cc 92.2x84.1mm, V8 cyl OHV, F/R
Max power 235bhp at 5500rpm
Max speed 125mph
Body styles coupé, convertible
Prod years 1965-67
Prod numbers c 12

PRICES
Difficult to quote, but probably in the range £10-25,000

Ghia 1500 GT

Flushed with success over the Fiat 2300 Coupé, for which they built the bodies, the Italian *carrozzeria* Ghia gambled that there was a market for a smaller coupé with Fiat parts and built the 1500 GT from 1962. This was a highly distinctive two-seater coupé with a long bonnet and all mechanical items taken from the Fiat 1500. These were largely unmodified, so performance and roadholding were not top-flight, but it did make the GT a practical car to own. Ghia OSI built them at the amazing rate of five per day. All but unknown in the UK.

Engine 1481cc 77x79.5mm, 4 cyl OHV, F/R
Max power 84bhp at 5200rpm
Max speed 105mph
Body styles coupé
Prod years 1962-67
Prod numbers c 925

PRICES
A 10,000 B 7000 C 5000

Gilbern GT

Llantwit Major may not have the ring of a Maranello or Detroit, but for a while it was the car manufacturing capital of Wales. The Gilbern GT was a rather plain 2+2 coupé with a tubular chassis and glassfibre body. Front suspension derived from the Austin A35, rear axle was BMC and the gearbox came from MG. Engines ranged from the 948cc Sprite (optionally supercharged), to the 1098cc Coventry-Climax and 1588/1622cc MGA (from '63, the 1798cc MGB). Good value in its day (sold in kit form) and high quality.

Engine 948cc 62.9x76.2mm/1098cc 72.4x66.6mm/1588cc 75.4x88.9mm/1622cc 76.2x88.9mm/1798cc 80.3x88.9mm, 4 cyl OHV, F/R
Max power 42-68bhp at 5500rpm/84bhp at 6900rpm/80bhp at 5600rpm/90bhp at 5500rpm/95bhp at 5400rpm
Max speed 85-107mph **Body styles** coupé **Prod years** 1959-67
Prod numbers c 400

PRICES
A 4000 B 2500 C 1600

Gilbern Genie/Invader

Smarter clothes for a strengthened version of the existing GT chassis made the Genie a good-looking grand tourer, moving Gilbern firmly up-market. Engines were V6 units from Ford (2.5 or 3.0 litres, some with fuel injection), and there was rather too much power for the chassis. The 1969 Invader remedied this fault, featuring a stiffer chassis; Invaders had flared arches and wider grilles. MkII model of '71 added an estate car body (offered 1971-72, 105 built), MkIII of '72 was fully-built only and had Cortina front suspension and rear axle. Fast (0-60 in 7.2 secs).

Engine 2495cc 93.7x60.3mm/2994cc 93.7x72.4, V6 cyl OHV, F/R
Max power 112bhp at 4750rpm/140-158bhp at 4750rpm
Max speed 110-125mph
Body styles coupé, estate
Prod years 1966-69/1969-76
Prod numbers 197/c 600

PRICES
Genie A 5000 B 3500 C 2000
Invader A 6000 B 4000 C 2500

Ginetta G4

Classic Ginetta, and the first volume-selling car the four Walklett brothers built. An attractive glassfibre body sat atop a tubular space frame, initially with front coil springs and live rear axle, optional front discs. Power came from the Ford 105E (997cc) and later the 1498cc Cortina. Short-tail Series II from '63 swapped Ford axle for BMC. Series III much revised: new chassis, Herald front wishbones, pop-up headlamps. Revived in 1981 and still made today as modernised G27. Like most Ginettas, the majority were sold in kit form.

Engine 997cc 81x48.4mm/1498cc 81x72.7mm, 4 cyl OHV, F/R
Max power 39bhp at 5000rpm to 85bhp at 6000rpm
Max speed 90-122mph
Body styles sports, coupé
Prod years 1961-69
Prod numbers c 500

PRICES
A 15,000 B 10,000 C 6000

Ginetta G10/G11

With an eye on the American market, Ginetta developed a TVR Griffith style sports car to be powered by Ford's 4.7-litre V8 engine. There was a new space frame chassis clothed with a smooth glassfibre body which incorporated MGB doors and, on the convertible, an MGB screen. Independent front suspension by coil springs and double wishbones, four-speed Ford 'box. Not eligible for SCCA racing in the USA, so few sold. G11 was G10 shape with MGB engine, transmission and rear axle. BMC parts supply difficulties killed the project. Only four G10 coupés, two G11 coupés.

Engine 4727cc 101.6x72.9mm/1798cc 80.3x88.9mm, V8/4 cyl OHV, F/R
Max power 275bhp at 6000rpm/95bhp at 5400rpm
Max speed 150/120mph
Body styles coupé, convertible
Prod years 1964-65/1966
Prod numbers 6/12

PRICES
Impossible to quote accurately, but likely to be similar to G4

Ginetta G15

The G15 took Ginetta to a new level of importance in the specialist car industry. Its formula of low price, small dimensions and keen road manners struck a chord with British buyers. Glassfibre body was bonded to a square tube ladder frame, front suspension/disc brakes were Herald-derived and the trailing rear suspension, engine and transaxle came as one from the Sunbeam Imp Sport. Almost straight away, Series II introduced with front-mounted radiator, Series III ('73) had improved trim, alloy wheels and larger rear quarter windows.

Engine 875cc 68x60.3mm, 4 cyl OHC, R/R
Max power 55bhp at 6100rpm
Max speed 95mph
Body styles coupé
Prod years 1967-73
Prod numbers 796

PRICES
A 6000 **B** 4000 **C** 2300

Ginetta G21

Increasing maturity of design was evident in the G21, although it took a long time to get it into production. A new backbone/tubular steel chassis employed coil spring and wishbone front suspension and a subframe for the suspension and final drive at the back. Conceived to use Ford 3-litre V6 power, but in fact most were fitted with the 1725cc four-cylinder engine from the Sunbeam Rapier, also available in 95bhp Holbay tune. Only two seats and fixed head coupé style with no opening boot lid, so not a practical MGB rival. Most were sold fully-built.

Engine 1725cc 81.5x82.5mm/2994cc 93.7x72.4mm, 4/V6 cyl OHV, F/R
Max power 79-95bhp at 5200rpm/128bhp at 4750rpm
Max speed 112-119/128mph
Body styles coupé
Prod years 1971-78
Prod numbers 170

PRICES
A 7000 **B** 4000 **C** 2500

Glas Goggomobil

Hans Glas's first attempt at motor car manufacture in 1955 was an unqualified success. The Goggomobil (named after his grandson, nicknamed 'Goggo') was a credible means of popular transport which significantly undercut the VW Beetle's price. Only 9ft 6in (290cm) long but capable of seating four adults. Original 250cc model was joined by T300 and T400 versions, which were quite quick. Pretty 2+2 TS coupé models (photo) followed in '56. Last ones badged as BMWs, still selling at a rate of 100 per month! Plenty still around.

Engine 247cc 53x56mm/296cc 58x56mm/392cc 67x56mm, 2 cyl TS, R/R **Max power** 14bhp at 5400rpm/15bhp at 5000rpm/20bhp at 5000rpm **Max speed** 47-62mph
Body styles saloon, coupé
Prod years 1955-69
Prod numbers Saloon: 214,198/Coupé: 66,511

PRICES
Saloon **A** 2000 **B** 1100 **C** 600
Coupé **A** 3000 **B** 1600 **C** 1000

Glas Isar T600/T700

Larger brothers of the Goggomobil (11ft 3in/343cm long), fitted with new air-cooled flat-twin four-stroke engines in 600 and 700 sizes. Sited in front and driving the rear wheels, the power units provided more power but were disappointingly unreliable. Claimed to seat five adults. Only T700 sold in Britain, as the Royal (saloon) and Esquire (estate), but never caught on here. Even now, they have virtually no following and are rare. Marketed as economy cars with the 'big car look'. Pretty S-35 Coupé version of 1959 was stillborn.

Engine 584cc 72x72mm/688cc 78x72mm, 2 cyl OHV, F/R
Max power 20bhp at 5000rpm/30bhp at 4900rpm
Max speed 60/70mph
Body styles saloon, estate
Prod years 1958-65
Prod numbers 87,575

PRICES
A 1800 **B** 1000 **C** 500

Glas 1004/1204/1304

Buoyed up with funds from the best-selling Goggomobil, Glas moved up-market in 1961 with the 1004, a four-seater coupé with styling by Frua of Italy. Its most interesting feature was its engine: a four-cylinder 993cc unit, it was the world's first engine to use a toothed belt for the camshaft in place of a chain. In TS guise, this developed no less than 64bhp. The range expanded to encompass a two-door saloon, three-door estate and convertible, while engines expanded to 1.2 litres (1204) and 1.3 litres (1304). Again TS versions were available.

Engine 993cc 72x61mm/1189cc 72x73mm/1289cc 75x73mm, 4 cyl OHC, F/R **Max power** 42bhp at 5000rpm to 85bhp at 5800rpm **Max speed** 80-105mph **Body styles** saloon, estate, coupé, convertible **Prod years** 1961-67/1963-65/1965-68 **Prod numbers** 9346/16,902/12,259
PRICES
Saloon/Kombi **A** 2500 **B** 1500 **C** 800
Coupé **A** 3500 **B** 2000 **C** 1200
Cabriolet **A** 6000 **B** 4000 **C** 2500

Glas 1300GT/1700GT/1600GT

Frua's hand was behind the attractive shape of the Glas 1300GT, first presented at the 1963 Frankfurt show. At first the engine was a 1.3-litre unit (soon to be used in the 1304 TS), but from 1965 there was also 1700 version which provided useful performance. Convertible was rare (364 built). Handling was a little suspect but *Road & Track* said the GT was 'one of the most attractive small GT cars we've driven'. After BMW's 1966 take-over, the GT was re-engineered for BMW's 1600TI engine, semi-trailing rear and BMW kidney grille, and was sold as the BMW 1600GT.

Engine 1289cc 75x73mm/1682cc 78x88mm/1573cc 84x71mm, 4 cyl OHC, F/R **Max power** 85bhp at 5800rpm/100bhp at 5500rpm/105bhp at 6000rpm **Max speed** 106/110/115mph **Body styles** coupé, convertible **Prod years** 1963-67/1965-67/1967-68 **Prod numbers** 3760/1802/1002
PRICES
Coupé **A** 7500 **B** 5000 **C** 3000
Cabriolet **A** 11,000 **B** 7500 **C** 5000

Glas 1700

Presented at the same time as the 1300GT was another Frua-styled Glas, a four-door saloon which competed with the BMW, which was a very bold move for the company to make. The show car had a 1.5-litre engine but the saloon entered production with a 1.7-litre unit. Again, the engine was the car's strong point: in basic tune it developed 80bhp and in TS form 100bhp, enough to get the five-seater from rest to 60mph in 11.5 secs – hot stuff for 1964. But it never sold as well as Glas hoped and was one of the major reasons for the company's collapse.

Engine 1682cc 78x88mm, 4 cyl OHC, F/R **Max power** 80bhp at 4800rpm/100bhp at 5500rpm **Max speed** 96/106mph **Body styles** saloon **Prod years** 1964-67 **Prod numbers** 13,789

PRICES
A 2500 **B** 1500 **C** 600

Glas 2600 V8/3000

The crowning glory of Glas's over-ambitious expansion turned out to be the final nail in its coffin. The spectacular 2600 V8 was based on the 1700 floorpan and was powered by an intriguing engine – a DOHC V8 consisting of two 1300GT fours joined together and sharing a single crank. De Dion rear end suspended by semi-elliptic springs and Boge automatic self-levelling. Semi-fastback styling by Frua led to it being nicknamed the 'Glaserati'. BMW continued making it after '66 with a stroked V8 as the BMW-Glas 3000, and it kept going until BMW's own 2800CS arrived.

Engine 2580cc 75x73mm/2982 78x78mm, V8 cyl DOHC, F/R **Max power** 140bhp at 5600rpm/160bhp at 5100rpm **Max speed** 121/125mph **Body styles** coupé **Prod years** 1965-67/1967-68 **Prod numbers** 300/71

PRICES
A 15,000 **B** 10,000 **C** 6000

Goliath GP700/GP900

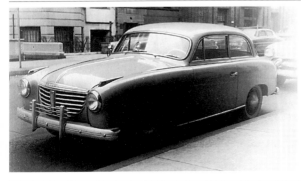

Goliath was the post-war junior relation of Borgward. The first car following the war was a modern-looking saloon, which unfortunately hid an antiquated tubular chassis and power train. The engine was a two-stroke twin, mounted transversely and driving the front wheels (who says the Mini was a pioneer?). Reasonable power output and performance offset by rough nature and tiresome petroil mixing. Early examples were carburetted, later ones had Bosch fuel injection – a real technical novelty for the time. GP900 from '55.

Engine 688cc 74x80mm/886cc 84x80mm, 2 cyl TS, F/F **Max power** 24-40bhp at 4000rpm **Max speed** 59-72mph **Body styles** saloon, estate, convertible **Prod years** 1950-57 **Prod numbers** 36,270

PRICES
Saloon **A** 2500 **B** 1300 **C** 800
Cabriolet **A** 5000 **B** 3500 **C** 2200

Goliath GP700 Sport

A sports car with a fuel-injected two-stroke engine and front-wheel drive? It could only be the Goliath GP700 Sport, although fuel injection was strictly optional (and rather unreliable). The Sport looked pretty racy for 1951 Germany. The cars were built for Goliath by coachbuilders Rometsch of Berlin, although a few appear to have been made in Munich by a firm called Rupflin. Lasted only a couple of seasons, so very few were made. That's why Goliath enthusiasts (and they do exist) lust after the Sport in a manner which embarrasses its true ability.

Engine 845cc 82x80mm, 2 cyl TS, F/F
Max power 32bhp at 4000rpm
Max speed 80mph
Body styles coupé
Prod years 1951-53
Prod numbers 26

PRICES
A 14,000 B 9000 C 6000

Goliath 1100

A water-cooled four-stroke flat-four engine spearheaded Goliath's 1957 change of direction. The wet-liner alloy head power unit still drove the front wheels but the column-change four-speed 'box was now all-synchro. Fuel injection was abandoned and the front suspension uprated from leaf springs to coils. The model was renamed Hansa after 1958, and was given little tail fins to celebrate. The so-called Coupé version was little more than a saloon with a shorter roof, but it offered an alternative to the VW Karmann-Ghia – and it was faster, too.

Engine 1094cc 74x64mm, 4 cyl OHV, F/F
Max power 40bhp at 4250rpm to 55bhp at 5000rpm
Max speed 78-86mph
Body styles saloon, estate, coupé, convertible
Prod years 1957-61
Prod numbers 42,659
PRICES
Saloon/Estate A 2500 B 1500 C 1000
Coupé/Cabriolet A 5500 B 4000 C 2500

Gordon-Keeble GK1/IT

John Gordon's project for an American engined four-seat GT car debuted at Geneva in 1960 as the Gordon GT. The styling was Giugiaro's first ever, as a 21-year old trainee at Bertone. However, production did not begin until 1964, renamed Gordon-Keeble GK1 (some also called IT, for International Touring). Chevy V8 engine, space frame, De Dion rear axle, discs all round, glassfibre bodywork. Thoroughly competent in virtually every area, but priced too low and company went bust. 1968 revival by an American company failed. Practical, unusual, and delightful.

Engine 5355cc 101.6x82.5mm, V8 cyl OHV, F/R
Max power 300bhp at 5000rpm
Max speed 135mph
Body styles coupé
Prod years 1964-66
Prod numbers 99

PRICES
A 21,000 B 16,000 C 12,000

GSM Delta

Sports car of South African origin, created by Bob van Niekirk and Vester de Witt, and built from 1958. UK manufacturing operation began in 1960, but lasted less than a year. Steel tube ladder chassis with tuned Ford Anglia power (optional Coventry-Climax for racing), subframes carrying transverse leaf spring front suspension and 100E rear axle with coil springs. Bodywork: open or coupé, the latter having a reverse-angle rear screen. South African production continued until 1964, where an improved version was marketed as the Flamingo.

Engine 997cc 81x48.4mm, 4 cyl OHV, F/R
Max power 57-90bhp at 7000rpm
Max speed 100-115mph
Body styles sports, coupé
Prod years 1960-61 in UK
Prod numbers 35 in UK

PRICES
A 5000 B 3000 C 1500

Gutbrod Superior

This German garden tool manufacturer built Standard Superiors from 1933-35. After the war, Gutbrod's own Superior arrived. This was a boxy two-seater roll-top coupé, based on a backbone chassis with swing axle rear suspension and advanced double wishbone/coil spring front end. The engine was a two-cylinder two-stroke unit which drove the front wheels. Also sold as a full convertible and an estate. Larger 663cc engine option from '51, plus very rare Wendler-built Sport Roadster. High quality for a small car.

Engine 593cc 71x75mm/663cc 75x75mm, 2 cyl TS, F/F
Max power 20bhp at 4000rpm/26-30bhp at 4300rpm
Max speed 60-71mph
Body styles saloon, estate, coupé, convertible, sports
Prod years 1950-54
Prod numbers 7726
PRICES
Saloon/Estate A 2000 B 1200 C 500
Convertible A 3000 B 2000 C 1200

Healey 2.4-Litre Westland/Elliot/Sportsmobile

There was no faster British four-seater just after the war than the Healey. Stiff box-section chassis, coils all round (with trailing arms at the front), hydraulic brakes. Streamlined alloy-over-wood bodies and only a ton in weight, so the high-camshaft Riley engine gave it a 105mph top speed. Westland (pictured) was the roadster, Elliot the saloon. Sportsmobile from '48 was an unpopular slab-sided drophead coupé with 104bhp. Some special bodies by Duncan and even Bertone. Rare and highly desirable in any form.

Engine 2443cc 80.5x120mm, 4 cyl OHV, F/R
Max power 90-100bhp at 4200rpm/104bhp at 4500rpm
Max speed 105-110mph
Body styles sports/saloon/drophead coupé
Prod years 1946-50/1946-50/1948-50
Prod numbers 64/101/23

PRICES
Westland **A** 24,000 **B** 18,000 **C** 12,000
Elliot **A** 15,000 **B** 11,000 **C** 8000

Healey 2.4-Litre Tickford/Abbott

1950's replacements for the Westland and Elliot were longer, heavier and better equipped, although they looked very like the old models. Same Riley-engined chassis (now with Girling brakes), but performance was blunted by the extra weight, despite lower gearing to compensate. More refined, however, and you got a proper boot. Tickford is the saloon, Abbott the drophead coupé. Attractive touring cars, forgotten after Nash- and Austin-Healeys. More common than the earlier types and less favoured by collectors. Even so, prices are high.

Engine 2443cc 80.5x120mm, 4 cyl OHV, F/R
Max power 106bhp at 4800rpm
Max speed 102mph
Body styles saloon/drophead coupé
Prod years 1950-54
Prod numbers 224/77

PRICES
Tickford **A** 14,000 **B** 9000 **C** 6000
Abbott **A** 22,000 **B** 16,000 **C** 10,000

Healey Silverstone

Pared-down cycle wing bodywork on a virtually unmodified chassis made this sporting Healey more of a lightweight: just over 2000lb (940kg). Front anti-roll bar, stiffer springing. Unusual features included narrow-set headlamps behind grille, spare wheel which doubled up as a rear bumper and a windscreen which retracted into the scuttle. Alloy bodywork by Abbey Panels. Well-priced new and very popular as a club racing and sprint machine which you could drive to each meeting. Superseded by Nash-Healey in 1951. Prices very high, and fakes are not unknown.

Engine 2443cc 80.5x120mm, 4 cyl OHV, F/R
Max power 106bhp at 4800rpm
Max speed 112mph
Body styles sports
Prod years 1949-51
Prod numbers 105

PRICES
A 40,000 **B** 25,000 **C** 16,000

Healey G-Type

A home market version of the Nash-Healey (see page 161), this model was officially known as the 3-Litre Sports Convertible, and commonly referred to as the Alvis-Healey or G-Type. Twin carb Alvis TB21 engine meant no bonnet bulge was necessary and the grille was smaller and neater than the Nash. Full-width body was rather heavy (2800lb/1270kg), so it always felt underpowered. Never really stood a chance against the cheaper, faster, prettier XK120. British tastes were satisfied by a walnut dash and superior upholstery. Rarity lends some enchantment.

Engine 2993cc 84x90mm, 6 cyl OHV, F/R
Max power 106bhp at 4200rpm
Max speed 99mph
Body styles convertible
Prod years 1951-54
Prod numbers 25

PRICES
A 28,000 **B** 20,000 **C** 16,000

Heinkel

Prof Ernst Heinkel's bubble car came out one year after BMW's Isetta. If that seems plagiaristic, at least the Heinkel represented an advance on the Isetta: it was better-looking, had seats for 2+2, and was lighter. Unitary construction, hydraulic brakes (front end only), single door at the front without the Isetta's folding steering column. Mostly three wheels, but some narrow-track four-wheelers were made. Irish production started when German production ceased (1958), made in UK by Trojan from 1961.

Engine 174cc 59x61mm/198cc 64x61.5mm/204cc 65x61.5mm, 1 cyl OHV, R/R
Max power 9-10bhp at 5500rpm
Max speed 52mph
Body styles saloon
Prod years 1956-65
Prod numbers c 29,000

PRICES
A 3700 **B** 2500 **C** 1500

Hillman Minx Phase I/II

When it was first seen in 1939, the half-unitary construction of the Minx was quite revolutionary. A large number were built for service use during the war, but private sales restarted in 1945. This was one of the first British cars with an alligator bonnet (the grille and sides rose as well). Side-valve engine, semi-elliptic springs, floor change, Bendix brakes. Two-door drophead coupé from '46, plus Commer van based estate. Phase II had integral headlamps, column change, hydraulic brakes, but was as dull as ever.

Engine 1185cc 63x95mm, 4 cyl SV, F/R
Max power 35bhp at 4100rpm
Max speed 65mph
Body styles saloon, estate, drophead coupé
Prod years 1939-47/1947-48
Prod numbers n/a

PRICES
Saloon **A** 2200 **B** 1500 **C** 700
DHC **A** 4500 **B** 3000 **C** 1500

Hillman Minx Phase III/IV/V/VI/VII/VIII/VIIIA

Full-width four-light body of the new '48 Minx featured full unibody construction. Coil-and-wishbone IFS but otherwise mechanically very similar to previous models. Bench front seat on saloons, split on drophead coupés (which now had rear quarterlights). 1949 Phase IV had larger engine and separate sidelamps. 1951 Phase V added extra brightwork. 1953 'Anniversary' Phase VI had new cylinder head, better interior, plus new Californian hardtop coupé model. '53 VII had restyled rear, '54 VIII had new OHV option, '55 VIIIA 'Gay Look' two-tone paint.

Engine 1185cc 63x95mm/1265cc 65x95mm/1390cc 76.2x76.2mm, 4 cyl SV/SV/OHV, F/R **Max power** 35bhp/37.5bhp/43bhp **Max speed** 65/70/75mph **Body styles** saloon, estate, drophead coupé, coupé
Prod years 1948-57
Prod numbers 378,705 (28,619/90,832/59,777/44,643/60,711/94,123)
PRICES
Saloon/Estate **A** 2000 **B** 1300 **C** 600
DHC **A** 4000 **B** 2700 **C** 1200
Californian **A** 3500 **B** 2400 **C** 1200

Hillman Husky

While the 1946-57 Commer Minx estate was basically a van with rear seats, the Husky was sold as an estate car with Hillman badges, and as a van under the Commer Cob name. However, it was ultra-basic, and its launch price of £599 undercut the '54 Minx estate by a large margin. The Husky was arrived at by sawing 9in out of a van's wheelbase, shortening overall length by 18.5in to just over 12ft. Sidevalve engine only, floor gearchange, fold-down rear bench, single rear door. Low gearing so fairly lively, but zero refinement and handling.

Engine 1265cc 65x95mm, 4 cyl SV, F/R
Max power 35-37.5bhp at 4200rpm
Max speed 65mph
Body styles estate
Prod years 1954-57
Prod numbers 41,898

PRICES
A 1700 **B** 1000 **C** 400

Hillman Minx Series I/II/III/IIIA-C

The all-new Minx was launched in 1956, essentially a four-door version of the Sunbeam Rapier announced one year previously. Larger dimensions than the old series, more space, more weight, and an extra 9bhp for 80mph-plus top speed. Special has floor change, separate front seats, De Luxe has column change and front bench. Estate from '57, SII of same year has shallow grille, auto clutch option, SIII of '58 gets 1494cc engine. SIIIA ('59) has fins, another new grille, bigger brakes and auto option. SIIIB ('60) has hypoid rear axle, SIIIC gains 1.6 engine.

Engine 1390cc 76.2x76.2mm/1494cc 79x76.2mm/1592cc 81.5x76.2mm, 4 cyl OHV, F/R **Max power** 51bhp at 4600rpm/53-57bhp at 4600rpm/57bhp at 4100rpm **Max speed** 80-84mph
Body styles saloon, estate, convertible **Prod years** 1956-57/1957-58/1958-59/1959-63
Prod numbers c 500,000

PRICES
Saloon/Estate **A** 1400 **B** 900 **C** 500
Convertible **A** 4000 **B** 2500 **C** 1200

Hillman Husky I/II/III

Replacement for the dreadfully basic '54 Husky. Based on the new '56 Minx, but 8in shorter in the wheelbase for an overall length of 12ft 6in (381cm), so still compact. Series II (1960) has lower roofline, larger front and rear screens, close ratio gearbox. Hypoid rear axle from September 1960. '63 Series III (pictured) has new grille incorporating sidelamps/indicators, lower bonnet, redesigned facia. All-synchro 'box and front anti-roll bar from late '64. Still the budget workhorse model of the range, basically not very pleasant but certainly dependable.

Engine 1390cc 76.2x76.2mm, 4 cyl OHV, F/R
Max power 51bhp at 4600rpm
Max speed 80mph
Body styles estate
Prod years 1958-60/1960-63/1963-65
Prod numbers 56,000/n/a/n/a

PRICES
A 1200 **B** 700 **C** 300

Hillman Minx Series V/VI

The quite different appearance of the '63 Minx belies the fact that very little had actually changed. Reshaped and now almost negligible rear fins, squared-off rear roof line without wrap-around screen, bigger windscreen, new grille, revised facia. No estates or convertibles. Underneath, you now benefited from front disc brakes and Borg Warner became the automatic gearbox supplier instead of Smiths. All-synchro gearbox from '64. Series VI of 1965 had five-bearing 1725cc engine, another new facia and self-adjusting rear brakes. Nobody loves them today.

Engine 1592cc 81.5x76.2mm/1725cc 81.5x82.5mm, 4 cyl OHV, F/R
Max power 57bhp at 4100rpm/59bhp at 4200rpm
Max speed 85mph
Body styles saloon
Prod years 1963-65/1967-67
Prod numbers n/a

PRICES
A 900 **B** 600 **C** 300

Hillman Super Minx

Minx mechanicals in new clothes: 5in longer wheelbase, central gear lever, Smiths Easidrive auto option. Trendy styling features included hooded headlamps, tail fins and wrap-around front and rear screens. Estate and attractive convertible versions from '62, though drop-top dies in '64. Technical updates as ordinary Minxes: Series II (1962) has front discs, Borg Warner auto option, individual front seats. Series III (1964) has sharper roofline, larger screens front and rear, wood veneer dash. Series IV (1965) has new 1725cc engine, optional overdrive. Singer and Humber versions also made.

Engine 1592cc 81.5x76.2mm/1725cc 81.5x82.5mm, 4 cyl OHV, F/R
Max power 58-62bhp at 4400rpm/65bhp at 4800rpm
Max speed 80-88mph
Body styles saloon, estate, convertible
Prod years 1961-67
Prod numbers c 135,000

PRICES
Saloon/Estate **A** 1100 **B** 700 **C** 400
Convertible **A** 3500 **B** 2200 **C** 1000

Hillman Imp/Husky

Rootes' belated response to the Mini was extraordinary in so many ways. Rear-engine format followed continental small car practice, but the all-alloy Coventry-Climax engine was sweet and could be very powerful. Sadly it was not 100% reliable, and there were transaxle problems too. Rear window hinged up for access to foldable rear seats. Four-wheel independent suspension. Super Imp from '65 added a few goodies and mock grille. Californian from '67 had pretty coupé roof-line and no hatch. '67 Husky was van-derived estate with 39bhp engine. Singer/Sunbeam derivatives more sought-after.

Engine 875cc 68x60.4mm, 4 cyl OHC, R/R
Max power 37bhp at 4800rpm/39bhp at 5000rpm
Max speed 81mph
Body styles saloon, coupé, estate
Prod years 1963-76/1967-70
Prod numbers 440,032

PRICES
Imp/Husky **A** 1200 **B** 750 **C** 300
Californian **A** 1600 **B** 1200 **C** 500

Hillman Hunter/Minx

For 1966, the new Hunter looked bang up-to-date, thanks to styling input from William Towns. Ultimately, however, it was a rather boring device, lacking dynamic flair. MacPherson strut front suspension, front discs, overdrive and automatic options. Standard engine was the 1725cc and, from January '67, the Minx came with the 1496cc engine (or 1725cc if automatic was specified). MkII of '67 was mild facelift. Servo brakes standard from '68. Minx name died in 1970, as 88bhp Hunter GT was launched. Even better was the '72 GLS with 93bhp, and Rostyle wheels.

Engine 1496cc 81.5x71.6mm/1725cc 81.5x82.5mm, 4 cyl OHV, F/R
Max power 60bhp at 4800rpm/61bhp at 4700rpm to 93bhp at 5200rpm
Max speed 86-110mph
Body styles saloon, estate
Prod years 1966-77/1967-70
Prod numbers c 470,000

PRICES
Minx/Hunter **A** 1000 **B** 500 **C** 200
Hunter GT/GLS **A** 1700 **B** 1200 **C** 800

Hillman Avenger

Under Chrysler ownership, this all-new saloon arrived in 1970, an early example of a 'world car'. In character, worthy but dull. New range of engines, MacPherson struts at front, beam axle at rear, auto option on 1500 engine. Twin carb GT version adds front discs, stiff suspension, better instrumentation. 1972 GLS shares GT engine, plus sports wheels, vinyl roof, servo brakes. Later two-door GT has three-quarter vinyl roof. 1250 and 1500 engines uprated to 1300 and 1600 in 1973. Badged Chrysler from '76 and Talbot from '79, also known as a Plymouth, Dodge and even VW in other countries. Almost no following.

Engine 1248cc 78.6x64.3mm/1295cc 78.6x66.7mm/1498cc 86.1x64.3mm/1598cc 87.3x66.7mm, 4 cyl OHV, F/R
Max power 53bhp at 5000rpm/57bhp at 5000rpm/63bhp at 5000rpm to 78bhp at 5600rpm/69bhp at 5000rpm to 81bhp at 5500rpm
Max speed 84-102mph
Body styles saloon, estate
Prod years 1970-81
Prod numbers 826,353

PRICES
A 800 **B** 400 **C** 200

Hillman Avenger Tiger

If you want an Avenger, make it a Tiger. This was Hillman's limited production Avenger intended for race/rally use as well as the road. Standard spec included the GT's twin carb engine, power bulge on the bonnet, rear aerofoil, alloy wheels and special decal treatment. More power was available for competition use. Tiger I lasted only five months (May-Oct 1972), so quite rare. Tiger II was around for only slightly longer, and is distinguishable by its twin headlamps and lack of bonnet bulge. With the right mods, could be made to exceed 110mph.

Engine 1498cc 86.1x64.3mm, 4 cyl OHV, F/R
Max power 78bhp at 5600rpm
Max speed 100mph
Body styles saloon
Prod years 1972-73
Prod numbers n/a

PRICES
A 3000 **B** 2000 **C** 1000

Hino Contessa I/II

Hino could have become a Nissan or Toyota, but it was absorbed by Toyota in 1966 and killed off in 1971. After assembling Renault 4CVs, it built the Contessa from 1961 using Renault-inspired mechanicals. Highly attractive coupé version of 1962 was styled by Michelotti and had an extra 10 horses (45bhp). The same design house redesigned the Contessa range in 1964 and the engine was expanded to 1251cc. The coupé version was not as pretty as the first edition, nor as collectable.

Engine 893cc 60x79mm/1251cc 71x79mm, 4 cyl OHV, R/R **Max power** 35-45bhp at 5000rpm/55-65bhp at 5000rpm **Max speed** 72-90mph
Body styles saloon, coupé
Prod years 1961-64/1964-71
Prod numbers n/a
PRICES
Contessa I **A** 4500 **B** 2800 **C** 1400
Contessa Coupé I **A** 8000 **B** 6000 **C** 4000
Contessa Coupé II **A** 3000 **B** 2000 **C** 1000

Honda S500/S600/S800

Motorbike maker Honda did not produce any cars until 1963, when the tiny S500 sports car was launched. And this was no ordinary sports car: just over 9ft (3m) long, fitted with a tiny 531cc DOHC engine (peak power at 8000rpm!), and motorbike technology evident in the chain final drive to the rear wheels. Engine grew to 606cc in the S600 of 1964, and to 791cc in the '66 S800. UK sales began in '66. Great looks, amazing ability, but needs understanding owners. Many have now returned to Japan.

Engine 531cc 54x58mm/606cc 54.5x65mm/791cc 60x70mm, 4 cyl DOHC, F/R **Max power** 44bhp at 8000rpm/57bhp at 8500rpm/70bhp at 8000rpm **Max speed** 85/90/100mph
Body styles coupé, convertible
Prod years 1963-64/1964-66/1966-70
Prod numbers n/a
PRICES
S600/S800 Coupé **A** 6000 **B** 4000 **C** 1800
S600/S800 Convertible **A** 12,000 **B** 7500 **C** 4500

Honda N360/N600

Japan's 'K' class of microcars was shaken up by the arrival of Honda's N360 in 1966. Subject to the rules, it used a 354cc OHC air-cooled twin mounted up front and driving the front wheels. Like its sporting sister, this engine could rev to 8500rpm. Great handling, fair performance, but a real buzz-box and no synchromesh at all. Auto option from '67. N400 and N500 offered in some export markets. In the UK, the '68 N600 is best known: healthy power from free-revving alloy 599cc twin. Never stood much chance here against the Mini, but technically interesting.

Engine 354cc 62.5x57.8mm/599cc 74x69.6mm, 2 cyl OHC, F/F
Max power 31bhp at 8500rpm/42bhp at 6600rpm
Max speed 71/85mph
Body styles saloon
Prod years 1966-74
Prod numbers n/a

PRICES
A 1000 **B** 600 **C** 300

Honda Z600

Idiosyncratic micro-coupé launched in Japan in 1970 as the Z Coupé with a 354cc engine, but we got only the Z600 export version with a detuned N600 engine. Novel features included 'TV screen' rear window, aircraft-style overhead console and lots of equipment. UK cars were any colour you liked, as long as it was orange (with optional black Starsky & Hutch stripes). In Japan, there was a pillarless Hard Top model from 1972, but we only ever got the standard version. Exceptionally economical (held *Motor*'s record at 136mpg at a constant 30mph). Lovable.

Engine 599cc 74x69.6mm, 2 cyl OHC, F/F
Max power 32bhp at 6000rpm
Max speed 75mph
Body styles coupé
Prod years 1970-74
Prod numbers n/a

PRICES
A 1400 **B** 800 **C** 400

Honda Civic

A dynasty was created when Honda tackled its first non-domestic non-microcar model, the Civic. By far the most important model in Honda's history. Transverse front-drive four-cylinder engine, front discs, independent rear, manual or dreadful Hondamatic semi-auto 'boxes. Hatchback or saloon styles, longer wheelbase 1500 model launched '72, UK-available from '75. Larger 1238cc base engine from '77. Replaced by evolution model in '79. Millions of shoppers discovered new levels of refinement, build quality and convenience. Honda has never looked back.

Engine 1169cc 70x76mm/1238cc 72x76mm/1488cc 74x86.5mm, 4 cyl OHC, F/F
Max power 54bhp at 5500rpm/60bhp at 5500rpm/75bhp at 5500rpm
Max speed 86-97mph
Body styles saloon, hatchback
Prod years 1972-79
Prod numbers n/a

PRICES
A 1000 **B** 600 **C** 300

Honda Accord

Another step up for Honda: larger dimensions than Civic, more equipment but basically the same mechanical package, albeit with larger engine. Independent suspension all round, five speeds or Hondamatic. Sophisticated styling for mid-1970s, in coupé-type three-door hatchback or four-door saloon versions, both on the same wheelbase. Lots of goodies, Prelude-type 1602cc engine from '79. Scored a phenomenal hit in the USA, but established Honda's bizarre blue-rinse following in Britain. Rust is the major enemy, but would you save one?

Engine 1599cc 74x93mm/1602cc 77x86mm, 4 cyl OHC, F/F
Max power 80bhp at 5300rpm
Max speed 100mph
Body styles saloon, hatchback
Prod years 1976-81
Prod numbers n/a

PRICES
A 900 **B** 500 **C** 300

Hotchkiss 686/2050

The French *grand routier* maker Hotchkiss returned in 1946 with its pre-war straight six 686, but sold only 117 examples. A somewhat revised version was relaunched in 1949 as the 686 S49, with coil-spring IFS and twin Zenith carbs. In 1950, it was renamed 2050 (or 1350 with a four-cylinder engine), when a facelift added a 'V' screen and recessed headlamps. The Anjou saloon had four doors, the Chapron-bodied Antheor was a convertible, but the 130bhp Grand Sport two-door is the one to have.

Engine 3485cc 86x100mm, 6 cyl OHV, F/R **Max power** 100-130bhp at 4000rpm **Max speed** 88-100mph
Body styles sports, saloon, limousine, convertible **Prod years** 1936-50/1950-54 **Prod numbers** 569/n/a

PRICES
686 Saloon **A** 12,500 **B** 9,000 **C** 5000
686 Cabriolet **A** 18,000 **B** 12,000 **C** 10,000
2050 Grand Sport **A** 22,000 **B** 17,500 **C** 14,000

Hotchkiss Grégoire

Hotchkiss squandered a lot of money on the development of the smaller Grégoire saloon, designed by the eponymous French engineer. Unorthodoxies included part-aluminium chassis construction, one-piece alloy dash, front-wheel drive, and a flat four-cylinder engine mounted ahead of the front wheels, resulting in front-heavy styling. Independent suspension all round, four-speed 'box with overdrive. Four-door saloon body was most common, but some two-door coupés and convertibles were made. Intriguing and idiosyncratic in a very French sort of way.

Engine 2180cc 89.9x85.9mm, 4 cyl OHV, F/F
Max power 70bhp at 4000rpm
Max speed 95mph
Body styles saloon, coupé, convertible
Prod years 1951-54
Prod numbers 180

PRICES
A 16,000 **B** 11,000 **C** 7000

HRG 1100/1500/1500 Aerodynamic

HRG's very traditional sports cars returned after the war with little change: running boards, upright grille, cutaway doors *et al*. Engines derived from Singer, in both 1.1 and 1.5-litre forms, although they were significantly modified by HRG. 1500 had slightly longer wheelbase. Specification varied according to the buyer's wishes. Last 12 1500s were known as WS and had Singer SM engines. 1500 Aerodynamic had full-width bodywork supported by outriggers, but gained little following. All HRGs sought-after.

Engine 1074cc 60x95mm/1496cc 68x103mm/1497cc 73x89.4mm, 4 cyl OHC, F/R
Max power 40bhp at 5100rpm/61bhp at 4800rpm/65bhp at 4800rpm
Max speed 80-95mph
Body styles sports
Prod years 1938-50/1939-56/1946-47
Prod numbers 49/138/35

PRICES
A 20,000 **B** 16,000 **C** 12,000

HRG Twin Cam

High hopes for HRG's brand new sports car were misplaced, as the company gave up on cars only a year after its 1955 presentation. Alloy bodywork clothed a twin tube chassis, which boasted all-independent suspension, four-wheel discs and magnesium alloy wheels. As its name indicated, the engine was a twin cam unit modified by HRG on a Singer SM block. Whether it would have made much impression on traditional HRG owners is difficult to say, but since virtually nobody was buying HRGs by the mid-1950s, the point is a delicate one. Very rare.

Engine 1497cc 73x89.4mm, 4 cyl DOHC, F/R
Max power 108bhp
Max speed 115mph
Body styles sports
Prod years 1955-56
Prod numbers 4

PRICES
Impossible to quote, but more than other HRGs

Hudson '48-'54

The pinnacle of Hudson's existence came in 1948, when its new 'Step-Down' was launched, so called because of the dropped floorpan construction. Safety, handling, ride and space were the Hudson's strong suits and its torpedo styling was revolutionary – but unitary build meant major annual changes were impossible, heresy in Detroit. Super Six, Super Eight and Commodore models were joined by budget Pacemaker in '50, the Carrera Panamericana-winning Hornet in '51, Wasp in '52. Hudson merged with Nash in 1954.

Engine 3310cc/3802cc/4293cc/5084cc 6 cyl SV, 4168cc 8 cyl SV, F/R
Max power 104-170bhp at 4000rpm
Max speed 88-100mph **Body styles** saloon, coupe, convertible **Prod years** 1948-54 **Prod numbers** 665,766

PRICES
Hudson saloon **A** 6500 **B** 4000 **C** 2500
Hudson convertible **A** 10,000 **B** 7000 **C** 5000
Hudson Hornet coupe **A** 10,000 **B** 7000 **C** 4000

Hudson Italia

When Hudson's chief stylist Frank Spring was allowed to design a two-door coupé style, he came up with the Italia, so called because the bodies were built and finished by Touring of Milan. The platform was that of the Hudson Jet, a compact saloon built 1953-54. Extraordinary styling details included doors which cut into the roof, wing cutaways which funnelled air to the brakes and rear lights that looked like rocket launchers. The aluminium bodywork was only 48in high, but too heavy for the feeble side-valve six engine. Nash killed it off post haste.

Engine 3310cc 76.2x120.6mm, 6 cyl SV, F/R
Max power 106-114bhp at 4000rpm
Max speed 92mph
Body styles coupe
Prod years 1954
Prod numbers 26

PRICES
A 30,000 **B** 20,000 **C** 15,000

Humber Hawk I & II/Snipe/Super Snipe I/Pullman I

Humber's first post-war output was a revival of the 1938-style Snipe, itself a badge-engineered 'luxury' Hillman 14. Hawk has Hillman four-cylinder engine, Snipe has 2.7-litre six. All post-war models have hydraulic brakes, IFS, projecting boots, sliding roofs. MkII Hawk has synchromesh and column change. Snipe has cross-braced chassis and disc wheels, Super Snipe boasts 100bhp 4086cc engine, but still not fast. Pullman has 127.5in wheelbase (over a foot more than Snipe), optional Mulliner sedanca bodywork. All are exceedingly comfortable.

Engine 1944cc 75x110mm/2731cc 69.5x120mm/4086cc 85x120mm, 4/6/6 cyl SV, F/R **Max power** 56bhp at 3800rpm/65bhp at 3500rpm/100bhp at 3400rpm **Max speed** 64/72/81mph **Body styles** saloon, limousine, landaulette **Prod years** 1945-48 **Prod numbers** c 8000/1240/3909/c 500
PRICES
Hawk/Snipe **A** 4000 **B** 2000 **C** 800
Super Snipe **A** 4500 **B** 2300 **C** 1000
Pullman **A** 6000 **B** 4000 **C** 1700

Humber Super Snipe II & III/Pullman & Imperial II, III & IV

New big Humbers still looked back in time for inspiration, with 'modern but dignified' styling, said Humber: ie. alligator bonnet, integral headlamps, near full-width bodywork but still running boards. Mechanically a little updated: column change, 'Evenkeel' transverse leaf IFS, but the same side-valve engine. Touring limo and Tickford two-door convertible from 1949. Pullman came on huge 131in wheelbase, Imperial was identical except for lack of division. Pullman/Imperial continued 1953-54 with OHV engine, improved suspension and brakes.

Engine 4086cc 85x120mm/4138cc 88.9x111.1mm, 6 cyl SV/OHV, F/R
Max power 100/113bhp at 3400rpm
Max speed 84/82mph **Body styles** saloon, touring limousine, drophead coupé, estate, limousine **Prod years** 1948-52/1948-54 **Prod numbers** 17,334/4140
PRICES
Super Snipe **A** 4500 **B** 2300 **C** 1000
Super Snipe DHC **A** 12,500 **B** 7500 **C** 4000. Pullman/Imperial **A** 6500 **B** 4000 **C** 1700

Humber Hawk III/IV/V/VI/VIA

Updated full-width Americanesque bodywork clothes essentially antique underpinnings: separate chassis, side-valve engine, column-change 'box. But you do get coil-spring IFS. Shorter wheelbase so less bulk: 1¼ tons (1250kg). Power improved in 1950's MkIV as capacity increased to 2267cc, plus higher geared steering and wider tyres. MkV (1952) has lower bonnet, extra front grilles, more chrome. '54 MkVI brings in OHV version of 2267cc engine, extended rear wings, bigger brakes, front anti-roll bar, optional overdrive. Estate from '55, duo-tone VIA from '56.

Engine 1944cc 75x110mm/2267cc 81x110mm, 4 cyl SV/OHV, F/R
Max power 56bhp at 3800rpm/58bhp at 3400rpm to 75bhp at 4000rpm
Max speed 69/71/80mph
Body styles saloon, estate, touring limousine
Prod years 1948-50/1950-52/1952-55/1954-56/1956-57
Prod numbers 59,282 (10,040/6492/14,300/18,836/9614)
PRICES
A 3500 **B** 2500 **C** 1300

Humber Super Snipe IV

The old Super Snipe was replaced some four years after the Hawk by a model which shared its central hull with the '48 Hawk. The front and rear ends were extended and you got plenty of chrome and usually duo-tone paint. First Humber to have OHV engine, a 4.1-litre seven-bearing unit from Commer. All-synchromesh 'box and optional walnut trim from '54, overdrive option from '55 (when Zenith carb increases output by 3bhp), auto option from '56. A big car: 16ft 5in (294cm) and 1¾ tons (1750kg), so heavy to drive and heavy on fuel. Rust record acceptable.

Engine 4138cc 88.9x111.1mm, 6 cyl OHV, F/R
Max power 113-116bhp at 3400rpm
Max speed 90mph
Body styles saloon, touring limousine
Prod years 1952-57
Prod numbers 17,993

PRICES
A 4500 **B** 3000 **C** 1500

Humber Hawk I/II/III/IV

A new generation of Humbers was launched in 1957 with the unitary construction Hawk. Styling was very American (wrap-around screens, 'kick' in the rear doors, whitewall tyres), but still roughly the same size. Saloon, estate and touring limousine versions on offer. Four speeds with column change, overdrive option, and auto available until 1960 only, except for export. SII (1960) adds servo front discs, improved gearbox, '62 SIII has restyled rear window, overdrive on third as well as top. Final SIV (1964) has sharper roof-line, all-synchro 'box, better suspension. Robustly built and durable.

Engine 2267cc 81x110mm, 4 cyl OHV, F/R
Max power 78bhp at 4400rpm
Max speed 81mph
Body styles saloon, estate, touring limousine
Prod years 1957-60/1960-62/1962-64/1964-68
Prod numbers 41,191 (22,352/7230/6109/5500)

PRICES
A 3200 **B** 1800 **C** 1000

Humber Super Snipe/Imperial

Arriving some 18 months after the four-cylinder Hawk, the square-six Super Snipe shared the Hawk's bodyshell but added adornments like a snipe bonnet mascot, fluted chrome grilles around the sidelamps and duo-tone paint schemes. Three-speed 'box, plus optional overdrive, automatic and power steering. 3.0-litre engine and front discs from '59 SII, four headlamps on '60 SIII. In 1963, the SIV pushed power up, added more chrome and the final SV had the '64 Hawk body changes, plus twin carbs and PAS. Imperial is luxury model SV with black roof and standard auto.

Engine 2651cc 82.5x82.5mm/2965cc 87.3x82.5mm, 6 cyl OHV, F/R
Max power 112bhp at 5000rpm/129bhp at 4800rpm to 133bhp at 5000rpm
Max speed 90-102mph
Body styles saloon, touring limousine
Prod years 1958-67/1964-67
Prod numbers 30,031

PRICES
A 2700 **B** 1500 **C** 800

Humber Sceptre I/II

The beginning of the end for Humber: the Super Snipe proved to be the last model designed specifically as a Humber, and the 1963 Sceptre was basically a hybrid Hillman Super Minx. However, it was fitted with the grille from the Sunbeam Rapier, and wrap-around side grilles and quad headlamps. Mechanically it was uprated, too: self-cancelling dual overdrive on close-ratio gearbox, servo front discs, and twin Zenith carbs (replaced by single Solex after six months). MkII of '65 had the 1725cc engine, automatic option and Super Minx frontal styling.

Engine 1592cc 81.5x76.2mm/1725cc 81.5x82.5mm, 4 cyl OHV, F/R
Max power 80bhp at 5200rpm/85bhp at 5500rpm
Max speed 88-92mph
Body styles saloon
Prod years 1963-67
Prod numbers 28,996 (17,011/11,985)

PRICES
A 1800 **B** 1200 **C** 700

Humber Sceptre

The final ignominy for the Humber name was this badge-engineered Hillman Hunter, replacing all existing Humber models. 'Luxury' spec included quad headlamps, a special grille, black vinyl roof, walnut veneer facia, more instruments and twin reversing lights. Mechanically, it always came with twin carb engine, servo front discs, and standard dual overdrive (or optional auto). Estate car version launched October '74, when gaudy chrome panel wraps around rear end. By no means collectable, although they are becoming rare these days.

Engine 1725cc 81.5x82.5mm, 4 cyl OHV, F/R
Max power 82-88bhp at 5200rpm
Max speed 100mph
Body styles saloon, estate
Prod years 1967-76
Prod numbers 43,951

PRICES
A 1000 **B** 600 **C** 300

Innocenti Spider/Coupé

Lambretta manufacturer Innocenti began making Austin A40s under licence in 1960, and the same year launched a rebodied Austin-Healey Sprite. The bodywork was designed by Tom Tjaarda of Ghia, which also built the bodies. A removable hardtop was offered from autumn 1961 and the larger Austin 1098cc engine arrived in 1963, alongside front disc brakes. The Spider left production in 1967, leaving only the C (Coupé) model, built by Ghia's neighbour OSI. Interesting alternative to a Spridget.

Engine 948cc 62.9x76.2mm/1098cc 64.6x83.7mm, 4 cyl OHV, F/R
Max power 43bhp at 5000rpm/58bhp at 5500rpm
Max speed 85/90mph
Body styles convertible, coupé
Prod years 1960-70
Prod numbers c 17,500

PRICES
A 6000 **B** 3500 **C** 1800

Innocenti Mini & Mini Cooper

Italian-built Minis became a reality in October 1965, when Innocenti produced its own 850. Up-market spec included opening rear quarter windows, lever-pull door handles and special trim. Distinguishing points included a unique grille and different lighting. Mk2 model (1968) had high compression engine, as did later 998cc 1001. Unique winding windows from '70. Mini-Cooper from '66, Mk2 version (faster than Longbridge's) from '68, Mk3 from '70, 1300 from '72, 1300 Export (the best version) from '73 to '76. Unique full-width facia from Cooper Mk2, with five dials at first, six from 1300.

Engine 848cc/998cc/1275cc, 4 cyl OHV, F/F **Max power** 34bhp at 5500rpm to 48bhp at 5800rpm/48-60bhp at 5500rpm/76bhp at 5800rpm
Max speed 75-95mph **Body styles** saloon, estate **Prod years** 1965-76
Prod numbers c 450,000

PRICES
Mini **A** 2000 **B** 1000 **C** 700
Mini-Cooper 1000 **A** 4000 **B** 2500 **C** 1500
Mini-Cooper 1300 Export **A** 5000 **B** 3500 **C** 2000

Innocenti Mini 90/120

This was a fascinating project. BL sponsored the creation of a new Bertone-styled suit of clothes for the Mini floorpan. Mechanically it was very Mini (only the exhaust, 12in wheels and front-mounted radiator differed), and engine choices were the 998cc or detuned 1275cc Cooper units. Great packaging included hatchback and folding rear seats, though very compact dimensions made it cramped and there were rigidity problems. UK production was planned but expense scotched that idea. De Tomaso bought Innocenti in '76 and made his own dressed-up version. Daihatsu-engined from '82.

Engine 998cc 64.6x76.2mm/1275cc 70.6x68.3mm, 4 cyl OHV, F/F
Max power 49bhp at 5600rpm/65bhp at 5600rpm to 77bhp at 6050rpm
Max speed 87/96-99mph
Body styles hatchback
Prod years 1974-82
Prod numbers c 220,000

PRICES
Mini 90/120 **A** 1700 **B** 1100 **C** 600
Mini De Tomaso **A** 2500 **B** 1700 **C** 1000

Intermeccanica Italia/Torino/IMX

Hungaro-American Frank Reisner built his Italia in – you guessed it – Italy. The project grew out of the stillborn TVR-based Griffith Omega, a striking coupé styled by ex-Bertone man Franco Scaglione. Reisner was left with about 150 body/chassis units when the Omega failed. He fitted these with Ford Mustang V8 engines and manual or optional auto transmission. Convertible Torino followed in '67. Last models were glassfibre-bodied IMX coupés. Suspect build quality, but attractive styling helps.

Engine 4949cc 101.6x76.2mm/5752cc 102.1x88.9mm, V8 cyl OHV, F/R
Max power 200bhp at 4400rpm to 290bhp at 4800rpm
Max speed 125-140mph
Body styles coupé, convertible
Prod years 1965-71
Prod numbers c 1000

PRICES
A 18,000 **B** 12,000 **C** 7000

Intermeccanica Indra

Erich Bitter represented Intermeccanica in Germany and it was he who first proposed that Reisner should use Opel drive trains (a precursor to his own marque, Bitter). A new model was thus readied for a debut at the 1971 Geneva show, based on shortened floorpans from the Opel Admiral/Diplomat and fitted with a choice of Opel-supplied 2.8-litre or V8 5.4-litre engines. The Indra was another Scaglione design, a very handsome machine realised in steel. Handling and equipment were praised, but quality was not. Reisner relocated to California in 1975.

Engine 2784cc 92x69.8mm/5354cc 101.6x82.5mm, 6/V8 cyl OHV, F/R
Max power 190bhp at 6200rpm/250bhp at 3400rpm
Max speed 127/140mph
Body styles coupé, convertible
Prod years 1971-75
Prod numbers n/a

PRICES
A 16,000 **B** 12,000 **C** 7000

Iso Rivolta

Fridge, motor scooter and bubble car maker Renzo Rivolta turned his hand to GT cars in 1962. The first Italian GT car to use an American V8, the Rivolta's credentials were impressive: a young Giotto Bizzarrini designed the chassis and an equally young Giugiaro did the bodywork. Four seats, Corvette engine, IFS, De Dion rear. Sales success based on price (35% less than the cheapest Ferrari), but it was fast and competent around corners, as well as being practical. Usual exotic rust problems.

Engine 5357cc 101.6x82.5mm, V8 cyl OHV, F/R
Max power 300-340bhp at 5000rpm
Max speed 135/142mph
Body styles coupé
Prod years 1962-70
Prod numbers 797

PRICES
A 15,000 **B** 10,000 **C** 5000

Iso Grifo 300 & 365/7-Litre

Iso's most celebrated car was the Grifo, a two-seater coupé based on a shortened version of the Rivolta chassis. The fabulous bodywork was styled by Giugiaro and built by his then boss Bertone. Series 2 front-end restyle by Gandini occurred in 1970. With a choice of Corvette engines (5.4 or 7.0 litres), the Grifo tends to upset the purists but it is more practical and usable than Italian thoroughbreds. In 7-litre form it could easily hold its own against a Ferrari Daytona or Lamborghini Miura in a straight line, but Grifo's handling is not their equal. Strangely, Coupés are preferred to Spiders.

Engine 5359cc 101.6x82.5mm/6997cc 108x95.5mm, V8 cyl OHV, F/R
Max power 300-365bhp at 5000rpm/406bhp at 5200rpm
Max speed 155/162/171mph
Body styles coupé, cabriolet
Prod years 1965-74/1969-74
Prod numbers 414/90

PRICES
300/365 **A** 35,000 **B** 25,000 **C** 15,000
7-Litre **A** 40,000 **B** 30,000 **C** 22,000

Iso Fidia

Not one of Giugiaro's best designs: the Fidia was a slightly ungainly attempt to create a four-door grand tourer in the Maserati Quattroporte class. Stretched Rivolta floorpan, familiar Corvette engine, though from '73 switched to Ford's 5.8-litre V8. Lots of equipment and good performance despite huge bulk (3950lb/1790kg), but not well enough made and too expensive (the same price as a Rolls-Royce Silver Shadow). Production ended in 1974 when Iso went under, but a handful more were built until 1979 under the Ennezeta name.

Engine 5359cc 101.6x82.5mm/5735cc 101.6x88.4mm/5752cc 101.6x88.8mm, V8 cyl OHV, F/R
Max power 300bhp at 4800rpm/355bhp at 5000rpm/325bhp at 5800rpm
Max speed 128-138mph
Body styles saloon
Prod years 1967-74
Prod numbers 192

PRICES
A 9000 **B** 6000 **C** 3000

Iso Lele

The Lele was Iso's belated replacement for the Rivolta. Bertone's Marcello Gandini was responsible for the coachwork. The same old Rivolta chassis sat underneath it, although the V8 was available with more horsepower than its predecessor. This didn't help to sell the car at a difficult time for large sports cars generally. Like all Isos, the Lele could be had with five-speed manual or automatic 'boxes, and post-'73 examples switched from Chevy to Ford power. Short-lived Marlboro special edition in 1973. 1970s styling somehow lacks charisma but the Lele does have the benefit of four seats.

Engine 5359cc 101.6x82.5mm/5735cc 101.6x88.4mm/5752cc 101.6x88.8mm, V8 cyl OHV, F/R
Max power 300bhp at 4800rpm/355bhp at 5000rpm/325bhp at 5800rpm
Max speed 140/150/145mph
Body styles coupé
Prod years 1969-74
Prod numbers 317

PRICES
A 15,000 **B** 10,000 **C** 6000

Isuzu 117 Coupé

In Britain, we know Isuzu only for its tank-like off-roaders, but its production life started in 1962 with the large Toyota-rivalling Bellel. Only one Isuzu is really of any interest: the 117 Coupé, launched at the end of 1968. Mechanically, it shared its underpinnings with the Florian saloon, but its stylish suit of clothes was designed by Giugiaro while at Ghia. Launched with a 1.6-litre engine, but upgraded in 1970 to a 1.8-litre twin cam, in XE guise having fuel injection. Not sold in UK, but some have been imported privately.

Engine 1584cc 82x75mm/1817cc 84x82mm, 4 cyl OHV/DOHC, F/R
Max power 90bhp at 5400rpm/100bhp at 5400rpm to 140bhp at 6400rpm
Max speed 100/106-122mph
Body styles coupé
Prod years 1968-81
Prod numbers n/a

PRICES
A 3000 **B** 2000 **C** 1000

Jaguar 1½-, 2½- & 3½-Litre

The first post-war Jaguars were broadly similar to the cars built in 1939 apart from the name plate, changed from SS to Jaguar, better braking and minor trim changes. The 1½-Litre had a four-cylinder engine, the 2½ and 3½ had sixes, all of them with overhead valves, which was advanced for those days, so the rigid front axle looked a bit out-of-date. Most went for export, especially the rare dropheads. Elegant all-steel bodies rust badly, rod-operated drum brakes not really up to it and 1½-Litre is terminally slow, otherwise a very attractive car. Drophead coupés (1947-48) worth 50% more.

Engine 1776cc 73x106mm/2664cc 73x106mm/3485cc 82x110mm, 4/6/6 cyl OHV, F/R **Max power** 66bhp at 4500rpm/104bhp at 4500rpm/126bhp at 4250rpm **Max speed** 70/85/94mph
Body styles saloon, drophead coupé
Prod years 1935-49
Prod numbers 13,046/7230/5424

PRICES
1½-Litre **A** 15,000 **B** 9000 **C** 5000
2½-Litre **A** 16,000 **B** 10,000 **C** 6000
3½-Litre **A** 22,000 **B** 15,000 **C** 8000

Jaguar Mark V 2½- & 3½-Litre

The Mark V of 1949 was an interim model, looking similar to the pre-war car but with all-new bodywork (swoopy roofline, headlamps built into the wings and spats over the rear wheels). All-new chassis became the basis for all Jaguar's 1950s big saloons. The Mark V was only built with six-cylinder engines and was extremely plush for its day. New features for Jaguar were independent front suspension and hydraulic brakes. Drophead models are very rare (only 1001 made) and, because of short production run, even the saloons are now scarce.

Engine 2664cc 73x106mm/3485cc 82x110mm, 6 cyl OHV, F/R
Max power 104bhp at 4500rpm/126bhp at 4250rpm
Max speed 87/94mph
Body styles saloon, drophead coupé
Prod years 1948-51
Prod numbers 1675/8791
PRICES
2.5 sal **A** 20,000 **B** 15,000 **C** 8000
2.5 DHC **A** 30,000 **B** 21,000 **C** 14,000
3.5 sal **A** 23,000 **B** 16,000 **C** 10,000
3.5 DHC **A** 32,000 **B** 24,000 **C** 15,000

Jaguar Mark VII/VIIM/VIII

New generation Jaguar saloons began with the 1950 Mark VII (photo); the Mark VI never existed as rival Bentley had already used the number. Six-cylinder XK engine had two overhead camshafts and could power it to a genuine 100mph. Dreadful two-speed auto option from '53, better overdrive option from '54. Follow-on Mark VIIM had more power, wrap-around rear bumpers. The Mark VIII of 1956 had revised grille, two-tone paint with chrome side strip, cut-away rear spats and one-piece windscreen. Major problems are killer rust and a gargantuan appetite for fuel.

Engine 3442cc 83x106mm, 6 cyl DOHC, F/R
Max power 162bhp at 5200rpm/192bhp at 5500rpm
Max speed 100/112mph
Body styles saloon
Prod years 1950-54/1954-57/1956-58
Prod numbers 20,908/10,061/6212

PRICES
Mark VII **A** 12,000 **B** 9000 **C** 5000
Mark VIIM **A** 13,000 **B** 10,000 **C** 5500
Mark VIII **A** 11,000 **B** 7500 **C** 3500

Jaguar Mark IX

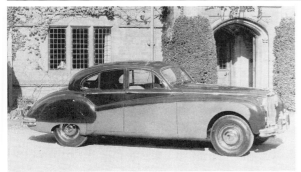

In October 1958, the Mark IX was introduced. From the outside the car differed from the Mark VIII only in the smallest details. The changes all happened under the skin: a bored-out twin carb 3.8-litre version of the XK engine, standard power steering and disc brakes on all four wheels. Almost all cars came with two-tone paintwork separated by a chrome swage line. Usually automatic, but manual and overdrive manual gearboxes were offered. Relatively plentiful so some bargains around and likely to rise in value. Rust again, but a much more capable cruising car than the 3.4-litre large saloons.

Engine 3781cc 87x106mm, 6 cyl DOHC, F/R
Max power 223bhp at 5500rpm
Max speed 115mph
Body styles saloon
Prod years 1958-61
Prod numbers 10,009

PRICES
A 15,000 **B** 10,000 **C** 5000

Jaguar

Jaguar XK120

The sensation of the 1948 London Motor Show, this two-seater roadster had a light alloy body and its new XK engine was the father of all six-cylinder Jaguar engines into the 1990s. After the first 240 cars had been delivered body construction switched to steel. Closed coupé (from '51) and comfortable drophead coupé (from '53) versions followed the roadster. All models were able to clock 120mph – hence the name. SE version boosted power to 180bhp. Rare rhd attracts a significant premium. Plentiful power but handling less satisfactory. Alloy roadsters nearly double the price of steel ones.

Engine 3442cc 83x106mm, 6 cyl DOHC, F/R **Max power** 160bhp at 5000rpm/180bhp at 5300rpm **Max speed** 122mph **Body styles** roadster, coupé, drophead coupé **Prod years** 1948-54 **Prod numbers** 12,055 (7612/2678/1765)

PRICES
Roadster **A** 40,000 **B** 30,000 **C** 20,000
Coupé **A** 30,000 **B** 20,000 **C** 14,000
DHC **A** 35,000 **B** 25,000 **C** 18,000

Jaguar XK140

You have to count the spokes in the grille and spot the overriders on the heavier bumpers to tell the difference between the XK120 and XK140. On the bootlid there was a logo celebrating the XK's victory at Le Mans. The 3.4-litre engine provided an extra 30bhp, while rack-and-pinion steering became standard and overdrive on the four-speed gearbox was offered (also a rare automatic). Drophead (photo) and Fixed-Head Coupés had marginal rear seats, the latter with an ugly extended roof. Again there was a faster SE version (210bhp) but this was very rare indeed.

Engine 3442cc 83x106mm, 6 cyl DOHC, F/R **Max power** 190bhp at 5500rpm/210bhp at 5750rpm **Max speed** 125mph **Body styles** roadster, coupé, drophead coupé **Prod years** 1954-57 **Prod numbers** 9051 (3354/2808/2889)

PRICES
Roadster **A** 38,000 **B** 28,000 **C** 20,000
Coupé **A** 27,000 **B** 18,000 **C** 12,000
DHC **A** 34,000 **B** 24,000 **C** 16,000

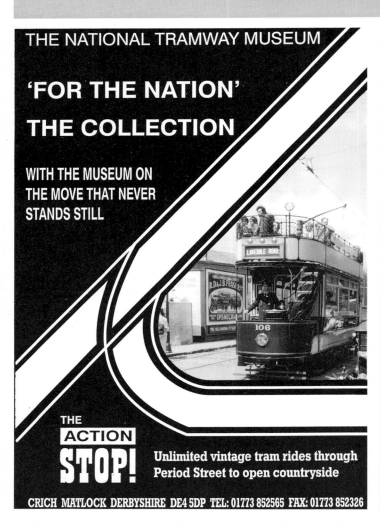

Jaguar XK150/XK150S

Despite Jaguar's factory fire in 1957, it was able to present a successor for the XK140 that year, the XK150. It compromised the XK's exotic styling with less pronounced wing lines and a one-piece windscreen but, in roadholding and comfort, the XK150 was a significant advance. The S versions had 250bhp (265bhp in 3.8 form) but were rare: only 924 S roadsters were made, plus 249 S fixed heads and 140 S dropheads. 220bhp 3.8 engine standard from 1959. Like all XKs, watch for underbody rust. Standard 3.8s about 10% dearer, S models of either engine size 15-20% more.

Engine 3442cc 83x106mm/3781cc 87x106mm, 6 cyl DOHC, F/R
Max power 3.4: 190bhp/210bhp/250bhp. 3.8: 220bhp/ 265bhp.
Max speed 124-138mph
Body styles roadster, coupé, drophead coupé **Prod years** 1957-61
Prod numbers 9398 (2265/4462/2671)
PRICES
Roadster 3.4 **A** 40,000 **B** 32,000 **C** 22,000
Coupé 3.4 **A** 22,000 **B** 16,000 **C** 10,000
DHC 3.4 **A** 35,000 **B** 28,000 **C** 18,000

Jaguar C-Type

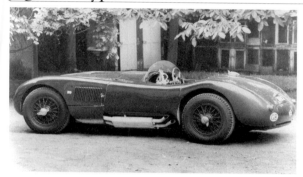

The XK120 C, or C-Type, was conceived specifically to win Le Mans – which it did in 1951. Private buyers stumping up £2300 often used C-Types as hairy road cars. New tubular steel chassis but mechanically it's basically XK120, except for an uprated cylinder head, new camshaft, 9:1 compression (so 204bhp), rack-and-pinion steering, uprated front suspension, transverse torsion bar rear end. Aerodynamic all-aluminium bodywork. Triple carbs, Panhard rod and four-wheel servo discs from 1953. Ultra-rare, although most of the 54 cars built survive. A favourite for replication.

Engine 3442cc 83x106mm, 6 cyl DOHC, F/R
Max power 204bhp at 5800rpm
Max speed 144mph
Body styles sports
Prod years 1951-54
Prod numbers 54

PRICES
Dependent on individual history, but usually £200-300,000

QUALITY + ORIGINALITY

VEHICLE RESTORATION COMPONENTS

THREADED FASTENERS

TRADE ONLY

MILLVALE HOUSE, SELSLEY HILL,
DUDBRIDGE, STROUD,
GLOUCESTERSHIRE GL5 3HF
TEL: 01453 751731   FAX: 01453 759630

Norman Motors Ltd

JAGUAR SPARES SPECIALISTS
SPARES FOR ALL MODELS

For all your Jaguar spares
from the XK120 to the XJS

If you are restoring or just require parts
for a routine service,
then contact us:

100 MILL LANE, LONDON NW6 1NF

Open: Monday-Friday 9.00am-6.00pm;
Saturday 9.00am-1.00pm

 Tel: 0171 431 0940
Fax: 0171 794 5034

Jaguar

Jaguar D-Type & XKSS

Next generation Le Mans Jaguar, styled by Malcolm Sayer around a central aluminium monocoque. Mechanically similar to C-Type, except for new synchro 'box, and some had Lucas fuel injection. The ultimate Jaguar road car is undoubtedly the XKSS (photo), essentially a D-Type racer modified for road use, ie trim, hood, bumpers, windows and no fin. The car had the chassis of the D-Type, plus its suspension, brakes and gearbox. Ultra-rare and if you ever find an XKSS for sale, be careful as many replicas were made using ordinary Jaguar parts or those of the D-Type.

Engine 3442cc 83x106mm, 6 cyl DOHC, F/R
Max power 250 bhp at 6000rpm
Max speed 149mph
Body styles convertible
Prod years 1957
Prod numbers 16

PRICES
Last XKSS sold fetched £500,000 and that could be higher today

Jaguar 2.4-Litre/3.4-Litre (Mark 1)

Flushed with the success of its large saloons, Jaguar muscled in on the smaller luxury car market with the sensational 1955 2.4-Litre. This relatively compact saloon had unitary construction, a short-stroke version of the twin cam six, and coil spring independent front suspension. Manual gearbox available with overdrive, or unpleasant automatic from 1957, when all-wheel disc brake option arrived. The 3.4-Litre was launched as a sister model in 1957: wider grille, cutaway rear spats, 3442cc engine, usually wire wheels. Later 2.4s gain wider grille, too. Hulls rust with ease, and Mark 2 preferred.

Engine 2483cc 83x76.5mm/3442cc 83x106mm, 6 cyl DOHC, F/R
Max power 112bhp at 5750rpm/210bhp at 5500rpm
Max speed 101/120mph
Body styles saloon
Prod years 1955-59/1957-59
Prod numbers 19,992/17,405

PRICES
2.4-Litre **A** 10,000 **B** 7000 **C** 4000
3.4-Litre **A** 13,000 **B** 9000 **C** 6000

Jaguar Mark 2 2.4/3.4/3.8 and 240/340

Major overhaul of the Mark 1 gave the '59 Mark 2 a bigger glass area, improved interior, wider rear track (so only semi-spats), new grille, fog lamps, standard four-wheel disc brakes. The 3.8-litre XK engine made the Mark 2 the fastest production saloon car in the world, and today a 3.8 fetches a significant premium over the two other models. All-synchro 'box from '65, plastic upholstery from '66, optional power steering from '67. Run-out 240/340 were cut-price Mark 2s, but had better performance, though officially no 3.8s (nine were made). 240/340 worth 20% less than the equivalent Mark 2.

Engine 2483cc 83x76.5mm/3442cc 83x106mm/3781cc 87x106mm, 6 cyl DOHC, F/R **Max power** 120-133bhp/210bhp/ 220bhp
Max speed 96-106/120-124/125mph
Body styles saloon **Prod years** 1959-67 and 1967-69 **Prod numbers** 83,980 (25,173/28,666/30,141) and 4446/2796

PRICES
Mk2 2.4 **A** 12,000 **B** 8500 **C** 4500
Mk2 3.4 **A** 18,000 **B** 11,000 **C** 7000
Mk2 3.8 **A** 20,000 **B** 12,500 **C** 8000

Jaguar E-Type 3.8

Unquestionably a landmark model and the star of the 1961 Geneva Salon, the E-Type redefined what a sports car should be. A William Lyons/Malcolm Sayer design masterpiece, achieved 150mph in road tests, and all for just over £2000. The 3.8, made 1961-64, has clean, unadulterated lines and ultimately better performance than the later 4.2. In open form, it was a cracking car, available with removable hardtop; coupé has practical rear hatch but is less highly regarded. Severe rust problems can decimate the hull, bonnets cost a bomb, prices high. Most ones for sale are re-imports.

Engine 3781cc 87x106mm, 6 cyl DOHC, F/R
Max power 265bhp at 5500rpm
Max speed 150mph
Body styles coupé, convertible
Prod years 1961-64
Prod numbers 7827/7669

PRICES
Roadster **A** 35,000 **B** 25,000 **C** 18,000
Coupé **A** 25,000 **B** 19,000 **C** 12,000

Jaguar E-Type 4.2 Series 1

New 4.2-litre Jaguar XK engine provided better mid-range torque, but no significant increase in power or ultimate performance. It made the E-Type more pleasant to drive, especially now that synchro was included for first gear as well, and a much better brake servo was fitted. A third body style arrived in 1966: the 2+2. This had an extra 9in in the wheelbase, a redesigned 2in higher roof-line, 2cwt (110kg) more weight, and optional auto. 2+2 not favoured as a collector's E. Series 1½ (1967-68) had uncowled headlamps, some had S2 interiors.

Engine 4235cc 92.1x106mm, 6 cyl DOHC, F/R
Max power 265bhp at 5400rpm
Max speed 136-149mph
Body styles convertible, coupé
Prod years 1964-68
Prod numbers 22,916 (Roadster: 9548/Coupé: 7770/2+2: 5598)

PRICES
Roadster **A** 32,000 **B** 22,000 **C** 16,000
Coupé **A** 20,000 **B** 15,000 **C** 10,000
2+2 **A** 16,000 **B** 11,000 **C** 7500

Jaguar E-Type Series 2

US Federal law was behind the S2 bodywork changes: uncowled headlamps, larger raised bumpers, bigger front and rear light clusters mounted below bumper level, redesigned bonnet. Twin electric fans, larger brakes, auto option on 2+2, plus power steering and even air con options. Rocker switches inside and choice of 'easy clean' wire or steel wheels. Avoid LHD US imports which have twin Stromberg carburettors, only 171bhp, and gutless performance. If there are E-Type bargains to be found, it will be from this period. Do make sure yours isn't a rusty heap waiting to gobble your money.

Engine 4235cc 92.1x106mm, 6 cyl DOHC, F/R
Max power 265bhp at 5400rpm
Max speed 136-149mph
Body styles convertible, coupé
Prod years 1968-70
Prod numbers 18,808 (Roadster: 8627/Coupé: 4855/2+2: 5326)

PRICES
Roadster **A** 28,000 **B** 18,000 **C** 12,500
Coupé **A** 18,000 **B** 14,000 **C** 9000
2+2 **A** 14,000 **B** 9500 **C** 6000

Jaguar E-Type Series 3

First use of Jaguar's brand new alloy V12 engine, obligatory on all S3s: a magnificent unit with torque galore, very fast (0-60 in 6.4 secs) and refined. Only two models: roadster and 2+2 coupé, both based on the long 2+2 wheelbase. Redesigned front end has ugly egg-crate grille, extended arches all round, quadruple exhaust tail-pipes. Standard power steering, auto option for both models, ventilated discs. 80% went to the USA. Run-out edition of 50 cars (photo), all RHD, painted black with special plaque. A good example of an S3 roadster is the most desirable E-Type of all.

Engine 5343cc 90x70mm, V12 cyl OHC, F/R
Max power 272bhp at 5850rpm
Max speed 146mph
Body styles convertible, coupé
Prod years 1971-75
Prod numbers 7990/7297

PRICES
Roadster **A** 36,000 **B** 25,000 **C** 18,000
Coupé **A** 18,000 **B** 13,000 **C** 9000

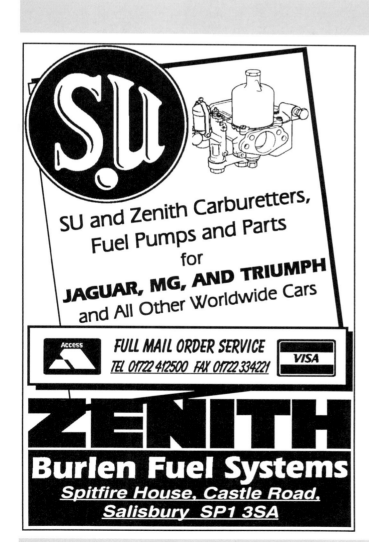

Jaguar

Jaguar Mark X/420G

At nearly 17ft long, 6ft 4in wide (the widest British car ever made until the XJ220) and two tons, the Mark X was quite a beast. Smooth styling hid its bulk admirably, though high sill line makes access awkward. Independent suspension all round (modified Mark 2 front, E-Type rear), four-wheel discs with Kelsey Hayes assistance, standard PAS. XK150 S spec 3.8-litre engine supplanted by new 4.2 from 1964, which gave superior mid-range pull, though no more power. All-synchro 'box from '64, always an auto option. 420G ('66) has detail styling changes. Limos rare (42 built). Structurally very strong but watch that rust.

Engine 3781cc 87x106mm/4235cc 92.1x106mm, 6 cyl DOHC, F/R
Max power 265bhp at 5500rpm/265bhp at 5400rpm
Max speed 120/122mph
Body styles saloon, touring limousine
Prod years 1961-66/1966-70
Prod numbers 25,212 (3.8: 12,678/4.2: 5680/420G: 6554)

PRICES
MkX 3.8 **A** 9000 **B** 6000 **C** 3000
MkX 4.2/420G **A** 8000 **B** 5500 **C** 2500

Jaguar S-Type 3.4/3.8

Shrewd niche model slotting in between Mark 2 and Mark X. Used a modified version of the platform and centre hull of the Mark 2, updated its frontal styling and added a Mark X-style rear end. Unlike Mark 2, this has IRS (therefore better handling) and inboard rear discs. All-synchro 'box standard from 1965, Borg Warner automatic optional. Only ever offered with the 3.4 and 3.8-litre engines. Last ones made came with Jaguar's contemporary cost-cutting measures like Ambla plastic upholstery. Generally overlooked, and so fairly cheap when compared with a Mark 2.

Engine 3442cc 83x106mm/3781cc 87x106mm, 6 cyl DOHC, F/R
Max power 210bhp at 5500rpm/220bhp at 5500rpm
Max speed 120/122mph
Body styles saloon
Prod years 1963-68
Prod numbers 10,036/15,135

PRICES
3.4 **A** 10,000 **B** 6500 **C** 4000
3.8 **A** 11,000 **B** 7500 **C** 4500

Jaguar 420/Daimler Sovereign

More up-market 4.2-litre version of the S-Type, sharing its bodywork from the windscreen back but boasting Mark X-inspired frontal treatment. Detuned 4.2-litre engine had only two SU carbs, not the triple carb system of the Mark X. Usual Jaguar four-speed 'box, with or without overdrive, or Borg Warner three-speed automatic. Marles Varamatic power steering a desirable option; dual-circuit Girling discs derived from the Mark X. Sovereign was the first pure badge-engineered Daimler, but had standard PAS and overdrive, fluted grille and 'D' hubcaps. An under-rated car.

Engine 4235cc 92.1x106mm, 6 cyl DOHC, F/R
Max power 245bhp at 5500rpm
Max speed 125mph
Body styles saloon
Prod years 1966-69
Prod numbers 9801/5829

PRICES
A 8000 **B** 5500 **C** 3000

Jaguar XJ6 Series 1

At one stroke, in 1968 Jaguar introduced a car which would replace all its saloon models within months. The XJ6 was so good, it didn't need a supporting cast. Well proportioned style (though lacking most traditional Jag elements) initially on short wheelbase – 4in longer lwb model from '72 (only 874 built). Mark X type suspension now had anti-dive geometry, was superbly refined. Rack-and-pinion power assisted steering, manual 'box with or without overdrive, and auto option. 2.8-litre version suffered from overheating and lack of power. Classic 4.2-litre is best. Gaining value as rarity increases.

Engine 2791cc 83x86mm/4235cc 92.1x106mm, 6 cyl DOHC, F/R
Max power 180bhp at 6000rpm/245bhp at 5500rpm
Max speed 117/124mph
Body styles saloon
Prod years 1968-73
Prod numbers 19,322/59,951

PRICES
2.8 **A** 4500 **B** 2700 **C** 1200
4.2 **A** 5700 **B** 3500 **C** 1300

Jaguar XJ12 Series 1

Four years after the XJ6 came the XJ12, fitted with the V12 engine first used in the E-Type. Fabulous quad carb V12 is gutsy yet refined, but a DIY nightmare and gas guzzler par excellence. Same basic suspension as XJ6, but higher rate springs, ventilated discs, wider wheels and tyres. Automatic only (Borg Warner at first, later GM 400), and standard air con, gold-plated V12 badging. Voted Car of the Year 1973. All Series 1 V12s are rare: 2474 short-wheelbase examples and just 754 XJ12L long wheelbases were built from October 1972. A good one is worth cosseting.

Engine 5343cc 90x70mm, V12 cyl OHC, F/R
Max power 253bhp at 6000rpm
Max speed 140mph
Body styles saloon
Prod years 1972-73
Prod numbers 3228

PRICES
A 6000 **B** 3700 **C** 1500

Jaguar XJ6/XJ12 Series 2

Re-engineered XJ bodyshell is instantly recognisable by its higher front bumper, smaller grille, and revised lighting, all to satisfy US safety laws. All-new dash with every gauge now sited in front of the driver, and column stalks for major controls. 2.8 engine for export only for a very short period, 4.2 engine's power drops thanks to exhaust-heated air intake system, V12 (photo) boasts fuel injection and 285bhp from 1975. New 3.4 model from April '75. Short wheelbase on 2.8 and 4.2 only, made only until September '74. Build quality could be dreadful and rust record is not impeccable either.

Engine 2791cc 83x86mm/3442cc 83x106mm/4235cc 92.1x106mm/5343cc 90x70mm, 6/V12 cyl DOHC, F/R **Max power** 180bhp at 6000rpm/161bhp at 5000rpm/170bhp at 4500rpm/253-285bhp at 5750rpm **Max speed** 117-147mph **Body styles** saloon **Prod years** 1973-79 **Prod numbers** 93,011 (2.8: 170/3.4: 6880/4.2: 69,951/5.3: 16,010)
PRICES
XJ6 **A** 4200 **B** 2500 **C** 1000
XJ12 **A** 5200 **B** 3000 **C** 1200

Jaguar XJ6C/XJ5.3C

The Coupé bodyshell was launched at the same time as the Series 2 saloon, but did not make production until 1975 because of window sealing problems. Based on a short wheelbase floorpan, the Coupés had a pillarless hardtop style, so all windows wound down out of sight. Reduced rear leg room and 4in longer doors. Slightly less weight, so performance subtly improved. But rigidity was not a strong suit and nor were wind noise and weather sealing; this, combined with poor build quality and aborted Broadspeed racing programme, killed it. All Coupés had black vinyl roofs.

Engine 4235cc 92.1x106mm/5343cc 90x70mm, 6/V12 cyl DOHC/V12 cyl OHC, F/R **Max power** 170bhp at 4500rpm/285bhp at 5750rpm **Max speed** 126/148mph **Body styles** coupé **Prod years** 1975-77 **Prod numbers** 6487/1855

PRICES
4.2 **A** 8000 **B** 5000 **C** 2500
5.3 **A** 9500 **B** 6000 **C** 3000

Jaguar XJ-S 5.3

All the ingredients were in place to make the XJ-S perhaps the greatest car in the world: although not a true successor to the E-Type, it was Jaguar's sports/grand touring model, and benefited from XJ12 suspension, brakes and fuel-injected engine, all on a short (102in) wheelbase. But the styling rubbed everyone up the wrong way: traditionalists hated its slab flanks, black rubber bumpers and the absence of the traditional grille, chrome and walnut. Sports car drivers questioned its flying buttresses and cramped interior. Manual 'box rare (352 built). 5.3 engine replaced by 6.0 in '93.

Engine 5343cc 90x70mm, V12 cyl DOHC, F/R **Max power** 285bhp at 5500rpm **Max speed** 153mph **Body styles** coupé **Prod years** 1975-93 **Prod numbers** 79,447

PRICES
XJ-S '75-'81 **A** 5000 **B** 2800 **C** 1500

Jensen PW

Immediate post-war production at Jensen was the merest of trickles. Its 1946 effort was the PW four-door sports saloon and convertible. This was a massive and expensive beast, but undoubtedly handsome. Tubular braced chassis, coil springs all round (independent at the front), hydraulic brakes. Original engine was Meadows designed and built straight eight, but this suffered from chronic vibration, so from 1949 Austin A135 engines were fitted (no Meadows powered cars survive). Only one convertible, not a lot more saloons.

Engine 3860cc 85x85mm/3993cc 87x111mm, 8/6 cyl OHV, F/R **Max power** 130bhp/130bhp at 3700rpm **Max speed** 95mph **Body styles** saloon, convertible **Prod years** 1946-52 **Prod numbers** 7

PRICES
Too rare to price

Jensen Interceptor

Based on a PW chassis shortened by 12in, the Interceptor was a drophead coupé of rather grand proportions, able to carry 5-6 passengers. Its slab-sided appearance was not especially handsome, but it could do 100mph, thanks to its Austin A135 straight six. Semi-elliptics at the rear and hydromech brakes on early cars, soon upgraded to full hydraulics. Hardtop saloon version followed in 1951, with a fabric-covered roof. Overdrive standard from '52, lower bonnet from '53. Drop-tops are rare, sedancas rarer still. Decidedly vintage in feel, and alluring for it.

Engine 3993cc 87x111mm, 6 cyl OHV, F/R **Max power** 130bhp at 3700rpm **Max speed** 100mph **Body styles** drophead coupé, sedanca, saloon **Prod years** 1949-57 **Prod numbers** 88 (32/4/52)

PRICES
Saloon **A** 9000 **B** 7000 **C** 4000
DHC **A** 10,500 **B** 8500 **C** 6000

Jensen 541/541R/541S

Although announced at Earls Court in 1953, it took over a year for Jensen's revolutionary new 541 to go into production – and with a pioneering glassfibre body, not steel as originally shown. Styling was distinctive, panoramic visibility featuring highly. Triple carb Austin engine; high compression 150bhp version offered from '56 had standard wire wheels and all-wheel discs (another advanced feature). 1957 541R also had 150bhp, plus rack-and-pinion steering and an opening boot lid. 1960 541S is wider, longer, usually has auto 'box and limited-slip diff. Under-rated – and no body rust!

Engine 3993cc 87x111mm, 6 cyl OHV, F/R
Max power 130-150bhp at 4000rpm
Max speed 105-115mph
Body styles coupé
Prod years 1955-59/1957-60/1960-63
Prod numbers 226/193/127
PRICES
541 **A** 15,000 **B** 10,000 **C** 6000
541R **A** 16,000 **B** 11,000 **C** 7500
541S **A** 12,000 **B** 8000 **C** 5000

Jensen CV8

Jensen's glassfibre sports saloon updated the heavy old PW chassis and added a controversially restyled body. Sloping quad headlamps are unmistakable and length increases by 6.5in. Chrysler 5.9-litre V8 engine and Torqueflite automatic transmission gave it a top speed in the mid-130s, and up to 140mph with the 6.3-litre V8 fitted from 1964. MkII of 1963 adds Selectaride dampers, MkIII has equal-size headlamps, improved brakes, reclining seats. Plus: no body rust to worry about. Minus: 12-14mpg, no power steering, few admirers.

Engine 5916cc 105x85.7mm/6276cc 107.9x85.7mm, V8 cyl OHV, F/R
Max power 305bhp at 4800rpm/335bhp at 4600rpm
Max speed 133/140mph
Body styles coupé
Prod years 1962-66
Prod numbers 499 (68/250/181)

PRICES
A 15,000 **B** 10,000 **C** 6000

Jensen Interceptor/SP

From relative obscurity, Jensen hit the jackpot with the Interceptor. A handsome body, built by Vignale in steel, incorporated a big curved glass tailgate. Chrysler V8 power again, servo discs and automatic standard (only 24 manuals built), as is PAS from 1969. MkII (Oct '69) has revised front/interior, MkIII (Oct '71) has cast alloy wheels, vented discs, restyled seats. '71 SP has 385bhp 7.2-litre V8, air con, vinyl roof, louvred bonnet (Interceptor gets 330bhp big lump from '72). Rare '74 convertible (267 made) and '75 notchback coupé (60). Big fuel bills, big rust.

Engine 6276cc 107.9x85.7mm/7212cc 109.8x95.5mm, V8 cyl OHV, F/R
Max power 325bhp at 5000rpm to 385bhp at 4700rpm **Max speed** 135-150mph **Body styles** hatchback coupé, convertible, coupé
Prod years 1966-76/1971-73
Prod numbers 6175/232
PRICES
Interceptor **A** 12,500 **B** 9000 **C** 4000
SP **A** 14,000 **B** 10,000 **C** 5000
Coupé/Convertible **A** 30,000 **B** 20,000 **C** 14,000

Jensen FF

Ahead of its time or a white elephant? It took other manufacturers until the 1980s to realise the benefits of permanent four-wheel drive and anti-lock brakes, but the FF had them both. FF stood for Ferguson Formula, and the ABS was Dunlop's Maxaret system, originally designed for aircraft. A complicated, expensive, thirsty and unreliable car, but supremely safe. 4in longer wheelbase, double air vents in front wings, and bonnet scoop identify. A daunting ownership prospect but a guaranteed classic: its historical importance is well appreciated.

Engine 6276cc 107.9x85.7mm/7212cc 109.8x95.5mm, V8 cyl OHV, F/R
Max power 325bhp at 5000rpm to 385bhp at 4700rpm
Max speed 135-145mph
Body styles hatchback coupé
Prod years 1966-71
Prod numbers 320

PRICES
A 25,000 **B** 17,000 **C** 8000

Jensen-Healey

A real hotch-potch: Jensen made it, Donald Healey designed and put his name to it, Lotus supplied the 16-valve engines, Vauxhall supplied the Viva suspension and steering and Rootes the Sunbeam four-speed gearbox. The result was a competent, reasonably quick sports car, which failed to be outstanding in any way. There were too many problems (engine reliability, scuttle shake, poor hood, build quality, anonymous shape, rust), and too few redeeming features. Oil crisis killed it – and Jensen. Getrag five-speed 'box from 1974. Later attempts are better buys. If you acquire one, you'll need to look after it.

Engine 1973cc 95.2x69.3mm, 4 cyl DOHC, F/R
Max power 140bhp at 6500rpm
Max speed 125mph
Body styles convertible
Prod years 1972-76
Prod numbers 10,926

PRICES
A 5000 **B** 3500 **C** 1800

Jensen GT

Reliant had made a big impact with its GTE, so Jensen tried its luck with an estate version of the Healey (Donald Healey had severed his links by this stage, so the car was called simply GT). Rear hatch, 1in higher roof, rather cramped 2+2 seating and luxury interior (burr walnut facia, electric windows, optional air con and leather seat inserts). All had five-speed Getrag gearboxes. Most of the teething troubles of the Healey had been sorted out, so the GT was actually a reasonable buy, although too small inside and too expensive to be a satisfying GTE/Volvo ES rival.

Engine 1973cc 95.2x69.3mm, 4 cyl DOHC, F/R
Max power 140bhp at 6500rpm
Max speed 120mph
Body styles estate
Prod years 1975-76
Prod numbers 473

PRICES
A 6000 **B** 4000 **C** 2200

Jowett Javelin

Jowett's approach was unusually brave in the post-war period: here was a brand new saloon, with a new flat-four engine, an aerodynamic body, room for six at a pinch (thanks to forward positioning of the engine), fine roadholding, good economy, smooth gearchange for a column shift, and 80mph plus top speed. Why then did it ultimately fail? Probably an unrefined engine and poor reliability – though remedied by 1952, two years after full hydraulic brakes were standardised. Jowett was eventually forced out of business because of body supply difficulties.

Engine 1485cc 72.5x90mm, 4 cyl OHV, F/R
Max power 50-52bhp at 4100rpm
Max speed 80-83mph
Body styles saloon
Prod years 1947-53
Prod numbers 23,307

PRICES
A 7000 **B** 5000 **C** 3000

Jowett Bradford

The putt-putting flat-twin engine which powered this utility Jowett dated all the way back to 1910. In fact, the Bradford was a true survivor from another era: cart springs, mechanical brakes, three speeds, alloy-over-ash body. The Bradford was essentially a commercial vehicle adapted for private use, in four- and six-light forms, with four or six seats. Hardly any performance and a board-hard ride, but ceaseless ability to keep going, especially up hills. Inspired a fanatical following in its day, which has never really gone away. High values considering it's only a vintage van with windows.

Engine 1005cc 79.4x101.6mm, 2 cyl SV, F/R
Max power 25bhp at 3500rpm
Max speed 54mph
Body styles estate
Prod years 1946-54
Prod numbers 38,241 (incl commercials)

PRICES
A 4500 **B** 3500 **C** 2500

Jowett Jupiter

A 1949 tubular chassis design by ERA's Prof Eberan von Eberhorst was taken up by Jowett and developed for its own luxury sports model, the Jupiter. The entire nose section of the aluminium body hinged forward, revealing Jowett's flat-four engine. Larger carbs meant 60bhp. Fairly heavy at 2000lb (910kg), but acceleration was brisk, while handling was surprisingly good. Scored class victories at Le Mans 1950-52. MkIA (1952) has opening boot lid, reshaped wings, more power. Several special bodies by Abbott, J.E. Farr, Pinin Farina et al. Plastic-bodied R4 of '53 never reached production.

Engine 1485cc 72.5x90mm, 4 cyl OHV, F/R
Max power 60-62bhp at 4500rpm
Max speed 85-88mph
Body styles convertible
Prod years 1950-54
Prod numbers 899

PRICES
A 15,000 **B** 11,000 **C** 7000

Kaiser Manhattan

'Anatomic design' was how Kaiser described its Dutch Darrin designed new series of 1950. No other car looked like it: more glass, a lower waist line and more vivid colours than anything else Detroit was making. It also offered advanced safety features, soft ride and sharp handling by American standards. Only saloons and coupes, and lack of a V8 option was a sales liability. The 1951 Dragon trim option included alligator-look 'dragon' vinyl; '53 Dragon had 'bambu' roof. Last ones had superchargers. Convertibles rare (only 131 built).

Engine 3707cc 84.1x111.1mm, 6 cyl SV, F/R
Max power 115-140bhp at 3650rpm
Max speed 86-96mph
Body styles saloon, coupe, convertible
Prod years 1950-55
Prod numbers 202,856

PRICES
A 6000 **B** 4000 **C** 2000

Kaiser-Darrin · Lada · Lagonda

Kaiser-Darrin

This was a last-gasp sports car from a company losing money by the hour. On the basis of the Kaiser-built Henry J compact saloon, Dutch Darrin designed this novel glassfibre body in 1952 – before the Corvette – and it featured patented doors which slid into the front wings. The soft top was a three-position item. The six-cylinder Willys engine came from the Henry J and its 90bhp was good for 0-60mph in 13 secs. When Kaiser's career came to an end in 1955, the remaining stock – about 100 cars – was bought by Darrin and fitted with Cadillac V8s, in which form a frightening 140mph was possible.

Engine 2638cc 79.4x88.9mm, 6 cyl SV, F/R
Max power 90bhp at 3800rpm
Max speed 98mph
Body styles convertible
Prod years 1954-55
Prod numbers 435

PRICES
A 16,000 **B** 10,000 **C** 7500

Lada 2100 series

Fiat signed a deal with the Russian government in 1966 to supply expertise for a new factory at Togliattigrad. The first product was the VAZ 2101 of 1969, basically a lightly modified Fiat 124 fitted with a different 1.2-litre OHC engine. Known as the Lada in export markets (including the UK from 1974), it quickly became Russia's most popular car. Badged Lada 1200/1300/1500/1600 in UK. Cheap prices outweighed lack of development, rough engines, crude detailing and poor build quality. The Lada Riva you can buy today is basically the same as that of 1969.

Engine 1198cc 76x66mm/1294cc 79x66mm/1452cc 76x80mm/1570cc 79x80mm, 4 cyl OHC, F/R
Max power 62bhp at 5600rpm/67bhp at 5700rpm/75bhp at 5600rpm/78bhp at 5200rpm
Max speed 87-96mph
Body styles saloon, estate
Prod years 1969-date
Prod numbers n/a

PRICES
A 500 **B** 250 **C** 100

Lagonda 2.6-Litre

A new smaller engine designed by W.O. Bentley powered Lagonda's post-war cars, but Lagonda was one of those firms which suffered in difficult post-war years, and was bought by Aston Martin in 1947. Bentley-designed chassis had inboard rear brakes, four-speed column shift 'box (though Cotal electric 'box with automatic clutch theoretically available), rack-and-pinion steering, and all-independent suspension. '52 MkII had minor interior and exterior changes. Luxurious, but you didn't get a heater!

Engine 2580cc 78x90mm, 6 cyl DOHC, F/R
Max power 105bhp at 5000rpm
Max speed 85mph
Body styles saloon, drophead coupé
Prod years 1946-53
Prod numbers 550

PRICES
Saloon **A** 16,000 **B** 10,000 **C** 7000
DHC **A** 22,000 **B** 15,000 **C** 10,000

Lagonda 3-Litre Saloon/DHC

A race-bred, larger bore version of the Bentley-designed engine made it into the Lagonda chassis in 1953, well before the Aston Martin. The chassis remained as before but the full-width body was all-new, based on a Tickford design. Two-door saloon and drophead coupé styles were offered, both weighing around 1½ tons. Expensive, luxurious, but not very fast and rather hard-riding. Four-door saloon arrived in 1954, displacing the two-door after a short period, and a brake servo was standardised at the same time. '56 Series II saloon had floor shift, but there were no longer any open cars.

Engine 2922cc 83x90mm, 6 cyl DOHC, F/R
Max power 140bhp at 5000rpm
Max speed 99mph
Body styles saloon, drophead coupé
Prod years 1953-58/1953-56
Prod numbers 430

PRICES
Saloon **A** 15,000 **B** 10,000 **C** 7000
DHC **A** 20,000 **B** 14,000 **C** 9000

Lagonda Rapide

Touring had done a fabulous styling job on Aston Martin's DB4, so why was the Rapide such a lash-up? Hideous frontal aspect and wrongly proportioned cabin area probably explained the Rapide's failure to capture a bigger share of the luxury car market. First use of the new Tadek Marek 4.0-litre engine (the DB5 got it three years later), platform chassis, dual-circuit servo disc brakes all round, De Dion rear axle on torsion bars, aluminium bodywork, automatic standard (optional synchro manual). Lack of interest then is mirrored today, and prices are low for a David Brown hand-built performance machine.

Engine 3995cc 96x92mm, 6 cyl DOHC, F/R
Max power 236bhp at 5000rpm
Max speed 125mph
Body styles saloon
Prod years 1961-64
Prod numbers 55

PRICES
A 25,000 **B** 18,000 **C** 11,000

Lamborghini 350GT

The myth goes that tractor builder Ferruccio Lamborghini was dissatisfied with his Ferrari and the treatment he received at Maranello, so he vowed to build a better car. The Superamerica-rivalling 350GT was the result. Bizzarrini did the chassis engineering and magnificent V12 engine, Franco Scaglione the aluminium bodywork, though it was tidied up by Touring. Independent suspension all round ensured pace-setting handling. Five-speed 'box standard, and the last 23 cars built with 4.0-litre engine. Strictly two seats.

Engine 3463cc 77x62mm/3929cc 82x62mm, V12 cyl QOHC, F/R
Max power 270bhp at 6500rpm/320bhp at 6500rpm
Max speed 150/155mph
Body styles coupé
Prod years 1964-67
Prod numbers 143

PRICES
A 75,000 B 50,000 C 30,000

Lamborghini 400GT

You'd never know it, but Touring completely restyled the bodywork for the 2+2 400GT, and made the bodies too (in steel, not aluminium). Four headlamps and fastback roof-line identify it. Convertible was a two-seater. The glorious 4.0-litre V12 had six twin-choke Webers, and 320bhp. Coupé's rear seats were, in fact, of negligible use, the noise levels were deafening and the build quality wasn't up to much. But if you savoured fine handling, the 400GT was the car to own in 1966. Steel body spells potential trouble, as does that V12 engine.

Engine 3929cc 82x62mm, V12 cyl QOHC, F/R
Max power 320bhp at 6500rpm
Max speed 155mph
Body styles coupé, convertible
Prod years 1966-67
Prod numbers 247

PRICES
A 65,000 B 45,000 C 30,000

Lamborghini Miura/S/SV

The car every rock star had to have. Stunning bodywork by Marcello Gandini of Bertone was only 43in (110cm) high and a classic of post-war styling, often painted in gaudy colours. Same Lambo quad cam V12 engine was mounted transversely in front of the rear axle in a unitary structure. Lambo claimed 180mph (170 nearer the mark), but '69 S model had 375bhp and could do much nearer that figure. SV of '71 had no less than 385bhp, plus wider wheelarches. Some concern over stability at high speed and poor build quality, but the Miura has long since passed into the book of motoring legends.

Engine 3929cc 82x62mm, V12 cyl QOHC, M/R **Max power** 350bhp at 7000rpm/375bhp at 7700rpm/385bhp at 7850rpm **Max speed** 170/178/180mph **Body styles** coupé
Prod years 1966-69/1969-71/1971-72
Prod numbers 475/140/150

PRICES
Miura A 75,000 B 55,000 C 40,000
Miura S A 85,000 B 60,000 C 45,000
Miura SV A 95,000 B 70,000 C 55,000

Lamborghini Islero/Islero S

The chassis and mechanicals are 400GT, but the suit of clothes is all-new, courtesy of Carrozzeria Marazzi, which also built the bodies. Pop-up headlamps, standard air con, power steering, LSD, electric windows and more interior space confirmed this as more of a grand tourer, though still hairy. V12 engine produced 325bhp in standard guise, while the S version had an extra 25 horses. Not as characterful as the earlier 400GT, but more discreet. Lamborghini enthusiasts tend to overlook the Islero and collectors shun it, but it's a fast, exciting car.

Engine 3929cc 82x62mm, V12 cyl QOHC, F/R
Max power 325bhp at 6500rpm/350bhp at 7000rpm
Max speed 155/160mph
Body styles coupé
Prod years 1968-70
Prod numbers 125/100

PRICES
A 30,000 B 20,000 C 13,000

Lamborghini Espada

Full four seats and Miura's reflected glory helped this executive's Lamborghini become a best-seller for the Sant'Agata manufacturer. Developed from Bertone's Marzal prototype, it had a 4in longer wheelbase than other front-engined Lambos, but mechanically was identical to the Islero. Lots of goodies – air con, power steering, later an automatic gearbox option – to justify its price tag as the most expensive Lambo. More power from 1970 Series II, even more (365bhp) on 1972 Series III. Cheap these days, but set aside plenty of money for engine maintenance and rust prevention.

Engine 3929cc 82x62mm, V12 cyl QOHC, F/R
Max power 325bhp at 6500rpm/350bhp at 7000rpm/365bhp at 7500rpm
Max speed 155/160mph
Body styles coupé
Prod years 1968-78
Prod numbers 1217

PRICES
A 22,000 B 16,000 C 10,000

Lamborghini Jarama/Jarama S

A shortened Espada chassis sat underneath the Jarama, whose Bertone-penned lines looked rather heavy as a result, with uncomfortably long overhangs. Bodies were built in steel at Marazzi. Power was spiralling ever higher: 350bhp initially, and 385bhp in the '73 S version, making the Jarama one of the quickest cars of its day, despite its substantial weight. Power steering offered in later cars, as was automatic transmission. Since no-one seems to like the Jarama, it's possible that this could be the holy grail: a genuine Lamborghini bargain.

Engine 3929cc 82x62mm, V12 cyl QOHC, F/R
Max power 350bhp at 7000rpm/385bhp at 7500rpm
Max speed 155/160mph
Body styles coupé
Prod years 1970-73/1973-78
Prod numbers 177/150

PRICES
A 28,000 **B** 20,000 **C** 14,000

Lamborghini Urraco P250/P300/P200

A brave but not entirely successful attempt to launch a lower-priced Lamborghini, aimed at the Ferrari Dino. Complicated steel body construction was a near-monocoque and a new V8 engine meant a delay of two years getting it into production. Engine sited centrally in a transverse position, but even so there were four seats – just. Front and rear suspension by MacPherson struts and lower wishbones. P250 was entry-level model, P300 had four camshafts and 265bhp, P200 was Italian market tax-break special. Somehow not exotic enough for supercar or Lamborghini enthusiasts.

Engine 2463cc 86x53mm/2996cc 86x64.5mm/1994cc 77.4x53mm, V8 cyl OHC/QOHC/OHC, M/R
Max power 220bhp at 7500rpm/265bhp at 7800rpm/182bhp at 7500rpm
Max speed 145/160/134mph
Body styles coupé
Prod years 1970-79
Prod numbers 520/190/66

PRICES
P250 **A** 17,000 **B** 10,000 **C** 6500
P300 **A** 25,000 **B** 18,000 **C** 12,000

Lamborghini Countach LP400/LP400S

One of the all-time supercar icons, the Countach supplanted the Miura in spectacular fashion. Extraordinary Bertone-styled body clothed a brand new multi-tube frame. The engine was now positioned lengthways (LP = Longitudinale Posteriore), and the gearbox squeezed in front of it, virtually between the seats. '400' tag referred to the unchanged 4.0-litre engine. Novel fold-forward doors, trapezoidal wheel arches. Pure shape gradually adulterated, starting with LP400S of '78 (optional rear wing, five-pot wheels, wider tyres, improved suspension). A pain in traffic, but glorious out in the open.

Engine 3929cc 82x62mm, V12 cyl QOHC, M/R
Max power 385bhp at 8000rpm
Max speed 180mph
Body styles coupé
Prod years 1973-79/1978-82
Prod numbers 150/466

PRICES
LP400 **A** 55,000 **B** 45,000 **C** 35,000
LP400S **A** 60,000 **B** 50,000 **C** 40,000

Lamborghini Silhouette

Somewhat marginal Lamborghini made to capture a slice of the open car market. Underneath all those Bertone-effected bulges and spoilers (a presage of what was about to happen on the Countach), the Silhouette was basically an Urraco P300, although it had only two seats. The structure was stiffened up to accommodate a removable targa roof panel. Lack of class, poor ergonomics, iffy build quality, leaky roof, and non-compliance with US safety laws confined it to the shadows. That means examples are fairly cheap today – if you can find one, that is.

Engine 2996cc 86x64.5mm, V8 cyl QOHC, M/R
Max power 265bhp at 7800rpm
Max speed 156mph
Body styles targa
Prod years 1976-77
Prod numbers 54

PRICES
A 25,000 **B** 18,000 **C** 12,000

Lanchester Ten

Following the war, Lanchester was able to introduce a model it had been preparing for launch in 1940. The Ten was utterly conventional in appearance, but advanced in specification, including coil spring IFS with anti-roll bar, an OHV engine which was among the most powerful of contemporary 10HP units, and fluid flywheel/preselector gearbox. A high overall weight meant dowdy performance and heavy braking, but solid construction, sound cornering and comfortable ride won it friends. Four-light aluminium Barker saloon from '49 (579 built).

Engine 1287cc 63.5x101.6mm, 4 cyl OHV, F/R
Max power 40bhp at 4200rpm
Max speed 68mph
Body styles saloon
Prod years 1946-51
Prod numbers 3030

PRICES
A 4500 **B** 2500 **C** 1200

Lanchester Fourteen/Leda

The Fourteen was basically what would become a Daimler Conquest body and chassis fitted with a Lanchester grille and a new 60bhp four-cylinder engine (the Conquest would have a six). Pressed steel body but virtually all hand-built: LJ200 for home market has wood framing, export LJ201 (Leda) is all-steel. Torsion bar and wishbone IFS, hydromechanical brakes, Luvax-Bijur central chassis lubrication, and still the fluid flywheel. Two follow-ups (Hooper-bodied Dauphin, unitary Sprite) never reached production. The Fourteen is basically a boring, solid, middle-of-the-road car.

Engine 1968cc 76.2x108mm, 4 cyl OHV, F/R
Max power 60bhp at 4200rpm
Max speed 75mph
Body styles saloon, drophead coupé
Prod years 1950-54
Prod numbers 2100

PRICES
A 5000 **B** 3500 **C** 2000

Lancia Ardea

Just before the outbreak of war, Lancia had introduced the little Ardea, which looked like a miniature version of the Aprilia. The wheelbase was a full foot shorter than the Aprilia's at 8ft, and the engine was a comparatively tiny four-cylinder OHV unit of 903cc. IFS but semi-elliptic rear. Series II (1941) added a boot lid, Series III (1948) an alligator bonnet and five gears, while the Series IV (1949) had an even shorter wheelbase and an alloy cylinder head. Displaced by the Appia.

Engine 903cc 65x68mm, 4 cyl OHV, F/R
Max power 29-30bhp at 4600rpm
Max speed 67mph
Body styles saloon
Prod years 1939-52
Prod numbers n/a

PRICES
A 4000 **B** 3000 **C** 2500

Lancia Aprilia

Way ahead of its time is the only way to describe the Aprilia. Launched in 1937, it boasted unitary construction, an overhead cam alloy-block V4 engine, all-independent suspension (sliding pillars up front, torsion bars out back), hydraulic brakes and pillarless aerodynamic styling. Performance and handling would not have disgraced a contemporary sports car, while its spaciousness was a revelation. All post-war Aprilias had 1.5-litre engines and steel disc wheels. Standard wheelbase was 108.3in, but two slightly longer versions were also offered to coachbuilders.

Engine 1486cc 74.6x85mm, V4 cyl OHC, F/R
Max power 48bhp at 4300rpm
Max speed 80mph
Body styles saloon, various coachbuilt
Prod years 1936-49
Prod numbers 27,642

PRICES
A 14,000 **B** 9000 **C** 5000

Lancia Aurelia B10/B15/B21/B22/B12

The Aprilia's replacement was no less innovative. Vittorio Jano designed the Aurelia and its new 60-degree hemi-head all-alloy V6 engine and four-speed all-synchro transaxle. Other technically interesting points were inboard rear brakes, semi-trailing IRS and hydraulic brakes. Initial B10 and replacement B22 had 1.8-litre engine, '51 B21 had 2.0 litres, '52 B15 was a six-light long wheelbase limousine,. The 1954 B12 (ie. second series) had a de Dion rear axle and a 2.3-litre engine. Interesting but expensive to restore and maintain.

Engine 1754cc 70x76mm/1991cc 72x81.5mm/2266cc 75x85.5mm, V6 cyl OHV, F/R **Max power** 56bhp at 4700rpm/64-90bhp at 4800rpm/85bhp at 4800rpm **Max speed** 80-96mph
Body styles saloon, estate, convertible, limousine **Prod years** 1950-53/1952-53/1951-53/1951-53/1954-56 **Prod numbers** 12,786
PRICES
B10/B15 **A** 9000 **B** 6000 **C** 3500
B21/B22 **A** 10,000 **B** 7000 **C** 4500
B12 **A** 12,000 **B** 8000 **C** 5000

Lancia Aurelia B20 GT

A grand tourer which would have made the late Vincenzo Lancia smile. Mechanics broadly as Aurelia saloon, except for the engine which was a twin Weber 2.0-litre V6 from the outset in 1951. Simple, handsome, full-width styling was by Pinin Farina. Handling and performance were excellent, enabling the B20 to score many race victories. S2, launched very soon after, upped power to 80bhp, S3 of '52 had 2.5-litre engine (and was often called GT 2500), 1954's S4 changed to a De Dion rear axle. '56 S5 has direct drive top gear, '57 S6 has uprated engine. Lovely but rust-infected.

Engine 1991cc 72x81.5mm/2451cc 78x85.5mm, V6 cyl OHV, F/R
Max power 75-80bhp at 4700rpm/110-118bhp at 5100rpm
Max speed 100-112mph
Body styles coupé
Prod years 1951-58
Prod numbers 3871
PRICES
B20 S1-S2 **A** 30,000 **B** 24,000 **C** 18,000
B20 S3-S6 **A** 24,000 **B** 16,000 **C** 10,000

Lancia Aurelia B24 Spider/GT 2500 Convertible

The B24 was much more than a convertible B20. Its Pinin Farina designed and built bodywork was all-new and on a shorter (96.5in) wheelbase. Launched in 1954, it shared the latest B20 spec of 2.5 litres and De Dion rear end. The first 240 Spiders had wrap-around windscreens; the 1956 GT 2500 Convertible (aka America) had no wrap-around but quarter-lights and winding windows. LHD cars can have floor-mounted gear levers, but rare RHD will be column change only. As with all Aurelias, the metalwork seems purpose designed to trap water: these Lancias love rust.

Engine 2451cc 78x85.5mm, V6 cyl OHV, F/R
Max power 110-118bhp at 5100rpm
Max speed 110-115mph
Body styles sports, convertible
Prod years 1955-56/1956-58
Prod numbers 240/521

PRICES
Spider **A** 50,000 **B** 42,000 **C** 30,000
Convertible **A** 40,000 **B** 28,000 **C** 18,000

Lancia Appia

Lancia always built a relatively cheap car to tempt upwardly mobile Fiat drivers. Its 1953 effort was the Appia, which looked like a scaled-down Aurelia, but underneath it was more utilitarian: beam rear axle, conventional column-change four-speed gearbox, outboard rear brakes. Doors, boot, bonnet and rear wings were aluminium. Unusual 1.1-litre V4 had only 10 degree angle between the 'V', and was uprated to 43.5bhp in S2 from '56, which also had notchback rear end and 1in longer wheelbase. S3 of '59 (pictured) had twin leading shoe front brakes, 53bhp and wider grille. A delight to drive.

Engine 1090cc 68x75mm, V4 cyl OHV, F/R
Max power 38bhp at 4800rpm to 53bhp at 5200rpm
Max speed 75-90mph
Body styles saloon
Prod years 1953-63
Prod numbers 98,006

PRICES
A 6000 **B** 4000 **C** 2500

Lancia Appia (coachbuilt)

Many coachbuilt bodies were based on the Appia Series 2. Vignale's angular convertible appeared in 1957, featuring a wide front grille. The Farina coupé looked surprisingly similar, but featured a wrap-around rear window and seats for children in the rear. Lombardi, Scioneri and Vignale did saloons, Viotti an estate and a coupé. After building around 50 pre-production cars, Zagato offered an alloy GT coupé from 1956, soon joined by a swb Sport. S3 GTE of 1959 is pictured ('E' is for export) and there were only 721 Zagatos. Always LHD and rarity depends on model.

Engine 1090cc 68x75mm, V4 cyl OHV, F/R **Max power** 53-60bhp at 5200rpm
Max speed 90mph
Body styles coupé, convertible, estate **Prod years** 1956-62
Prod numbers 9239

PRICES
Vignale Convertible **A** 10,000 **B** 7500 **C** 4500
Farina Coupé **A** 8000 **B** 6000 **C** 4000
Zagato Coupé **A** 25,000 **B** 20,000 **C** 15,000

Lancia Flaminia

Impressive new Lancia bought by industrialists, bankers and even the Pope. Designed and built by Farina, the bodywork was expansive (191in/485cm) and heavy at 1½ tons. The De Dion rear end and transaxle recalled the Aurelia, but the sliding pillar front suspension was abandoned in favour of wishbones and coil springs. Also new was the 2458cc alloy V6 engine. Disc brakes were incorporated after a short while, and the first cars even had a wiper for the rear window. Optional triple carbs from '62, 2.8-litre engine from '63, and production ended as late as 1970. Yet to fire collectors' imaginations.

Engine 2458cc 80x81.5mm/2775cc 85x81.5mm, V6 cyl OHV, F/R
Max power 98-110bhp at 5000rpm/129bhp at 5000rpm
Max speed 95/102mph
Body styles saloon
Prod years 1956-70
Prod numbers 3943

PRICES
A 7000 **B** 4500 **C** 2500

Lancia Flaminia Coupé 2.5/2.8

Farina's lines for the coupé version of the Flaminia were much more harmonious, although in reality this was more like a handsome five-seater two-door saloon. Electric windows and lots of other goodies justified a much higher price tag than the saloon, but even so more customers bought the two-door. More powerful engines than the saloon, too: 119bhp for the early 2.5-litre models and 136bhp for the post-'63 2.8-litre ones. RHD available from '60, though LHD offered in UK before then. Still quite cheap, as the usual rust and parts problems dissuade collectors.

Engine 2458cc 80x81.5mm/2775cc 85x81.5mm, V6 cyl OHV, F/R
Max power 119bhp at 5100rpm/136bhp at 5400rpm
Max speed 105/117mph
Body styles coupé
Prod years 1959-67
Prod numbers 5284 (4151/1133)

PRICES
A 11,000 **B** 7000 **C** 4000

Lancia Flaminia GT/GTL/Convertible

The true Flaminia coupé was the GT, based on a wheelbase shortened by 14in. It started life as a one-off but was pressed into production from 1959 in coupé and open-top forms. Touring of Milan was responsible for the design and execution of the bodywork, which featured double headlamps set in hooded cowls. All had front disc brakes and limited slip diff. Even more powerful than the Farina coupé and therefore popular for club racing as well as road use. Larger 2.8-litre engine from 1963, when the longer wheelbase GTL was also launched. LHD only and even more spares headaches.

Engine 2458cc 80x81.5mm/2775cc 85x81.5mm, V6 cyl OHV, F/R
Max power 119-136bhp at 5100rpm/146bhp at 5400rpm
Max speed 105-120mph
Body styles coupé, convertible
Prod years 1959-67
Prod numbers 1718/300/847

PRICES
GT/GTL **A** 20,000 **B** 15,000 **C** 10,000
Convertible **A** 35,000 **B** 25,000 **C** 15,000

Lancia Flaminia Sport/Supersport Zagato

Zagato's creations on the Flaminia were probably the prettiest of all. The first of these was the Sport, a strict two-seater fastback coupé on the same short floorpan as Touring's GT. Styling features included a 'double bubble' roof, squared-off wheelarches and a prominent bonnet scoop. 3C version from 1961 (pictured) had triple Webers. Supersport model had distinctive cowled headlamps. Power options were the hairiest of all Flaminias, up to 154bhp being quoted, while some special semi-race engines were built with 175bhp. The most desirable Flaminia of them all.

Engine 2458cc 80x81.5mm/2775cc 85x81.5mm, V6 cyl OHV, F/R
Max power 140bhp at 5600rpm/154bhp at 5600rpm
Max speed 118-130mph
Body styles coupé
Prod years 1959-67
Prod numbers 443/150

PRICES
Sport **A** 50,000 **B** 35,000 **C** 25,000

L&M LANCIA LTD

SPECIALISTS IN NEW AND USED SPARES

FULL WORKSHOP FACILITIES
MAIL ORDER
ALL LANCIAS BOUGHT AND SOLD

Currently Breaking
Dedra 2000ie - Thema 8v and 16v
Delta Integrale 16v, Delta HF and GTie
Beta Coupé, HPE, Spyder and VX

New Turbo Units Available
Lancia 8V integrale £325+VAT
Lancia Delta HFie £275+VAT
Lancia Thema 8V £325+VAT
All with 12 months guarantee
All other makes supplied

UNIT 2, THE EBOR WORKS, CHAPEL LANE
HIGH WYCOMBE, BUCKS, ENGLAND, HP12 4BS
TEL: (01494) 538899 OR 0860 614703

Classic Panels

SPECIALISTS IN THE RE-MANUFACTURE OF OBSOLETE CAR BODY PANELS

Suppliers of Exhaust Systems, Spare Parts and some Original Body Panels especially for Lancia Cars

LANCIA BODY PANELS
FULVIA FLAVIA AURELIA B20 AND BETA

With over 30 years of experience, we produce the most comprehensive range of post production panels for Lancia cars. We make them from 1mm C.R.C.A. steel and they are of the highest possible quality-guaranteed. Full catalogue and price list available on request.

We also carry a large stock of mechanical and service parts especially for Fulvia.

As well as original equipment exhaust systems in mild or stainless steel.

Huge amounts of second hand panels and spares.

OVERSEAS ORDERS WELCOME

"Sunnyside", 10 Rhosnesni Lane, Wrexham,
Clwyd L12 7LY
Tel: 01978 357776
Fax: 01978 357776
Mobile: 0378 795303

Lancia

Lancia Flavia

In the early 1960s, Lancia's range was polarised: the Appia and Flaminia were effective opposites. To plug the gap, Lancia offered its first-ever front-wheel drive car, the Flavia. A new light-alloy 1.5-litre flat-four engine was developed by Antonio Fessia, expanded to 1.8 litres in 1963 and 2.0 litres in 1969. Servo disc brakes all round, beam rear axle, column change. Boxy styling not the last word in elegance, but '67 restyle improved things and offered floor change. Fuel injection option from '65, '69 LX version has PAS. Another restyle in '69 and five speeds from '73.

Engine 1488cc 82x71mm/1800cc 88x74mm/1991cc 89x80mm, 4 cyl OHV, F/F
Max power 78bhp at 5200rpm/92-115bhp at 5200rpm/131bhp at 5400rpm
Max speed 94-118mph
Body styles saloon
Prod years 1960-74
Prod numbers 79,764
PRICES
1500/1800 **A** 3000 **B** 2000 **C** 800
2000 **A** 4000 **B** 2500 **C** 1000

Lancia Flavia Coupé/Convertible

Pininfarina designed the coupé body for the Flavia, but compromises were necessary to accommodate the long nose/front-drive layout. The platform was shortened by 5.5in, but there were still four seats. Twin carbs on the earliest 1.5s (photo) pushed the power output up slightly, but thereafter the power outputs remained the same as the saloon, including the fuel-injection option (2150 built). Convertible built by Vignale from 1962 to 1969. Coupé restyled in '69: lower bonnet line, full-width grille incorporating twin headlamps, revised interior. Five speeds from '71.

Engine 1488cc 82x71mm/1800cc 88x74mm/1991cc 89x80mm, 4 cyl OHV, F/F
Max power 80bhp at 5400rpm/92-115bhp at 5200rpm/131bhp at 5400rpm
Max speed 95-120mph
Body styles coupé, convertible
Prod years 1961-73/1962-69
Prod numbers 26,084/n/a
PRICES
Coupé **A** 5000 **B** 3000 **C** 1500
Convertible **A** 10,000 **B** 8000 **C** 5000

Lancia Flavia Sport Zagato

Zagato has never been a coachbuilder which follows fashion. It proved this in spectacular style with its Flavia Sport of 1963, which looked like nothing else, ever. Its light-alloy bodywork – which shared the Flavia Coupé's shortened floorpan – featured bulbous wings, chiselled out back end and rear windows which curved into the roof. Briefly offered with a 1500 engine and two twin-choke Solex carbs, more commonly in 1800 form. The 3.818:1 axle ratio made it suitable for club racing. RHD imports were available from '64, but only 27 British enthusiasts bought one.

Engine 1488cc 82x71mm/1800cc 88x74mm, 4 cyl OHV, F/F
Max power 90bhp at 5800rpm/100-115bhp at 5800rpm
Max speed 106-116mph
Body styles coupé
Prod years 1963-67
Prod numbers 628

PRICES
A 22,500 **B** 17,000 **C** 8000

Lancia Fulvia

New base junior Lancia for 1963 boasted a new V4 OHC engine mounted at an angle under the bonnet and driving the front wheels. Four-wheel discs, transverse-leaf IFS, four-speed all-synchro 'box with column change. 2C (twin carb) option from '64, standard on 1100 from '66. GT saloon from 1967 has 1200 engine, optional floor gearchange, lasts until 1969. 95bhp 1300 engine from '68 in GTE model, standard 1300 model from '70 (revised styling and longer wheelbase at same time). Five speeds standard from '71. Not much of a following, and most have now rusted away.

Engine 1091cc 72x67mm/1216cc 76x67mm/1298cc 77x69.7mm, V4 cyl OHC, F/F
Max power 58bhp at 5800rpm to 95bhp at 6000rpm
Max speed 88-105mph
Body styles saloon
Prod years 1963-72
Prod numbers 192,097

PRICES
Fulvia **A** 2000 **B** 1500 **C** 800
Fulvia GT/GTE **A** 3000 **B** 2000 **C** 1200

Lancia Fulvia Coupé/HF

With over two feet chopped out of the floorpan, the in-house designed coupé version of the Fulvia looked very compact and purposeful. Always four headlamps, floor-mounted gear lever, all-wheel discs. 1200 version initially, joined by Rallye 1.3 from 1967 (including light alloy doors, boot and bonnet); Rallye S has 103bhp. HF models have chassis mods, tuned engines, no bumpers, alloy panels and plastic windows (Lusso comes without these). 101bhp 1.3 joined in '68 by 1.6, which has five speeds, alloy wheels and up to 132bhp. All get five speeds from '71. A wonderful machine in all respects, except rust prevention.

Engine 1216cc/1298cc/1584cc, V4 cyl OHC, F/F **Max power** 80bhp at 6200rpm/87-103bhp at 6000rpm/115-132bhp at 6200rpm **Max speed** 100-120mph **Body styles** coupé
Prod years 1965-76/1966-72
Prod numbers 134,035/6419

PRICES
Coupé **A** 5000 **B** 3000 **C** 1200
HF 1300/1600 **A** 20,000 **B** 15,000 **C** 10,000
HF 1600 Lusso **A** 10,000 **B** 7000 **C** 4000

Lancia Fulvia Sport Zagato

Lancia's love affair with Zagato continued when it took up an alternative coupé body style proposed by the coachbuilder. Unlike Lancia's own coupé, the Sport Zagato was a two-seater but it was both longer and wider (but lighter thanks to the use of aluminium in its bodywork). Odd front end treatment, dominated by rectangular headlamps, made the Zagato look ill-proportioned. But it was a great driver's car, on road and track. 1300 version was sold in UK 1967-69, when the Series 2 was launched. 1600 version also offered (in UK 1972-73). Not often seen over here.

Engine 1298cc 77x69.7mm/1584cc 82x75mm, V4 cyl OHC, F/F
Max power 87-103bhp at 6000rpm/115-132bhp at 6200rpm
Max speed 104-120mph
Body styles coupé
Prod years 1967-73
Prod numbers 7102

PRICES
1300 S1 **A** 10,000 **B** 7500 **C** 3500
1300 S2 **A** 9000 **B** 6500 **C** 3000
1600 **A** 12,000 **B** 8000 **C** 4000

Lancia Stratos

In 1970, Fiat took control of Lancia. The same year, Bertone displayed the Stratos, a striking mid-engined show car using Lancia parts. With a few changes, that car made it into production in 1973, Bertone building them. A steel monocoque was clothed with glassfibre panels; suspension was independent by coils and wishbones, and there were four-wheel discs. The Ferrari Dino donated its V6 engine. Works cars won the World Rally Championship twice. It took seemingly forever to sell the required number of these cramped and nervous-handling machines for homologation, but they are now avidly enjoyed.

Engine 2418cc 92.5x60mm, V6 cyl QOHC, M/R
Max power 190bhp at 7000rpm
Max speed 142mph
Body styles coupé
Prod years 1973-75
Prod numbers 492

PRICES
A 70,000 **B** 45,000 **C** 30,000

TECHNILOCK
and Welding Services Ltd.

CAST-IRON & ALLOY REPAIR SPECIALISTS

COLD METAL STITCHING & GAS FUSION WELDING

TO CRACKED CYLINDER HEADS, BLOCKS, MANIFOLDS, GEARBOXES, SUMPS ETC.

VETERAN & PRE-1920 ENGINES A SPECIALITY,

THROUGH TO MODERN-DAY CLASSICS

TOP-QUALITY REPAIRS FOR 20 YEARS

ALL WORK GUARANTEED

CALL: **DAVID BAKER**

TEL: (01283) 222202.
FAX: (01283) 222203

UNIT 7, VIKING BUSINESS CENTRE, HIGH STREET, WOODVILLE, DERBYSHIRE DE11 7EH

Richard Thorne
CLASSIC CARS

UNIT 1, BLOOMFIELD HATCH, MORTIMER, READING RG7 3AD
Tel: 01734 333633
Fax: 01734 333715

LANCIA SPECIALIST

We are leading specialists in the sales, service and restoration of Lancia cars of all ages. Our team of seven have extensive experience in the preparation of Lancias for historic motorsport and are proud to be 1995 HSCC Roadsport Overall Champions with Richard Thorne's Lancia Beta Coupé. We have our own in house bodyshop which is happy to carry out accident repairs to classics as well as restoration work. We have our own low bake paint oven on the premises. If you are interested in buying or selling a Lancia please contact me.

Lancia

Lancia Beta

The Beta was Fiat's idea of what a Lancia should be. Fiat supplied the basis for the engine, a twin cam unit with Weber carburation. Fitted transversely, it drove the front wheels, and you got four disc brakes as standard. Styling was smart, but there was no hatchback. Engine sizes were initially 1400, 1600 and 1800, with 1300 from '75. '76 Series 2 had bigger glass area and revised engine range: old 1300, new 1600 and 2000. ES model is highly specified. Good cars but destroyed by their appalling propensity for rusting just about everywhere. Even the best ones (with Ziebart protection) are cheap.

Engine 1297cc 76x71.5mm/1438cc 80x71.5mm/1592cc 80x79.2mm/ 1585cc 84x71.5mm/1756cc 84x79.2mm/1995cc 84x90mm, 4 cyl DOHC, F/F
Max power 82bhp at 5800rpm to 122bhp at 5500rpm
Max speed 100-112mph
Body styles saloon
Prod years 1972-82
Prod numbers 194,916
PRICES
A 1200 **B** 700 **C** 100

Lancia Beta Coupé/Spider

Chop 7.5in out of the floorpan of the Beta and clothe it with a pretty new 2+2 coupé body from Zagato and you have a recipe for an entertaining little car. Always five speeds, disc brakes, MacPherson strut suspension. 1300/1600/2000 engines joined by 1367cc unit in 1980s. 1975 Spider had similar bodywork to the coupé below the waist, but featured a 'convertible' roof (a fixed roll-over bar splitting a targa top and a drop-down rear section). Spider died before coupé, and fuel injection 2000 from 1982 was only model left. Supercharged 135bhp Volumex from '83 on coupé only.

Engine 1297cc 76x71.5mm/1367cc 78x71.5mm/1592cc 80x79.2mm/ 1585cc 84x71.5mm/1756cc 84x79.2mm/1995cc 84x90mm, 4 cyl DOHC, F/F **Max power** 82bhp at 6200rpm to 135bhp at 5500rpm
Max speed 103-125mph
Body styles coupé/convertible
Prod years 1973-84/1975-83
Prod numbers 111,801/9390
PRICES
Coupé **A** 2000 **B** 1200 **C** 400
Spider **A** 4000 **B** 2500 **C** 1000

Lancia Beta HPE

Lancia tried the sports estate theme on the standard Beta saloon floorpan in 1975 and it proved a long-lasting and attractive design. Coupé front end was mated to a sharply-styled rear end with a hatchback and individually folding rear seats. The HPE acronym stood for High Performance Executive. Only two engine sizes (1600/2000), but a high standard of trim and an automatic option. Like the Beta Coupé, fuel injection standard on 2000 from 1981, Volumex supercharged HPE from '83. 1600 version dies '84, 2000 the following year. Volumex versions are becoming sought after.

Engine 1585cc 84x71.5mm/1995 84x90mm, 4 cyl DOHC, F/F
Max power 100bhp at 5800rpm/115-135bhp at 5500rpm
Max speed 108-125mph
Body styles estate
Prod years 1975-85
Prod numbers 71,258

PRICES
HPE 1600/2000 **A** 2500 **B** 1200 **C** 500
HPE Volumex **A** 3800 **B** 2200 **C** 1000

Lancia Beta Montecarlo

Conceived as a bigger sister for the X1/9 (and planned to be sold as a Fiat), the Lancia Montecarlo plundered the Beta parts bin to create a good value sports car. Pininfarina's lines were purposeful but the reality was less sparkling: 120bhp engine not really powerful enough, roadholding dangerous in the wet, and severe front-wheel lock-up under braking forced production to halt in 1978. Revised version relaunched in '80: new grille, bigger wheels, non-servo front brakes, better suspension. Dreadful rust traps, and avoid US-only Scorpions with 1800 engines and 80bhp.

Engine 1756cc 84x79.2mm/1995cc 84x90mm, 4 cyl DOHC, M/R
Max power 80bhp at 5900rpm/120bhp at 6000rpm
Max speed 112/120mph
Body styles coupé, targa
Prod years 1975-78 and 1980-84
Prod numbers 7595

PRICES
A 6500 **B** 4500 **C** 2500

Lancia Gamma

Lancia had been without an executive class car since the demise of the Flaminia in 1970, but Fiat revived the idea with the Gamma in 1979. Pininfarina created a not entirely happy fastback saloon body, but the mechanical package – front-wheel drive, all-independent suspension, disc brakes all round, five speeds/automatic option – sounded competitive. Yet the flat-four engines (the 2.5-litre export unit was the largest 'four' then being made) were not powerful or refined enough for 1980s luxury car buyers. Apart from orders from the Italian diplomatic corps, the Gamma was a disastrous sales flop.

Engine 1999cc 91.5x76mm/2484cc 102x76mm, 4 cyl DOHC, F/F
Max power 115bhp at 5500rpm/140bhp at 5400rpm
Max speed 115-121mph
Body styles saloon
Prod years 1976-84
Prod numbers 15,296

PRICES
A 1800 **B** 1000 **C** 500

Lancia Gamma Coupé

If the Gamma saloon was frumpy, the Coupé was a stunning polar opposite. Pininfarina's elegant notchback lines fitted a Gamma platform shortened by 4.5in. Mechanically, the Coupé was identical to the saloon: 2-litre engine for Italy only, carburetted 2.5-litre unit at launch and fuel-injected version from '81. Like the saloon, did not arrive in UK until 1978. A good one is rewarding in the extreme, but good ones are rare. Typically, a Gamma has gasket problems, rust appearing everywhere and a large appetite for swallowing good money. First-rate credentials for future classic status.

Engine 1999cc 91.5x76mm/2484cc 102x76mm, 4 cyl DOHC, F/F
Max power 115bhp at 5500rpm/140bhp at 5400rpm
Max speed 115-121mph
Body styles coupé
Prod years 1976-84
Prod numbers 6789

PRICES
A 3000 **B** 2000 **C** 700

Lea Francis 12/14

Modernised pre-war engineering for the 1946 Lea Francis, sold in the 14HP class but available on request, very briefly, with a 12HP 1496cc engine. Standard engine had 'hemi' head and two camshafts. Aluminium body panels were semi-stressed and mated to a steel floor. Traditional, sturdy, and rather expensive. Rare two-door coupé made 1947-48. 1947 MkIII and MkIV: latter has radio and standard heater. Torsion bar IFS on the MkV/VI from '48 (except for the large 'woody' estate car which got it from 1950), Girling hydromech brakes in '49. Recessed headlamps from Four-Light of '50.

Engine 1496cc 69x100mm/1767cc 75x100mm, 4 cyl OHV, F/R
Max power 50-55bhp at 4800rpm/56-65bhp at 4700rpm
Max speed 70/75mph
Body styles saloon, estate, coupé
Prod years 1946-47/1946-54
Prod numbers 13/2133

PRICES
A 10,000 **B** 7000 **C** 2500

Lea Francis 14/70/18

The restyled, six-light version of the 14 was intended mainly for export. It was sold in the UK 1950-51 as the 14/70, with flowing front wings extending back to the rear wheels, non-suicide front doors, and narrow grille. The 18 (confusingly also known as the 2½-Litre, MkVII or Six-Light Saloon), was the same car with a 95bhp 2.5-litre engine. Full hydraulic brakes from 1952. 'Pillar of society' character, rather ponderous in looks and performance and some problems from chassis rust and body stress. But there is an excellent club and strong parts support.

Engine 1767cc 75x100mm/2496cc 85x110mm, 4 cyl OHV, F/R
Max power 70bhp at 4800rpm/95bhp at 4000rpm
Max speed 78/90mph
Body styles saloon
Prod years 1948-51/1949-54
Prod numbers 162/69
PRICES
14/70 **A** 10,000 **B** 7000 **C** 3000
18 **A** 12,000 **B** 8000 **C** 3500

Lea Francis 12/14/2½-Litre Sports

LeaF's sports car was certainly individual: flush headlamps, cutaway doors, humped scuttle and rear wheel spats. Launched in 1947 with the 1.5-litre 12HP engine, but the 14HP unit – in twin carb form – was standardised the following year. Chassis was the same as the saloon (though shortened by a foot), so you still got beam axles and mechanical brakes – hardly pace-setting stuff. 2½-Litre model of '49 has same chassis but 3in wider body with fuller doors and winding perspex windows, IFS and hydromech (later hydraulic) brakes, plus 100mph performance. Occasional rear seats useful.

Engine 1496cc 69x100mm/1767cc 75x100mm/2496cc 85x110mm, 4 cyl OHV, F/R **Max power** 64bhp at 5000rpm/77-87bhp at 5100rpm/105bhp at 4000rpm to 125bhp at 5200rpm **Max speed** 75-100mph
Body styles sports
Prod years 1947-48/1947-49/1949-54
Prod numbers n/a/118/77

PRICES
12/14 **A** 15,000 **B** 11,000 **C** 8000
2½-Litre **A** 20,000 **B** 14,000 **C** 10,000

Leyland P76

Sometimes described as 'Australia's Edsel', the Leyland P76 was a genuine catastrophe. Conceived by Leyland Australia to rival the Holden and Ford Falcon, it was a conventional affair with styling by Michelotti. Straight six engine (derived from BL E-series) was standard, but expanded Rover V8 was optional. Production delays, poor quality control and the oil crisis squashed sales and forced Leyland to abandon Aussie manufacture. Force 7 coupé version was stillborn (9 complete cars were made). A tiny handful of saloons were imported to the UK, but do any survive?

Engine 2622cc/4416cc, 6/V8 cyl OHC/OHV, F/R
Max power 130-192bhp at 4100rpm
Max speed 105-120mph
Body styles saloon
Prod years 1973-74
Prod numbers c 22,000

PRICES
A 2000 **B** 1200 **C** 800

Ligier JS2

French millionaire racing driver and ex-rugby international Guy Ligier decided to build his own sports car in 1970. The JS designation was in memory of his friend, Jo Schlesser, killed racing in 1968. A pressed steel platform was fitted with a glassfibre coupé body. As presented at the 1970 Paris Salon, it had a mid-mounted Ford Capri engine, but production cars had Citroën SM V6 engines turned through 180 degrees (Ligier also built the SM). Servo discs all round, all-independent suspension. Ligier went on to make F1 cars and microcars. JS2 never sold in UK, but in-the-know French enthusiasts appreciate them.

Engine 2670cc 87x75mm/2965cc 91.6x75mm, V6 cyl QOHC, F/R
Max power 170bhp at 5500rpm/195bhp at 5500rpm
Max speed 150/153mph
Body styles coupé
Prod years 1970-77
Prod numbers c 150

PRICES
A 15,000 **B** 10,000 **C** 5000

Lincoln Continental '41-'48

With its 1936 Zephyr V12, Lincoln created a monolith in American culture. The Continental went one better: long bonnet, huge spare wheel on the boot, and an imposing presence. Club coupe and convertible styles were offered, the latter with an optional power top from '41. Larger 5018cc engine from '41. Dropped in 1942, the model was revived after the war with minor styling changes and the original smaller engine. Ford V8-derived sidevalve engine's power output was nothing to write home about.

Engine 4378cc 70x95mm/5018cc 74.6x95.2mm, V12 cyl SV, F/R
Max power 120/130bhp at 3600rpm
Max speed 90mph
Body styles coupe, convertible
Prod years 1941-42/1946-48
Prod numbers Coupe: 2993/Convertible: 1927

PRICES
Coupe **A** 20,000 **B** 14,000 **C** 8000
Convertible **A** 40,000 **B** 30,000 **C** 20,000

Lincoln Continental MkII

In 1955 the Continental name returned as a marque in its own right when Lincoln displayed what was one of the biggest stars of the 1955 Paris Salon. The new coupe instigated a deluge of orders, even though the asking price was $10,000 exactly – over five times the price of a contemporary Ford Mainline. That huge price tag was partly justified by the hand-built nature of much of the car. As a result, very few were actually built. Only two prototypes were made of the attractive 1957 convertible version. A latter-day Duesenberg, but Ford lost $1000 on every one made.

Engine 6031cc 101.6x93mm, V8 cyl OHV, F/R
Max power 285-300bhp at 4800rpm
Max speed 110mph
Body styles coupe
Prod years 1956-57
Prod numbers 1767

PRICES
A 25,000 **B** 18,000 **C** 12,000

Lincoln Continental MkIII/IV/V

The '58 Continental marked an about-change in thinking: its sticker price plummeted to around $6000, as it was basically an up-market edition of the Lincoln Capri/Premiere. A monstrous machine: 18ft 9in (572cm) long, and 2¼ tons. Unsubtle in the extreme, with its diagonal quad headlamps, reverse-angle rear screen, hefty bumpers, scalloped wings, the largest engine then being made and goodies dripping from every quarter. Became a Lincoln again for 1959, with additional limo and Town Car styles. 1960 MkV had styling retouches which made it appear slimmer.

Engine 7045cc 109.2x94mm, V8 cyl OHV, F/R
Max power 315-375bhp at 4100rpm
Max speed 116-120mph
Body styles saloon, hardtop saloon, coupe, convertible, limousine
Prod years 1957-58/1958-59/1959-60
Prod numbers 12,550/11,126/11,086
PRICES
Saloon **A** 10,000 **B** 6000 **C** 4000
Coupe **A** 17,000 **B** 10,000 **C** 6000
Convertible **A** 25,000 **B** 18,000 **C** 10,000

Lincoln Continental '61-'69

The Continental became Lincoln's only offering in 1961, as a brand-new 'downsized' pair of models bowed in: a four-door saloon and a convertible with the unusual fitment of four doors, arranged in 'clap hands' format. It needed to be rigid – and it was. Simple lines, sound engineering and an endorsement from the White House gave the Continental an almost unstoppable momentum. Convertible had an electric hood. Extended wheelbase from '64, two-door coupe and bigger engine from '66. A surprising best-seller and one of the all-time great American car designs, and the market recognises this.

Engine 7045cc 109.2x94mm/7565cc 111.25x97.3mm, V8 cyl OHV, F/R
Max power 300-365bhp at 4600rpm
Max speed 120-125mph
Body styles saloon, convertible, coupe
Prod years 1961-69
Prod numbers 342,781
PRICES
Saloon '61 **A** 10,000 **B** 7000 **C** 3500
Convertible **A** 20,000 **B** 16,000 **C** 12,000
Coupe **A** 8000 **B** 5000 **C** 2000

Lloyd 300/400/250

Because of severe post-war metal shortages, the 1950 Lloyd was introduced by the parent Borgward group with fabric body panels over a wooden frame. This helped keep its price down to 25% less than a VW Beetle. Since the Lloyd looked at least superficially like a real car and could seat four, it sold strongly. Air-cooled two-stroke twin engine, choice of bodies (LP = saloon, KS/LS = estate, LC = coupé). The 400 models arrived in 1953 and graduated to all-steel construction. A super-economy 250 model arrived in 1956, aimed at holders of a German Group 4 (motorbike) licence.

Engine 293cc 54x64mm/386cc 62x64mm/250cc 50x64mm, 2 cyl TS, F/F
Max power 10bhp at 4000rpm/13bhp at 3750rpm/11bhp at 5000rpm
Max speed 47/52/47mph
Body styles saloon, estate, coupé, rolltop saloon
Prod years 1950-52/1953-57/1956-57
Prod numbers 18,087/109,878/3768

PRICES
A 3500 B 2200 C 1000

Lloyd 600/Alexander

The LP600 joined its smaller-engined brethren in 1955. Under the familiar bonnet sat an all-new four-stroke two-cylinder overhead cam engine. The Alexander version from '57 had a four-speed synchro 'box, while the '58 Alexander TS had a spine-tingling 25bhp on tap, independent rear suspension and the option of Saxomat automatic transmission. It could out-accelerate a Beetle, but it now cost a fraction more, which explains the model's rapid decline after the late 1950s. There was even a miniature MPV version called the LT600 with three rows of seats and a van-like appearance.

Engine 596cc 77x64mm, 2 cyl OHC, F/F
Max power 19bhp at 4500rpm to 25bhp at 5000rpm
Max speed 62-70mph
Body styles saloon, estate, rolltop saloon
Prod years 1955-62/1957-62
Prod numbers 176,524

PRICES
LP600 A 2800 B 1800 C 800
Alexander TS A 3200 B 2000 C 900

Lloyd Alexander Frua

One of Piero Frua's most unlikely projects was the task of rebodying the LP600 with a sports car body. Its stylised coupé bodywork (L-shaped bumpers, headlamps recessed in chrome nacelles, wrap-around rear screen) resembled a shrunken Renault Floride, but its performance could never match its looks. It was also fiendishly expensive new, and although it had lots of goodies like a rev counter and wooden steering wheel, this was scant compensation for the price and few were sold. Today, Lloyd collectors chase each other to secure the opportunity to buy one of the rare survivors.

Engine 596cc 77x64mm, 2 cyl OHC, F/F
Max power 25bhp at 5000rpm
Max speed 67mph
Body styles coupé
Prod years 1959
Prod numbers 49

PRICES
A 14,000 B 10,000 C 8000

Lloyd Arabella

Faced with the realisation that the microcar boom would not last forever, Lloyd moved up-market with the larger Arabella. This had a four-cylinder engine, independent front suspension by double wishbones and coil springs and a column gearchange. Sadly the new engine was put into production before being developed properly and the resulting warranty claims helped tip the Borgward combine towards financial disaster. Styling was transatlantic miniature, with fins, a strong swage line and wrap-around rear screen. The last Arabellas were badged as Borgwards, but the nasty taste lingered.

Engine 897cc 69x60mm, 4 cyl OHV, F/F
Max power 38bhp at 4800rpm to 45bhp at 5300rpm
Max speed 75-80mph
Body styles saloon
Prod years 1959-63
Prod numbers 47,042

PRICES
A 2000 B 1200 C 900

LMX 2300 HCS

The LMX was one of those innumerable Italian sports car projects of the 1960s, but this one did actually make it to production. Glassfibre bodywork (designed by Eurostyle of Turin) hid a forked backbone chassis, Ford Zodiac MkIV suspension and disc brakes, and a Ford Taunus 2.3-litre V6 engine in various states of tune, up to – would you believe? – a 210bhp turbocharged lump. Although it was, and is, an extremely rare bird, if you find one it might make a good-looking and usable sports car which is cheap to run – and it's just about the only Italian car which doesn't rust.

Engine 2293cc 90x60.1mm, V6 cyl OHV, F/R
Max power 108-210bhp at 5100rpm
Max speed 123-145mph
Body styles coupé, convertible
Prod years 1969-74
Prod numbers 43

PRICES
Too rare to value accurately, but will be in the £3000-£12000 range

Lotus MkVI

Amateur builder/racer Colin Chapman was persuaded to enter production with the MkVI after winning just about everything he entered. An ultra-basic sports machine, it had a space frame, aluminium body panels of the starkest nature, split Ford front beam axle, IFS and Ford live rear axle. Wide choice of engines, but mostly Ford Ten or MG TC. Always kit-built, the package was quite comprehensive but totally unfinished. Its excellent competition record makes it a favourite of historic racers today.

Engine 1172cc to 1500cc, 4 cyl SV/OHV, F/R
Max power 40bhp to 75bhp
Max speed 75-90mph
Body styles sports
Prod years 1952-56
Prod numbers c 110

PRICES
A 15,000 B 12,000 C 8000

Lotus Seven S1/S2/S3

Chapman said he knocked the Seven off in a week; it has now lasted almost 40 years and become a British sports car institution. It was an updated MkVI, having a new space frame, Lotus Eleven Club type rear suspension, modified Lotus Twelve front end, stressed aluminium body panels, Ford hydraulic drums. Power from Ford sidevalve, BMC A-series, or Coventry-Climax (Super Seven). S2 of 1960 had flared GRP front wings, new nose, 105E power option. '61 Super Seven had 95bhp Cosworth 109E engine. S3 had Cortina 1600 engine, Escort back axle. Ultimate '69 Twin Cam SS had Lotus power (13 built).

Engine 948cc to 1599cc, 4 cyl SV/OHV/DOHC, F/R
Max power 37bhp at 4800rpm to 115bhp at 5500rpm
Max speed 80-108mph
Body styles sports
Prod years 1957-60/1960-68/1968-70
Prod numbers 242/1350/350
PRICES
S1 A 14,000 B 11,000 C 9500
S2/S3 A 13,000 B 10,000 9000
S2 Cosworth A 15,000 B 12,000 C 10,000

Lotus Seven S4

Lotus hoped the Seven S4 would capture some of the Midget/Spitfire market and, to some extent, it succeeded, but the company was not geared up to satisfying MG buyers' expectations of after-sales service. The S4 was a total departure from earlier Sevens: an Alan Barrett styled all-glassfibre body bonded on to a new chassis, and an appearance which bordered on beach buggydom. Lotus Europa-type front suspension, Watt linkage at the rear. Cortina 1300/1600 and Lotus twin cam engine options. Soft and civilised when compared to other Sevens, but still fun. Enthusiasts loathe the compromises.

Engine 1297cc 81x63mm/1599cc 81x77.6mm/1558cc 82.55x72.75mm, 4 cyl OHV/OHV/DOHC, F/R
Max power 70bhp at 5700rpm/84bhp at 6500rpm/115bhp at 5500rpm to 125bhp at 6200rpm
Max speed 100-116mph
Body styles sports
Prod years 1970-73
Prod numbers c 900
PRICES
1300/1600 A 8000 B 6000 C 4000
Twin Cam A 9000 B 6500 C 5000

Lotus Eleven

While the XI (Eleven) was primarily a racing machine, it was conceived in the days when racers drove their cars to the event they would compete in. As such, the Eleven was offered with full lighting equipment and even a basic hood. Fabulous aerodynamic bodywork, and one of the best space frames ever seen. Three versions were offered: the Le Mans (Coventry-Climax power in 75, 83 or 100bhp forms, De Dion rear end), Club (75bhp Climax power and live axle/coil springs), and Sport (Ford sidevalve engine and live rear axle). Wishbone IFS after 1956. A1 collectable.

Engine 1172cc 63.5x92.5mm/1098cc 72x67mm/1462cc 76x80mm, 4 cyl SV/DOHC/DOHC, F/R
Max power 36bhp at 4500rpm/75-83bhp at 6800rpm/100bhp at 6200rpm
Max speed 120-160mph
Body styles sports
Prod years 1956-60
Prod numbers 426

PRICES
A 38,000 B 30,000 C 25,000

Lotus Elite

Chapman applies his ingenuity to a true road car, looking for profit. In fact the Elite lost lots of money, but it was a technical triumph. As the world's first competent glassfibre monocoque, it was a daring design, and handsome too. IFS by wishbones and coils, Chapman struts at the rear, discs all round (inboard at the rear). Coventry-Climax alloy engines were light and powerful (up to 95bhp in the '62 Super 95 and more in the Super 100/105), leading to accusations that the plastic body cracked under stress. SE of 1960 had close-ratio ZF 'box. Noisy, harsh-riding but gloriously balanced.

Engine 1216cc 76.2x66.7mm, 4 cyl OHC, F/R
Max power 71bhp at 6100rpm to 105bhp at 7250rpm
Max speed 110-130mph
Body styles coupé
Prod years 1957-63
Prod numbers 998

PRICES
A 26,000 B 20,000 C 15,000

Lotus Elan S1/S2/S3/S4/Sprint

The Elan is still regarded, even today, as a benchmark in handling. Having learned its lesson about monocoque road cars, Lotus gave the Elan a steel backbone chassis, but stuck with a GRP body, independent suspension and all-round discs. Pop-up headlamps a novelty and overall dimensions are extremely compact. S2 of '64 has improved brakes, better dashboard. S3 has higher final drive, comes as coupé as well, SE has 115bhp and brake servo. S4 has flared arches, standard servo. Delectable Sprint boasts 126bhp big-valve engine, two-tone paint, may have five speeds.

Engine 1558cc 82.5x72.75mm, 4 cyl DOHC, F/R **Max power** 106bhp at 5500rpm to 126bhp at 6500rpm **Max speed** 112-124mph **Body styles** sports, coupé **Prod years** 1962-64/1964-66/1965-68/1968-71/1970-73 **Prod numbers** c 9150 (900/1250/2650/3000/1350)
PRICES
S1 **A** 15,000 **B** 12,000 **C** 7500
S2/S3/S4 **A** 13,000 **B** 10,000 **C** 6000
Sprint **A** 15,000 **B** 12,000 **C** 8500

Lotus Elan +2

The market was demanding a four-seater Lotus, so the Elan was given a new shell, stretched by a foot and made wider to allow room for two small seats in the rear. Coupés only, and not now offered in kit form, unlike all other Lotus models to this time. Standard 118bhp engine, S models have better trim and foglamps. 126bhp big-valve unit and two-tone body on post-'70 cars (called +2S/130), and a few were made to +2S/130-5 spec with five speeds (from '72). Not as sharp handling or as fast as the ordinary Elan, but still a great sports car. On all Elans, watch for chassis rust.

Engine 1558cc 82.55x72.75mm, 4 cyl DOHC, F/R **Max power** 118bhp at 6250rpm/126bhp at 6500rpm **Max speed** 112-120mph **Body styles** coupé **Prod years** 1969-74 **Prod numbers** c 3300

PRICES
+2 **A** 7500 **B** 5000 **C** 4000
+2S/130 **A** 9000 **B** 7000 **C** 5500

Lotus Europa S1/S2/Twin Cam

New 'cheap' Lotus, originally intended to oust the Seven but redirected to become Lotus's 'car for Europe' – hence the name. Modified Renault 16 engine/transaxle sat in the middle of an Elan-type backbone chassis, to which was bonded a very low glassfibre body. There were unkind comments about its bread van styling, its cockpit was claustrophic (the windows didn't even open), and visibility was poor. But nothing handled as well as a Europa. S2 had bolt-on body, electric windows, improved cockpit. Underpowered, so Lotus twin cam engine from '71 a better bet. S2 UK-available from '69 on, 1565cc unit for USA only.

Engine 1470cc 76x81mm/1565cc 77x84mm/1558cc 82.55x72.75mm, 4 cyl OHV/OHV/DOHC, M/R **Max power** 78-82bhp at 6000rpm/83-87bhp at 6000rpm/105-126bhp at 6500rpm **Max speed** 105-125mph **Body styles** coupé **Prod years** 1966-68/1968-71/1971-75 **Prod numbers** 9230
PRICES
S1 **A** 5000 **B** 4000 **C** 2700
S2 **A** 6000 **B** 5000 **C** 3000
Twin Cam **A** 11,000 **B** 8500 **C** 6000

Lotus Elite/Eclat S1

Classic backbone chassis format remained for the otherwise all-new Elite of 1974, which took Lotus firmly up-market. Sharply styled glassfibre hatchback body could seat four but was controversial. New 2-litre 16-valve 907 engine, first seen in Jensen-Healey, gave tremendous performance for a 'four'. Five speeds, independent suspension, four-wheel discs. Several models: base 501, 502 (air con), 503 (PAS), 504 (auto). Fastback Eclat (pictured) followed in '75, had rare four-speed 'box option, better aerodynamics, cheaper price. Build quality and reliability were Achilles' heels.

Engine 1973cc 95.2x69.3mm, 4 cyl DOHC, F/R **Max power** 160bhp at 6200rpm **Max speed** 126/130mph **Body styles** sports hatchback/coupé **Prod years** 1974-80/1975-80 **Prod numbers** 2398/1299

PRICES
Elite **A** 3500 **B** 2000 **C** 1000
Eclat **A** 4000 **B** 2500 **C** 1200

Lotus Esprit S1/S2/S2.2

Giugiaro excelled himself when he designed the Esprit, first shown as a concept car as early as 1972. Dramatic two-seat wedge shaped plastic body clothed backbone chassis, independent suspension and disc brakes all round. Transaxle came from Citroën's SM and 907 engine was mounted centrally. S1s had cooling problems, sorted by '79 S2 with its wider wheels, better interior, integrated spoiler. S2.2 of 1980 had the larger 917 2174cc engine (same power but more torque). A brilliant supercar in terms of handling and image, but flawed in so many ways. Can be picked up very cheaply.

Engine 1973cc 95.2x69.3mm/2174cc 95.2x76.2mm, 4 cyl DOHC, M/R **Max power** 160bhp at 6200rpm/160bhp at 6500rpm **Max speed** 135mph **Body styles** coupé **Prod years** 1976-78/1978-80/1980-81 **Prod numbers** 994/980/88

PRICES
S1 **A** 5000 **B** 4000 **C** 2500
S2 **A** 6000 **B** 4500 **C** 3000
S2.2 **A** 7500 **B** 5500 **C** 3500

Marcos GT

'Wooden Wonder' was the natural nickname for Jem Marsh and Frank Costin's incredibly ugly gullwing coupé: it had a wooden monocoque construction. Indeed, the only non-wooden part was the GRP nose cone. Triumph Herald front coils, Ford live axle located by leading arms and Panhard rod at the rear, Ford 105E/Classic engines. Very successful on the track (eg Bill Moss, Jackie Stewart), simply too hideous for much commercial success. Two restyles by Dennis Adams, increasing the use of glassfibre.

Engine 997cc 81x48.4mm/1340cc 81x65.1mm/1498cc 81x72.7mm, 4 cyl OHV, F/R **Max power** 41bhp at 5000rpm to 64bhp at 4600rpm
Max speed 95-110mph
Body styles coupé
Prod years 1960-64
Prod numbers 39

PRICES
Too rare for accurate valuation, but examples have fetched in excess of £20,000

Marcos 1800/1500/1650/1600

The classic Marcos design was created by Dennis Adams. The chassis was laminated plywood, but the sculpted body was glassfibre. Triumph front wishbones, semi-De Dion and coil sprung rear (switched to Ford live axle after first year), Volvo P1800 engine. Sold mostly in kit form. Cortina 1500 power from 1965, optionally bored out by Lawrencetune to 1650cc. Ford crossflow 1600 engine from '67. Shape still looks futuristic today – indeed, you can still buy a Marcos which looks broadly the same. Wooden chassis can rot, in which case repair can be expensive.

Engine 1780cc 84.1x80mm/1498cc 81x72.7mm/1599cc 81x77.6mm/ 1650cc, 4 cyl OHV, F/R
Max power 96-114bhp at 5800rpm/82bhp at 5200rpm/120bhp/88bhp at 5400rpm
Max speed 115/110/125/112mph
Body styles coupé
Prod years 1964-65/1965-67/1967/ 1967-69
Prod numbers 99/82/32/192
PRICES
A 9000 B 6000 C 4000

Marcos 3-Litre/3-Litre Volvo/2½-Litre/2-Litre

More power was what the Marcos chassis had always called out for and it got it in 1968 when Jem Marsh fitted a Ford 3-litre V6 engine under the bonnet. Early 3-Litres had the old wooden chassis, but steel replaced it after about 100 cars. To please the US market, Marcos beat a path to Volvo, fitting the 164 3-litre power unit in 1970. There was a short-lived 2½-Litre model in 1971 with a Triumph TR6 engine. The last 'classic shape' Marcos was the 2-Litre, fitted with a Ford V4 engine, but lacking the overdrive of the larger-engined models. Marcos relaunched production in 1981.

Engine 2994cc/2978cc/2498cc/ 1996cc, V6/6/6/V4cyl OHV, F/R
Max power 140bhp at 5700rpm/ 130bhp at 5000rpm/150bhp at 5500rpm/85bhp at 5000rpm
Max speed 112-130mph
Body styles coupé **Prod years** 1968-71/1970-71/1971/1970-71
Prod numbers 100/250/11/40
PRICES
2-Litre A 8000 B 6000 C 4000
2½-Litre/3-Litre A 11,000 B 8000 C 5000

Marcos Mini-Marcos

The only Marcos not to have had some design input by Dennis Adams, the Mini-Marcos was born out of the DART prototype (as was the similar Minijem). Of glassfibre monocoque construction, it used complete front and rear Mini subframes bolted through metal plates. Basic kits started at just £199, so it was no surprise that this was a crude machine. The only British finisher at Le Mans in '66. Improved slightly until the firm's demise in 1971. Rob Walker took it on and made it with a rear hatch and winding windows; from 1975-81 it was the precursor of Harold Dermott's Midas. Recently relaunched in MkV form.

Engine 848cc 62.9x68.3mm to 1275cc 70.6x81.3mm, 4 cyl OHV, F/F
Max power 34bhp at 5500rpm to 76bhp at 5800rpm
Max speed 80-105mph
Body styles coupé
Prod years 1965-81
Prod numbers c 1200

PRICES
A 3500 B 2200 C 1200

Marcos Mantis

The Mantis has had a lot of bad publicity over the years: it has been slated for its 'rashers of bacon' styling and blamed for Marcos's crash of 1971. It's true that the shape did not come out quite as intended, but it was not the main cause of the financial trouble. Marcos's first four-seater, it had a new semi-spaceframe chassis, TR6 engine, Triumph GT6 IFS, coil-sprung live rear axle, and a luxurious interior. Surprisingly good ride and reasonable performance, but expensive in its day (even in kit form). Very rare, it has its own eccentrics-only following.

Engine 2498cc 74.7x95mm, 6 cyl OHV, F/R
Max power 150bhp at 5500rpm
Max speed 125mph
Body styles coupé
Prod years 1970-71
Prod numbers 32

PRICES
A 6000 B 4000 C 2000

Maserati A6/A6G/ A6G/2000

The first ever Maserati road car was the 1946 A6/1500. This had a 1.5-litre straight six engine derived from the supercharged 6CM racer of 1936. The simple ladder frame had coil spring and wishbone front suspension, live rear axle, four-speed 'box. Maserati sold only the rolling chassis; most were bodied by Pinin Farina with a full-width coupé or convertible body, plus some by Zagato. A6G of '51 had a 1954cc engine, while the A6G/2000 of 1954 had a detuned 1986cc F2 twin cam six. Large range of bodies, and all are hugely sought-after.

Engine 1488cc 66x72.5mm/1954cc 72x80mm 1986cc 76.5x72mm, 6 cyl OHC/OHC/DOHC, F/R **Max power** 65bhp at 4700rpm/100bhp at 5500rpm/120-140bhp at 6000rpm **Max speed** 94/105/125-130mph **Body styles** coupé, sports, convertible **Prod years** 1946-51/1951-53/1954-57 **Prod numbers** 61/16/61

PRICES
A 140,000 B 110,000 C 90,000

Maserati 3500GT/GTI

With the 3500GT, Maserati became a more serious supercar player and true Ferrari competitor. Chassis was derived from the A6G, though it had semi-elliptic rear suspension and a powerful 3.5-litre straight six twin cam engine with three twin-choke Webers (or Lucas fuel injection from '61, which was a maintenance nightmare). All-synchro 'boxes were initially four speeds; from '61, five-speed (even automatic from '64). Braking evolved from drums to disc/drums and eventually all-wheel discs. Touring built the coupé bodies and Vignale the shorter Spider convertibles.

Engine 3485cc 86x100mm, 6 cyl DOHC, F/R **Max power** 220-235bhp at 5500rpm **Max speed** 135-142mph **Body styles** coupé, convertible **Prod years** 1957-64 **Prod numbers** 1981/242

PRICES
Coupé A 28,000 B 20,000 C 15,000
Spider A 40,000 B 27,000 C 19,000

Maserati 5000GT

While the 3500GT derived its engine from the 350S racer, the 5000GT got its quad cam V8 from the 450S. This included gear-driven camshafts and four Webers and was mated to a four-speed 'box (later ZF five-speeders). The chassis was shared with the 3500GT. Disc brakes all round and new chain-driven camshaft engine with Lucas fuel injection from '61. Incredibly powerful machines: 170mph and 0-60mph in 6.5 secs in 1959 – and even more for the three customers who ordered full race engines! Most came with Allemano built 2+2 coupé bodies; others came from Touring, Frua, Ghia, etc., including at least one convertible.

Engine 4935cc 98.5x81mm/4941cc 94x89mm, V8 cyl QOHC, F/R **Max power** 330bhp at 5700rpm/ 370bhp at 6000rpm **Max speed** 170-175mph **Body styles** coupé, convertible **Prod years** 1959-64 **Prod numbers** 32

PRICES
A 110,000 B 90,000 C 65,000

Maserati Sebring I/II

By using the short wheelbase 3500GTI convertible chassis and clothing it with a smart 2+2 coupé body by Vignale, Maserati created a useful extra model, the Sebring. Five speeds, disc brakes, fuel injection (though many have now been converted to less troublesome carbs). Series II had 3.7-litre or 4-litre engines, hooded headlamps and a more restrained bonnet scoop. The main target market was America, so air conditioning, a radio and three-speed automatic transmission appeared on the options list. All had four headlamps, many had wire wheels. Excellent value today.

Engine 3485cc 86x100mm/3694cc 86x106mm/4014cc 88x110mm, 6 cyl DOHC, F/R **Max power** 235bhp at 5500rpm/ 245bhp at 5200rpm/255bhp at 5200rpm **Max speed** 142/145/150mph **Body styles** coupé **Prod years** 1962-65/1965-66 **Prod numbers** 346/98

PRICES
A 25,000 B 18,000 C 12,000

Maserati Mistral Coupé/Spider

An even shorter (94.5in) and reinforced version of the old chassis, now with square tubes, sat under the new Mistral of 1963. Better handling resulted, while performance was enhanced by lighter weight. Engines identical to the Sebring, though the 4.0-litre unit does not make it until '67. Pretty bodywork was by Frua: the coupé version had a large curved glass opening tailgate, while the '64 Spider could be ordered with a factory steel hardtop. If it looks like an AC 428, it's no surprise, because Frua designed that too – and even had the cheek to use some Mistral panels!

Engine 3485cc 86x100mm/3694cc 86x106mm/4014cc 88x110mm, 6 cyl DOHC, F/R **Max power** 235bhp at 5500rpm/ 245bhp at 5200rpm/255bhp at 5200rpm **Max speed** 145/150/155mph **Body styles** coupé, convertible **Prod years** 1963-70 **Prod numbers** 828/120

PRICES
Coupé A 30,000 B 20,000 C 10,000
Spider A 40,000 B 30,000 C 22,000

Maserati Quattroporte

Quattroporte means 'four doors', and Maserati started a new Italian fashion with its high-performance luxury saloon. It was the fastest four-door car in the world, at 130mph when launched in 1963. Frua designed and built the bodywork, though it was not especially elegant. Early cars had a complicated De Dion rear axle, but after 1966 semi-elliptics were standardised. Gearbox choices were ZF five-speed manual or Borg Warner auto. Original 4.1-litre V8 superseded by 4.7 in '67 and the following year air conditioning was fitted. Not much fancied thanks to its looks and rear doors.

Engine 4136cc 88x85mm/4719cc 93.9x85mm, V8 cyl QOHC, F/R
Max power 260bhp at 5200rpm/ 290bhp at 5000rpm
Max speed 130/140mph
Body styles saloon
Prod years 1963-71
Prod numbers 759

PRICES
A 9000 **B** 7000 **C** 4000

Maserati Mexico

Basically a Quattroporte mechanically, on a shorter wheelbase and with rather bland notchback coupé bodywork by Vignale. However, there was a live rear axle and semi-elliptic springs from the outset in 1965. Full four seats, and a bulky, thirsty machine (well over 1½ tons and 15mpg), but a good performer thanks to that lusty four-Weber quad cam V8 (0-60mph in 7.5 secs). 4.7-litre version was offered from 1969, when air conditioning became standard. PAS optional. Not a strong seller, particularly in America, and one of the cheaper Maseratis today.

Engine 4136cc 88x85mm/4719cc 93.9x85mm, V8 cyl QOHC, F/R
Max power 260bhp at 5200rpm/ 290bhp at 5000rpm
Max speed 140/150mph
Body styles coupé
Prod years 1965-72
Prod numbers 250

PRICES
A 23,000 **B** 17,000 **C** 11,000

Maserati Ghibli

In his capacity as chief designer at Ghia, Giugiaro penned the lines of the Ghibli, one of the designs of which he is most proud today. Same chassis as the Mexico (albeit shortened by 3.5in) so archaic live axle and leaf springs remained. Same mechanical specification as Mexico: i.e. five-speed/auto 'boxes, disc brakes, but more powerful 4.7-litre quad cam V8. From '70, the SS had a 4.9-litre engine with only 5bhp extra but a huge amount of torque. Very pretty Spider available from '69, also with factory hardtop; but beware of coupés which have been converted to Spiders. SS models worth 10% extra.

Engine 4719cc 93.9x85mm/4930cc 93.9x89mm, V8 cyl QOHC, F/R
Max power 330bhp at 5500rpm/ 335bhp at 5500rpm
Max speed 165mph
Body styles coupé, convertible
Prod years 1966-73
Prod numbers 1247 (Coupé: 1149; Spider: 125)

PRICES
Coupé **A** 26,000 **B** 20,000 **C** 14,000
Spider **A** 45,000 **B** 30,000 **C** 22,000

Maserati Indy

Successor to the Sebring, based on the Quattroporte type chassis (102.5in wheelbase), but with its body welded to the frame to make a semi-unitary whole. Design and manufacture of the body was by Vignale, and the shape was compromised by the requirement for four seats. Four-Weber 4.1-litre V8 at first, then 4.7 in 1970 and finally the 4.9 in 1973. ZF five-speed or optional auto, pop-up headlamps and electric windows standard, LSD option. A good seller for Maserati, but lacking the panache and appeal of the two-seater Ghibli. Fast and good value, but rust-ridden.

Engine 4136cc 88x85mm/4719cc 93.9x85mm/4930cc 93.9x89mm, V8 cyl QOHC, F/R
Max power 260bhp at 5200rpm/ 290bhp at 5000rpm/335bhp at 5500rpm
Max speed 152/156/162mph
Body styles coupé
Prod years 1966-74
Prod numbers 1136

PRICES
A 22,000 **B** 17,000 **C** 11,000

Maserati Bora

Another Giugiaro masterpiece for Maserati's first ever mid-engined road car, a straight competitor for the Lamborghini Miura. Steel unitary construction, Alfieri designed all-independent suspension, familar 4.7-litre V8 (after '76, uprated to the 4.9-litre), mounted longitudinally on a subframe and mated to a five-speed transaxle. Citroën had taken over Maserati by this stage, so olio-pneumatic brakes standard, as were hydraulically operated seats, pedal adjustment and pop-up headlamps. High speed and stunning beauty make for firm prices, but inexpensive compared with a Miura.

Engine 4719cc 93.9x85mm/4930cc 93.9x89mm, V8 cyl QOHC, M/R
Max power 310bhp at 6000rpm/ 335bhp at 5500rpm
Max speed 160/165mph
Body styles coupé
Prod years 1971-78
Prod numbers 571

PRICES
A 35,000 **B** 25,000 **C** 18,000

Maserati Merak 2000/3000/SS

Same structure as Bora and virtually the same body, too, but a smaller V6 engine from the Citroën SM made the Merak more of a Dino rival. Only slightly lighter than the Bora but much less powerful, so not nearly as much raw energy, but still plenty of entertainment value in handling terms. Early examples had Citroën SM instruments and steering wheel, swapped after 1975 for conventional items; also Citroën five-speed 'box ditched in 1976 for ZF unit. 2+2 seating arrangement was marginal in the extreme. SS version of '74 had 220bhp, while the Merak 2000 was an Italy-only tax break special.

Engine 1999cc 80x66.3mm/2965cc 91.6x75mm, V6 cyl QOHC, M/R
Max power 159bhp at 7000rpm/ 190bhp at 6000rpm/220bhp at 6500rpm
Max speed 127/145/153mph
Body styles coupé
Prod years 1972-83
Prod numbers 1140 (102/814/224)

PRICES
A 20,000 **B** 14,000 **C** 9000

Maserati Khamsin

Marcello Gandini, stylist of the Lamborghini Miura, was also responsible for the Maserati Khamsin. Unitary construction by Bertone, 4.9-litre V8 mounted up front, standard rack-and-pinion steering with Citroën's power assistance (plus power brakes), Ghibli-derived suspension. Five-speed manual gearbox or automatic option. A pretty heavy car (3500lb/1550kg), so not as fast as many previous front-engined Maseratis, but you couldn't call it slow. 2+2 seating was optimistic and the interior was not especially comfortable. Not the most sought-after Maserati.

Engine 4930cc 93.9x89mm, V8 cyl QOHC, M/R
Max power 320bhp at 5500rpm
Max speed 160mph
Body styles coupé
Prod years 1973-82
Prod numbers 421

PRICES
A 19,000 **B** 14,000 **C** 8000

Maserati Kyalami

After De Tomaso bought Maserati in 1975, he decided to fit the De Tomaso Longchamp (which had a Ford V8 engine) with a Maserati 4.1-litre engine and badge it as the Kyalami. There were other differences: Ghia restyled the front end to incorporate double headlamps and a trident grille and added interior trim changes. 4.9-litre engine offered from '78 as an option, which made the Kyalami a quick machine, but somehow the idea of a badge-engineered Maserati never caught the buying public's imagination. De Tomaso pulled the plug on it long before the Longchamp.

Engine 4136cc 88x85mm/4930cc 93.9x89mm, V8 cyl QOHC, F/R
Max power 255bhp at 6000rpm/ 280bhp at 5600rpm
Max speed 147/152mph
Body styles coupé
Prod years 1977-83
Prod numbers 150

PRICES
A 12,000 **B** 9000 **C** 7000

Matra Djet

René Bonnet's pioneering (but expensive) mid-engined Djet was taken on in 1964 by aerospace giant Matra, which was already building the bodies. Matra Sports was thus created and the Djet was relaunched in pretty much the same form as before under the name Matra-Bonnet Djet 5. That meant a narrow plastic body (now restyled at the front), and Renault 8 power (uprated in '66 to a 1255cc Gordini engine, when the car was renamed Djet 6). LHD only and not marketed in the UK, but a small following now exists here.

Engine 1108cc 70xm72m/1255cc 74.5x72mm, 4 cyl OHV, M/R
Max power 70bhp at 6000rpm/ 103bhp at 6750rpm
Max speed 109/126mph
Body styles coupé
Prod years 1964-68
Prod numbers 1681 (all Djets)

PRICES
A 9000 **B** 5000 **C** 3000

Matra 530

Matra developed a somewhat larger machine for 1967, utterly bizarre and very French in appearance. Still mid-engined, it turned to the Ford V4 for its motive power. Unitary steel underbody, glassfibre 2+2 bodywork, independent suspension, discs all round, gearbox mounted behind rear axle line. Original 530 LX had pop-up headlamps and a targa roof panel, but the more down-market 530 SX, offered from 1971, had a fixed roof and a curious arrangement of four free-standing 'bug-eye' headlamps. Great to drive, not so great to look at. Virtually unknown in Britain.

Engine 1699cc 90x66.8mm, 4 cyl OHV, M/R
Max power 75bhp at 5000rpm
Max speed 100mph
Body styles coupé, targa
Prod years 1967-73
Prod numbers 9609

PRICES
A 5000 **B** 3000 **C** 1800

Matra-Simca · Mazda

Matra-Simca Bagheera

In 1969, Matra switched its allegiance to Simca and developed the Bagheera which was launched in 1973. Now it was a Simca 1.3-litre engine which was mounted transversely amidships, though its lowly power output hardly made the Bagheera an exciting sports car; the '75 1442cc Bagheera S was slightly better. Space frame, Philippe Guedon-designed glassfibre body, torsion bar independent suspension, four-wheel discs, LHD. Most unusual feature was three-abreast seating layout, and there was a glass hatchback. Became a Talbot-Matra from '79. Check for chassis rust and engine problems.

Engine 1294cc 76.7x70mm/1442cc 76.7x78, 4 cyl OHV, M/R
Max power 84bhp at 6000rpm/ 90bhp at 5800rpm
Max speed 102/115mph
Body styles coupé
Prod years 1973-80
Prod numbers 47,802

PRICES
A 4000 **B** 2500 **C** 1000

Mazda Cosmo 110S

An intriguing car in virtually every respect: Mazda's first sports car and also its first Wankel-engined car. World's first production twin-rotor engine – just beating NSU's Ro80 – had a healthy output but an unhealthy reliability record. De Dion back axle/leaf springs, coils and wishbones up front, front disc brakes. L10B model from '68 had longer wheelbase, larger under-bumper grille. Five-speed 'box and more power from '69. Was briefly sold in the UK with RHD – and there are survivors. Handling not up to much and chronic parts problems.

Engine 2 x 491cc, twin-rotor Wankel, F/R **Max power** 110bhp at 7000rpm/ 128bhp at 7000rpm
Max speed 115/125mph
Body styles coupé
Prod years 1967-72
Prod numbers 1176

PRICES
A 15,000 **B** 10,000 **C** 4000

Mazda Luce 1500/1800/R130

This should have become the Alfa Romeo 1750: Bertone styled the saloon body for Alfa but when it turned it down, the design was sold to Mazda. Despite this fillip, the Luce was hardly an exciting car. Spearheaded the Japanese firm's drive towards exports (known here simply as the 1800 – only a handful of 1500s were imported). Overhead cam engines were punchy, and it boasted twin headlamps, servo front discs and a lot of equipment. More interesting R130 rotary coupé version of 1969 was never imported. It had a twin-rotor engine, front-wheel drive and hard-top styling. Mazda rotaries had sealing problems.

Engine 1490cc 78x78mm/1796cc 78x94mm/2 x 655cc, 4 cyl OHC/ OHC/twin-rotor Wankel, F/R / F/F
Max power 92bhp at 5800rpm/ 104bhp at 5500rpm/126bhp at 6000rpm
Max speed 100/103/118mph
Body styles saloon, estate, coupé
Prod years 1966-72/1968-72/1969-72
Prod numbers 121,804/39,041/n/a
PRICES
1500/1800 **A** 2500 **B** 1300 **C** 700
R130 **A** 6000 **B** 4000 **C** 1800

Mazda 616/818/RX2/RX3

Mazda's first non-sporting rotary car was the 1968 R100, a twin-rotor version of the boring 1200. More successful was the RX2, the Wankel version of the equally boring 616 (1500/1600/ 1800), plus the RX3, the rotary version of the smaller 818 (1300/1500/1600). Known as Capella and Savanna in Japan. Conventional engineering: rigid rear axle, MacPherson front, front discs, recirculating ball steering. At least the RXs had some performance, but even so only the coupés are of any interest. Cosmo-type Wankel engine on RX3, and bigger version for the RX2. Most have now rusted into orange puddles.

Engine 1272cc/1490cc/1586cc/ 1796cc/2 x 451cc/2 x 573cc, 4 cyl OHC/twin-rotor Wankel, F/R
Max power 75bhp/92bhp/100bhp/ 104bhp/110bhp/120bhp
Max speed 100-118mph
Body styles saloon, estate, coupé
Prod years 1970-78 **Prod numbers** 254,919/625,439/225,004/286,685
PRICES
616 **A** 1000 **B** 500 **C** 100
RX2 Coupé **A** 2500 **B** 1500 **C** 500
RX3 Coupé **A** 2400 **B** 1300 **C** 400

Mazda 929/RX4

The last and biggest of Mazda's 1970s rotary saloons (excepting the Japan-only Roadpacer) was the RX4. Based on the 1.8-litre 929, a car which reached new levels of boredom in the upper-medium class. The Wankel unit was the same as that in the RX2, although it was expanded in 1974 to a nominal capacity of 2.6 litres, in which case a five-speed gearbox was standard. In UK '73-'76. Coupé is the only vaguely interesting version, but people have enough trouble getting excited about the much more engaging NSU Ro80, let alone this rusty, thirsty, unreliable '70s porridge-mobile.

Engine 1796cc 78x94mm/2 x 573cc/2 x 654cc, 4 cyl OHC/twin-rotor Wankel, F/R
Max power 104bhp at 5500rpm/ 120bhp at 6500rpm/135bhp at 6500rpm
Max speed 103/112/118mph
Body styles saloon, estate, coupé
Prod years 1970-77
Prod numbers 213,988
PRICES
929 **A** 1000 **B** 600 **C** 200
RX4 Coupé **A** 2700 **B** 1500 **C** 600

Mercedes-Benz 170V/170S/170SV/170D/170SD

The last design of Mercedes-Benz's great Hans Nibel was the 1936 170V. It returned after the war almost unchanged, which meant side valves, transverse leaf IFS, swing axle IRS, divided trackrod steering, upright four-light styling. Numerous coachbuilt bodies – estates, taxis, convertibles, etc. Bigger engine in the 170S of 1949, plus a very sluggardly diesel model (170D). Hypoid axle from '52. SV and SD models of '53 had revised front suspension and bodywork. Extremely durable but heavy, wallowy and not much fun. Proper UK imports resumed from 1953 SV/SD.

Engine 1697cc 73.5x100mm/1767cc 75x100mm, 4 cyl SV & diesel, F/R
Max power 38bhp at 3200rpm to 52bhp at 4000rpm
Max speed 60-75mph
Body styles saloon
Prod years 1936-53/1949-53/1953-55/1949-53/1953-55
Prod numbers 143,612 (49,367/42,413/3122/33,823/14,887)

PRICES
A 8000 B 5000 C 2500

Mercedes-Benz 170S Cabriolet

The first post-war convertible from Stuttgart was the 1949 170S Cabriolet. Two body styles were offered: the Cabriolet A with 2+2 seating, or the Cabriolet B with full five-seat accommodation and rear passenger windows. The same basic mechanical package and bodywork as the 170S saloon, but two doors only. Build quality was high, with leather and walnut standard inside. No Cabriolets ever came officially to the UK, but some can be found. Very sought-after on the continent because the total production run was so low. Rust can be an expensive problem.

Engine 1767cc 75x100mm, 4 cyl SV, F/R **Max power** 45bhp at 3600rpm to 52bhp at 4000rpm
Max speed 70-75mph
Body styles convertible
Prod years 1949-52
Prod numbers 2433

PRICES
A 27,000 B 20,000 C 15,000

Mercedes-Benz 220/Cabriolet/Coupé

Six cylinders and an overhead camshaft made this a much better all-round sister of the 170S. Styling was identical apart from the headlamps, which were integrated into the front wings, and fixed bonnet sides. Decent performance and reasonable economy (22mpg), and popular with new generation of German industrialists. Convertible versions (A and B) followed 170S practice, too, though with integrated headlamp treatment. The coupé (available from 1954) was nothing more than a convertible with a fixed steel roof, and easily the rarest of the bunch (only 85 made).

Engine 2195cc 80x72.8mm, 6 cyl OHC, F/R **Max power** 80bhp at 4850rpm
Max speed 90mph
Body styles saloon, convertible, coupé
Prod years 1951-55/1951-55/1954-55
Prod numbers 16,154/2275/85
PRICES
220 A 12,000 B 9000 C 5000
220 Cabriolet A 32,000 B 23,000 C 17,000
220 Coupé: Rarity means pricing difficult, but similar to cabrio

Mercedes-Benz 180/180D/190/190D

First unitary construction Mercedes boasted full-width bodywork, alligator bonnet, integrated headlamps, subframes for suspension and drive train. However, the engine remained the same old 170V sidevalve, updated in 1957 to a larger OHC unit. A diesel 180D was offered from 1954. The 190 looked precisely the same as the 180, except for additional chromework, but it had a 1.9-litre engine, 220-type single pivot swing axle, and a servo brake option. 190D diesel from '58. Lower bonnet and wider grille from '59. Estate cars to special order only.

Engine 1767cc 75x100mm/1897cc 85x83.6mm/1988cc 87x83.6mm, 4 cyl SV & diesel/OHC & diesel/diesel, F/R
Max power 40bhp at 3200rpm to 68bhp at 4400rpm **Max speed** 75-84mph **Body styles** saloon, estate
Prod years 1953-62/1954-62/1956-61/1958-61
Prod numbers 118,234/152,983/89,808/81,938

PRICES
A 4000 B 2500 C 1000

Mercedes-Benz 220A/220S/220SE/219

Basically a six-cylinder bigger sister of the 180, the 220A was distinguishable by its longer bonnet and 111in wheelbase, rear quarterlights and pierced disc wheels. But the bodywork was almost identical, as was the suspension. Brake servo from '55, twin carb 220S from '56, short (108in) wheelbase 219 also from '56. Hydrak auto clutch option from '57, fuel-injected SE from '58. Convertible and coupé appeared at Frankfurt 1955 and went into production with the S engine in '56, and SE in '58. These had shortest wheelbase at 106in. Rare, opulent and sought-after. Expect some rust. SE worth 25% more.

Engine 2195cc 80x72.8mm, 6 cyl OHC, F/R **Max power** 85bhp at 4800rpm to 120bhp at 5000rpm **Max speed** 92-105mph **Body styles** saloon, coupé, convertible **Prod years** 1954-56/1956-59/1958-59/1956-59
Prod numbers Saloon: 25,937/55,279/1974/27,845.
Coupé & Cabriolet: 5371
PRICES
Saloon A 9000 B 6500 C 4000
Coupé A 15,000 B 10,000 C 7000
Cabriolet A 18,000 B 12,000 C 9000

Mercedes-Benz

Mercedes-Benz 300A/300B/300C

A big car which was launched at the same time as the BMW 501, and the ultimate German prestige car of the 1950s. Alloy-head overhead cam twin carb straight six engine, all-synchro four-speed 'box with column change, IRS with swing axle and selectable auxiliary springs, first German use of hypoid drive, tubular cruciform separate chassis. 300B has tuned engine and servo brakes, 300C has the same engine but an automatic option. Six-light four-door styling is well proportioned, if stately, and the 300 became a favourite of heads of state. Four-door Cabriolet is an impressive beast, made 1952-56.

Engine 2996cc 85x88mm, 6 cyl OHC, F/R
Max power 115bhp at 4600rpm/125bhp at 4500rpm
Max speed 100/104mph
Body styles saloon
Prod years 1951-54/1954-55/1955-57
Prod numbers Saloon: 7746/Cabriolet: 642
PRICES
300A/B/C **A** 15,000 **B** 12,000 **C** 8000
300 Cabriolet **A** 40,000 **B** 30,000 **C** 18,000

Mercedes-Benz 300S/300SC

For the wealthy, sporting motorist (almost inevitably Americans), the 300S was a highly attractive device. Chassis was shortened by 4in and fitted with a triple Solex engine. Three versions were offered: a fixed-head coupé, a convertible, and a roadster without landau irons and a top which folded out of sight (pictured). All had only two doors and two-light windows. For 1956, the engine received direct fuel injection and dry sump lubrication, and the model was renamed SC. Launched as a stop-gap after the 300SL had been delayed, and it cost even more new than an SL.

Engine 2996cc 85x88mm, 6 cyl OHC, F/R
Max power 150bhp at 5000rpm/175bhp at 5400rpm
Max speed 108-115mph
Body styles coupé, landau convertible, convertible
Prod years 1951-55/1955-58
Prod numbers 760

PRICES
Coupé/Cabriolet/Roadster **A** 80,000 **B** 60,000 **C** 35,000

Mercedes-Benz 300D/300D Cabriolet

The 300D was built in limited numbers between August 1957 and March 1962. The wheelbase was longer, the body (now 17ft long) was largely redesigned, with longer rear wings, more upright headlamps and a larger glasshouse incorporating a wrap-around rear screen. All six windows could be wound down out of sight for completely pillarless motoring. Always fitted with the fuel-injected engine, and three-speed automatic was usual (manual was offered but in LHD form only). Optional power steering. High prices reflect rarity; the Cabriolet is impossible to value.

Engine 2996cc 85x88mm, 6 cyl OHC, F/R
Max power 160bhp at 5300rpm
Max speed 105mph
Body styles saloon, convertible
Prod years 1957-62
Prod numbers 3008/65

PRICES
300D **A** 20,000 **B** 16,000 **C** 11,000
300D Cabriolet Seldom offered for sale, but probably well in excess of £50,000

Mercedes-Benz 300SL Coupé

The classic 'Gullwing' SL is one of those designs which stand out from everything else. It grew out of the 1952 race car which won Le Mans and the Carrera Panamericana. Much of the mechanical side derived from the 300 saloon, including the suspension, engine (though this was fuel-injected, dry sumped and mounted at an angle), and gearbox (though with floor shift). The chassis was a space frame design and gullwing doors were fitted to maintain rigidity. It cost as much as a small house, oversteered, had mere drum brakes, and today represents a severe restoration challenge. 29 were built with alloy bodies.

Engine 2996cc 85x88mm, 6 cyl OHC, F/R
Max power 215bhp at 5800rpm to 240bhp at 6100rpm
Max speed 145-155mph
Body styles coupé
Prod years 1954-57
Prod numbers 1400

PRICES
A 140,000 **B** 100,000 **C** 80,000

Mercedes-Benz 300SL Roadster

Took over from the Gullwing as a more practical, less claustrophobic and frankly better sports car, more suited to the climate of California (the Gullwing had no winding windows). Classic car collectors aren't as interested, however, because it lacks the sporting edge of the Gullwing. As on the coupé, steel bodywork with alloy doors, bonnet and boot lid. More powerful fuel-injected engine, restyled headlamps, low-pivot rear swing axle, proper doors and soft top were the main changes. Removable hardtop offered from '58, and all-wheel discs from '61.

Engine 2996cc 85x88mm, 6 cyl OHC, F/R
Max power 225bhp at 5900rpm to 250bhp at 6200rpm
Max speed 140-155mph
Body styles sports
Prod years 1957-63
Prod numbers 1858

PRICES
A 100,000 **B** 75,000 **C** 55,000

Mercedes-Benz

Mercedes-Benz 190SL

The 190SL looked a bit like a scaled-down 300SL but that was all the cars had in common. It traded off its big sister's reputation and was plainly aimed at American boulevards, not circuits. Engine was a twin carb version of the standard 190 1.9-litre unit, and the four-speed all-synchro 'box had a floor-mounted lever. Slight styling revisions in '55, servo brakes from '56. Very expensive for what it was – a shortened 190 saloon floorpan, only mildly fast, not especially pretty and no sharp handler. Coupé (non-gullwing) and removable hardtop for the roadster were offered.

Engine 1897cc 85x83.6mm, 4 cyl OHC, F/R
Max power 105bhp at 5700rpm
Max speed 105mph
Body styles sports, coupé
Prod years 1954-63
Prod numbers 25,881

PRICES
A 20,000 **B** 14,000 **C** 9000

Mercedes-Benz 220/220S/220SE & 230/230S

All-new W111 series arrived in 1959, but only in 2.2-litre form. Angular styling by Karl Wilfert contrasted strongly with previous rounded contours and was inevitably called the 'fintail'. Wrap-around front and rear screens were *de rigueur*. Mechanically, the new series was very similar to the old 219/220: even the wheelbase was identical. S had twin carbs, SE was fuel-injected; whole range slightly more powerful than before. Front discs from '62. SE's Bosch injection system is complicated and thirsty. Replaced in '65 by the twin carb 230/230S (latter has cowled headlamps). Estate from '66.

Engine 2195cc 80x72.8mm/2281cc 82x72.8mm, 6 cyl OHC, F/R
Max power 95bhp at 5000rpm/110bhp at 5200rpm/120bhp at 5000rpm/105-110bhp at 5400rpm
Max speed 95-109mph
Body styles saloon, limousine, estate
Prod years 1959-65 & 1965-68
Prod numbers 69,691/161,119/66,086 & 40,258/41,107

PRICES
A 10,000 **B** 7000 **C** 4000

Mercedes-Benz 190/190D & 200/200D

Two years after the 220 arrived, the 190 replaced the old rounded 180/190 series. Basically the same shape as the 220, but a shorter nose to accommodate a four-cylinder engine, plus single round headlamps. More spacious but heavier than the old series, but a stronger seller especially in 55bhp 190D diesel form – this was a slow but doggedly reliable machine capable of returning over 35mpg. Replaced by larger-engined twin carb 200 in '65, which also boasted dual-circuit servo disc/drum brakes and choice of floor or column gearchanges. Lwb saloon 1967-68, Belgian-built estate 1966-67.

Engine 1897cc 85x83.6mm/1988cc 87x83.6, 4 cyl OHC/OHC & diesel, F/R
Max power 80bhp at 5000rpm/55bhp at 4200rpm/95bhp at 5200rpm/60bhp at 4200rpm
Max speed 80-92mph
Body styles saloon, limousine, estate
Prod years 1961-65 & 1965-68
Prod numbers 130,554/225,645 & 70,207/161,618

PRICES
A 4500 **B** 2600 **C** 1200

Mercedes-Benz 300SE/300SEL

The ultimate fintail was the 3-litre 300SE, introduced in 1961 alongside the 220. It could be distinguished from its lowlier brethren by its extra chromework and high spec, but the big change occurred underneath. The engine was the familiar Mercedes all-alloy six fitted with Bosch fuel injection, the brakes were discs all round, steering was power-assisted and you also got air suspension (derived from the 600) which was self-levelling at the rear. '63 SEL saloon had an extra 4in in the wheelbase. 300s cost almost twice as much as a 220, which explains their rarity – and their high values today.

Engine 2996cc 85x88mm, 6 cyl OHC, F/R
Max power 160bhp at rpm to 170bhp at 5400rpm
Max speed 111-118mph
Body styles saloon
Prod years 1961-65/1963-65
Prod numbers 5204/1546

PRICES
A 14,000 **B** 9000 **C** 5000

Mercedes-Benz 220SE/300SE Coupé/Cabriolet

The Coupé W111 arrived just before the Cabriolet, both in 1961. Ironically, these two-door 'fintails' had their fins removed. The early engine was the 220SE fuel-injected unit, but this was joined in '62 by the 300SE, which shared its mechanical spec with the saloon and was vastly expensive – double the price of the 220SE versions. 220SE replaced by 250SE in '65 (see page 142), but 300SE continued until 1967. All had floor-mounted gearchanges or automatic. These are now highly prized machines, worth large sums. Tracking down a 300SE might prove difficult.

Engine 2195cc 80x72.8mm/2996cc 85x88mm, 6 cyl OHC, F/R **Max power** 120bhp at 4800rpm/170bhp at 5400rpm
Max speed 106/118mph **Body styles** coupé, convertible **Prod years** 1961-65/1962-67 **Prod numbers** 220SE: 16,902/300SE: 3127

PRICES
200SE Coupé/Cabriolet **A** 15,000/22,000 **B** 9000/16,000 **C** 5000/10,000
300SE Coupé/Cabriolet **A** 20,000/30,000 **B** 12,000/22,000 **C** 8000/14,000

Mercedes-Benz

Mercedes-Benz 250S/250SE, 280S/280SE/280SEL and 300SE

More conservative styling for the revised '65 W108 medium-class Mercedes. Gone were the little fins, the bodywork was lower and slightly longer, though the wheelbase was shorter. New seven-bearing 2.5-litre engine at first, replaced in '68 by the 2.8-litre 280S/280SE and long-wheelbase 280SEL. The '65 300SE was less interesting than the model it succeeded, substituting mere coil springs for air suspension, though the lwb 300SEL (below) retained it. It differed from the 250SE in having standard PAS and better equipment. All SEs came with fuel injection and more power.

Engine 2496cc 82x78.8mm/2778cc 86.5x78.8mm/2996cc 85x88mm, 6 cyl OHC, F/R **Max power** 128bhp at 5400rpm to 148bhp at 5500rpm/140bhp at 5200rpm to 158bhp at 5500rpm/168bhp at 5400rpm **Max speed** 112-124mph **Body styles** saloon **Prod years** 1965-69, 1968-72, 1965-68 **Prod numbers** 250: 74,677/55,181. 280: 93,666. 300: 2737
PRICES
250/280 **A** 6500 **B** 4500 **C** 3000
300 **A** 7500 **B** 5500 **C** 3500

Mercedes-Benz 280SE & 280SEL 3.5/4.5 and 300SEL 2.8/3.5/4.5/6.3

The 300SEL was the long wheelbase derivative of the W108 family, dubbed W109. It retained the old air suspension system abandoned by the short wheelbase 300SE, but in other respects was identical. Until 1968, that is, when its 3-litre engine was downsized to 2.8 litres (from the 280SL) – confusingly, it was still called 300SEL. 3.5-litre V8 engine arrives in '69 in the 300SEL 3.5, and in '71 in the 280SE/SEL 3.5. The 4.5 was offered in USA only. Ultimate version is '68 300SEL 6.3, using the 600's enormous V8 engine, air suspension and power disc brakes.

Engine 2778cc/2996cc 6 cyl OHC, 3499cc/4520cc/6332cc V8 OHC, F/R **Max power** 168bhp at 5400rpm to 250bhp at 4000rpm **Max speed** 110-140mph **Body styles** saloon **Prod years** 1965-72 **Prod numbers** 280SE/SEL 3.5: 12,260/4.5: 21,700. 300SEL 2519/3.5: 9583/4.5: 2553/6.3: 6526.
PRICES
280SE 3.5 **A** 6500 **B** 5400 **C** 3000
300SEL **A** 7500 **B** 5500 **C** 3500
300SEL 6.3 **A** 10,000 **B** 7000 **C** 5000

Mercedes-Benz 250SE/280SE/280SE 3.5 Coupé/Cabriolet

This starts to get confusing: although Mercedes' late-1960s coupés and convertibles were called 250/280 and had the seven-bearing sixes from the contemporary W108 saloons, the bodywork remained the same old 1961 W111 style. Discs all round, swing axle load compensator, optional PAS and automatic. As expensive as ever. 280SE replaced the 250SE in 1968. Most attractive variant is the 280SE 3.5 offered from 1969-71, capable of 125mph on the autobahn, and boasting standard PAS and auto 'box. A cheap alternative to the Rolls-Royce Corniche? Coupés worth 40% less.

Engine 2496cc/2778cc 6 cyl OHC, 3499cc V8 cyl OHC, F/R **Max power** 148bhp at 5500rpm/200bhp at 5400rpm **Max speed** 113-125mph **Body styles** coupé, convertible **Prod years** 1965-68/ 1968-71/1969-71 **Prod numbers** 6213/5187/4502 **PRICES** 250SE Cabriolet **A** 20,000 **B** 15,000 **C** 12,000
280SE Cabriolet **A** 24,000 **B** 18,000 **C** 14,000
280SE 3.5 Cabriolet **A** 32,000 **B** 25,000 **C** 19,000

Mercedes-Benz 600/600 Pullman

The best car in the world? In 1963, perhaps it was. A leviathan of a car, in standard form 18ft long and 5500lb (2475kg); if you ordered the long-wheelbase Pullman nine-seater (with a choice of four or six doors), it was fully 20ft 5in and 5850lb (2650kg). Giant 6.3-litre V8, air suspension, obligatory automatic, air conditioning, hydraulic control for the seats, windows, locks and partition. Special-order landaulette very rare. It cost as much new as a Rolls-Royce Phantom, but at one stage plummeted in value as a classic. Now it's back on a pedestal, in Germany at least.

Engine 6332mm 103x95mm, V8 cyl OHC, F/R
Max power 250bhp at 4000rpm
Max speed 123mph
Body styles saloon, limousine, landaulette
Prod years 1963-81
Prod numbers 2190/487

PRICES
A 35,000 **B** 25,000 **C** 15,000

Mercedes-Benz 230SL/250SL/280SL

Because of the concave roof-line of the hardtop model, the new SL series appropriately became known as the 'pagoda roof'. Styling was by Paul Bracq and the basis was a truncated 220 floorpan, complete with swing axle rear and double wishbone front. Four-speed manual or auto, disc front brakes on 230SL, upgraded to four-wheel discs on the larger-engined 250SL of 1966 (plus optional five-speed 'box). Like 250SL, later 280SL had improved torque. Cars come with hood only, hard-top and hood (pictured), or hardtop on its own with a rear seat (transverse for one occupant on 230SL, folding bench for two on 250/280SL).

Engine 2306cc 82x72.8mm/2496cc 82x78.8mm/2778cc 86.5x78.8mm, 6 cyl OHC, F/R **Max power** 148bhp at 5500rpm/150bhp at 5600rpm/170bhp at 5750rpm **Max speed** 120-127mph **Body styles** coupé, convertible **Prod years** 1963-67/1966-68/1968-71 **Prod numbers** 19,831/5186/23,885

PRICES
230SL **A** 14,000 **B** 10,000 **C** 7500
250SL **A** 15,000 **B** 12,000 **C** 9000
280SL **A** 18,000 **B** 15,000 **C** 11,000

Mercedes-Benz 200/220/230.4/230.6/240/240 3.0/250/250 2.8/280E

New generation entry-level Mercs, codenamed W114/W115. All-new but very traditional unitary body, semi-trailing link rear suspension, bewilderingly huge range of engines: base 2.0-litre petrol and diesel, 2.2-litre petrol and diesel, 2.3-litre four, 2.3-litre six, 2.4-litre diesel, 2.5-litre six, 2.8-litre six and, on 280E, a new twin cam straight six. Five-cylinder 3-litre diesel was not offered in UK. Optional auto, PAS. 17ft 6in long-wheelbase limo available in 220D, 240D and 230.6 guises. Archetypal durable Merc, beloved of taxi drivers and hire trade, still loads around.

Engine 1988cc/2197cc/2307cc/2292cc/2404cc/3005cc/2496cc/2778cc/2746cc, 4/5/6 cyl OHC/diesel/DOHC, F/R
Max power 55bhp at 4200rpm to 185bhp at 6000rpm
Max speed 80-124mph
Body styles saloon, limousine
Prod years 1968-76
Prod numbers 1,326,099

PRICES
A 5000 **B** 3000 **C** 1000

Mercedes-Benz 250C/250CE/280C/280CE

W114 coupé model shared just about everything in common with its saloon brethren except for the two-door styling. Slightly shorter and 2in lower than the saloon, but also much more cramped for rear seat passengers, and significantly more expensive than the saloon. C versions have carburettors, CEs have fuel injection and more power. 250CE was sold in UK 1968-72, though it continued elsewhere with the 2778cc engine as late as 1976. The twin cam 2746cc 280 models arrived in 1971. Power steering standard from '75. Not very exciting, but better value than previous Merc coupés.

Engine 2496cc 82x78.8mm/2778cc 86.5x78.8mm/2746cc 86x78.8mm, 6 cyl OHC/OHC/DOHC, F/R **Max power** 130bhp at 5400rpm/140bhp at 5400rpm/150bhp at 5000rpm/160bhp at 5500rpm/185bhp at 6000rpm
Max speed 115-125mph
Body styles coupé
Prod years 1968-76/1971-76
Prod numbers 30,611/24,669
PRICES
250CE **A** 6000 **B** 4000 **C** 2000
280CE **A** 7000 **B** 4800 **C** 2500

Mercedes-Benz 280SL/350SL/450SL

The W107 SL was an absolute classic Mercedes which lasted in production for almost 20 years. Expanded in every way over the 'pagoda roof' SL, it was initially powered by Merc's 3.5-litre V8, or the 4.5-litre V8. Modernised suspension featured semi-trailing arm rear end, and there was standard power steering, and optional automatic. From '74, the 280SL was launched on the continent (but not UK), sporting the twin cam six-cylinder engine from the 280E. Like its predecessor, the optional removable hardtop had a dip in the roof. Range revamped in 1980 to become 280SL/380SL/500SL.

Engine 2746cc 86x78.8mm/3499cc 92x65.8mm/4520cc 92x85mm, 6/V8/V8 cyl DOHC/OHC/OHC, F/R
Max power 185bhp at 6000rpm/200bhp at 5800rpm/225bhp at 5000rpm
Max speed 127/130 /134mph
Body styles coupé, convertible
Prod years 1974-85/1971-80/1971-80
Prod numbers 25,436/15,304/66,298
PRICES
280SL **A** 12,500 **B** 9000 **C** 6000
350/450SL **A** 14,000 **B** 10,000 **C** 6000

Mercedes-Benz 280SLC/350SLC/450SLC

Insertion of a sizeable 14in into the wheelbase created a four-seater fixed-head variant on the SL theme, launched one year after the roadster. Not as pretty as the SL, but at least you no longer have the controversial dippy hardtop roof-line. Mechanically all but identical to the SL, but quite a bit heavier (450SLC was 3600lb/1635kg). 280SLC arrived in '74 but not sold in Britain until 1980. Mechanical fuel injection from 1975. Splendidly refined, long-legged cruisers with sports car handling and impressive pace. As with the SL, new engines appeared in the 1980 line-up.

Engine 2746cc 86x78.8mm/3499cc 92x65.8mm/4520cc 92x85mm, 6/V8/V8 cyl DOHC/OHC/OHC, F/R
Max power 185bhp at 6000rpm/200bhp at 5800rpm/225bhp at 5000rpm **Max speed** 127/130/134mph **Body styles** coupé
Prod years 1974-81/1972-80/1972-80
Prod numbers 10,666/13,925/31,739
PRICES
280SLC **A** 8000 **B** 6500 **C** 4000
350SLC **A** 10,000 **B** 7500 **C** 5000
450SLC **A** 10,000 **B** 7500 **C** 5000

Mercedes-Benz 280S/280SE/280SEL/350SE/350SEL/450SE/450SEL

One magazine called the new W116 S-Class Mercedes the best car in the world: certainly its powerful combination of fine engines, superb handling, comfort, ride and refinement made it a formidable all-rounder. Semi-trailing arm rear suspension from SL, auto 'box obligatory (though some 280s did have manual), PAS standard. 2.8-litre six not really man enough to pull 3550lb/1610kg, but glorious 350 and 450 V8s provided effortless power. SEL denoted 4in wheelbase stretch, all used to increase rear passenger space, but the UK only got the 450SEL. 300SD for USA only. Sure to appreciate.

Engine 2746cc 86x78.8mm/3499cc 92x65.8mm/4520cc 92x85mm, 6/V8/V8 cyl DOHC/OHC/OHC, F/R
Max power 155-185bhp at 5500rpm/200bhp at 5800rpm/225bhp at 5000rpm **Max speed** 118-130mph
Body styles saloon **Prod years** 1972-80 **Prod numbers** 122,848/150,775/7032/ 51,140/4266/41,604/59,575
PRICES
280S/SE **A** 4500 **B** 3000 **C** 1600
350SE **A** 6000 **B** 4000 **C** 2000
450SE/SEL **A** 7000 **B** 5000 **C** 2500

Mercedes-Benz · Messerschmitt

Mercedes-Benz 450SEL 6.9

This car deserves a whole entry to itself, as it was probably the best car money could buy in the late 1970s. And you needed a lot of money to own one: it cost half as much again as a 450SEL and twice as much as a Jaguar XJ12, while you could expect no more than 16mpg. The engine was a bored-out version of the 600 V8, and it delivered storming performance for such a bulky car (0-60mph in 7.3 secs), while hydropneumatic self-levelling suspension provided superlative ride and masterful cornering. Many examples have been maintained to exceptional standards – as they need to be.

Engine 6834cc 106.9x95mm, V8 cyl OHC, F/R
Max power 286bhp at 4250rpm
Max speed 140mph
Body styles saloon
Prod years 1975-80
Prod numbers 7380

PRICES
A 10,000 **B** 7500 **C** 4500

Mercedes-Benz 200/230/240/250/280/300

Another best-selling Mercedes, despite its conventional approach. Codenamed W123, the mechanical package remained very much as before and the shape evolved discreetly (horizontally-arranged double headlamps and rounded arches were ID features). Confusing spread of engines again: 2.0-litre and 2.3-litre engines as before (but replaced in 1980 by all-new engines), 2.0 and 2.4-litre diesels, new 2.5-litre six, 2.8 and, from '76, the five-cylinder 3-litre diesel first seen in the W115. Also Mercedes' first estates, badged 'T'. Basically boring, but finely engineered and long-lasting.

Engine 1988cc/1997cc/2307cc/2299cc/2399cc/2402cc/2525cc/2746cc/3005cc/2998cc, 4/5/6cyl OHC/DOHC/diesel, F/R
Max power 55bhp at 4200rpm to 185bhp at 5800rpm
Max speed 80-121mph
Body styles saloon, estate
Prod years 1975-84
Prod numbers 1,162,230

PRICES
A 4000 **B** 2500 **C** 1300

Mercury Cyclone

Cyclone began as a sports version of the Comet for the '64 model year and was progressively upgraded in line with the muscle car splurge. The '67 Cyclone reached a peak of 425bhp. The new '68 Cyclone coupe was given a fastback roof or, for one year only, a GT notchback. 1969's Cyclone CJ had the Ford 428 Cobra Jet V8 and had already been sprouting spoilers and a bonnet scoop when optional Ram-Air induction was specified. This muscle car mayfly bowed out after 1971. Collectors like the rare '68 Spoiler Dan Gurney and '70/'71 Spoiler editions.

Engine 4736cc 101.6x72.9mm/6391cc 103x96mm/6997cc 107.4x96mm/7014cc 104.9x101.1mm/7030cc 110.7x91.2mm, V8 cyl OHV, F/R
Max power 200bhp at 4400rpm to 425bhp at 5600rpm
Max speed 109-133mph
Body styles coupe, convertible
Prod years 1963-71
Prod numbers 90,236

PRICES
'66 **A** 6000 **B** 4000 **C** 1800
'68 **A** 8000 **B** 6000 **C** 3000

Mercury Cougar

Mercury dealers felt short-changed that they had no Mustang to sell, so FoMoCo developed the Cougar. It debuted for 1967 and was slightly larger than the Mustang and only ever offered with V8 engines. As it also had better equipment, it's not surprising that it cost about 10% more. Styling was crisper than the contemporary Ford's but became blander, especially after the '70 facelift, and positively hideous after a neo-classic grille was plastered on in '71. XR-7 was the leather-and-fake-wood luxury model. GTE handling package and 428-powered '69 Eliminator are desirable.

Engine 4736cc/4949cc/5752cc/6391cc/6997cc/7014cc/7030cc V8 cyl OHV, F/R **Max power** 168bhp at 4000rpm to 390bhp at 5600rpm **Max speed** 105-125mph **Body styles** coupe, convertible **Prod years** 1966-73 **Prod numbers** 614,225

PRICES
'67 Coupe **A** 8000 **B** 5000 **C** 2200
'67 Convertible **A** 12,000 **B** 8000 **C** 5000
'73 Convertible **A** 8000 **B** 5500 **C** 3000

Messerschmitt KR175

Aircraft engineer Fritz Fend found himself jobless after the war and turned to microcar manufacture with the Flitzer and other small three-wheelers. The most promising was the Kabinenroller, which was taken on by the Messerschmitt ex-aeroplane works. This bizarre projectile used a tubular space frame, handlebar steering, compressed rubber suspension and a Fichtel & Sachs 174cc engine sited in the tail. The KR175 had semi-enclosed front wheels which severely limited its turning circle. Messerschmitts have acquired a certain kudos and are among the most sought-after micros.

Engine 174cc 62x58mm, 1 cyl TS, R/R
Max power 9bhp at 5250rpm
Max speed 55mph
Body styles coupé
Prod years 1953-55
Prod numbers 19,668

PRICES
A 5000 **B** 3500 **C** 2500

Messerschmitt · Metropolitan · MG

Messerschmitt KR200/KR201

In 1955, the little Kabinenroller was uprated with a 191cc Sachs engine. The engine was now reversible, allowing the possibility of reverse in four gears! As before, cable brakes, transverse front arms and rear swinging arm, but torsional rubber bushes gave a better ride and there was now a curved windscreen. You still got in by lifting the Plexiglass canopy sideways to reveal a tandem seating arrangement for 2/3 people. On the '56 convertible KR201, you could simply step over the side if you wished. Surprisingly nippy, easy to restore, and lots of support from an enthusiastic club scene.

Engine 191cc 65x58mm, 1 cyl TS, R/R
Max power 10bhp at 5250rpm
Max speed 62mph
Body styles coupé, convertible
Prod years 1955-64/1956-64
Prod numbers 41,190

PRICES
A 5000 **B** 3500 **C** 2500

Messerschmitt Tg500

This little bombshell was intended to be called Tiger, but truck maker Krupp already had the name. People always called it the Tiger anyway. Possibly this was the best microcar ever built, boasting four wheels, a 493cc Sachs engine and a claimed top speed of 85mph (but more like 75mph in reality – still better than a contemporary Opel Rekord). Launched in dome-top form, it was joined by a very pretty low-roof Roadster model in 1959. Hydraulic brakes helped you slow from those hair-raising speeds. Almost mythical status at odds with its true worth: concours ones have sold for £30,000 plus!

Engine 493cc 67x70mm, 2 cyl TS, R/R
Max power 20bhp at 5000rpm
Max speed 75mph
Body styles coupé, sports
Prod years 1958-64
Prod numbers c 450

PRICES
A 20,000 **B** 16,000 **C** 12,000

Metropolitan

Nash and Austin were introduced by Donald Healey and they struck a deal whereby Austin would build a Nash-designed car using Austin parts to sell in the USA. The Nash/Hudson Metropolitan resulted, a gracelessly squashed pastiche of the US continental style. 1200cc A40 engine swapped for 1489cc B-series in '56. A puzzled British public was presented with it in 1957. Two-tone paint for UK, opening boot from '59. Lurches round corners, struggles up to speed, rusts like nothing else. For all that, it does have lots of character.

Engine 1200cc 65.5x88.9mm/1489cc 73x89mm, 4 cyl OHV, F/R
Max power 42bhp at 4500rpm/47-51bhp at 4500rpm
Max speed 75-78mph
Body styles coupé, convertible
Prod years 1953-61
Prod numbers 104,368

PRICES
Coupé **A** 4500 **B** 2500 **C** 1200
Convertible **A** 6000 **B** 4000 **C** 2000

MG TC

Cecil Kimber's 21st anniversary MG model of 1945 was the TC, essentially a wider pre-war TB with a part-synchromesh gearbox. Twin carb OHV engine in leaf-sprung chassis gave nimble performance. 19in wire wheels, fold-flat screen and slab tank at the rear evoke vintage charm, as do hard ride and heavy steering. Hydraulic drum brakes were a nod to modernity. US servicemen took them home and secured MG's future in America. Wooden body frames prone to rotting, but not a single parts difficulty. All TCs were built with right-hand drive.

Engine 1250cc 66.1x90mm, 4 cyl OHV, F/R
Max power 54.4bhp at 5200rpm
Max speed 78mph
Body styles sports
Prod years 1945-49
Prod numbers c 10,000

PRICES
A 15,000 **B** 11,000 **C** 8000

MG YA/YB

This sporting saloon used many Morris 8 body pressings to keep its price down, but at £672 it was still expensive. TC chassis underneath, fitted with a detuned TC engine, floor-change four-speed 'box, rack-and-pinion steering, coil spring and wishbone IFS. Luxuriously trimmed in leather and wood, but cramped. YB of '51 substituted a hypoid rear axle for the earlier spiral bevel, plus 15in wheels and twin leading-shoe brakes. Hard ride, not particularly quick with its single SU carb, but light and direct steering and good roadholding. The cheapest way into 1940s MG motoring.

Engine 1250cc 66.1x90mm, 4 cyl OHV, F/R
Max power 46-48bhp at 4800rpm
Max speed 72mph
Body styles saloon
Prod years 1947-51/1951-53
Prod numbers 6158/1301

PRICES
A 6500 **B** 5000 **C** 3000

MG YT

The Y Tourer was the last of a long line of four-seater open MGs, but its dumpy style was the least elegant. YA chassis but twin SU carbs (the same spec as the later TD), therefore more of a sporting machine. Cutaway doors, more square-cut rear and folding hood identify it. Not marketed in Britain, but you do find quite a few still on the road here. In the main target market (America, naturally), MG said: 'It looks, it rides, it drives like sports car'. Almost true, but not a patch on the TD. LHD most likely, but some RHD examples were built for export to the colonies.

Engine 1250cc 66.1x90mm, 4 cyl OHV, F/R
Max power 54bhp at 5200rpm
Max speed 75mph
Body styles convertible
Prod years 1948-51
Prod numbers 877

PRICES
A 7500 **B** 5000 **C** 3000

MG TD

Same chassis as the TC, but the coil spring IFS and rack-and-pinion steering of the YA. Gear ratios also lower and more widely spaced to compensate for increased weight. Bulkier appearance, 15in steel disc wheels and front and rear bumpers had enthusiasts questioning if this was a 'real' MG. It was certainly roomier, and boasted a higher top speed, especially if optional tuning kits were ordered (ranging from 60bhp right up to 100bhp with Shorrock supercharging). MkII model also available from '52 had a higher compression engine rated at 57.5bhp. Most were exported to the USA.

Engine 1250cc 66.1x90mm, 4 cyl OHV, F/R
Max power 54-57.5bhp at 5200rpm
Max speed 80mph
Body styles sports
Prod years 1949-53
Prod numbers 29,664

PRICES
A 13,000 **B** 9000 **C** 6000

MG TF 1250/1500

The modernisation of the TD was very tentative: MG buyers were ardent about traditionalism. Thus the TF had a lower bonnet, headlamps integral with the wings, sloping grille and fuel tank, central instrument panel and separate, not bench, seats. The engine was the same as the TD MkII. On the one hand, it had to battle with increased weight; on the other, it had a higher axle ratio, so top speed was unaffected. Bored out to 1.5 litres in 1954, the old pre-war Nuffield engine restored a competitive edge to what was becoming an outclassed sports car. But TFs are the most sought-after T-types today.

Engine 1250cc 66.1x90mm/1466cc 72x90mm, 4 cyl OHV, F/R
Max power 57.5bhp at 5200rpm/ 63bhp at 5000rpm
Max speed 80/86mph
Body styles sports
Prod years 1953-54/1954-55
Prod numbers 6400/3400

PRICES
TF 1250 **A** 16,000 **B** 12,000 **C** 8500
TF 1500 **B** 19,000 **B** 13,000 **C** 9500

MG Magnette ZA/ZB

The first 'non-MG': basically a Wolseley 4/44 fitted with a BMC B-series engine and adorned with MG grille, straight front bumper and chrome strips around the front wheelarches. However, it did work as a car: the B-series gave it lively performance, roadholding was distinctly un-BMC like, the close-ratio gearbox was effective and the leather-and-walnut interior luxurious. ZB (photo) has more power, higher axle ratio, straight-through chrome strip, optional Manumatic two-pedal control. Magnette Varitone was a ZB spin-off with wrap-around rear window and (usually) duo-tone paint.

Engine 1489cc 73x89mm, 4 cyl OHV, F/R
Max power 60bhp at 4600rpm/ 68bhp at 5400rpm
Max speed 75/80mph
Body styles saloon
Prod years 1953-56/1956-58
Prod numbers 18,076/18,524

PRICES
A 4000 **B** 2500 **C** 1200

MGA 1500/1600/1600 MkII

Syd Enever's classically modern MG had a difficult birth, since Leonard Lord preferred the Austin-Healey. But the A brought about a new era of sports car motoring. Conventional chassis, coil spring IFS, rack-and-pinion steering. Robust, yet sweet-handling and good for nearly 100mph. Factory hardtop joined in '56 by a fixed-head coupé model. From '59, the 1600 MkI had engine uprated to 1588cc, plus front discs, separate rear indicators; optional De Luxe (all-wheel discs and centre-lock wheels). MkII of '61 has 1622cc engine, higher final drive, recessed grille slats, horizontal rear light clusters.

Engine 1489cc 73x89mm/1588cc 75.4x88.9mm/1622cc 76.2x88.9mm, 4 cyl OHV, F/R **Max power** 68-72bhp at 5500rpm/80bhp at 5600rpm/86bhp at 5500rpm **Max speed** 99/102/105mph
Body styles sports, coupé
Prod years 1955-59/1959-61/1961-62
Prod numbers 58,750/31,501/8719
PRICES
Roadster **A** 12,000 **B** 10,000 **C** 8000 Coupé **A** 8500 **B** 7000 **C** 5000
1600 Roadster **A** 13,000 **B** 10,500 **C** 8500

metex car DUSTCOVERS
British made covers at very affordable prices.

KEEP YOUR CLASSIC CLEAN AND PROTECTED WITH ONE OF OUR POPULAR COVERS
Made from a hardwearing yet very soft fabric the cover will keep your car dust free, damp free and in the pristine condition you would expect.

LOTUS / MINI	**£21.50**
SMALL SPORTS CARS eg Austin Healey, MG, TR	**£25.50**
SMALL SALOONS eg Ford Escort, Triumph Herald Morris Minor etc	**£29.50**
LARGE SPORTS CARS eg Ferrari, Jaguar (E Type) Aston Martin ect.	**£32.50**
MEDIUM SALOONS eg Ford Cortina, Triumph Dolomite	**£32.50**
LARGE SALOONS eg Rover, Jaguar	**£42.50**
ROLLS ROYCE / BENTLEY	**£45.50**

CANVAS STORAGE BAGS £4.00
Covers available in black or white
Please state colour when ordering

THIS IS A SMALL SAMPLE OF METEX COVERS. IF YOU DO NOT SEE YOUR CAR ABOVE PLEASE PHONE. WHATEVER YOUR MAKE / MODEL WE'LL HAVE ONE TO FIT YOURS.
All the prices include VAT. PLEASE ADD £4.00 PER COVER FOR POST & PACKING
Send cheques payable to METEX (Darwen) Ltd. or telephone with credit card details.
If you require any assistance or any other size than the ones above please telephone us and one of our highly qualified staff will deal with you personally.
(Trade and overseas enquiries most welcome).

metex car DUSTCOVERS
DEPT WOOD ST. MILL
DARWEN, LANCS BB3 1LS

CREDIT CARD HOTLINES-
01254 704625 / 703893
OR WHY NOT FAX
01254 776927

MG OWNERS' CLUB

The World's Largest one-marque car club with 50,000 members offers:

★ Superb colour magazine
★ Large MG Classified section
★ Special Insurance Scheme (UK only)
★ Free Technical Advice
★ Super Spares Deals & Special Offers
★ List of MG Parts Suppliers and Workshops
★ Over 140 Areas for Social and Sporting Activities
★ Fully equipped MG workshop for Club members
★ Exclusive Special Offers, Regalia and much more.

BUYING AN MG?
To get our booklet and current MG Magazine with lists of MGs for sale send £2 to cover postage & packing and don't buy a rotten MG by mistake.

For details write to: MGOC Freepost, Swavesey, CAMBRIDGE, CB4 1BR (No stamp required) or Telephone 01954 231125 (24hr service); Fax: 01954 232106

◆ **NEW SPARES**
More than 25,000 MGB & Midget quality parts & accessories "on the shelf". Massive Stocks!!!

◆ **RECON PARTS**
Gearboxes, racks, stubs, calipers, shox, starters, axles, gauges, etc.

◆ **SECONDHAND**
Thousands of parts from 100s of once loved MGBs & Midgets.

Visa
Mastercard
Access
Switch

ORDER LINE: 01522 703774
FAX LINE: 01522 704128

DRINSEY NOOK, GAINSBOROUGH RD, SAXILBY, LINCOLN, LN1 2JJ.

BRITBITS
MG & MINI SPARES

10% Discount.. Quote this ad.

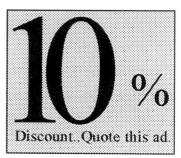

Mail Order
MG Spares
Worldwide

MGA Twin Cam

Harry Weslake developed a new twin cam head for the B-series 1588cc engine which, when fitted in the A chassis, produced a very rapid machine. Unfortunately, the engine quickly acquired a reputation for burning pistons and oil, but sound maintenance and fettling should dispel any worries. Chassis mods included four-wheel disc brakes and special lightweight steel wheels with centre-lock hubs, plus a front anti-roll bar from '59. BMC was not really geared up to mass-produce such a sophisticated engine, so the Twin Cam was short-lived. That makes it a desirable property in classic terms.

Engine 1588cc 75.4x88.9mm, 4 cyl DOHC, F/R
Max power 108bhp at 6500rpm
Max speed 110mph
Body styles sports, coupé
Prod years 1958-60
Prod numbers 2111

PRICES
Roadster **A** 16,000 **B** 12,000 **C** 10,000
Coupé **A** 13,000 **B** 10,000 **C** 8000

MGB

Unitary body, roomy interior and wind-up windows – MG heresy, but lapped up in America. 1.8-litre B-series development offered more power, higher final drive for elevated top speed, optional overdrive. Glassfibre hardtop available '64 on, as was stronger five-bearing crank; front anti-roll bar '66. MkII of '67 has all-synchro 'box, improved rear axle, automatic option. Rostyle wheels, matt black grille and revised interior from '69. From '74, ugly US safety rubber bumpers and higher ride height, plus ever-decreasing power outputs. Enormous following ensures parts abundance.

Engine 1798cc 80.3x88.9mm, 4 cyl OHV, F/R
Max power 95bhp at 5500rpm (later 62-84bhp)
Max speed 106mph
Body styles sports
Prod years 1962-80
Prod numbers 387,259

PRICES
'62-'74 **A** 8000 **B** 5000 **C** 3000
'74-'80 **A** 5500 **B** 3700 **C** 2500

MGB GT

Three years after the roadster, MG launched its fixed-head B. The fastback styling (with some Pininfarina input) was a success and the incorporation of a tailgate made it much more practical; you also had two small rear seats for children. From the start, it had the five-bearing engine and quieter back axle. Mechanically identical to the roadster. A little heavier and not as sharp handling as the roadster, but you can't discount it because it's cheaper, prettier and more practical. US versions had progressively more strangled power outputs from 1967 on.

Engine 1798cc 80.3x88.9mm, 4 cyl OHV, F/R
Max power 95bhp at 5500rpm (later 62-84bhp)
Max speed 106mph
Body styles hatchback coupé
Prod years 1965-80
Prod numbers 125,621

PRICES
'65-'74 **A** 5000 **B** 2800 **C** 1500
'74-'80 **A** 4000 **B** 2400 **C** 1200

MGC

The long-in-the-tooth Austin-Healey was discontinued in 1968 and its replacement was the MGC. BMC did not install the Healey's straight six, but a new seven-bearing unit whose only other use was in the Austin 3-Litre saloon. Fitting this engine meant revising the front suspension with torsion bars. Other changes were 15in wheels in place of 14in, servo brakes and a bonnet bulge. Automatic and overdrive optional. Handling was wayward and performance was well down on the Healey, so the C got pasted in the press, not entirely justified. Spares can be problematic.

Engine 2912cc 83.4x88.9mm, 6 cyl OHV, F/R
Max power 145bhp at 5250rpm
Max speed 120mph
Body styles sports, hatchback coupé
Prod years 1967-69
Prod numbers 8999 (4542/4457)

PRICES
MGC **A** 9000 **B** 6000 **C** 4000
MGC GT **A** 6500 **B** 4500 **C** 3000

MGB GT V8

Several tuning companies fitted Rover's V8 into the MGB shell before BL did it in 1973. The result was better conceived than the MGC: the light alloy V8 weighed little more than the B-series, so the car was not as nose-heavy as the C and handled far better. Unique alloy wheels, different gearing, modified suspension, standard overdrive and servo brakes. Straddles chrome/rubber-bumper periods. Perhaps it should have looked more special, because sales were disappointing (the V8 could not be sold in America because of emissions difficulties). But it's an attractive classic: smooth, fast and torquey.

Engine 3528cc 88.9x71.1mm, V8 cyl OHV, F/R
Max power 137bhp at 5000rpm
Max speed 125mph
Body styles hatchback coupé
Prod years 1973-76
Prod numbers 2591

PRICES
'73-'74 **A** 8000 **B** 6000 **C** 3500
'74-'76 **A** 7000 **B** 5000 **C** 3000

MG Midget MkI/II/III/1500

Badge-engineered Austin-Healey Sprite, fitted with MG grille and chrome side strip. Mechanically identical, they were even built on the same production line. Bigger 1098cc engine and front discs from '62. MkII of '64 has curved screen, winding windows, door locks, more power and semi-elliptic rear springs. MkIII moves up to 1275cc, has fixed hood. '69 sees matt black grille and sills, Rostyle wheels, slimmer bumpers. Round rear wheelarches from '72, returned to square for the unloved '74 1500, with its huge black safety bumpers and tip-toe ride height. Enduring starter sports car.

Engine 948cc/1098cc/1275cc/1493cc, 4 cyl OHV, F/R **Max power** 42bhp at 5500rpm/56-59bhp at 5750rpm/65bhp at 6000rpm/ 65bhp at 5500rpm **Max speed** 86/90/95/98mph **Body styles** sports **Prod years** 1961-64/1964-66/1966-74/1974-79 **Prod numbers** 226,427 (25,681/26,601/100,246/73,899)
PRICES
MkI/II **A** 4000 **B** 2600 **C** 1300
MkIII '66-'74 **A** 4000 **B** 2500 **C** 1200
1500 **A** 3500 **B** 2300 **C** 1000

MG Magnette MkIII/IV

Supposedly sporting member of the decidedly unsporting BMC Farina range. Visual differences over the Austin Cambridge were MG grille, separate indicator/side light units, plus MG wheel trims and badges. Under the bonnet there was a twin carb B-series engine and inside you got leather trim and a wood-covered dash. MkIV of 1961 had 1622cc twin carb engine, front anti-roll bar, rear stabilizer bar, automatic option but – unlike other Farina saloons – the same size rear fins as before. Duo-tone paint schemes from '62. Best you can say is that it's a cheap MG, but not an enthusiast's machine.

Engine 1489cc 73x88.9mm/1622cc 76.2x88.9mm, 4 cyl OHV, F/R **Max power** 66.5bhp at 5200rpm/ 68bhp at 5000rpm **Max speed** 85-88mph **Body styles** saloon **Prod years** 1959-61/1961-68 **Prod numbers** 16,676/14,320

PRICES
A 2500 **B** 1500 **C** 700

MG 1100/1300

As the Magnette is to the Farina, so these are to the BMC ADO16 range. Launched in 1962 with the usual Hydrolastic suspension and front discs, but MG grille, hub caps and badges. More powerful 1100 engine, with optional 1275cc from June '67. 1100 two-door for export only. Official 1300 MkI model arrives in October '67 in two and four-door forms, though latter lasts only six months. Meanwhile the 1100 progresses to MkII in '67 (smaller tail fins, auto option) but is axed in '68. 65bhp twin carb engine for 1300 in April '68, MkII launched in September '68 (70bhp, two-door only). The Japanese like 'em.

Engine 1098cc 64.6x83.7mm/1275cc 70.6x81.3mm, 4 cyl OHV, F/F **Max power** 55bhp at 5500rpm/58-70bhp at 6000rpm **Max speed** 89/92-101mph **Body styles** saloon **Prod years** 1962-68/1967-73 **Prod numbers** 124,860/32,549

PRICES
A 2500 **B** 1500 **C** 800

Mini 850/1000

Issigonis' masterpiece is a strong candidate for most significant car of the 20th century. Transverse engine/sump-mounted 'box and front-wheel drive allowed miracle packaging. MkI is purest. Early ones badged Austin Seven/Morris Mini-Minor, and marque distinction continued until '69 MkIII. Hydrolastic suspension (1964-69) and auto option ('65 on) are demerits. '67 MkII has squarish grille, bigger rear screen, 998cc option. '69 MkIII has concealed door hinges, winding windows, and reverts to dry cone suspension. Little development thereafter except for 12in wheels/front discs from '84.

Engine 848cc 62.9x68.3mm/998cc 64.6x76.2mm, 4 cyl OHV, F/F **Max power** 33-34bhp at 5500rpm/ 38bhp at 5250rpm to 42bhp at 5250rpm **Max speed** 74/82-84mph **Body styles** saloon **Prod years** 1959-80/1967-92 **Prod numbers** 3,152,489
PRICES
MkI '59-'61 **A** 3500 **B** 2200 **C** 1000
MkI '61-'67 **A** 3000 **B** 2000 **C** 800
MkII/III **A** 1500 **B** 900 **C** 400

Mini Countryman/Traveller

Based on the extended floorpan of the Mini van, the load-lugging Mini was some 10in longer than the saloon. Austin Countryman and Morris Traveller launched with glued-on wood (non-structural), and later all-metal as an alternative. Double swing load doors are a pain, but capacity is amazing. Stiffer rear springs, never Hydrolastic. MkII has 998cc engine only. Separate Austin/Morris identities died when replaced by the Clubman Estate in 1969 (see p151); this also marked the end for the 'woody' estate. Stick-on wood seems to strike a chord with collectors today.

Engine 848cc 62.9x68.3mm/998cc 64.6x76.2mm, 4 cyl OHV, F/F **Max power** 34bhp at 5500rpm/ 38bhp at 5250rpm **Max speed** 72/75mph **Body styles** estate **Prod years** 1960-69 **Prod numbers** c 207,000

PRICES
Wood **A** 3800 **B** 2500 **C** 1100
Metal **A** 2500 **B** 1700 **C** 700

Mini

Mini-Cooper 997/998

Mini's inherent roadholding lent itself to sporting applications, and John Cooper was quick to realise this. His long stroke/narrow bore twin carb engine produced 55bhp, and the gear ratios were lengthened to produce a remarkable 87mph. Front discs, remote gearchange, duo-tone paint, special grilles for Austin and Morris versions, luxury trim. 1964 saw the Riley Elf's 998cc engine fitted, but in twin carb form; also Hydrolastic suspension the same year. MkII versions shared the standard Mini's bodywork changes, but still had a different grille. Even non-'S' Coopers are valuable.

Engine 997cc 62.4x81.3mm/998cc 64.6x76.2mm, 4 cyl OHV, F/F
Max power 55bhp at 6000rpm/55bhp at 5800rpm
Max speed 87/90mph
Body styles saloon
Prod years 1961-64/1964-69
Prod numbers 24,860/55,760

PRICES
997 **A** 5000 **B** 3200 **C** 2200
998 **A** 4000 **B** 2800 **C** 1800

Mini-Cooper 'S' 1071/970/1275

Downton developed the Cooper 'S', which immediately became the stuff of which legends are made. 1071cc engine had a new crankshaft, revised valves, block and head, plus twin carbs. Power and torque were unheard of in such a small car. Two 'boxes and two final drives to choose from, bigger front discs, ventilated disc wheels. 970cc version was a short-stroke homologation unit, now almost extinct. Tall-block 1275 was the definitive 'S'. Hydrolastic from '64 on. Twin tanks on '67 MkII, concealed hinges on '69 MkIII. Do your research on chassis numbers to avoid buying a fake.

Engine 1071cc 70.6x68.3mm/970cc 70.6x61.9mm/1275cc 70.6x81.3mm, 4 cyl OHV, F/F
Max power 70bhp at 6200rpm/65bhp at 6500rpm/76bhp at 5800rpm
Max speed 91/89/96mph
Body styles saloon
Prod years 1963-64/1964-65/1964-71
Prod numbers 4031/963/40,153

PRICES
MkI **A** 7500 **B** 5500 **C** 3000
MkII/III **A** 6500 **B** 5000 **C** 2700

GEARBOX REPAIR KITS

Why pay prices of £250 or more when you can **easily repair your own gearbox in one weekend** for a fraction of the normal cost and know it's been well done!
Automatics or Manuals ● Axles or Overdrives

With "TRANS-FIX KITS"® it's possible!

Kits contain: Gasket Set, Ball Bearings, Needle Roller Bearings, Oil Seals, Synchromesh Baulk Rings
SUBJECT TO GEARBOX TYPE

Model	Price	Model	Price
MG Midget 1275/Minor 1098	£57.50	Triumph Herald/Spitfire 1/2	£47.50
MG Midget/Spitfire 1500 MkV	£58.50	Vitesse Mk1/Spitfire Mk3	£51.50
MGB 3-synchro Pre-69 + O/Dr	£107.50	Vitesse Mk2/Spitfire Mk4/GT6	£57.50
MGB 3-synchro Pre-69 Non O/Dr	£117.50	TR6/2000/2500/Sprint	£138.50
MGB 3-synchro 69-on + O/Dr	£137.50	Stag/2500 late models	£120.50
MGB 3-sync 69-on Non O/Dr	£123.50	Jensen-Healey 1725	£57.50
MGB 4-synchro Non O/Dr	£83.50	Reliant Scimitar 3.0 1972	£72.50
MGB/GT V8/MGC + O/Dr	£112.50	Reliant Scimitar 3.0 1972-on	£79.50
MG Montego 5-spd (VW)	£110.50	Ford Cortina/Lotus 2000E	£75.50

*Subject to VAT at current rate.

For fast delivery – telephone for mail order by Access or Visa
PARTS DIVISION:
AUTO-TECH TRANSMISSIONS LTD
UNIT 7, BALMORAL TRADING ESTATE,
STUART ROAD, BREDBURY,
STOCKPORT, CHESHIRE SK6 2SR

Mini · Mitsubishi

Mini Clubman

A lesson in how to milk more money out of a best-seller: stick an extended nose on it, give it special wheel trims, new instruments and up-market trim and charge a hefty premium. That was the Mini Clubman, in every respect a dreadful move. More length, more weight, worse aerodynamics, less economy, more ugliness – in short, the antithesis of Issigonis. Estate version replaced the old woody estates and was also less lovable: it had fake wood trim pasted on the side. 1098cc after '75, unless you wanted auto, in which case a 998cc engine was fitted specially. Avoid unless you want a cheap Mini.

Engine 998cc 64.6x76.2mm/1098cc 64.6x83.7mm, 4 cyl OHV, F/F
Max power 38-41bhp at 5250rpm/45bhp at 5250rpm
Max speed 75/82mph
Body styles saloon, estate
Prod years 1969-80
Prod numbers 473,189

PRICES
A 1000 **B** 500 **C** 200

Mini 1275 GT

Supposedly a Mini-Cooper replacement, but it was in fact a step backwards. Single carb 1275cc engine was nothing like a Cooper 'S', being taken directly from the BMC 1300. Low final drive initially tried to compensate, but dreadful fuel economy forced a return to standard. Sadly, it had the Clubman nose. Disc front brakes, rev counter, Rostyle wheels and body stripes attempted to redress the balance. Rubber cone suspension from '71, 12in wheels from '74, run-flat Denovo wheels and tyres from '77 (but do any still have them?). If you are a 1970s decal fetishist and can't afford a Cooper, buy one.

Engine 1275cc 70.6x81.3mm, 4 cyl OHV, F/F
Max power 54-59bhp at 5300rpm
Max speed 86-90mph
Body styles saloon
Prod years 1969-80
Prod numbers 110,673

PRICES
A 2000 **B** 1300 **C** 600

Mini Moke

Having failed to sell the idea of a Mini-based 'Buckboard' to the armed forces, BMC tried to sell it to the public. Moke is a type of packhorse donkey and the Mini Moke was marketed as the same: a basic carry-all with stark, open bodywork and Mini mechanicals. Fold-flat screen, rear-mounted spare, rudimentary hood, optional passenger seats, only one paint scheme (green), very cheap price. Carnaby Street took a fancy to it, but that's about all. Australia built them with bigger engines 1966-81, Portugal 1980-94. British-built (1964-68) Mokes are most sought-after, but rust, noise and discomfort overwhelming.

Engine 848cc 62.9x68.3mm/998cc 64.6x76.2mm/1098cc 64.6x83.7mm/1275cc 70.6x81.3mm, 4 cyl OHV, F/F
Max power 34bhp at 5500rpm/38-40bhp at 5250rpm/50bhp at 5100rpm/54-65bhp at 5250rpm **Max speed** 65-70mph **Body styles** utility
Prod years 1964-94
Prod numbers c 51,000

PRICES
'64-'68 **A** 3500 **B** 2200 **C** 1500
'68-'81 **A** 3000 **B** 1800 **C** 1100

Mitsubishi Galant GTO

Mitsubishi arrived in the UK in 1973 under the name Colt, with typically Japanese products. Galant was a Morris Marina competitor with about as much appeal. The GTO Coupé was the only interesting one: 2-litre engine, five speeds, four headlamps, front spoiler, bolt-on wheelarch extensions, jazzed-up interior. Hardtop coupé bodywork looked very like the Toyota Celica. We got the top-model GSR (sold here to '78), but there were cheaper 1600/1700 and SL/GS versions in Japan. Must be pretty rare now, as rust record was abysmal.

Engine 1995cc 84x90mm, 4 cyl OHC, F/R
Max power 115bhp at 6000rpm
Max speed 112mph
Body styles coupé
Prod years 1970-76
Prod numbers n/a

PRICES
A 2500 **B** 1200 **C** 400

Mitsubishi Celeste

This was basically a Lancer – Mitsubishi's base-model Escort competitor – dolled up with a fancy fastback body. Styling was therefore a little top-heavy, but more than a hint of Mustang II can be seen. Tailgate helps practicality. UK got the 1600 version in ST and GS forms, the latter having twin carbs and five speeds, plus 2000 GT (engine derived from Galant), which could be had with five speeds or optional auto. Essentially, not a very exciting car, but quite a few were sold here before it was withdrawn in 1981. A rust-free one would be a miracle.

Engine 1597cc 76.9x86mm/1995cc 84x90mm, 4 cyl OHC, F/R
Max power 92bhp at 6000rpm/105bhp at 5400rpm
Max speed 99/103mph
Body styles coupé
Prod years 1975-81
Prod numbers n/a

PRICES
A 1300 **B** 600 **C** 200

Mitsubishi Sapporo

Although it was called Sapporo here, its Japanese name of Galant Lambda provided the evidence of its origin: this was based on the new '76 Galant saloon series (which we called the Sigma). Floorpan, inner panels, drive train and running gear all from the saloon, but new body panels. Hardtop coupé style was nothing special, but you got standard PAS and overhead console. Launched in UK in '78, new 1997cc engine in '80, rectangular headlamps '81, bespoilered Turbo version here from '82. Japanese stopped developing it in '83, but it was sold here till '85.

Engine 1995cc 84x90mm/1997cc 85x88mm, 4 cyl OHC, F/R
Max power 105bhp at 5400rpm/120bhp at 6000rpm to145bhp at 5500rpm
Max speed 106-112mph
Body styles coupé
Prod years 1976-85
Prod numbers n/a

PRICES
A 1700 **B** 700 **C** 200

Monica GT

Fascinating Anglo-French project, funded by Jean Tastevin and engineered by Chris Lawrence. Work began in 1967 but it took until 1974 to start making 'la Jaguar française'. Engine choice in production cars was a modified Chrysler V8, sitting in a tubular chassis with a De Dion rear end. Hand-built steel bodywork was impressive rather than handsome. Vastly expensive and killed by mid-1970s economic wasteland. Panther planned to revive it, but didn't. You're more likely to find a prototype than one of only 10 true production cars.

Engine 5560cc 102.6x84.1mm, V8 cyl OHV, F/R
Max power 305bhp at 5500rpm
Max speed 145mph
Body styles saloon
Prod years 1974-75
Prod numbers 35 (incl prototypes)

PRICES
Very rarely seen for sale, but expect to pay £5000-£15,000

Monteverdi High Speed/375/400/Berlinetta

Swiss BMW dealer and racing driver Peter Monteverdi built his own GT in 1967. Called the 375S, it had a huge tubular chassis, De Dion/Watt linkage rear, big Chrysler V8 engine, auto or manual transmission, discs all round, and PAS. 400SS had an extra 15bhp. Coupé body designed by Frua and built by Fissore. 1969 375L had longer wheelbase and 2+2 seats. 375C Cabrio from '71. 1972 Berlinetta had the 450bhp Hemi engine from the Hai. '75 Palm Beach was a short-wheelbase three-seater convertible. Very rare in any form.

Engine 6974cc 108x95.25mm/7212cc 109.7x95.25mm, V8 cyl OHV, F/R
Max power 305-375bhp at 4600rpm/390bhp at 5000rpm/450bhp at 5000rpm **Max speed** 140-160mph
Body styles coupé, convertible
Prod years 1967-77
Prod numbers n/a
PRICES
375 Coupé **A** 35,000 **B** 24,000 **C** 15,000
375 Convertible/Palm Beach worth 50% extra

Monteverdi 375/4

Encouraged by surprisingly high sales levels for his graceful GT cars, Peter Monteverdi launched a limousine version in 1971. Called the 375/4 to indicate it had four doors, it used a 375 chassis stretched by 26in, creating a car which measured 17ft 5in (531cm) long and weighed two tons. There was plenty of room for rear seat passengers and opulent fittings, including an optional division. Usual Chrysler V8 engines and very high claimed top speed. Like all Monteverdis, rust can be widespread and build quality suspect. But at least you'll have a genuine rarity.

Engine 6974cc 108x95.25mm/7212cc 109.7x95.25mm, V8 cyl OHV, F/R
Max power 305-375bhp at 4600rpm
Max speed 135-145mph
Body styles saloon, limousine
Prod years 1971-77
Prod numbers n/a

PRICES
A 25,000 **B** 16,000 **C** 9000

Monteverdi Hai

Switzerland had its Lamborghini Miura rival as early as 1970, but the Hai could hardly be described as having finesse. The tubular chassis was crude and the mid-mounted engine such a tight squeeze that it protruded in between the driver and passenger. Monteverdi himself did the styling, though Fissore built it. Lots of equipment, including air conditioning. Handling was reportedly frightening, especially in the wet, considering you had 390bhp under your foot in the 450 GTS and 450bhp in the 450 SS. It cost roughly twice the price of a 375L coupé. It is believed only two were built.

Engine 6974cc 108x95.25mm, V8 cyl OHV, M/R
Max power 390bhp at 5000rpm/450bhp at 5000rpm
Max speed 165-180mph
Body styles coupé
Prod years 1970-77
Prod numbers n/a

PRICES
Almost never comes up for sale, but worth more than 375 series

Moretti 1000

There is no Italian manufacturer or coachbuilder which built such a diversity of cars post-war as Moretti. Moretti was unusual in that it made its own overhead cam engines, chassis and components. Only after 1957 did it begin making cars based on Fiat floorpans. The 1000 pictured is typical of Moretti's output in this period. It was launched at the 1960 Turin Motor Show and was a 2+2 coupé using a Fiat 1100 engine. Other small 1960s Morettis included coupés and sports cars based on Fiats 500, 600 and 850.

Engine 1089cc 72x75mm, 4 cyl OHV, F/R
Max power 55bhp at 5200rpm
Max speed 92mph
Body styles saloon, estate, coupé, convertible
Prod years 1960-62
Prod numbers n/a

PRICES
Too rare to price accurately, but probably in the £5-10,000 range

Morgan F Super/F4

Before 1935, Morgan's entire output had consisted of three-wheelers; after the war, it consisted almost solely of four-wheelers. The 1934 F Type trike was the only three-wheeler offered by Morgan – or indeed any British maker – when hostilities ceased. Pressed steel chassis, Ford sidevalve engine, three-speed 'box, chain drive to the single rear wheel, IFS, quarter-elliptic rear springing, brakes on front wheels only. F Super was a two-seater, F4 was a longer wheelbase four-seater. Pre-war V-twins are preferred.

Engine 1172cc 63.5x92.5mm, 4 cyl SV, F/R
Max power 39bhp at 5000rpm
Max speed 70-75mph
Body styles sports
Prod years 1934-52
Prod numbers 265/367 (post-war)

PRICES
A 8000 **B** 5000 **C** 3000

Morgan 4/4

The first four-wheeled Morgan was little more than a development of the existing three-wheeler. It returned post-war in almost identical form, except for dropping the old Climax engine for Standard's OHV 1.3-litre unit. Both sports and drophead coupé styles were listed, the latter having sliding windows, squarer rear body and vertically-slatted grille. From '46, a four-seat Tourer was also available. Classic Morgan layout of tubular chassis, sliding pillar IFS, live rear axle on semi-elliptic springs, mid-mounted gearbox, worm-and-peg steering. Larger (but still mechanical) front brakes in '48.

Engine 1267cc 63.5x100mm, 4 cyl OHV, F/R
Max power 40bhp at 4300rpm
Max speed 80mph
Body styles sports, drophead coupé, tourer
Prod years 1936-39/1945-50
Prod numbers 824/1428

PRICES
A 15,000 **B** 10,000 **C** 7500

Morgan 4/4 Series II/III/IV/V/1600

The 4/4 was relaunched in 1955 after a five-year gap to offer low-cost Morgan ownership. Price kept down by Ford 100E sidevalve engine and three-speed 'box with push-pull scuttle-mounted lever. Only one body style now (sports two-seater, though Morgan called it a 'tourer'), but always with sloping cowled grille. SIII has Anglia engine and four speeds, SIV has Ford Classic engine and front discs, SV has Cortina engine (83bhp GT unit on the 'Competition') and all-synchro 'box. Post-68 cars called 1600 and have Ford crossflow engines, four seat option from late '68. Morgans are incomparable.

Engine 1172cc/997cc/1340cc/1498cc/1599cc, 4 cyl SV/OHV, F/R
Max power 36bhp/39bhp/54bhp/65bhp-83bhp/74-96bhp
Max speed 75/80/90/90-100/100-110mph **Body styles** sports, tourer
Prod years 1955-60/1960-61/1961-63/1963-68/1968-82 **Prod numbers** 387/59/206/646/c 3480

PRICES
SII **A** 10,000 **B** 8500 **C** 6500
SIII/IV/V **A** 12,000 **B** 9000 **C** 7000
1600 **A** 13,000 **B** 10,000 **C** 7000

Morgan Plus 4

The Plus 4 replaced the original 4/4 in 1950, offering greatly improved performance thanks to its Standard Vanguard engine. Slightly longer wheelbase, stronger chassis, better legroom, hydraulic brakes, four-speed 'box, but more weight (1cwt/51kg). Four-seater offered from 1951. Curved, cowled grille from '53 on roadster, from '54 on DHC and four-seat tourer. Triumph TR2 engine became optional in 1954 (90bhp TR3 unit from '55, 106bhp TR4 from '61). Front discs from '60. Desirable Super Sports (101 built between 1962-68) has twin Weber carb Lawrencetune engine, alloy body.

Engine 2088cc/1991cc/2138cc, 4 cyl OHV, F/R **Max power** 68bhp/90bhp/105bhp-125bhp **Max speed** 85/96/105-120mph **Body styles** sports, drophead coupé, tourer
Prod years 1950-58/1954-69/1960-68
Prod numbers 799 (Vanguard)/3642 (TR)/101 (Super Sports)

PRICES
Vanguard **A** 12,000 **B** 9000 **C** 7000
TR **A** 15,000 **B** 10,000 **C** 7500
Super Sports **A** 23,000 **B** 18,000 **C** 13,000

MORGAN SPECIALISTS

Mike Parkes
John Dangerfield Garages
Staple Hill Road
Fishponds
Bristol
BS16 5AD
Tel: 0117 956 6525/
0117 956 6373

Deborah White
Allon White & Son Ltd
The Morgan Garage
High Street
Cranfield
Bedfordshire
MK43 0BT
Tel: 01234 750205

Richard Thorne
Richard Thorne Classic Cars
Bloomfield Hatch, Mortimer
Reading, RG7 3AD
Tel: 01734 333633

RTCC are Morgan agents for West Berkshire, South Oxfordshire, North Hampshire and East Wiltshire. In addition to the normal sales, service and parts facilities we offer a full restoration service.

John Carter
Cliffsea Car Sales
654 Sutton Road
Southend-on-Sea
Essex
S52 5PX
Tel: 01702 602042

Rick Bourne
Brands Hatch Morgans
Maidstone Road,
Borough Green
Kent, TN15 8HA
Tel: 01732 882017

Suppliers of new and secondhand Morgans. Huge stocks of Morgan parts and accessories. Mail order. Specialists in classic car MOTs.

David Randall
Lifes Motors Ltd
West Street, Southport
Lancashire, PR8 1QN
Tel: 01704 531375

We have been Morgan agents since 1926, the oldest in the world, and usually carry up to 22 Morgans in stock.

David Gilbertson
Thomson & Potter Ltd
High Street, Burrelton
Blairgowrie, Perthshire
PH13 9NX
Tel: 01828 670247

New for '96: hire a Morgan, sample classic motoring in a fabulous part of the world!

A. M. Parker
Parkers of Stepps
Mayston Garage
38 Glasgow Road
Kirkintrillach
Glasgow
G66 1BJ
Tel: 0141 776 1708

Mr Burnett
Burlen Services
Spitfire House
Castle Road
Salisbury
Wiltshire
SP1 3SA
Tel: 01722 412500

Mr Andreas Adamou
Autorapide Ltd
The Wells Road, Latcham
Wedmore, Somerset, BS28 45B
Tel: 01934 712170

Specialists in restoration, standard V8 and modified engines, suspension tuning, modified exhaust systems, CVH fast road cams and cylinder heads.

Techniques
Porter Brothers Garage
Station Road, Radlett
Hertfordshire, WD7 8JX
Tel: 01923 853600

Offering comprehensive Morgan repairs/service to the highest standard. Recognised as the specialists for: L/H drive conversions and chassis replacement.

Morgan Plus 4 Plus

The one and only occasion on which Morgan tried to update its traditional bodywork, and a complete flop. The problem was more the method of construction than the controversial closed coupé styling: a heretical glassfibre body and (shock) winding windows! Plus Four chassis, usual ash frame, plastic body by EB, front discs, TR4 engine, some with Armstrong Selectaride dampers. Nominally an export car and Lotus Elite competitor, but compromised by vintage chassis. Morgan enthusiasts disapproved and negligible sales ensured the experiment never recurred. Rarity keeps prices high.

Engine 2138cc 86x92mm, 4 cyl OHV, F/R
Max power 105bhp at 5500rpm
Max speed 110mph
Body styles coupé
Prod years 1963-67
Prod numbers 26

PRICES
A 15,000 **B** 10,000 **C** 7000

Morgan Plus 8

One of the greatest of all British sports cars was created when Morgan judiciously installed a Rover V8 engine under the bonnet of its Plus 4. Track 2in wider from the start, further 2in increments in '73 and '76; chassis also lengthened by 2in in '69. Phenomenal acceleration (0-60mph in 6.5 secs). Moss four-speed 'box replaced by Rover 3500S all-synchro unit in '72 and a five-speeder from late '76. Steel bodywork or, from 1977, optionally in aluminium; alloy wheels always standard. Fuel injection from 1984, rack-and-pinion steering standard in '86, 3.9-litre engine '90. Apart from factory-built one-off, no four-seaters.

Engine 3528cc 88.9x71.1mm, V8 cyl OHV, F/R
Max power 160bhp at 5200rpm to 190bhp at 5280rpm
Max speed 125mph
Body styles sports
Prod years 1968-date
Prod numbers over 4000 to date

PRICES
'68-'86 (carb) **A** 17,000 **B** 13,000 **C** 9000

Morris Eight

Morris's 1938 Series E design was revived post-war with no change. Chassis frame has boxed-in side members (i.e. semi-unitary), beam front axle, leaf springs all round, hydraulic brakes, four speeds, old sidevalve engine. Both two- and four-door saloons were offered. Although no open tourers were made post-war, all but the very last saloons had standard sliding roofs. Styling was thought radical in 1938, with waterfall grille and faired-in headlamps which, incidentally, gave a dismal spread of light. Cheap, plentiful and worthy.

Engine 919cc 57x90mm, 4 cyl SV, F/R
Max power 30bhp at 4400rpm
Max speed 60mph
Body styles saloon
Prod years 1938-48
Prod numbers 120,434

PRICES
A 2000 **B** 1200 **C** 700

Morris Ten

The sister model to the Eight post-war was the Series M Ten, having benefited from much testing in wartime service. It was Morris's first unitary car but shared the Eight's rather wallowy suspension layout, helped slightly by a front anti-roll bar. Overhead valve engine is virtually unburstable and is an antecedent of the MG TC unit. Six-light styling more upright and traditional than the Eight's, but you still get a sliding roof (until early 1948 at least) and, from 1946, there was a new, curved radiator grille. Basically unexciting in just about every area.

Engine 1140cc 63.5x90mm, 4 cyl OHV, F/R
Max power 37bhp at 4600rpm
Max speed 63mph
Body styles saloon
Prod years 1938-48
Prod numbers 80,990

PRICES
A 2200 **B** 1300 **C** 700

Morris Minor MM/Series II

Issigonis' advanced post-war masterpiece, which became Britain's best-selling car. Smooth unitary bodywork (four-door option from '50 on), IFS by wishbones and torsion bars (but rigid rear axle and leaf springs), rack-and-pinion steering. Light and responsive, with legendary roadholding, its amazing abilities were compromised by its Series E pre-war sidevalve engine. Following the Austin/Morris merger, this was addressed by installing an Austin A30 803cc engine in the four-door from '52 and in the two-door from '53: better acceleration and top speed. Headlamps raised on to wings from '50.

Engine 918cc 57x90mm/803cc 58x76mm, 4 cyl SV/OHV, F/R
Max power 27.5bhp at 4400rpm/ 30bhp at 4800rpm
Max speed 60/62mph
Body styles saloon
Prod years 1948-53/1952-56
Prod numbers 176,002/269,838 incl Tourer/Traveller

PRICES
MM **A** 3000 **B** 2000 **C** 1000
Series II **A** 2200 **B** 1500 **C** 900

Morris

Morris Minor 1000

The definitive Minor arrived in 1956, fitted with the 948cc A-series engine, shared with the Austin A35 (except for its SU carburettor), which at last gave enough performance to suit the capable chassis. Close-ratio remote-control gearbox with higher final drive. Single-piece curved windscreen, larger rear screen, minor interior improvements. Wider-opening doors from '59, semaphore indicators replaced by flashing lights in '61, 1098cc 48bhp engine, bigger brakes and stronger gearbox from '62, but otherwise virtually no changes until the end – Minor fans demonstrated, to no avail.

Engine 948cc 63x76mm/1098cc 64.6x83.7mm, 4 cyl OHV, F/R
Max power 37bhp at 4750rpm/48bhp at 5100rpm
Max speed 70/78mph
Body styles saloon
Prod years 1956-62/1962-71
Prod numbers 544,048/303,443 incl Tourer/Traveller

PRICES
A 2200 **B** 1300 **C** 600

Morris Minor Tourer

The unitary body of the Minor was so strong that hardly any modification was required to strengthen it when it came to producing a convertible version. Early ones had removable rear side curtains, but from 1951 a glass window with a fixed frame was substituted. Mechanical and bodywork changes as per the saloon, although only a two-door body style was ever offered. MM Tourers are now extremely rare, especially the ones with headlamps incorporated into the grille. Even later ones have a value which exceeds almost all other post-war popular convertibles.

Engine 918cc 57x90mm 4 cyl SV, 803cc 58x76mm/948cc 63x76mm/1098cc 64.6x83.7mm, 4 cyl SV/OHV/OHV, F/R **Max power** 27.5bhp at 4400rpm/30bhp at 4800rpm/37bhp at 4750rpm/48bhp at 5100rpm
Max speed 60/62/70/78mph
Body styles convertible **Prod years** 1948-69 **Prod numbers** 74,960

PRICES
MM **A** 5000 **B** 3000 **C** 1500
Series II **A** 3500 **B** 2500 **C** 1200
1000 **A** 3200 **B** 2300 **C** 1100

MORRIS MINORS

We are manufacturers and suppliers of a large range of Morris Minor Parts, including our World famous 'Tourer Conversion Kits'.

Our 1996 colour illustrated Parts Catalogue is now available. Call us for your free copy now.

The Morris Minor Company, Dept CCBG, Unit 3, Woodfield Road Industrial Centre, Balby, Doncaster, DN4 8EP, England
Tel: (01302) 859331 Fax: (01302) 850052

MORRIS MINOR CENTRE (BIRMINGHAM)

WE STOCK OVER 2000 LINES OF PARTS

BODY ETC		SPECIAL OFFERS	
Wings (all Cars, Late) from	£57.50	Bulkhead Brake Reservoir Kit	£28.50
Front (Grille) Panel	£36.00	Swivel Pins (new)	£57.50
Rolls-Royce Quality Bumpers	£42.50	Swivel Pin Assemblies	£75.00
Trav Rear Bumper	£16.95	Lead Free Cylinder Head (exch)	£89.50
Top (Grille) Chrome	£22.50	Front and Rear Screen Rubbers	£19.95
Assembled Side Ash Frames	£245.00	Road Spring (saloon)	£28.00
Over-Riders (Rolls-Royce quality)	£17.95	Hockey Sticks (late)	£13.50
Carpet Sets	£60.77	Indicator Switch Assemblies	£32.00

IMPROVEMENTS	
Servo kit	£134.33
5-speed Gearbox Fitting Kit	£285.00
5-speed Ford Sierra Recon G/box	£295.00
Auto Seat Belts (from)	£27.50
Heated Rear Window (Saloon)	£119.80
Heated Rear Window (Traveller)	£86.00
Alternator Kit (inc alt)	£69.50
High Output Heater Core (for std heater)	£33.50

THE MORRIS MINOR CENTRE (B'HAM)

2 CAMDEN STREET,
PARADE,
BIRMINGHAM B1 3BN
Tel: 0121 236 1341
Fax: 0121 236 1342

Prices do not include Carriage and VAT

Morris Minor Traveller

It took until the 1953 Series II Minor for Morris to develop an estate car variant, which it called Traveller. Behind the front doors, the bodywork was replaced by aluminium panels and a wooden framework reminiscent of 1940s American 'woodies' (the woodwork is structural, so rotten wood means substantial repairs). Split van-type rear doors, sliding windows for rear seat passengers, no four-doors. Updated as per the saloon, and lasts for the same amount of time (bowing out in '71, though the van and pick-up remained in production for slightly longer).

Engine 803cc 58x76mm/948cc 63x76mm/1098cc 64.6x83.7mm, 4 cyl OHV, F/R
Max power 30bhp at 4800rpm/37bhp at 4750rpm/48bhp at 5100rpm
Max speed 62/70/78mph
Body styles estate
Prod years 1956-71
Prod numbers 215,328

PRICES
A 3000 **B** 2000 **C** 1200

Morris Oxford MO

Looking like an enlarged Minor both inside and out, the Oxford was, however, a much less satisfying package than its smaller sister. The 1.5-litre engine was still an archaic and gutless sidevalve, the four-speed gearbox had an unpleasant column change and the handling was poor by comparison, despite the fitment of IFS. That improved in 1950 after telescopic dampers were fitted. New chrome grille in '52. Traveller two-door estate, available from 1952, had structural woodwork and twin opening rear doors; it is very rare.

Engine 1476cc 73.5x87mm, 4 cyl SV, F/R
Max power 41bhp at 4200rpm
Max speed 70mph
Body styles saloon, estate
Prod years 1948-54
Prod numbers 159,960

PRICES
Saloon **A** 2000 **B** 1300 **C** 700
Estate **A** 2500 **B** 1800 **C** 1200

MORRIS SPARES SOUTH EAST

Approved stockists of Duckhams Heritage Oil & Dinitrol

WE CAN SUPPLY THE FOLLOWING:

Brakes, Body – Floor – Chassis Panels, Chrome, Carburation, Radiators, Water Pumps (incl 803), Hoses, Full range of Trim, Upholstery & Carpets, Electrical & Wiring Assemblies & Components, Engines/gearboxes & Components, Gaskets, Exhaust systems, Lighting, Literature, Paint, Rear Springs & Fittings, Rust Proofing, Steering, Suspension, Clutches, Traveller-Woodwork, Windscreens, Rubbers, Seals & Wipers and a range of Accessories.

Unit 9, Bassett Business Centre
Hurricane Way, North Weald, Epping
CM16 6AA
Tel: 01992 524249 (Out of hours: 01799 526278)
Fax: 01992 524542
FOR A NEW PRICE LIST, SEND 3 (loose) 1ST CLASS STAMPS

Morris Minor South West

Specialist in all aspects of Morris Minors
Service Workshop Facilities
Callers Welcome
FREE FRIENDLY ADVICE

FAST & EFFICIENT MAIL ORDER
Send SAE for our parts price list.
All your requirements are just a phone call away.

Phone/Fax 01872 70210
Morris Minor South West
Unit 7, Willowgreen Farm,
Threemilestone, Truro,
Cornwall TR4 9AL

Morris Six

The third new Morris of 1948 was the Six. In terms of raw appeal it was a rather insipid creature, sharing the looks and fittings of the down-market Oxford but costing £799, into executive territory. Large, traditional grille and long bonnet hid the car's best feature: its engine. This 66bhp overhead cam unit was mounted in a separate subframe and was smooth and a good cruiser, though its top speed was not high and vertical valves were troublesome (replaced by inclined valves in '52). De Luxe version offered from 1953. Not popular when new and very few survivors today.

Engine 2215cc 73.5x87mm, 6 cyl OHC, F/R
Max power 66bhp at 4800rpm
Max speed 70mph
Body styles saloon
Prod years 1948-54
Prod numbers 12,464

PRICES
A 2200 **B** 1300 **C** 1000

Morris Oxford II/III/IV/ Cowley

New unitary body and BMC B-series 1.5-litre engine brought decent performance and class-leading roominess to the new Morris Oxford in 1954. It was no beauty, had basically the same suspension as the old Oxford and suffered from a badly offset driving position and column change 'box; but it was a good seller. Series III had fluted bonnet and small tail fins, plus optional automatic transmission. Series IV was an all-metal estate. Cowley was an austerity model, initially with a 1200cc Austin A40 engine, and from '56 with the same engine as the Oxford. Very little interest.

Engine 1489cc 73x89mm/1200cc 65.5x89mm, 4 cyl OHV, F/R
Max power 50-55bhp at 4400rpm/42bhp at 4600rpm
Max speed 75/70mph
Body styles saloon, estate
Prod years 1954-56/1956-59/1957-60/1954-59
Prod numbers Oxford: 145,458/Cowley: 22,036

PRICES
A 1500 **B** 1000 **C** 400

Charles Ware's world famous
MORRIS MINOR CENTRE

- The specialist for all your needs since 1976.
- Free advice for restorers, buyers and sellers.
- Best agreed Value Insurance scheme in the U.K. – also available for young drivers.
- Cars for sale – largest stock of minors in the U.K.
- Parts, our famous catalogue – £5 (inc postage).
- Incredible value panels – from our Sri Lanka factory – up to 30% off normal U.K. prices!
- Agents in Kent, Essex, Berkshire

Avon House, Lower Bristol Road, Bath BA2 1ES
- Restoration and free advice Tel: (0225) 315449
- Parts Tel: (0225) 482747   Fax: (0225) 444642
- Ask for our free brochure.

KING STREET MOTOR SERVICES
An approved associate of The Morris Minor Centre, Bath

Morris Minor Specialists
*

Car Sales:
We stock a variety of all models.

Restoration:
Complete restorations and accident repairs undertaken.

Mechanical:
MOTs, servicing, general repairs.

Woodwork:
Replacement and care of wood carried out.

Parts:
One of the largest selection of parts.

Valuations for insurance.

K.A. Motors Ltd, 40 King Street, West Malling, Kent. ME19 4QT
Telephone: West Malling (01732) 843135

Morris Isis

Replacement for the Six, fitted with the new Austin A90 C-series straight six. Main hull derives from contemporary Oxford, but longer bonnet, fatter tyres, larger brakes, cam steering and stronger suspension differentiated it. Not a great handler but better performance than any Morris to date. Traveller estate was a graceless all-steel affair which could seat eight thanks to a third row of rear-facing seats. Series II of '56 has 90bhp, chrome-tipped rear wings, mesh grille and floor-mounted gear lever, plus optional overdrive or Borg Warner auto. Virtually extinct.

Engine 2639cc 79.4x88.9mm, 6 cyl OHV, F/R
Max power 80-90bhp at 4500rpm
Max speed 82-90mph
Body styles saloon, estate
Prod years 1954-58
Prod numbers 12,155

PRICES
A 2500 **B** 1500 **C** 1000

Morris Oxford V/VI

The Morris-badged version of the Farina Austin Cambridge/Wolseley 15-60 series arrived in March 1959, sharing every detail with the Cambridge except for a different grille (with horizontal bars) and badging. 1.5-litre engine on initial Series V model, 1622cc unit on Series VI of August 1961 (this also has cut-down rear fins, new grille, automatic option). Traveller estate version from September 1960. Duo-tone paint from late '62. 1489cc diesel saloon and estate made from 1962 to 1971, but are virtually unknown in the UK. Just as dull as all the rest of the BMC Farina range.

Engine 1489cc 73x88.9mm/1622cc 76.2x88.9mm, 4 cyl OHV & diesel/OHV, F/R
Max power 52bhp at 4350rpm/40bhp at 4000rpm/61bhp at 4500rpm
Max speed 81/66/84mph
Body styles saloon, estate
Prod years 1959-61/1961-71
Prod numbers 87,432/208,823

PRICES
A 2000 **B** 1300 **C** 700

MORRIS MINOR MANIA LTD
1-3 HALE LANE
MILL HILL, LONDON, NW7 3NU
Tel: 0181 959 0818 Fax: 0181 959 0819

We have been established since 1981 and we specialize in the repair and restoration of all types of Morris Minor, as well as normal day to day running repairs, welding and MOT testing. We supply and fit disc brakes and suspension modifications and have large stocks of new and used parts, and we now stock Heritage oil. We offer a mail order service and also free advice over the 'phone should you need any help. We have a Fiat 1600/2000cc fitting kit if you need that extra bit of power, or a Marina/Ital fitting kit which also sells quite well. Please call Nick or Alex if we can help you.

Restoring a Morris Minor?
Looking to buy a good one?
Or just a proud owner?
If your answer to any of these questions is yes, you ought to have a copy of our book.

ORIGINAL MORRIS MINOR

by Ray Newell

It's a complete guide to authenticity and original factory specification for all Minors – saloons, tourers, travellers, vans and pick-ups. The author gives full details of all production changes and modifications, and there are 250 colour photographs of outstanding cars, so you can check what's right and what's wrong for your Minor, from door pulls to air cleaner.

The book is an A4 size hardback with 128 pages.
Price £19.95

Available from good bookshops, but in case of difficulty phone us, Bay View Books, on 01237 479225/421285.

Morris 1100/1300

Morris was the first of BMC's badges to go on Issigonis' new small saloon, arriving in August 1962, a full year before Austin. Advanced set-up for 1962: Hydrolastic interconnected suspension, front discs, transverse engine and front-wheel drive. Two-door saloon for export only initially, automatic option from '65, three-door Traveller estate from '66. 1100 MkII from '67 has minor trim improvements, while 1300 is launched at the same time. All-synchro 'box from '68. MkIII of '71 in 1300 Traveller form only. Four-door 1300 GT (1969-71) with twin carbs and 70bhp is the only interesting one.

Engine 1098cc 64.6x83.7mm/1275cc 70.6x81.3mm, 4 cyl OHV, F/F
Max power 48bhp at 5100rpm/ 58bhp at 5250rpm to 70bhp at 6000rpm
Max speed 78/92/96mph
Body styles saloon, estate
Prod years 1962-71/1967-73
Prod numbers 801,966

PRICES
1100/1300 **A** 1500 **B** 900 **C** 500

Morris 1800/2200

By now, Austin and Morris were pure badge engineering partners. The 1800 was simply a rebadged Austin, which arrived some two years later than its Longbridge stablemate. Grille (with round Morris badge) the only difference. Walnut veneer and power steering option from '67, MkII of '68 has more power, S version has twin carbs and 96bhp. '72 MkIII has new grille, rod-operated gear change, floor-mounted handbrake. 2200 launched in '72 has new E6 six-cylinder engine, new facia and yet another new grille. No 'Landcrab' is very exciting, and the Morris version is perhaps the least remarkable.

Engine 1798cc 76.2x95.75mm/2227cc 76.2x81.3mm, 4/6cyl OHV/OHC, F/F
Max power 80bhp at 5000rpm/96bhp at 5700rpm/110bhp at 5250rpm
Max speed 90/99/108mph
Body styles saloon
Prod years 1966-75
Prod numbers 95,271/c 10,000

PRICES
A 1500 **B** 800 **C** 400

Morris Marina 1300/1800/1700/Diesel

The Marina ranks alongside the Austin Allegro in the high book of BL mediocrity. Hastily conceived, it was an utterly conventional, rear-wheel drive car, which ended up lasting (latterly as Ital) for 13 years. Early ones were dreadful understeerers and gained a reputation for shoddy build quality. 1.3 A-series and 1.8 B-series engines to start. Only twin carb TC Coupé even mildly interesting (replaced by GT in '75). MkII from '78 has 'fireplace' facia style, adds O-series 1.7 engine. Export-only diesel version (1977-80) has prehistoric 1489cc B-series engine. Tedious – but there's now a club!

Engine 1275cc/1798cc 80/1695cc/ 1489cc, 4 cyl OHV/OHV/ OHC/diesel, F/R **Max power** 57-60bhp/82bhp – 94bhp/78bhp/40bhp
Max speed 82/95-100/95/70mph
Body styles saloon, estate, coupé
Prod years 1971-80/1971-78/1978-80/1977-80 **Prod numbers** 953,576 (515,888/374,711/59,107/3870)

PRICES
1.3/1.8/1.7 **A** 700 **B** 400 **C** 100
1.8 TC Coupé **A** 1100 **B** 600 **C** 300

Morris 1800/2200

Almost forgotten member of the Harris Mann designed ADO71 family, which became the Princess after only a matter of months. Alongside Austin and Wolseley versions, there was a Morris too. The Morris one was identifiable by its raised bonnet centre and quad headlamps (like the Wolseley) and its own unique grille. Direct replacement for the old 1800/2200, but controversially styled. Fine ride (courtesy of Hydragas suspension) and big interiors were its strongest features. Very few Morris survivors, but don't expect to pay much.

Engine 1798cc 80.3x88.9mm/2227cc 76.2x81.3mm, 4/6cyl OHV/OHC, F/F
Max power 82bhp at 5250rpm/ 110bhp at 5250rpm
Max speed 96/105mph
Body styles saloon
Prod years 1975
Prod numbers c 6000/1650

PRICES
A 1500 **B** 900 **C** 400

Moskvich 400/401

When the Red Army pulled out of Germany after the war, it removed the tooling for the pre-war Opel Kadett from the Rüsselsheim works as reparations, and proceeded to set up a production line in Moscow to make it as the Moskvich 400 ('son of Moscow'). Virtually every part was duplicated exactly, and finish was reportedly very good. The 1954 401 had an extra 4bhp. Exports to Western Europe began as early as 1949 (where it was sometimes badged as the MKV or Moskvitch), but it was always pretty marginal outside the motherland.

Engine 1074cc 67.5x75mm, 4 cyl SV, F/R
Max power 23-27bhp at 3600rpm
Max speed 58-61mph
Body styles saloon
Prod years 1946-54/1954-56
Prod numbers n/a

PRICES
A 2500 **B** 1500 **C** 1000

Moskvich 402, 410 & 423/407 & 403

Russia's 'people's car' was substantially updated in 1956. The full-width bodywork was all-new and quite modern, and the old Opel unit was bored out to 1.2 litres and given overhead valves. Coil springs at the front, semi-elliptics at the back, three speeds on the column. Four-wheel drive 410 (1956-60), and 423 estate car from '57. 1358cc 407 arrived in 1958. Limited UK imports began in 1961. 403 of 1963 had the merest of styling changes. Assembly also took place in Belgium by the firm Sobimpex, cars being marketed on the continent under the name Scaldia. Sturdy and stolid.

Engine 1220cc 72x75mm/1358cc 76x75mm, 4 cyl OHV, F/R
Max power 35bhp at 4200rpm/45bhp at 4500rpm
Max speed 65-72mph
Body styles saloon, estate
Prod years 1956-58/1958-64
Prod numbers n/a

PRICES
A 1800 **B** 1000 **C** 400

Moskvich 408 & 426/412 & 427

Another major overhaul of the Moskvich occurred in 1964 when the 408 (saloon) and 426 (estate) were launched: new body style featuring twin headlamps, plus bigger 1.4-litre OHV engine, four-speed all-synchromesh 'box, same IFS/leaf spring rear suspension set-up. 1968 412/427 had a new 1.5-litre five-bearing overhead cam engine, rectangular headlamps. Sobimpex assembled some under the Scaldia name, fitted with British-made 1760cc Perkins diesel engines. Terrible reputation in Britain, so withdrawn in '75. Renamed 2140 in '75 and they still make them in Moscow!

Engine 1357cc 76x75mm/1478cc 82x70mm/1760cc 79.4x88.9mm, 4 cyl OHV/OHC/diesel, F/R
Max power 60bhp at 4750rpm/80bhp at 5800rpm/62bhp at 4000rpm
Max speed 75-90mph
Body styles saloon, estate
Prod years 1964-68/1968-75
Prod numbers n/a

PRICES
A 1000 **B** 500 **C** 100

Muntz Jet

Frank Kurtis began building his interesting Sports model in 1948: a two-seater convertible with unitary aluminium body panels, glassfibre bonnet (very unusual for the time) and 'any engine you wanted'. Project sold to used car dealer Earl 'Madman' Muntz in 1950. He made it as the Muntz Jet with Cadillac V8 power. 1952's longer, four-seater, steel-bodied version used a modified version of Lincoln's sidevalve V8 engine (later an overhead valve Lincoln V8 and glassfibre wings). Of the 394 built, around 50 still survive.

Engine 5424cc 96.8x92.1mm/5520cc 88.9x111.3mm/5203cc 96.5x88.9mm, V8 cyl OHV/SV/OHV, F/R **Max power** 160bhp at 4000rpm/152bhp at 3600rpm/205bhp at 4200rpm
Max speed 112/108/116mph
Body styles coupe, convertible
Prod years 1950-54
Prod numbers 394
PRICES
Coupe **A** 20,000 **B** 13,000 **C** 8000
Convertible **A** 30,000 **B** 20,000 **C** 12,000

Nash Healey

Donald Healey astutely formed an alliance with Nash of America whereby he would build sports cars for the American market using mechanicals from the Nash Ambassador. Started with Healey-designed bodywork (also sold in UK with Alvis engine – see page 102), but Pinin Farina restyled it with rather more flair for '52 (photo). Also in '52, the larger engine bowed in. Longer wheelbase Le Mans coupé version arrived for '53. The cost of Healey shipping completed chassis to Italy and then to America was terminal. Roadster died '53, coupé '54. Character is more boulevard than sporting.

Engine 3848cc 86x111.1mm/4138cc 88.9x111.1mm, 6 cyl OHV, F/R
Max power 125bhp at 4000rpm/140bhp at 4000rpm
Max speed 105mph
Body styles sports, coupé
Prod years 1951-54
Prod numbers Healey: 104/Farina: 402

PRICES
A 25,000 **B** 18,000 **C** 12,000

Nissan Sports Fairlady

Officially this was called the Datsun Fairlady, essentially Japan's answer to the MGB and Fiat Spider. Very European in conception, it was a three-seater sports car with a ladder chassis, steel body, 1.5-litre engine, IFS, semi-elliptic rear. Twin carb engine from '63, 1.6 engine from '65, OHC 2-litre/five-speed option and front discs from '67. Priced strategically below the MGB in the USA, Americans lapped them up. Not sold in Britain, but some US imports are now appearing. Surprisingly accomplished for such an early Japanese effort.

Engine 1488cc 80x74mm/1595cc 87.2x66.8mm/1982cc 87.2x83mm, 4 cyl OHV/OHV/OHC, F/R
Max power 77-85bhp at 5600rpm/96bhp at 6000rpm/135bhp at 6000rpm
Max speed 87-95/100/118mph
Body styles sports
Prod years 1962-69
Prod numbers n/a

PRICES
A 7000 **B** 5000 **C** 2500

Nissan 240Z

With the 240Z, Nissan slayed the American market for sports cars. Launched after considerable market research, it had everything going for it: a stunning body by Count Albrecht Goertz, a punchy 150bhp twin carb six-cylinder engine, nimble handling, lots of equipment and build quality which embarrassed MG and Triumph – all at $3500, the same price as a GT6. It quickly became the world's best-selling sports car. All-synchro 'box, all-independent MacPherson strut suspension, front discs. Early ones are now appreciating, but you'll need to look carefully for rust in the structure.

Engine 2393cc 83x73.7mm, 6 cyl OHC, F/R
Max power 130-150bhp at 6000rpm
Max speed 122mph
Body styles coupé
Prod years 1969-74
Prod numbers 150,076

PRICES
'70-'71 **A** 7500 **B** 5000 **C** 2500
'71-'74 **A** 6000 **B** 3500 **C** 1500

Nissan 260Z

Larger-engined version of 240Z arrived in 1974, identifiable by bigger rear light clusters and new interior. Also launched in '74 was 260Z 2+2 (photo), wheelbase lengthened by 12in and roofline modified to allow head room for two small rear passengers; slow two-seater sales meant this model was temporarily dropped in UK between 1975-77. 260Zs generally do not fetch nearly as much as earlier 240Z, despite extra power; 2+2 models especially are regarded as watered-down and compromised Z-cars. Just about every body panel rusts. 2-litre version for Japan and 2.8 litres (280Z) for America from '75.

Engine 2565cc 83x79mm, 6 cyl OHC, F/R
Max power 162bhp at 5600rpm
Max speed 125mph
Body styles coupé
Prod years 1974-79
Prod numbers 80,369

PRICES
260Z **A** 4000 **B** 2800 **C** 1200
260Z 2+2 **A** 3000 **B** 2300 **C** 1000

NSU Prinz I, II & II/30, III & III/30

NSU built Fiats under licence before presenting its own car in 1957. This was the little Prinz, shown at Frankfurt in 1957 and put into production the following year. A very compact four-seater saloon, it had a rear-mounted two-cylinder air-cooled ohc engine, and all-independent suspension. Prinz I was basic version with a crash gearbox, Prinz II had better trim and 23bhp engine, Prinz 30 had highest output of the lot. All replaced by Prinz III and III/30 in 1960 (front anti-roll bar, reclining seats). Noisy, uncomfortable, but characterful.

Engine 583cc 75x66mm, 2 cyl OHC, R/R
Max power 20bhp at 4600rpm/23bhp at 5000rpm/30bhp at 5500rpm
Max speed 68-75mph
Body styles saloon
Prod years 1958-62
Prod numbers 94,549

PRICES
A 1800 **B** 1000 **C** 500

NSU Prinz 4

Classic NSU shape emerged with the Prinz 4 in 1961, although it looked more frog than prince. 1.5in longer wheelbase but well over a foot longer overall, the 4 had a slightly bored-out version of the air-cooled twin engine and a standard all-synchro gearbox. Standard, De Luxe, Special Equipment and L trim levels available, plus a short-lived Super Prinz with a dummy front grille. Quite nippy considering its tiny engine, and better-handling than most small continental rear-engined cars. Sold over half a million, but surprisingly few examples of this base-model NSU left around.

Engine 598cc 76x66mm, 2 cyl OHC, R/R
Max power 30bhp at 5600rpm
Max speed 75mph
Body styles saloon
Prod years 1961-73
Prod numbers 576,023

PRICES
A 1200 **B** 700 **C** 400

NSU Sport Prinz

Amazingly mature coupé version of the original Prinz was effected by no less a man than Franco Scaglione, then at Bertone. Wherever you stand around the car, it always looks right – the sign of a classic design. It arrived as early as 1958 and was initially built by Bertone; later ones were built by Drauz of Heilbronn. Very light at just 1250lb (565kg), which was just as well because the tiny 583cc engine pumped out only 30bhp. Even the larger 598cc engine from September 1962 boasted the same output, so performance is not its strongest suit. Front discs from '65.

Engine 583cc 75x66mm/598cc 76x66mm, 2 cyl OHC, R/R
Max power 30bhp at 5500rpm/30bhp at 5600rpm
Max speed 75mph
Body styles coupé
Prod years 1958-68
Prod numbers 20,831

PRICES
A 2500 **B** 1700 **C** 800

NSU Wankel Spider

The fact that this was the world's first ever rotary engined production car would justify its place in every car museum. Under a chop-top version of the Sport Prinz's bodywork sat a tiny single-rotor engine designed by Felix Wankel. At the time, it created quite a stir: here was a revolutionary little sports car with a 50bhp engine capable of nearly 100mph. But reliability problems plagued it, and the public was suspicious of the new technology; also it was quite an expensive little beast. Front discs, trailing wishbone rear end. Attractive and engaging, but fraught with problems.

Engine 497cc, single-rotor Wankel, R/R
Max power 50bhp at 6000rpm
Max speed 98mph
Body styles sports
Prod years 1964-67
Prod numbers 2375

PRICES
A 3500 **B** 2000 **C** 1200

NSU 1000/TT/TTS/1200TT/110/110SC/1200C/1200CS

Grown-up brother of the little Prinz, with four-cylinder air-cooled engine, 6in longer wheelbase, expanded styling, and oval headlamps. 1000L/LS are base models (later called C/CS), TT has more powerful 996/1085cc engine and twin headlamps, TTS has twin carb 996cc engine, oil cooler and front discs (and qualifies as hairy transport in anyone's book), 1200TT has expanded 1.2-litre engine. 110 joins 1000 as 8in longer bigger sister (1.1-litre engine), from '66 as 110SC (1.2-litre engine), then almost immediately renamed 1200C/CS. Automatic clutch option from '71.

Engine 996cc/1085cc/1177cc, 4 cyl OHC, R/R **Max power** 40-51bhp/55-78bhp/70-102bhp/65bhp/66bhp/69bhp/73bhp **Max speed** 80-115mph **Body styles** saloon **Prod years** 1963-72/1965-72/1967-71/1967-72/1965-67/1966-67/ 1967-73 **Prod numbers** 196,000/ 14,942/2404/49,327/all 110 & 1200: 230,688

PRICES
1000/110/1200 **A** 1200 **B** 700 **C** 400
1000TT/1200TT **A** 2000 **B** 1100 **C** 600
TTS **A** 4000 **B** 2200 **C** 1000

NSU Ro80

Intellectually, the NSU Ro80 was a triumph. Dynamically, it was also a triumph. Why then was it an unmitigated flop? Essentially, it was short engine life and massive fuel consumption. No amount of smoothness, refinement, handling, aerodynamicism, comfort, performance, or Car of the Year awards could make up for the early sealing problems of the Wankel engine. Interesting spec included front-wheel drive, semi-automatic torque converter, all-independent suspension. Losses led to a takeover by VW/Audi. Today's technology can make the Wankel work, so avoid Ford V4/V6 conversions.

Engine 2 x 497.5cc, twin-rotor Wankel, F/F
Max power 115bhp at 5500rpm
Max speed 112mph
Body styles saloon
Prod years 1967-77
Prod numbers 37,398

PRICES
A 3500 **B** 1800 **C** 900

Ogle SX1000/Fletcher GT

David Ogle's design company began making cars in 1960 with the Riley-based 1.5, of which only eight were built. More successful was the Mini-based SX1000. A Mini van floorpan was clothed with a smooth, if bulbous, glassfibre body and fitted out with a luxurious interior. Alternatively, a customer's own Mini could be converted. It weighed 300lb (130kg) less than a Mini, so performance was excellent. David Ogle's death at the wheel of an SX1000 stopped production, though it was restyled/relaunched in 1966 by boat-builders Norman Fletcher.

Engine 848cc to 1275cc, 4 cyl OHV, F/F
Max power 34bhp at 5500rpm to 76bhp at 5800rpm
Max speed 80-110mph
Body styles coupé
Prod years 1961-63/1966-67
Prod numbers 66/4

PRICES
A 4000 **B** 2500 **C** 1200

Oldsmobile Toronado

A complete break for GM: front-wheel drive. And not in half measures, either: the Toronado was 17ft 7in (536cm) long, over two tons and had a 7-litre V8 engine! Handling was in a different league to other Detroit iron, thanks to forward positioning of the engine (it connected via a chain drive to the Hydra-Matic 'box), and performance was high. Bill Mitchell's styling was a success, too. Front disc option for '67, even bigger engine for '68, semi-notchback rear for '69. One of the most engaging Yanks of all time. Start saving now for your first tank-full.

Engine 6965cc 104.8x101mm/7446cc 104.8x108mm, V8 cyl OHV, F/F
Max power 385bhp at 4600rpm/375-400bhp at 4800rpm
Max speed 125-135mph
Body styles coupe
Prod years 1966-70
Prod numbers 143,134

PRICES
A 10,000 **B** 6000 **C** 3500

Oldsmobile 4-4-2

The Olds interpretation of the muscle car was essentially a Cutlass F85 platform and typical late 1960s body, plus big horsepower engines. The 4-4-2 moniker stood for 400 cubic inches, four-barrel carb and two exhausts. The 1968 4-4-2 was available as a convertible, a Sports coupe and a Holiday hardtop coupe. There were also limited-production Hurst/Olds 455 coupe models (1429 built in total) with up to 400bhp and bonnet scoops galore. All 4-4-2s boasted special paint schemes rich in stripes and decals. Well and truly trodden on by the early 1970s power squeeze.

Engine 6551cc 98.3x108mm/7446cc 104.8x108mm, V8 cyl OHV, F/R
Max power 290-360bhp at 4800rpm/340-400bhp at 4800rpm
Max speed 120-145mph
Body styles coupe, convertible
Prod years 1967-71
Prod numbers 86,883

PRICES
Holiday **A** 6000 **B** 3500 **C** 1500
Convertible **A** 12,000 **B** 7000 **C** 3000
Hurst 455 **A** 9000 **B** 5000 **C** 3000

Opel Olympia '47-'52

In 1946, Opel restarted truck production but was unable to relaunch the obvious post-war model, the Kadett, since the Russians had removed all the tooling for it (it resurfaced as the first Moskvich). It was also forbidden from making a car with more than a 1.5-litre engine, so in 1947 it revived the pre-war Olympia. Quite advanced for its day, with unitary construction, IFS, overhead valve engine and alligator bonnet. Ugly restyle for '50 (new nose, wings and boot), but cheap prices, so large production run.

Engine 1488cc 80x74mm, 4 cyl OHV, F/R
Max power 37bhp at 3500rpm to 43bhp at 4000rpm
Max speed 68-74mph
Body styles saloon
Prod years 1947-52
Prod numbers 187,055

PRICES
A 4500 **B** 2500 **C** 1200

Opel Kapitän '48-'53

When Allied rules about large-engined cars relaxed, Opel began building its pre-war Kapitän again. It had hardly changed since it was first seen in the 1930s: American influenced styling, six-cylinder overhead valve engine, and hydraulic drum brakes. Column shift arrived in 1950 and, in 1951, it got a facelift in the form of an unpleasantly American new grille, smaller wheels and a lengthened tail. It's amazing how many Kapitäns are still on the roads in Europe, a testament to their high build quality. However, Britain would not see Opels for some years yet.

Engine 2473cc 80x82mm, 6 cyl OHV, F/R
Max power 55-58bhp at 3700rpm
Max speed 75-78mph
Body styles saloon
Prod years 1948-53
Prod numbers 78,993

PRICES
A 5000 **B** 3000 **C** 1500

Opel Olympia Rekord

Everything American was good, according to the design office at Opel. Hence the 1953 Olympia Rekord was full of Chevy-esque Detroit clichés: high wing lines, prominent body feature lines, mouth-like grille, plus a curved windscreen. Same old engine, but compression increased in '55 along with a new oval mesh grille. Column shift, 13in wheels. Estate version was called Caravan. Cabriolet rare (12,504 built). Another new front end arrived in '56 (hooded headlamps, vertical bar grille). Very popular in its native Germany, less so in export markets. Again, Britain never received this Rekord.

Engine 1488cc 80x74mm, 4 cyl OHV, F/R
Max power 40bhp at 3600rpm to 45bhp at 3900rpm
Max speed 75mph
Body styles saloon, estate, convertible
Prod years 1953-57
Prod numbers 556,500

PRICES
Saloon/Estate **A** 3500 **B** 1800 **C** 1000
Cabriolet **A** 5000 **B** 3000 **C** 1500

Opel Kapitän '53-'58

All change for the big Opel in 1953: full-width bodywork which looked like an enlarged Rekord, IFS by coil springs and wishbones, half-elliptic rear springs with telescopic dampers, uprated engine. 1955 facelift did away with 'hawk's teeth' grille in favour of an oval surround. All-synchro 'box arrived in '56 (still column changed). Basically a gormless big car, the Teutonic equivalent of a Vauxhall Velox, but has a far stronger following on the continent than 1950s Vauxhalls do here. That said, no-one is really very interested in 1950s Opels in Britain.

Engine 2473cc 80x82mm, 6 cyl OHV, F/R
Max power 68bhp at 3700rpm to 75bhp at 3900rpm
Max speed 85-88mph
Body styles saloon
Prod years 1953-58
Prod numbers 154,098

PRICES
A 4000 **B** 2000 **C** 1000

Opel Kapitän '58-'59

In 1958, the Kapitän was updated according to the passing fads of Detroit. The wheelbase was increased by a couple of inches and a new body was fitted, featuring wrap-around front and rear screens. Of course, this meant cracking some part of your anatomy when you got in and out of the front seats, but hey, it looked cool. There was also less head room in the back. However, you got a more powerful engine and the suspension was firmer, giving improved roadholding. A luxury (L) version was also marketed. Germans call this one the 'Keyhole Kapitän' after its distinctive tail-lights.

Engine 2473cc 80x82mm, 6 cyl OHV, F/R
Max power 80bhp at 4100rpm
Max speed 90mph
Body styles saloon
Prod years 1958-59
Prod numbers 34,842

PRICES
A 4000 **B** 2000 **C** 1000

Opel Kapitän '59-'63

Although it was still inescapably transatlantic in looks and feel, the new '59 Kapitän was much more what the public wanted. The bodywork was substantially revised, featuring a higher rear roof-line, lower bonnet, full-width grille, new rear light treatment and a prominent feature line above the rear wheels. The engine was entirely new: a 2.6-litre straight six with more power, plus there were larger drum brakes and a stronger rear axle. An overdrive or Hydra-Matic transmission could be specified. Again, this model is all but unknown in Britain.

Engine 2605cc 85x76.5mm, 6 cyl OHV, F/R
Max power 90bhp at 4100rpm
Max speed 92mph
Body styles saloon
Prod years 1959-63
Prod numbers 145,618

PRICES
A 4000 **B** 2000 **C** 1000

Opel Olympia/Rekord P1/1200/P2

Opel themselves even described the new Olympia P1 as 'German made, American style'. Wrap-around screens and chrome did not look right on a car only 14ft 6in (174cm) long, while the Caravan estate was possibly the ugliest thing around in 1957. Rekord has more trim and a chrome body strip, Olympia is the cheapie (which only lasted until '59, when it was replaced by the smaller-engined 1200 as the base model). Olymat semi-automatic 'box and 1.7-litre engine options from '59. P2 had new grille, new glasshouse with less wrap-around and four-speed 'box.

Engine 1488cc 80x74mm/1205cc 72x74mm/1680cc 85x74mm, 4 cyl OHV, F/R **Max power** 45-50bhp at 4000rpm/40bhp at 4400rpm/55-60bhp at 4000rpm **Max speed** 72-84mph **Body styles** saloon, estate, convertible **Prod years** 1957-59/1959-62/1960-63
Prod numbers 787,835/67,952/755,658

PRICES
A 2700 **B** 1300 **C** 700

Opel Rekord A/B

At last some home-grown styling influence for the Rekord! It started to look much more like Vauxhalls, with clean-cut lines, no wrap-around, slightly larger dimensions and a new coupé body option. Underneath, it was all exactly the same as before: 1.5 or 1.7-litre engines, four speeds, rigid rear axle on semi-elliptics. From '64, the Rekord 6 came with the six-cylinder engine from the Kapitän. Rekord B arrived in '65, having front discs, three new OHC engines, rectangular headlamps, and a new grille with horizontal bars. The B model lasted only one year.

Engine 1488cc 80x74mm/1680cc 85x74mm/2605cc 85x76.5mm/1492cc 82.5x69.8mm/1698cc 88x69.8mm/1897cc 93x69.8mm, 4/4/6/4/4/4 cyl OHV/OHV/OHV/OHC/OHC/OHC, F/R **Max power** 55bhp at 4500rpm to 100bhp at 4600rpm **Max speed** 80-105mph **Body styles** saloon, estate, coupé, convertible **Prod years** 1963-65/1965-66 **Prod numbers** 864,197/288,627

PRICES
A 1800 **B** 1000 **C** 500

Opel Kadett A

After years churning out vast numbers of medium class cars, Opel turned its attention to taking on the VW Beetle. The Kadett A was the secret weapon, built at a new factory at Bochum from 1962. It had a new 1-litre OHV engine, all-synchromesh four-speed gearbox, rack-and-pinion steering and transverse-leaf IFS. Two-door saloon and estate versions were joined by a coupé model, which was nothing more than a different roof and grille on the standard Kadett shell, plus a high compression 48bhp engine. Strangely, collectors seem to like the estate car.

Engine 993cc 72x61mm, 4 cyl OHV, F/R
Max power 40bhp at 5000rpm/48bhp at 5400rpm
Max speed 75-80mph
Body styles saloon, estate, coupé
Prod years 1962-65
Prod numbers 649,512

PRICES
Saloon **A** 1800 **B** 1000 **C** 400
Estate **A** 2000 **B** 1200 **C** 600
Coupé **A** 2300 **B** 1500 **C** 900

Opel

Opel Kadett B

The 1965 restyle for the Kadett is recognisable by headlamps in square surrounds and featureless body sides. Also the range of engines expanded, in 1967, from one to four: a bored-out base unit, now measuring 1.1 litres, and the 1.5, 1.7 and 1.9 units shared with the Rekord (plus 1196cc from '71). At launch, the Coupé versions (L and S) had 54bhp, the S also having disc front brakes. Most exciting was the Coupé Rallye (photo) with its twin carb engine, dual-circuit brakes, front discs, stripes and matt black bonnet. From '67 it had a more powerful 1.9-litre engine. An Opel collector's favourite.

Engine 1078cc 75x61mm/1196cc 79x61cc/1492cc 82.5x69.8mm/1698cc 88x69.8mm/1897cc 93x69.8mm, 4 cyl OHV/OHV/OHC/OHC/OHC, F/R **Max power** 44bhp at 5000rpm to 90bhp at 5100rpm **Max speed** 75-102mph **Body styles** saloon, estate, coupé **Prod years** 1965-73 **Prod numbers** 2,649,501
PRICES
Saloon/Estate **A** 1200 **B** 600 **C** 200
Coupé **A** 2000 **B** 1000 **C** 600
Rallye Coupé **A** 2500 **B** 1600 **C** 900

Opel Olympia

Confusingly, Opel revived the old Olympia name in 1967 to plug the gap between the Kadett and the Rekord. It was confusing because 'Olympia' had been the name for the Rekord in the 1950s; now it was merely an upmarket Kadett. It was offered in two- or four-door saloon styles, a coupé, and two- and four-door fastbacks. This latter style was also available as the Kadett LS, incidentally. Olympias were better trimmed, had unique wrap-around grilles, vinyl roofs, fancy wheel trims and twin carb engines. Not very exciting really, and all unitary 1960s Opels have rust problems.

Engine 1078cc 75x61mm/1196cc 79x61cc/1492cc 82.5x69.8mm/1698cc 88x69.8mm/1897cc 93x69.8mm, 4 cyl OHV/OHV/OHC/OHC/OHC, F/R **Max power** 60bhp at 5200rpm to 90bhp at 5100rpm **Max speed** 90-102mph **Body styles** saloon, coupé, fastback coupé **Prod years** 1967-70 **Prod numbers** 80,637

PRICES
A 2000 **B** 1000 **C** 600

Opel Kapitän/Admiral/Diplomat

Opel's new 'big three' had an inescapably American feel to them. Indeed the styling was almost identical to the Buick Special of the period. Not only that, but the top-of-the-range Diplomat actually had an American V8 in the form of the Chevrolet Chevelle unit. The base Kapitän and mid-range Admiral had straight sixes. Spaciousness, refinement and luxury were the models' strong suits. They were also good value, costing much less than the BMW 2500 and Mercedes. Never available in Britain. Larger engines optional. If you must have one, find a Karmann-built Diplomat Coupé (304 built).

Engine 2605cc 85x76.5mm/2784cc 92x69.8mm/4638cc 98.4x76.2mm/5354cc 101.6x82.6mm, 6/6/V8/V8 cyl OHV, F/R **Max power** 100bhp at 4600rpm to 230bhp at 4700rpm **Max speed** 100-130mph **Body styles** saloon, coupé **Prod years** 1964-68 **Prod numbers** 89,277
PRICES
Saloon **A** 3000 **B** 2000 **C** 1000
Coupé **A** 9000 **B** 6000 **C** 4000

Opel Kapitän/Admiral/Diplomat

The new big Opels arrived in 1969. Under smartened-up bodywork (recognisable by its thick, vertical C-pillars) sat the same floorpan and virtually the same mechanical package, which would make do right up until 1977. Kapitän lasts only two seasons, Admiral available in various states of tune (up to 165bhp E). Diplomat is also offered for the first time with a six-cylinder engine as well as the 5.4-litre Chevy V8, in which form Hydra-Matic auto is standard. LWB Diplomat V8 offered from 1973 (extra 6in in wheelbase) is Mercedes 450SEL competitor.

Engine 2784cc 92x69.8mm/5354cc 101.6x82.6mm, 6/V8 cyl OHV, F/R **Max power** 129bhp at 5200rpm to 165bhp at 5600rpm/230bhp at 4700rpm **Max speed** 109-126mph **Body styles** saloon **Prod years** 1969-77 **Prod numbers** 4976/35,622/21,021

PRICES
A 3500 **B** 2000 **C** 700

Opel Rekord C/Commodore

Coke bottle styling inevitably pervaded Opel and the '66 Rekord C had it in full measure. Dual-circuit brakes (including servo front discs), coil springs for the live rear axle, choice of 1.5, 1.7 and 1.9-litre fours and a 2.2-litre six. Estate version was new, but coupé and convertible continued to be built by Deutsch of Cologne. Commodore arrived in '67, sharing the same bodywork but a higher level of trim and only six-cylinder engines. Sprint has 115bhp, GS has twin carbs and 140bhp, and GS/E has fuel injection, 150bhp and front spoiler. Big Commodore GS2800 was the sybarite's choice.

Engine 1492cc/1698cc/1897cc/ 2239cc/2490cc/2784cc, 4/4/4/6/6/6 cyl OHC, F/R **Max power** 58bhp at 4800rpm to 150bhp at 5800rpm **Max speed** 84-121mph **Body styles** saloon, estate, coupé, convertible **Prod years** 1966-71/1967-71 **Prod numbers** 1,276,681/156,330
PRICES
Rekord **A** 1800 **B** 1000 **C** 400
Commodore GS/E **A** 4000 **B** 2500 **C** 1000

Opel GT 1100/1900

The public's first sight of Opel's sports car was at the 1965 Frankfurt show, where it was displayed as a concept car. Production of this 'mini-Corvette' began in 1968, with bodywork made in France by Brissoneau & Lotz. Corvette influence was from GM's Clare MacKichan: manual rotating headlamps, Kamm tail with four round lights, all on a shortened Kadett floorpan. 1100 available to 1970 only, and very rare, 1900 more usual; '71 GT/J (GT Junior) has less chrome and poverty trim. Front discs, auto option, but no RHD. Convertible Aero GT was a 1969 one-off.

Engine 1078cc 75x61mm/1897cc 93x69.8mm, 4 cyl OHV/OHC, F/R
Max power 60bhp at 5200rpm/ 90bhp at 5100rpm
Max speed 95/115mph
Body styles coupé
Prod years 1968-70/1968-73
Prod numbers 103,373 (3573/99,800)

PRICES
A 4500 **B** 3200 **C** 2000

Opel Ascona/Manta

The Ascona replaced the Olympia in Opel's range, providing it at long last with a rival for the Ford Taunus. Conventional engineering (as ever with Opel) means rigid rear axle, coils all round, front discs, choice of engines. Manta shares the same floorpan, but has its own, larger coupé body, and was a strong rival to Ford's Capri. 1.2-litre model underpowered, but not sold in UK. 1.6 and 1.9 litres more usual for Manta; SR has 90bhp. Berlinetta is luxury special. Manta GT/E (1973-75, pictured), has Bosch injection (again not for UK, where a turbocharged version was listed).

Engine 1196cc 79x61mm/1584cc 85x69.8mm/1897cc 93x69.8mm, 4 cyl OHV/OHC/OHC, F/R
Max power 60bhp at 5400rpm to 105bhp at 5400rpm
Max speed 90-117mph
Body styles saloon, estate, coupé
Prod years 1970-75
Prod numbers 641,438/498,553
PRICES
Ascona **A** 900 **B** 400 **C** 100
Manta **A** 1900 **B** 1000 **C** 500
Manta GT/E **A** 3000 **B** 1700 **C** 800

Opel Rekord/Commodore

For the first time, an Opel shared the same floorpan as a model from its GM counterpart in Britain, in this case the Vauxhall Victor FE. As conventional as they come, the Rekord was a very successful million-seller, also available in coupé and estate forms. 2.0 litres for Italy only, plus 2.1 diesel. More interesting was the Commodore (saloon or coupé), identical to the Rekord except for its 2.5 and 2.8-litre six-cylinder engines and superior equipment. Attractive top-of-the-range Commodore GS/E (1974-77) had fuel injection, standard auto and PAS, front spoiler (plus alloy wheels from '75).

Engine 1698cc/1897cc/1998cc/ 2068cc/2490cc/ 2784cc, 4/4/4/4/6/6 cyl OHC/diesel, F/R **Max power** 60bhp at 4400rpm to 160bhp at 5400rpm **Max speed** 82-124mph **Body styles** saloon, estate, coupé **Prod years** 1972-77
Prod numbers 1,128,196/ 140,827
PRICES
Rekord **A** 1000 **B** 500 **C** 100
Commodore Coupé **A** 1800 **B** 1000 **C** 500
Commodore GS/E Coupé **A** 2700 **B** 1500 **C** 800

Opel Kadett

Extremely successful member of the GM 'T' Car range, produced all over the world, including in Britain as the Vauxhall Chevette. All stemmed from the 1973 Opel. Wide choice of bodies, standard Opel engine range, rigid rear axle, front discs on most. Several interesting variations developed. The first was the fastback coupé, going up to Rallye and fuel-injected 115bhp GT/E versions (latter post-'75 and LHD only). Unique Aero model (1976-77) had fold-down rear roof section, small rear quarter lights and removable targa panel, but is very rare (1242 built) and was never sold in the UK.

Engine 993cc 72x61mm/1196cc 79x 61mm 4 cyl OHV, 1584cc 85x69.8mm/ 1897cc 93x69.8mm/1979cc 95x69.8mm, 4 cyl OHV/OHC, F/R **Max power** 40bhp at 5400rpm to 115bhp at 5600rpm **Max speed** 76-118mph **Body styles** saloon, hatchback, estate, coupé, convertible **Prod years** 1973-79
Prod numbers 1,701,075
PRICES
Kadett **A** 800 **B** 400 **C** 100
Kadett GT/E **A** 2500 **B** 1500 **C** 1000
Kadett Aero **A** 4500 **B** 2500 **C** 1500

Opel Ascona/Manta

Another storming success for Opel, and the car which sired the Vauxhall Cavalier (albeit with a different nose). The Ascona saloon was a bigger car than the 1970 version, but no more technically advanced. SR has more power, stiffer suspension; Berlina has luxury equipment. Manta (photo) is coupé model, available as a conventional coupé or with a hatchback from '78 on. Manta sold to 1988 as a cheap-and-cheerful old school coupé. Ascona and Manta 400 (1979-81/1981-83) were rally homologation specials with Cosworth twin cam 16V 2.4-litre 144bhp engines, some glass-fibre panels and special paint (448/236 built).

Engine 1196cc 4 cyl OHV, 1584cc/ 1796cc/1897cc/1979cc 4 cyl OHC, 1988 diesel, 2410cc, 4 cyl DOHC, F/R
Max power 60bhp at 5400rpm to 144bhp at 5200rpm **Max speed** 88-125mph **Body styles** saloon, coupé, hatchback coupé **Prod years** 1975-81/ 1975-88 **Prod numbers** 1,512,971/ 603,000
PRICES Ascona **A** 700 **B** 400 **C** 100
Manta **A** 2500 **B** 1300 **C** 800
Ascona 400/Manta 400 **A** 10,000 **B** 6500 **C** 3000

OSCA · Packard · Panhard

OSCA 1600

Ten years after the forced sale of their eponymous sports car firm, the Maserati brothers had to sever their contracted arrangement with the new owner, Adolfo Orsi, and formed OSCA in 1947 to build sports and racing cars. Its 1.5-litre twin cam engine was adopted by Fiat in the 1959 1500S/1600S (see page 83). That engine was capable of high power outputs, and OSCA built its own interpretation on the Fiat 1600S floorpan. Most were bodied by Zagato (pictured), but Fissore and Vignale also did some. OSCA made tiny numbers of other sports cars, including the 1050 coupé and coupés using Ford V4 engines.

Engine 1568cc 80x78mm, 4 cyl DOHC, F/R
Max power 95bhp at 6000rpm to 140bhp at 7200rpm
Max speed 115-130mph
Body styles coupé
Prod years 1960-67
Prod numbers n/a

PRICES
A 30,000 **B** 25,000 **C** 17,500

Packard Caribbean

Post-war Packard never revived the glory years of the 1930s, losing ground to Cadillac. The one exception was the '53 Caribbean, a true rival to the new Eldorado. Glamorous styling by Richard A. Teague on a short-wheelbase Packard chassis, the old straight-eight engine, 'Continental' rear end, chrome wire wheels, lots of equipment and a high price (over $5000). Limited to only 750 examples in its first year and 400 in its second, it had snob appeal but, ultimately, far less cachet than the Caddy.

Engine 5358cc 88.9x108mm, 8 cyl SV, F/R
Max power 180-210bhp at 4000rpm
Max speed 100-108mph
Body styles convertible
Prod years 1952-54
Prod numbers 1150

PRICES
A 20,000 **B** 12,000 **C** 8000

Panhard Dyna

Fascinating little French car designed by Jean-Albert Grégoire. Very modern unitary body and light alloy doors/bonnet to keep weight down to barely over half a ton (550kg). Small and highly efficient air-cooled alloy flat-twin engine incorporated torsion bar valve springs. Drive was via a four-speed-and-overdrive 'box to the front wheels, which were independently sprung. 745cc engine option and torsion bar rear end from 1950 (Type 120), and 850cc option from 1951 (Type 130). Always hydraulic brakes, too.

Engine 610cc 72x75mm/745cc 79.5x75mm/848cc 84.8x75mm, 2 cyl OHV, F/F
Max power 24bhp at 4000rpm/35bhp at 5000rpm/38bhp at 5000rpm
Max speed 62/73/77mph
Body styles saloon, rolltop saloon, estate, convertible
Prod years 1946-54
Prod numbers c 55,000

PRICES
A 4000 **B** 2200 **C** 1000

Panhard Dyna Junior

This was the sports version of the little Dyna. It used the same engines as the Dyna 120/130, but tuned up and boasting twin carbs, while it had its own box-section chassis with tubular cross-members. Synchromesh was fitted on the top two gears. When launched at the Paris Salon in 1952, one journalist called it a bread basket; certainly it was no beauty, although it did win a concours d'élégance just after its debut. Optional supercharger developed no less than 60bhp. Was available in the UK as Junior Sprint, but virtually unknown outside France. Some coach-built bodies by Allemano, Ghia-Aigle, Wendler et al.

Engine 745cc 79.5x75mm/848cc 84.8x75mm, 2 cyl OHV, F/F
Max power 40bhp at 5000rpm to 60bhp at 5300rpm
Max speed 77-90mph
Body styles sports, convertible
Prod years 1952-56
Prod numbers c 2000

PRICES
A 6000 **B** 4000 **C** 2500

Panhard Dyna 54/57/58

It looked like the aliens had landed when the Dyna 54 drove down the Champs Elysées in 1953: close-set headlamps, an extraordinary grille and streamlined body. Welded light alloy panels kept body weight down to just 202lb (92kg). Interior was spacious, though claimed six seats was optimistic; much use was made of plastic and nylon. Mechanically, it was a Dyna 130. Hydraulic tappets and electric auto clutch option in '55. Powerful Tigre engine optional. Dyna 57 gets coil spring/damper units at the front, Dyna 58 has steel panelling and bigger brakes. RHD from '58. Unusual to say the least.

Engine 848cc 84.8x75mm, 2 cyl OHV, F/F
Max power 35bhp at 5000rpm to 50bhp at 5750rpm
Max speed 75-92mph
Body styles saloon, estate, convertible
Prod years 1953-59
Prod numbers c 155,000

PRICES
Saloon **A** 3500 **B** 2200 **C** 1000
Convertible **A** 10,000 **B** 7000 **C** 3000

Panhard · Panther

Panhard PL17

The centre hull of the Dyna was retained in the 1959 PL17, but the head and tail lamps gained quirky hoods and new smoothly curved bumpers were designed. The cheap 35bhp engined version disappeared, leaving 42bhp and twin-choke 50bhp 'Tigre' models. Synchromesh on top three gears from '60 (all-synchro from '62), front-hinged front doors from '61, separate front seats from '62, luxury Tigre S model from '62. Convertible was very expensive (nearly twice as much as a Morris Minor Tourer), and prices today reflect its rarity. Saloons can still be found quite cheaply in France.

Engine 848cc 84.8x75mm, 2 cyl OHV, F/F
Max power 42-50bhp at 5750rpm
Max speed 80-90mph
Body styles saloon, convertible
Prod years 1959-65
Prod numbers c 130,000

PRICES
Saloon **A** 3500 **B** 2200 **C** 1000
Convertible **A** 10,000 **B** 7000 **C** 3000

Panhard 24 series

The world's most expensive two-cylinder car: did the world really want a car which sounded like a 2CV for nearly the price of a Jaguar Mk2? But a fascinating classic: unusual Gallic style with double headlamps behind glass cowls, grille-less front, same old air-cooled twin engine providing impressive economy and amazing performance. 24C is 7ft 6in (230cm) wheelbase 2+2 coupé (CT has 60bhp Tigre engine), while 24B is full four-seater on 8ft 2in (249cm) wheelbase. Cheaper 24BA and Tigre-engined 24BT not sold in UK. Engaging little car, but the parts situation is awkward.

Engine 848cc 84.8x75mm, 2 cyl OHV, F/F
Max power 50bhp at 5750rpm/ 60bhp at 5800rpm
Max speed 90-100mph
Body styles saloon, coupé
Prod years 1963-67
Prod numbers 23,245 (B: 12,848/C: 10,397)

PRICES
A 4500 **B** 3000 **C** 1500

Panhard CD

After Charles Deutsch had dissolved his partnership with René Bonnet, he built the Panhard CD racing car to compete in long distance events like Le Mans (in which race it won its class in 1962). In 1963, Panhard decided to put the car into production as the most exotic road-going machine it had ever made. The CD had a steel tube chassis clothed in a dramatically rakish and low (46.5in/118cm) glassfibre body, with seating strictly for two. It was light (1280lb/580kg) and aerodynamic, so the standard Tigre engine gave excellent performance. Expensive new, expensive now.

Engine 848cc 84.8x75mm, 2 cyl OHV, F/F
Max power 60bhp at 5800rpm
Max speed 105mph
Body styles coupé
Prod years 1963-65
Prod numbers 160

PRICES
A 12,000 **B** 8000 **C** 5000

Panther J72

With his first production car, Robert Jankel became a British version of Brooks Stevens (Mr Excalibur). Style was undeniably SS 100, though it was not a replica. Tubular chassis, rigid front and rear axles (Jaguar IFS from '77), four-wheel discs, auto option from '76. Jaguar 3.8 or 4.2 XK engines, or V12 from '74. Attracted glitterati buyers when new (eg Liz Taylor), since it cost twice as much as an E-Type. The trade regards the J72 as brazenly vulgar. That means some bargains are to be found for what is a hand-built quality car.

Engine 3781cc 87x106mm/4235cc 92.1x106mm/5343cc 90x70mm, 6/6/V12 cyl DOHC/DOHC/OHC, F/R
Max power 190-265bhp at 5500rpm/180-265bhp at 5400rpm/253-285bhp at 6000rpm
Max speed 114-130mph
Body styles sports
Prod years 1972-81
Prod numbers 426

PRICES
A 15,000 **B** 10,000 **C** 7000

Panther FF

The second Panther came about because Swiss coachbuilder Willy Felber asked Jankel to create a Ferrari 125S-style body fitted with, unbelievably, Ferrari 330 GTC mechanicals. This he did with painstaking quality, although again it was not intended to be a replica. A square-tube frame supported the unstressed aluminium bodywork. At least one was built on a butchered Daytona and painted metallic purple (Ferrari *aficionados* may well faint) and most, if not all, went to Felber in Switzerland, although he is believed to have had more built by the Italian coachbuilder Michelotti.

Engine 3967cc 77x71mm, V12 cyl QOHC, F/R
Max power 300bhp at 7000rpm
Max speed 160mph
Body styles sports
Prod years 1974-75
Prod numbers 7

PRICES
No sale recorded, but will be less than a Ferrari 330 GTC

Panther De Ville

This gigantic car was a pastiche of the Bugatti Royale on Jaguar XJ mechanicals. Massive ladder chassis, hand-formed alloy body on space frame, complete Jaguar subframes front and rear, reskinned Austin 1800 doors. V12s much more common than the 4.2. Two-door convertible of 1976 was Britain's most expensive new car, offered with manual or power hood and a removable hardtop (a single fixed hardtop was also built). Panther went into liquidation at the end of 1979, but was sold to Korean Young C. Kim, and production restarted.

Engine 4235cc 92.1x106mm/5343cc 90x70mm, 6/V12 cyl DOHC/OHC, F/R
Max power 180-265bhp at 5400rpm/ 253-285bhp at 6000rpm **Max speed** 110-128mph **Body styles** saloon, convertible, coupé, limousine
Prod years 1974-85 **Prod numbers** 62 (Saloon: 54/Convertible: 7/Coupé: 1)

PRICES
Saloon **A** 30,000 **B** 18,000 **C** 10,000
Convertible **A** 45,000 **B** 30,000 **C** 22,000

Panther Rio

The concept of Rolls-Royce luxury in a fuel crisis-friendly suit of clothes was quite sound. But the Rio suffered from comparisons with the car it was based on – the Triumph Dolomite. The body was completely reskinned in hand-beaten aluminium, a 'traditional' grille went on the front and the interior was trimmed in leather and walnut. Despite this, its cramped interior and price (three times as much as a Dolomite) sunk it. Two models offered: standard and alloy-wheeled Especial, the latter based on the Dolomite Sprint with its 16-valve engine.

Engine 1854cc 87x78mm/1998cc 90.3x78mm, 4 cyl OHC/DOHC, F/R
Max power 92bhp at 5500rpm/ 129bhp at 5700rpm
Max speed 100/115mph
Body styles saloon
Prod years 1975-77
Prod numbers 34

PRICES
A 5000 **B** 3500 **C** 2500

Panther Lima

'Panther for the masses' brought volume to Jankel's firm, but also its demise. 1930s pastiche on strengthened Vauxhall Magnum floorpan, so IFS, front discs, 2.3-litre engine. Non-stressed glassfibre body based around MG Midget doors/screen featured controversial cheese-grater grille and spoiler. Sold through selected Vauxhall dealers, which was a mixed blessing (dealers were unused to sports cars, customers expected mass-produced reliability). 1979 Mk2 Lima had separate box-section chassis. Engine options included tuning by Vauxhall's DTV and, from '78, a rare 178bhp turbocharged version.

Engine 2279cc 97.5x76.2mm, 4 cyl OHC, F/R
Max power 108bhp at 5000rpm to 178bhp at 6000rpm
Max speed 105-125mph
Body styles sports
Prod years 1976-82
Prod numbers 918

PRICES
A 6000 **B** 3500 **C** 2200

Peerless/Warwick

The Peerless was essentially a good idea, and it nearly succeeded. First shown in 1957, its production began in 1958. A glassfibre body was bonded and riveted to the space frame chassis with Triumph TR3 mechanicals mated to a De Dion rear end. Optional triple overdrive and wire wheels. 1959 Phase 2 had one-piece body and recessed headlamps. Bankruptcy came in 1960, but project was revived under the name Warwick: new grille, reshaped rear windows/fins, Buick V8 option from '61. Interesting GT for Triumph TR fans.

Engine 1991cc 83x92mm/3528cc 88.9x71.1mm, 4/V8 cyl OHV, F/R
Max power 100bhp at 5000rpm/ 155bhp at 4600rpm
Max speed 117/135mph
Body styles coupé
Prod years 1958-60/1960-63
Prod numbers 290/45

PRICES
A 9000 **B** 6000 **C** 3500

Pegaso Z102

Spaniard Don Wilfredo Ricart harboured a grudge against ex-boss Enzo Ferrari after being described by him as wearing thick rubber-soled shoes so that his brain didn't get any shocks. The Pegaso was his revenge. The engine was a Formula 1 spec quad cam V8. With optional twin superchargers, the output could be 275bhp. Interestingly, the five-speed gearbox was mounted *behind* the differential to obtain the best weight distribution. No-cost-spared factory coupés supplemented by numerous coachbuilt bodies.

Engine 2474cc 75x70mm/2816cc 80x70mm/3178cc 85x70mm, V8 cyl QOHC, F/R
Max power 140bhp at 6000rpm/ 179bhp at 6500rpm/195-275bhp at 6500rpm
Max speed 120-160mph
Body styles coupé, convertible
Prod years 1951-56
Prod numbers 112

PRICES
A 60-100,000 **B** 48-70,000 **C** 35-50,000

Pegaso Z103

At the 1955 Paris Salon, Pegaso announced the Z103 with the claim that it was 'the fastest and highest quality car built to date'. However, the enthusiast must have felt short-changed by the new V8 engines: they were larger and more powerful (up to 350bhp with a supercharger), but had mere overhead valves. The firm which had done most of the Z102 bodies, Touring, also made the Z103 bodies on the same semi-monocoque basis as the Z102. Vastly expensive, the Pegaso was also just about the hairiest drive around. The parent company became bored and concentrated on its truck business.

Engine 3988cc 88x82mm/4450cc 93x82mm/4681cc 95x82.5mm, V8 cyl OHV, F/R
Max power 230-350bhp at 5500rpm
Max speed 140-170mph
Body styles coupé, convertible
Prod years 1955-58
Prod numbers incl in Z102

PRICES
A 70,000 **B** 50,000 **C** 38,000

Peugeot 202

Peugeot was able to restart manufacture after the wartime break as early as February 1945 with its pre-war 202, virtually unchanged. This was basically a scaled-down 402, boasting IFS, quarter-elliptic rear, three-bearing OHV engine, all-synchro three-speed 'box and close-set headlamps mounted behind the grille. Saloons all have sliding roofs. Not as pretty as the 402, nor as popular, though short post-war run of over 40,000 examples helped push its total up to more than 100,000. Can still be found for low prices in France.

Engine 1133cc 68x78mm, 4 cyl OHV, F/R
Max power 30bhp at 4000rpm
Max speed 60mph
Body styles saloon, rolltop saloon, estate, convertible
Prod years 1938-48
Prod numbers 104,126

PRICES
Saloon **A** 5000 **B** 3000 **C** 1200
Décapotable **A** 9000 **B** 6000 **C** 4000

Peugeot 203

Peugeot's first all-new post-war design was ahead of its time. The styling may have been American influenced, but the over-square engine with its light alloy head and hemispherical combustion chambers was a delight. You also got hydraulic brakes, coil spring IFS, rack-and-pinion steering and, from '54, an all-synchro four-speed 'box. A rugged rally winner as well as dependable family transport, a long time in production with virtually no changes. UK sales consisted of the saloon and long wheelbase estate only. In France, the 203 now has a similar following to the Traction Avant.

Engine 1290cc 75x73mm, 4 cyl OHV, F/R
Max power 42-45bhp at 4500rpm
Max speed 75mph
Body styles saloon, estate
Prod years 1948-60
Prod numbers 685,828 incl coupé/convertible

PRICES
A 2700 **B** 1800 **C** 1200

Peugeot 203 Coupé/Découvrable/Décapotable

Alongside the staid family saloons, Peugeot built a series of more interesting body styles. The first of these was the Découvrable, basically a four-door saloon with a canvas roof which could roll right back. The true drophead coupé (Décapotable) had only two doors, two seats and a full folding roof (photo), and was offered from October 1951 to 1956. The two-door Coupé was the rarest of this trio, less than 1000 being built in a short production run. A growing following in France ensures that prices are bound to rise, although the British 203 contingent is still very marginal.

Engine 1290cc 75x73mm, 4 cyl OHV, F/R
Max power 42-45bhp at 4500rpm
Max speed 75mph
Body styles coupé/convertible/drophead coupé
Prod years 1948-60
Prod numbers 955/11,129/2567
PRICES
Coupé **A** 12,500 **B** 10,000 **C** 7000
Découvrable **A** 6000 **B** 4000 **C** 2500
Décapotable **A** 18,000 **B** 14,000 **C** 10,000

Peugeot 403

Although it shared many parts in common with the 203, the new 1955 403 joined it as a sister model rather than replaced it. Full-width bodywork was designed with help from Pinin Farina and the engine was a larger (1.5-litre) unit. The wheelbase was longer, especially on the estate (which, at over 15ft long, was probably the largest 1.5-litre car at the time). The Familiale had eight seats arranged in three rows. 1300 version sold only in France. Jaeger electro-magnetic clutch option '58-'60, 403B with better gearbox from '60. 1959 1.8-litre diesel uncommon in UK.

Engine 1290cc 75x73mm/1468cc 80x73mm/1816cc 85x80mm, 4 cyl OHV/OHV/diesel, F/R
Max power 54bhp at 4500rpm/58-65bhp at 4900rpm/50bhp at 4000rpm
Max speed 75-85mph
Body styles saloon, estate
Prod years 1955-66
Prod numbers 1,119,460

PRICES
A 2300 **B** 1700 **C** 1000

Peugeot 403 Décapotable

By far the most attractive of the Farina 403 range, the Décapotable drophead coupé was launched at the 1956 Paris Salon. Based on the same wheelbase, the drop-top had only two doors, but – unlike the 203 DHC – a full four seats. Mechanically, it was identical to the 403 saloon, with the exception of a higher-compression engine. We know the Décapotable because TV detective Colombo drives a shabby example, but that is the exception: most examples have been painstakingly cared for by doting French owners. Very seldom seen in the UK, which seems a shame.

Engine 1468cc 80x73mm, 4 cyl OHV, F/R
Max power 65bhp at 4750rpm
Max speed 85mph
Body styles convertible
Prod years 1956-63
Prod numbers 2050

PRICES
A 15,000 **B** 9000 **C** 5000

Peugeot 404

It looked bigger than the 403, but the new Pininfarina-designed 404 was actually a couple of inches shorter, narrower *and* lower than its predecessor. The old 1.5 and new 1.6-litre engines were mounted at an angle of 45 degrees under the bonnet (plus a 2-litre diesel from '63); the 1.6 unit could be had with fuel injection as an option, and came with five bearings from '63. Same live rear axle and worm drive, auto option from '67, servo front discs from '68. Again there was a standard estate and a Familiale eight-seater estate. Much smarter than that other Farina saloon, the Austin/Morris Cambridge/Oxford.

Engine 1468cc 80x73mm/1618cc 84x73mm/1816cc 85x80mm/1948cc 88x80mm, 4 cyl OHV/OHV/diesel/diesel, F/R
Max power 53bhp at 5000rpm to 96bhp at 5700rpm
Max speed 75-100mph
Body styles saloon, estate
Prod years 1960-75
Prod numbers 2,769,361

PRICES
A 1800 **B** 1200 **C** 500

Peugeot 404 Coupé/Décapotable

Pininfarina created a quite different shape for the coupé and convertible versions of the 404 in 1961/62. Like the 403, they were proper four-seaters and as practical as they were elegant. UK imports started in June 1965, but you'd be hard pressed to find an original right-hand drive example today. Choice of 85bhp carburettor or 96bhp fuel-injection 1.6-litre engines. Mechanical modifications as 404 saloon, though both these models ducked out well before the *berline* (being replaced by the 504 coupé/convertible). Again, all the following is in France.

Engine 1618cc 84x73mm, 4 cyl OHV, F/R
Max power 85-96bhp at 5500rpm
Max speed 94-104mph
Body styles coupé, convertible
Prod years 1961-69/1962-69
Prod numbers 6837/3728

PRICES
Coupé **A** 6000 **B** 4000 **C** 1700
Décapotable **A** 9000 **B** 5500 **C** 3500

Peugeot 204

Breaking with tradition, the new small Peugeot featured a transverse overhead cam engine and front-wheel drive, IRS, MacPherson strut front, and servo disc front brakes. Column gear change, updated to floor shift in '69. Coupé and cabriolet versions sold from '66 to '70 were much abbreviated (measuring only 12ft 3in/230cm long) and very pretty. They are becoming popular as stylish and cheap-to-run classics even over here. 1.3-litre diesel engine available in estates from '68, though not in UK until '75. 1127cc engine from '75. So much better than a stodgy old Escort or BMC 1100.

Engine 1130cc 75x64mm/1127cc 78x59mm/1255cc 75x71mm, 4 cyl OHC/OHC/diesel, F/F **Max power** 58bhp at 5800rpm/59bhp at 5500rpm/45bhp at 5000rpm **Max speed** 87/78mph **Body styles** saloon, estate, coupé, convertible **Prod years** 1965-77 **Prod numbers** 1,387,473 (incl Coupé: 42,765/Cabriolet: 18,181)

PRICES
Saloon/Estate **A** 1000 **B** 700 **C** 300
Coupé **A** 3500 **B** 2500 **C** 1000
Cabriolet **A** 5000 **B** 3500 **C** 1500

Peugeot 304

Although it looked almost identical to the 204, the 304 was several inches longer in the wheelbase and lankier overall. The diecast alloy engine of the 204 was bored and stroked so that it became a much more potent 1.3-litre unit. The gear change was column mounted. Attractive coupé and convertible models arrived in 1970 and lasted until 1975, and are surprisingly common in RHD. 80bhp S engine arrived in '72, when a dual-circuit braking system was introduced, along with larger front discs. New 1127cc and 1290cc engines from '75, late 1970s 1357cc diesel engine not sold in UK.

Engine 1127cc 78x59mm/1288cc 76x71mm/1290cc 78x67.5mm/1357cc 78x71mm, 4 cyl OHC/diesel, F/F
Max power 45bhp at 5000rpm to 80bhp at 6100rpm **Max speed** 80-100mph **Body styles** saloon, estate, coupé, convertible **Prod years** 1969-80
Prod numbers 1,334,309 (incl Coupé: 60,186/Cabriolet: 18,647)

PRICES
Saloon/Estate **A** 1000 **B** 700 **C** 300
Coupé **A** 3500 **B** 2500 **C** 1000
Cabriolet **A** 5000 **B** 3500 **C** 1500

Peugeot 504

Usual Peugeot overlap period means the 504 didn't replace the 404 directly. Understated Pininfarina shape was larger than 404 and the 504 carved out for itself a best-seller slot as a dependable saloon beloved of the French police and middle classes. Hypoid bevel rear axle at last, plus four-wheel servo discs, and IRS. Fuel-injected engines available from outset, as was auto transmission. 2-litre petrol and diesel engines supplement 1.8 from '70; 2.1-litre diesel from '78, 2.3-litre diesel from '77. Usual estate (Commerciale) and eight-seater (Familiale) models.

Engine 1796cc 84x81mm/1971cc 88x81mm/1948cc 88x80mm/2112cc 90x83mm/2304cc 94x83mm, 4 cyl OHV/diesel, F/R
Max power 50bhp at 4500rpm to 97bhp at 5000rpm
Max speed 80-115mph
Body styles saloon, estate
Prod years 1968-82
Prod numbers 2,836,237

PRICES
A 1200 **B** 800 **C** 300

Peugeot 504 Coupé/Cabriolet

The Coupé and Cabriolet versions of the 504 shared only their mechanicals with the saloon. The floorpan was shorter, Pininfarina's bodywork was all-new and very elegant, and fuel injection was fitted from the outset, initially on the 1.8-litre four, and from '71 on the 2-litre four. Automatic available from '72. In 1974, this pair became the first cars to be fitted with joint Peugeot-Renault-Volvo V6 engine, but RHD UK imports ceased at this juncture. Five speeds from '78. Quite a few around, but like all Peugeots of this period, rust can be disastrously bad. V6 is a very desirable car.

Engine 1796cc 84x81mm/1971cc 88x81mm/2664cc 88x73mm, 4 cyl OHV/V6 cyl OHV, F/R **Max power** 103bhp at 5600rpm/110bhp at 5600rpm/136-144bhp at 5750rpm
Max speed 105-117mph
Body styles coupé, convertible
Prod years 1969-82
Prod numbers 26,629/8135
PRICES
Coupé **A** 4000 **B** 2200 **C** 1000
Cabriolet **A** 8000 **B** 5500 **C** 3500
Cabriolet V6 **A** 10,000 **B** 6500 **C** 4000

Peugeot 104

A hatchback ahead of its time, and the car which spawned a whole generation of Peugeot/Citroën/Talbot models in the 1970s/80s. Likeable Pug family styling; four doors to '76, hatchback thereafter. New range of inclined light alloy engines, all-independent suspension by coils, front-wheel drive. The only models of any lasting interest are the short-wheelbase three-doors available from '73. These were sold with Z badging and were just 10in longer than a Mini. ZS has 66bhp 1.1-litre engine (later 72-80bhp and 1.4 litres). Range pared right back after launch of 205 (and dropped in UK).

Engine 954cc 70x62mm/1124cc 72x69mm/1219cc 75x69mm/1361cc 75x77mm, 4 cyl OHC, F/F
Max power 45bhp at 6000rpm to 80bhp at 5800rpm
Max speed 80-102mph
Body styles saloon, hatchback
Prod years 1972-88
Prod numbers n/a

PRICES
A 1000 **B** 600 **C** 200

Peugeot 604

This should have been a sure-fire hit: PRV's new all-alloy overhead cam V6 engine, independent suspension all round, four disc brakes, and plenty of room inside. But it wasn't: the 604 was as dull as they come. Not exciting to look at, even less so to drive, it didn't even capture the imagination of French buyers. GM automatic or four-speed manual until '77, when the TI was launched (five speeds and fuel injection). Turbodiesel engine option from '79 (bigger 2.5-litre oil-burner from '84). 2849cc GTI launched in late '83. Rusted heavily and today are nothing more than banger material.

Engine 2664cc 88x73mm/2849cc 91x73mm/2304cc 94x83mm/2498cc 94x90mm, V6/V6/4/4 cyl OHC/OHC/diesel/diesel, F/R
Max power 136-144bhp at 5750rpm/155bhp at 5750rpm/80-95bhp at 4150rpm
Max speed 97-118mph
Body styles saloon
Prod years 1975-86
Prod numbers 240,100
PRICES
A 1000 **B** 500 **C** 100

Plymouth Barracuda '64-'66

Plymouth launched its own ponycar in the same year as the Ford Mustang, but it was destined to be an also-ran. The reason was that it was nothing more than a Valiant saloon with a new superstructure. Massive curved rear screen was distinctive, while the fold-down rear seat was useful for surf boards and romance. Six-cylinder or V8 engines on offer; V8's high-lift cam option developed 235bhp, and Rallye suspension pack helped handling. Shared none of the Mustang's roaring success, but was a useful 'character' model for Plymouth.

Engine 3682cc 86.4x104.8mm/4490cc 92.2x84.1mm, 6/V8 cyl OHV, F/R
Max power 145bhp at 4200rpm/180-235bhp at 5200rpm
Max speed 94-112mph
Body styles coupe
Prod years 1964-66
Prod numbers 126,068

PRICES
A 5000 **B** 3200 **C** 1600

Plymouth Barracuda '67-'69

A completely fresh approach for the '67 'Cuda saw it move more into ponycar territory. Wheelbase was 2in longer, overall dimensions were expanded and there was now a choice of three body styles: a Sports fastback hardtop coupe, a rather ugly notchback coupe and a convertible. Formula S pack helped handling, which was dire with the new and heavy four-barrel 383 V8. Together with the small-block 340, the 383 was Plymouth's muscle car for the late 1960s, offering up to 330bhp in the '69 model year. Various stripes, decals and adornments identified them.

Engine 3682cc 6 cyl OHV, 4490cc/5210cc/5563cc/6276cc /V8 cyl OHV, F/R **Max power** 145bhp at 4200rpm to 330bhp at 5200rpm **Max speed** 92-118mph **Body styles** fastback coupe, notchback coupe, convertible **Prod years** 1966-69 **Prod numbers** 139,933 (Fastback: 70,473/Notchback: 60,950/Convertible: 8510)

PRICES
Coupe **A** 6000 **B** 4000 **C** 1500
Convertible **A** 9000 **B** 6000 **C** 4000

Plymouth Barracuda '70-'74

The 1970 'Cuda was the sister model of the Dodge Challenger, with which it shared its body and mechanicals. Lower, wider and shorter than the previous incarnation, it looked less distinctive: you had to look twice to check it wasn't a Camaro. Wheelbase was the same, but the range of engines expanded right up to the 440 (7.2-litre) V8. Both the 440 and 426 Hemi had 'shaker' bonnets – a scoop attached directly to the carb housing. AAR (All American Racer) had stripes, glassfibre bonnet and rear spoiler (c 1500 made). Much reduced power after '72.

Engine 3682cc 6 cyl OHV, 5210cc/5563cc/6276cc/6974cc/7212cc, V8 cyl OHV, F/R
Max power 145bhp at 4200rpm to 425bhp at 5000rpm
Max speed 92-145mph
Body styles coupe, convertible
Prod years 1969-74
Prod numbers 125,886

PRICES
Coupe **A** 7000 **B** 4000 **C** 1500
Convertible **A** 12,000 **B** 8000 **C** 5000

Plymouth Road Runner/Superbird

Warner Bros allowed Plymouth to use images of its cartoon character on this most curious of muscle cars – you even got a 'meep-meep' horn. In marketing terms, the Road Runner was a success, combining cuteness with Hemi power, all at budget prices. Dummy bonnet scoops, lots of cosmetic add-on options, convertible from '69 model year. Startling 1970 Superbird had the same droop-snoot treatment and massive rear wing as the Dodge Charger Daytona, and it too had plenty of NASCAR success. Road Superbirds had a 440 engine and standard auto, though a manual 'box and 426 Hemi and 440 'Six Pak' were options.

Engine 6276cc 108x85.9mm/6974cc 108x95.3mm/7212cc 109.7x92.5mm, V8 cyl OHV, F/R
Max power 335bhp at 5200rpm to 425bhp at 5000rpm
Max speed 113-145mph
Body styles coupe, convertible
Prod years 1968-70
Prod numbers 125,904/1920
PRICES
Coupe **A** 7500 **B** 4500 **C** 2500
Convertible **A** 12,000 **B** 8000 **C** 5000
Superbird **A** 18,000 **B** 14,000 **C** 11,000

Pontiac Bonneville '57-'60

Like other Detroit makers, Pontiac needed a glamour model to get customers into its showrooms. So chief stylist Semon 'Bunkie' Knudsen created a flashy new convertible based on the existing Star Chief. Its engine was bored out to 370cu in, got a cam and fuel injection, boosting output up as far as 310bhp (or 348bhp on '60 389 cu in engine). Feebler outputs were also on offer. But this was a bulky car at 17ft 8in (538cm) long and nearly two tons, so it was no great sports car. It was by far the most expensive Pontiac then made.

Engine 6063cc 103.1x90.4mm/6377cc 103.1x95.2mm, V8 cyl OHV, F/R
Max power 255bhp at 4400rpm to 348bhp at 4800rpm **Max speed** 100-115mph **Body styles** saloon, estate, coupe, convertible **Prod years** 1956-60 **Prod numbers** 180,531
PRICES
'57 Convertible **A** 18,000 **B** 12,000 **C** 8000
'59 Convertible **A** 15,000 **B** 9000 **C** 7000
Saloon/Estate **A** 6000 **B** 4000 **C** 1500

Pontiac Tempest GTO '64-'67

The '64 GTO really began the whole muscle car craze. From its humble basis (the intermediate Tempest), the GTO managed to squeeze out unheard-of performance for a six-passenger car. The spec included a 325bhp V8 (later 360bhp), floor-mounted gear lever, quick rack and stiff suspension. The GTO tag invited comparisons with Ferrari's own GTO; amazingly, the Pontiac was almost as fast. Options included four speeds, metal brake linings, LSD and extra power. Price was also on its side: here was a cheap way to go obscenely fast. GTOs sold consistently well until replaced for '68.

Engine 6377cc 103.1x95.2mm/6555cc 104.6x95.3mm, V8 cyl OHV, F/R
Max power 255bhp at 4400rpm to 360bhp at 5200rpm
Max speed 100-114mph
Body styles coupé, convertible
Prod years 1963-67
Prod numbers 286,470

PRICES
Coupé **A** 6000 **B** 4000 **C** 2000
Convertible **A** 10,000 **B** 7500 **C** 4000

Pontiac Tempest GTO '68-'70

New A-body GTO also had a bigger 400cu in (6.6-litre) V8 as standard, with optional Ram Air (bonnet scoop) for 360bhp. Headlamps were hidden behind a new energy-absorbing impact bumper which was made of 'dent-resistant' Endura plastic and colour-coded. Among the '69 GTOs was The Judge – an option package consisting of stripes, bright paint, 366bhp V8 and three-speed manual floorshift. 370bhp available '69. Nasty Firebird-style front ends for '70. Convertibles also offered, but rare. '71 models and on were not true GTOs, but pale option packs on the Le Mans (and even the lowly Ventura in '74).

Engine 6555cc 104.6x95.3mm/7457cc 105.4x106.8mm, V8 cyl OHV, F/R
Max power 265bhp at 4400rpm to 370bhp at 5500rpm
Max speed 105-124mph
Body styles coupe, convertible
Prod years 1967-70
Prod numbers 200,120
PRICES
GTO **A** 5000 **B** 3000 **C** 1500
GTO Judge **A** 6500 **B** 4000 **C** 2000
GTO Convertible **A** 10,000 **B** 7000 **C** 3500

Pontiac Firebird

The first Firebird was sired by Chevrolet's Camaro and differed only in terms of trim, chrome and nose job. Restyled below the waist in '69, and no convertibles after mid-1970. '69 Trans Am had special paintwork, rear spoiler and 335-345bhp Ram Air power. All-new Firebird arrived for '71 with a distinctive style which lasted until 1981. High-performance Trans Am was a marginal seller to start with but became the most popular version (there were also standard, Esprit and Formula variants). Power dropped steadily through the 1970s, though there was a turbocharged version for 1980.

Engine 3769cc V6 cyl OHV, 4093cc 6 cyl OHV, 4340cc/4930cc/5001cc/5340cc/6555cc/7457cc V8 cyl OHV, F/R **Max power** 100bhp at 3800rpm to 345bhp at 5300rpm
Max speed 95-125mph
Body styles coupe, convertible
Prod years 1967-70/1970-81
Prod numbers 277,381/1,061,719
PRICES
'68 Coupe **A** 7000 **B** 4500 **C** 2000
'69 Trans Am **A** 10,000 **B** 7000 **C** 3000
'75 Trans Am **A** 3500 **B** 2000 **C** 800

Porsche 356

The number 356 was the next in line after Ferdinand Porsche's 1939 streamlined racer which was cancelled due to the war. The first examples had aluminium bodies and were made in Austria, then in 1950 Porsche relocated to Stuttgart and switched to steel. Steel platform with modified VW Beetle engine out back, plus VW-derived suspension. Cabriolet from '51, 1500 and one-piece (but still 'bent' in the middle) screen from '52, 1290cc engine and anti-roll bar from '54. 1500 Super has Hirth crank, high compression, sports cam, bigger carbs. Earliest alloy-bodied cars are too rare to price.

Engine 1086cc 73.5x64mm/1286cc 80x64mm/1290cc 74.5x74mm/1488cc 80x74mm, 4 cyl OHV, R/R **Max power** 40bhp at 4000rpm/44bhp at 4200rpm to 60bhp at 5700rpm/60bhp at 5000rpm/55bhp at 4400rpm to 70bhp at 5400rpm **Max speed** 87-109mph
Body styles coupé, convertible
Prod years 1948-55
Prod numbers 7627
PRICES
'50-'55 **A** 20,000 **B** 13,000 **C** 9000
Cabriolet **A** 30,000 **B** 20,000 **C** 10,000

Porsche 356A

356 shape became less dumpy after the 1955 restyle, instantly recognisable by new curved windscreen. Softer ride due to fewer leaves at the front and longer torsion bars at the back, plus a steering damper and some extra instruments, but the real news was the availability of a new 1600 engine (the 1300 was continued in Germany only, until 1958). This could be ordered in 60bhp standard or 75bhp Super tune, the latter having roller-bearing crankshaft, making it a very quick sports car. Bodies were made by Reutter of Stuttgart; also available was a hardtop for the cabriolet.

Engine 1290cc 74.5x74mm/1582cc 82.5x74mm, 4 cyl OHV, R/R
Max power 44bhp at 4200rpm to 60bhp at 5500rpm/60bhp at 4500rpm to 75bhp at 5000rpm
Max speed 90-109mph
Body styles coupé, convertible
Prod years 1955-59
Prod numbers 21,045

PRICES
Coupé **A** 15,000 **B** 10,000 **C** 6000
Cabriolet **A** 25,000 **B** 18,000 **C** 10,000

Porsche 356B/356C

Higher bumpers, more steeply sloping bonnet and new headlamps distinguished the new 356B of 1959. There were two new body styles: the Roadster (descended from the Speedster, see page 176), and the notchback coupé made by Karmann. A new Super 90 engine (90bhp) was also offered. Bigger rear window and Koni dampers in '61. 356C arrived in '63: four-wheel discs, ZF steering, flat hub caps. Previous Super sold as 1600C, 1600S or Super 75 and old Super 90 became 1600SC and gained an extra 5bhp. Like all 356s, fun around corners can turn suddenly into a foray off-road.

Engine 1582cc 82.5x74mm, 4 cyl OHV, R/R **Max power** 356B: 60bhp at 4500rpm to 90bhp at 5500rpm. 356C: 75bhp at 5200rpm/95bhp at 5800rpm
Max speed 100-116mph **Body styles** coupé, hardtop coupé, convertible
Prod years 1959-63/1963-65
Prod numbers 30,963/16,668
PRICES
356B Coupé **A** 18,000 **B** 12,000 **C** 7500
356B Cabriolet/Roadster **A** 28,000 **B** 20,000 **C** 15,000
356C worth 10% more

Porsche 356 Speedster

America lapped up Porsches from the earliest days, though that market always craved better performance. The Speedster was a model designed specifically for American buyers. Launched in September 1954, it was a lightweight roadster version of the 356A which also happened to look rakishly handsome. Windscreen was cut down by 3.5in and the winding windows were ditched; there was also a very low-cut soft top. 1500 and 1500 Super engines uprated to 1600cc in '56. In top level of tune, you could expect 120mph and 0-60mph in 10 secs. Replaced in '58 by Cabriolet D.

Engine 1488cc 80x74mm/1582cc 82.5x74mm, 4 cyl OHV, R/R
Max power 55bhp at 4400rpm to 75bhp at 5000rpm
Max speed 100-120mph
Body styles sports
Prod years 1954-58
Prod numbers 4854

PRICES
A 40,000 **B** 30,000 **C** 15,000

Porsche 356A/B Carrera

356s scored one competition victory after another and, following success in the Carrera Panamericana, produced a hairy road car called Carrera. A very special engine was developed by Dr Ernst Fuhrmann: a compact 1.5-litre flat-four engine, each bank having twin overhead camshafts driven by shaft and bevel gears, plus roller bearing crank, two spark plugs per cylinder and dry sump lubrication. That meant 100bhp, or 115bhp from '58 1.6-litre unit (now with plain bearing crank), but also daunting and expensive maintenance. All were lighter thanks to use of alloy doors and front/rear lids.

Engine 1498cc 85x86mm/1587cc 87.5x66mm, 4 cyl QOHC, R/R
Max power 100bhp at 6200rpm/115bhp at 6500rpm
Max speed 120-130mph
Body styles coupé, hardtop coupé, convertible
Prod years 1955-59
Prod numbers 700

PRICES
A 50,000 **B** 40,000 **C** 27,000

Porsche 356B/C Carrera 2

The Carrera 2 appeared in spring 1962 (so 356B-based for its first year, then 356C) with an enlarged 2-litre version of the amazing four-cam engine. Plain bearing crank retained, twin Solex carbs, power up to 130bhp. All-round disc brakes from the start, ahead of their appearance on pushrod-engined 356C. Popular with race/rally drivers (especially in America), for whom GS (140bhp) and GT (155bhp) versions were available, both with twin Weber carbs. A very complicated engine, and few specialists can reliably rebuild one. Double-check authenticity and, as on all 356s, look for rust.

Engine 1966cc 92x74mm, 4 cyl DOHC, R/R
Max power 130bhp at 6200rpm
Max speed 130mph
Body styles coupé, convertible
Prod years 1960-64
Prod numbers 506

PRICES
A 65,000 **B** 45,000 **C** 30,000

Porsche 911/911N/911L

One of the all-time classic cars was first shown at Frankfurt in 1963. Designed by Ferry Porsche's eldest son, Butzi, it retained the rear-engined air-cooled layout but moved up to six cylinders and two triple-choke carbs. 4.5in longer wheelbase than the 356, dual-circuit discs all round, five-speed 'box, IFS by wishbones and torsion bars, rear by trailing arms and torsion bars. 911N was stripped-out German base model, 911L replaces both in '67 and adds choice of Sportomatic four-speed semi-auto (not well liked). The original 911, and therefore plenty of appeal. Sportomatics and Targas (from '66) worth 10% less.

Engine 1991cc 80x66mm, 6 cyl DOHC, R/R
Max power 130bhp at 6100rpm
Max speed 130mph
Body styles coupé, targa
Prod years 1964-69
Prod numbers 22,333

PRICES
A 13,000 **B** 10,000 **C** 7500

Porsche 912/912E

The elevated price of the 911 sent many Porsche 356 buyers to other marques. Hence the 912 was launched: a 911 with the old four-cylinder 356 engine. There was also a four-speed gearbox (five-speed optional), no automatic availability and a plainer interior. Dual-circuit brakes from '67, longer wheelbase in '68. 912E was brief US-market revival in 1975, having the four-cylinder VW fuel injection engine of the 914. Targa and coupé body styles available. If ever there was an unloved Porsche, this is it: if you want a cheap 911, you could do worse than start with a 912.

Engine 1582cc 82.5x74mm/1971cc 94x71mm, 4 cyl OHV, R/R
Max power 90bhp at 5800rpm/80bhp at 4900rpm
Max speed 110mph
Body styles coupé, targa
Prod years 1965-69/1975-76
Prod numbers 30,300/2099

PRICES
A 5000 **B** 4000 **C** 2700

Porsche 911S

New high-performance 911 was instantly recognisable by its five-spoke alloy wheels, later found on other 911s. The engine was given a higher compression ratio and, from late '68, fuel injection. From late '69, the engine expanded to 2.2 litres, then 2.4 litres and chin spoiler from '71. All S models have ventilated discs, the rear pair mounted directly adjacent to the differential. From September 1968, in common with all 911/912 models, the wheelbase was stretched by 2.5in and the wheelarches were flared. Also rear screen wiper, and optional self-levelling front gas struts for improved handling.

Engine 1991cc 80x66mm/2195cc 84x66mm/2341cc 84x70.4mm, 6 cyl OHC, R/R
Max power 160bhp at 6600rpm to 170bhp at 6800rpm/180bhp at 6500rpm/190bhp at 6500rpm
Max speed 140-146mph
Body styles coupé, targa
Prod years 1966-73
Prod numbers 14,841
PRICES
2.0/2.2 **A** 15,000 **B** 12,000 **C** 8000
2.4 **A** 18,000 **B** 14,000 **C** 10,000

Porsche 911T

'T' stood for Touring, an indication that this was the junior member of the 911 family (at least after the demise of the 912). The de-tuned engines had carburettors rather than fuel injection and were the least powerful of the sixes (2-litre to '69, 2.2-litre '69-'71, 2.4-litre '71-'73). Four speeds standard, but optional five-speeder or Sportomatic semi-auto. Like other 911s, also available in Targa guise, which meant a fixed roll-over bar and removable roof panel; removable plastic rear screen at first, but fixed glass optional from '68, obligatory from '71. Performance edge not really there.

Engine 1991cc 80x66mm/2195cc 84x66mm/2341cc 84x70.4mm, 6 cyl OHC, R/R
Max power 110bhp at 5800rpm/ 125bhp at 5800rpm/130bhp at 5600rpm
Max speed 117-132mph
Body styles coupé, targa
Prod years 1967-73
Prod numbers 38,333
PRICES
2.0 **A** 12,500 **B** 8000 **C** 6000
2.2/2.4 **A** 13,000 **B** 8500 **C** 6000

Porsche 911E

This model superseded the 'luxury' 911L in 1968, at the time when all 911s were updated with slightly longer wheelbases and flared wheelarches. In its role as the comfortable member of the 911 family, the E boasted standard self-levelling front suspension, alloy wheels and a fuel injected engine, although the power output was never as high as that of the 911S. As with other 911s, bigger engines arrived in '69 and '71, and the E got an S-style chin spoiler in late '72. Rust does take its toll on 911s and spares, though plentiful, are almost always expensive.

Engine 1991cc 80x66mm/2195cc 84x66mm/2341cc 84x70.4mm, 6 cyl OHC, R/R
Max power 140bhp at 6500rpm/ 155bhp at 6200rpm/165bhp at 6200rpm
Max speed 130-140mph
Body styles coupé, targa
Prod years 1968-73
Prod numbers 12,159
PRICES
2.0/2.2 **A** 14,000 **B** 10,000 **C** 7500
2.4 **A** 15,000 **B** 11,000 **C** 8500

Porsche 911 Carrera RS

The first 911 with the 2.7-litre six was the 1972-73 Carrera RS. This extremely special machine was made as a homologation car for Group 4 racing, and it proved popular, selling much more than the minimum 500 required. Two road versions were Sport (basic interior, fewer than 200 made) and Touring (trimmed like 911S, heavier), both sharing 210bhp engine, stiffer suspension, flared rear arches. Lighter steel bodywork normal, but extended production meant standard panels on late cars. Most have 'ducktail' spoiler, bold 'Carrera' side decals optional. Late on, 59 road models had 3.0/230bhp engine. Fakery abounds.

Engine 2687cc 90x70.4mm/2993cc 95x70.4mm, 6 cyl OHC, R/R
Max power 2.7: 210bhp at 6300rpm 3.0: 230bhp at 6200rpm **Max speed** 150-155mph **Body styles** coupé
Prod years 1972-73
Prod numbers 1590

PRICES
RS Sport **A** 40,000 **B** 30,000 **C** 20,000
RS Touring **A** 35,000 **B** 26,000 **C** 17,000

Porsche 911 2.7

In 1973, all 911s received the 2.7-litre engine. Still a three-tier range: basic 911 (150bhp), mid-ranking 911S (175bhp) and top-spec Carrera (210bhp). Although 210bhp unit is from previous RS, new Carrera is a mainstream model, beginning Porsche's progressive cheapening of the name. All have US impact-absorbing bumpers, front spoiler; rear spoiler optional on Carrera ('ducktail' at first, 'whaletail' from '75). Targa and Sportomatic available on all versions. Top model becomes Carrera 3.0 in '75, with bigger engine but less power (200bhp). Big bumpers and weak performance mean the lesser 2.7s are the cheapest of all 911s.

Engine 2687cc 90x70.4mm/2993cc 95x70.4mm, 6 cyl DOHC, R/R
Max power 150bhp at 5700rpm to 210bhp at 6300rpm **Max speed** 130-150mph **Body styles** coupé, targa
Prod years 1973-77
Prod numbers 37,737

PRICES
911/911S **A** 8000 **B** 6000 **C** 4000
Carrera 2.7 **A** 18,000 **B** 13,000 **C** 10,000
Carrera 3.0 **A** 15,000 **B** 12,000 **C** 9500

Porsche · Princess · Puma · Reliant

Porsche 911 Turbo

After many years of planning, Porsche finally felt confident enough to use its racing turbo technology on a road car. The Turbo was presented at the Paris Salon in 1974 and went into production the following year. You got 260bhp and monster acceleration thanks to the KKK compressor, but had to pay twice as much as a standard 911. Special four-speed 'box, big brakes, stiffer floorpan. Leather and air con inside and, on the outside, distinctive whale-tail rear spoiler and massive alloy wheels in extended arches. Walloping turbo lag and usual on-the-limit snap-away made it a handful.

Engine 2993cc 95x70.4mm, 6 cyl OHC, R/R
Max power 260bhp at 5500rpm
Max speed 155mph
Body styles coupé
Prod years 1975-77
Prod numbers 2873

PRICES
A 25,000 **B** 18,000 **C** 12,500

Porsche 924

This was originally going to be an Audi, but Porsche took it on and made it into its best-seller. Dutchman Harm Lagaay styled it, Audi built it, and VW supplied most of the mechanicals, including the 2-litre 'four' from the Volkswagen Transporter. Porsche's first front-engined and first water-cooled car got a thumbs-down from the traditionalists, but people were attracted by the lowish price and Porsche badge. Rear-mounted transaxle, initially with four speeds (five from '78), or rare auto. Not very quick, but lots around, longevity is assured, and quite practical too.

Engine 1984cc 86.5x84.4mm, 4 cyl OHC, F/R
Max power 125bhp at 5800rpm
Max speed 125mph
Body styles hatchback coupé
Prod years 1976-85
Prod numbers 122,304

PRICES
A 5000 **B** 3300 **C** 1500

Princess 1700/1800/2000/2200

What started life as a range of seven models with Austin, Morris and Wolseley badges was pared down to just four under a new marque name, only six months after launch. The new front end style was basically the same as the old Austin brand, though the 1800 and 1800HL had twin headlamps and the six-cylinder 2200HL and HLS had single trapezoidal lamps. Series II of '78 had new 1700 and 2000 O-series four-cylinder engines, though the 2200 remained. Compared to the Cortina, a sales flop. And they rust. Superseded in 1981 by the reworked hatchback Austin Ambassador.

Engine 1695cc 84.4x75.8mm/1798cc 80.3x88.9mm/1993cc 84.4x89mm/2227cc 76.2x81.3mm, 4/4/4/6 cyl OHC/OHV/OHC/OHC, F/F
Max power 87bhp at 5200rpm/82bhp at 5250rpm/93bhp at 4900rpm/110bhp at 5250rpm **Max speed** 90-106mph
Body styles saloon **Prod years** 1975-81 **Prod numbers** 205,842 (41,134/72,867/35,498/56,343)

PRICES
A 1000 **B** 600 **C** 100

Puma GT

In 1964, Italian immigrant Genaro Malzoni began building sports cars in Sao Paolo, Brazil. He began with two Fissore designs based on the DKW 3=6 and graduated to building the attractively-styled Puma in 1967. This was based on a VW Karmann-Ghia floorpan (later VW Beetle) and could be serviced through VW dealers. Plastic bodies, VW engines. A handful were imported to Britain in the early 1970s. From 1974, Puma also made the GTB, a bigger coupé based on a locally-built Chevy chassis.

Engine 1493cc 83x69mm/1584cc 85.5x69mm/1795cc 93x66mm, 4 cyl OHV, R/R
Max power 44bhp at 4000rpm to 72bhp at 4800rpm
Max speed 85-100mph
Body styles coupé, sports, convertible
Prod years 1967-85
Prod numbers n/a

PRICES
A 3500 **B** 1800 **C** 800

Reliant Sabre/Sabre Six

Having made three-wheelers dating back to before the war, Reliant became a sports car maker in 1961. It had helped develop the Sabra for Israel (basically an LMB chassis, modified Ashley glassfibre body and Ford Consul engine), and made it in Tamworth for British consumption. Novel leading arm and coil spring front suspension, front discs, and curious overriders. Sabre Six had TR4 front suspension, Ford Zephyr 6 power, rounded arches, restyled front end, Kamm rear. Too expensive, too basic and a curiosity today.

Engine 1703cc 82.55x79.5mm/2553cc 82.55x79.5xmm, 4/6 cyl OHV, F/R
Max power 73-90bhp at 4600rpm/106-120bhp at 5000rpm
Max speed 90-100/110-115mph
Body styles coupé, sports
Prod years 1961-63/1962-66
Prod numbers 208/77

PRICES
Sabre 4 **A** 5500 **B** 3700 **C** 2200
Sabre 6 **A** 7000 **B** 5000 **C** 3300

Reliant Scimitar GT SE4a/SE4b/SE4c

Ogle Design built a handsome one-off called the SX250 on a Daimler SP250 chassis and Reliant had the foresight to buy the rights. It based the car mechanically on the Sabre Six, including its Ford-derived engine and gearbox (though a ZF was optional). Trailing arm rear suspension from '65 improved handling. SE4b of '66 had extra chassis bracing, 3-litre Ford V6, steel disc wheels in place of wires and much better performance. 'Poverty special' 2.5-litre V6 (SE4c) was unpopular. Glassfibre body doesn't rust, mechanicals are simple, and the GT is seriously underrated in classic terms.

Engine 2553cc 82.55x79.5mm/2994cc 93.7x72.4mm/2495cc 93.7x60.3mm, 6/V6/V6 cyl OHV, F/R
Max power 120bhp at 5000rpm/146bhp at 4750rpm/121bhp at 4750rpm
Max speed 108-121mph
Body styles coupé
Prod years 1964-66/1966-70/1967-70
Prod numbers 1003 (296/590/117)

PRICES
A 4000 **B** 2500 **C** 1000

Reliant Scimitar GTE SE5/SE5a

The 'rightness' of Reliant's Ogle-designed sports estate remains true even today. The first sports estate to make production, it was favoured by Princess Anne (she has owned nine). New steel chassis, pretty glassfibre four-seat body with a glass rear hatch, folding rear seats. Mechanicals Ford, as ever: V6 engine, servo front discs, manual, overdrive or (from '69) auto gearbox. SE5a of '71 has no chrome waist strip, revised dash, reversing lights. Much copied but never bettered. Deservedly one of the most popular and practical classic cars – no worries about body rust, and few about the chassis.

Engine 2994cc 93.7x72.4mm, V6 cyl OHV, F/R
Max power 135bhp at 5500rpm to 144bhp at 4750rpm
Max speed 120mph
Body styles estate
Prod years 1968-75
Prod numbers 9416

PRICES
A 4500 **B** 3000 **C** 1200

Reliant Scimitar GTE SE6/SE6a/SE6b

By 1975, the GTE needed replacement and some thought an opportunity had been missed when all Reliant could do was expand the original theme: the old shape was religiously kept, though the car was 4in longer in the wheelbase and 3in wider. The chassis was new, but the handling suffered and weight was up. Revised SE6a launched soon after, featuring stiffer scuttle and bigger dual-circuit brakes. Ford Cologne 2.8-litre V6 and front spoiler from December 1979, galvanised chassis from '81. More stiffening on SE6b of '82. A much more practical all-round car than the SE5.

Engine 2994cc 93.7x72.4mm/2792cc 93x68.5mm, V6 cyl OHV, F/R
Max power 135bhp at 5500rpm to 150bhp at 5200rpm
Max speed 120mph
Body styles estate
Prod years 1975-79/1979-82/1982-86
Prod numbers 543/3877/437

PRICES
A 5000 **B** 3500 **C** 1500

Reliant Regal MkI-VI

Reliant made Raleigh Safety Seven vans after the war, but their first passenger car was the Regal, first shown in 1951. Van origins very discernible: pressed steel chassis, Austin Seven derived 747cc engine, long stub front axle with torsion bar, rear semi-elliptic springs and live axle. Hydraulic brakes and four speeds were unexpectedly sophisticated. Aluminium-over-ash open body, new grille on '54 MkII, Hardtop model from '55, all-new rounded glassfibre body from '56 MkIII. Improved chassis and damping on '58 MkIV, saloon only from '59 MkV. 1960 MkVI was the last of the 'old-style' Regals.

Engine 747cc 56x76mm, 4 cyl SV, F/R
Max power 16bhp at 4000rpm
Max speed 60mph
Body styles saloon, tourer
Prod years 1951-62
Prod numbers n/a

PRICES
A 2500 **B** 1200 **C** 400

Reliant Regal 3/25 and 3/30

3/25 means three wheels and 25bhp, hinting at the extra reserves of power offered by the all-new 1962 Regal. That was courtesy of Reliant's own Austin Seven-based 598cc diecast all-alloy OHV engine – one of the very first aluminium engines built in Britain. Angular styling reminiscent of Ford's Anglia, only uglier. Same old perimeter frame chassis and gearbox. Amazingly popular (selling up to 20,000 a year). 1967 3/30 had an expanded 701cc version of the OHV engine with 30bhp, plus an estate body option. You might not believe it, but there is a devoted owners' club for old Regals.

Engine 598cc 55.9x61mm/701cc 60.5x61mm, 4 cyl OHV, F/R
Max power 25bhp at 5250rpm/30bhp at 5000rpm
Max speed 65/70mph
Body styles saloon, estate
Prod years 1962-67/1967-73
Prod numbers c 100,000

PRICES
A 900 **B** 400 **C** 100

Reliant Rebel

The Rebel was little more than a four-wheeled version of the Regal, sharing its drive train and rear end but a different, and totally characterless, glassfibre body. A pretty crude device which had an impossible time competing with the Mini. Economy was the main reason you'd buy one: light weight meant it could average 60mpg, though it cost fifty quid more than a Mini. Estate car had squared-up rear and side-opening rear door. Bigger engines in '67 and '72. Its only attraction must be as a runaround in seaside towns where your average rust-bucket has more holes than a pitch-and-putt.

Engine 598cc 55.9x61mm/701cc 60.5x61mm/748cc 62.5x61mm, 4 cyl OHV, F/R
Max power 27bhp at 5250rpm/31bhp at 5000rpm/32bhp at 5500rpm
Max speed 65-73mph
Body styles saloon, estate
Prod years 1964-73
Prod numbers n/a

PRICES
A 800 **B** 400 **C** 100

Reliant Robin

Although the 'plastic pig' has become the infamous butt of car-jokers everywhere, the Robin was an excellent seller, if only for the reason that no other tax-busting trikes were available in the 1970s. Under a modern hatchback body styled by Ogle sat a new twin-rail chassis fitted with the familiar alloy four-pot engine, now bored out to 748cc. An estate was also available. For the first time, a Reliant trike had an all-synchro gearbox and the rear suspension was by single-leaf semi-elliptic springs. 848cc engine from '75. Ex-bikers still search the classifieds for them.

Engine 748cc 62.5x61mm/848cc 62.5x69.1mm, 4 cyl OHV, F/R
Max power 32bhp at 5500rpm/40bhp at 5500rpm
Max speed 73-80mph
Body styles saloon, estate
Prod years 1973-81
Prod numbers n/a

PRICES
A 800 **B** 400 **C** 100

Reliant Kitten

This really was nothing but a Robin with the correct number of wheels. An 848cc engine was standard from the start. The Kitten got quite a lot of publicity at launch because it had a tighter turning circle than a London taxi and was probably the most economical four-wheeler you could then buy: it could reach 60mpg without going too light on the accelerator, while being able to exceed 80mph. For all that, it was a rather unpleasant car: noisy, cramped, plasticky and poorly built. The glass hatch was a pain when compared with the average supermini. A fringe product at best, and now extremely rare.

Engine 848cc 62.5x69.1mm, 4 cyl OHV, F/R
Max power 40bhp at 5500rpm
Max speed 80mph
Body styles saloon, estate
Prod years 1975-82
Prod numbers 4074

PRICES
A 800 **B** 400 **C** 100

Renault Juvaquatre

As a stop-gap model, pending the launch of the 4CV, Renault reintroduced its Juvaquatre 8HP in 1945. It was only sold with four doors post-war, its appearance strongly resembling the old Opel Kadett. Sidevalve engine, independent front suspension, transverse leaf springs all round, hydraulic brakes. Saloon was dropped in 1950 but the van and estate continued until 1960, latterly with engines from the 4CV ('53) and Dauphine ('56), in which guise it was called Dauphinoise. Some RHD cars were built by Renault at its UK headquarters in Acton.

Engine 1003cc 58x95mm/747cc 54.5x80mm/845cc 58x80mm, 4 cyl SV/OHV/OHV, F/R
Max power 24bhp at 3500rpm/23bhp at 4000rpm/26bhp at 4200rpm
Max speed 56-60mph
Body styles saloon, estate
Prod years 1937-60
Prod numbers 40,681 (saloon)

PRICES
A 2500 **B** 1800 **C** 1000

Renault 4CV

Alloy-bodied prototypes of the new Renault were running during the war, but steel was employed from launch in 1946. Rear-mounted OHV engine, all-independent suspension, three-speed constant mesh gearbox, four doors. Reduced-bore 747cc engine from '50, revised grille '53, Ferlec automatic clutch option from '55, more power from '58. R1052 Sport model launched in 1952 had light alloy head, twin-choke carb, stiffer crank and 42bhp, known in UK as the 'Competition' (some even have five-speed crash 'boxes). Open Découvrable is rare. The French Morris Minor, with a strong following in France.

Engine 760cc 55x80mm/747cc 54.5x80mm, 4 cyl OHV, R/R
Max power 18bhp at 4000rpm/23bhp at 4000rpm to 42bhp at 6000rpm
Max speed 58-87mph
Body styles saloon, convertible
Prod years 1946-61
Prod numbers 1,105,543

PRICES
4CV **A** 3000 **B** 2000 **C** 1000
4CV Sport **A** 5000 **B** 3300 **C** 1800
4CV Découvrable **A** 6000 **B** 3500 **C** 2000

Renault

Renault Dauphine/Dauphine Gordini

Hugely popular development of the 4CV, sharing most of its mechanicals but having an all-new and much roomier four-door body and a bored-out engine. Coil spring and wishbone IFS, Ferlec option for finger-touch gear changes, Aerostable semi-pneumatic variable-rate suspension from '59, optional four-speed 'box from '63 on standard model, four-wheel discs from '64. Gordini is the one to have: 38-49bhp, four speeds and 'interesting' handling. 1093 rally model ('61 on) is a real scorcher, but only 2500 were built. All Dauphines rust appallingly.

Engine 845cc 58x80mm, 4 cyl OHV, R/R
Max power 26bhp at 4200rpm to 49bhp at 5600rpm
Max speed 70-90mph
Body styles saloon
Prod years 1956-68
Prod numbers 2,120,220

PRICES
Dauphine **A** 1800 **B** 1100 **C** 500
Gordini **A** 2500 **B** 1500 **C** 1000

Renault Frégate

Uninspiring big Renault. Unitary construction, four-wheel independent suspension, four-speed all-synchro 'box with column change, automatic option (or, from '57, Transfluide semi-auto drive). 2-litre models deleted in '57, 2.1-litre models available from '55. Confusing range of names for trim levels: base Affaires, de luxe Amiral, direct top gear Caravelle (all 2-litre), Amiral and Grand Pavois (2.1-litre), plus Domaine estate car and Manoir 6/8 seater estate. Almost zero chic level, not a good rust record and very little following outside France.

Engine 1996cc 85x88mm/2141cc 88x88mm, 4 cyl OHV, F/R
Max power 56-60bhp at 3800rpm/80bhp at 4000rpm
Max speed 75-90mph
Body styles saloon, estate
Prod years 1951-60
Prod numbers 177,686

PRICES
A 2000 **B** 1100 **C** 700

Renault Colorale Prairie

Decades before Nissan came up with its people carrier/estate cross-over, Renault had its own Prairie estate. This was basically a stop-gap model launched at around the same time as the Frégate, giving Renault a presence in the station wagon market until the Domaine arrived in '55. Also available in panel van and truck forms. It had nothing to do with the Frégate, having its own rather dumpy body on a shorter wheelbase and a 2.4-litre sidevalve engine first seen in 1936 (or, from '52, the Frégate's 2-litre engine). A bit of a slogger, all but unknown in Britain. Why would you want to own one?

Engine 2383cc 85x105mm/1996cc 85x88mm, 4 cyl SV/OHV, F/R
Max power 56bhp at 5800rpm/ 56bhp at 3800rpm
Max speed 62/66mph
Body styles estate
Prod years 1951-55
Prod numbers n/a

PRICES
A 1500 **B** 800 **C** 400

Renault Floride/Floride S

To call this a sports car would be stretching the definition, but Renault's first post-war coupé was certainly stylish and could be fun to drive. Basically a Dauphine Gordini under the Frua-designed skin. Therefore you get a rear-mounted 845cc engine, three speeds (optionally four speeds or the Ferlec electro-magnetic clutch), drum brakes, swing axles and Aerostable suspension. The cabriolet could be ordered with a removable hardtop and all models were 2+2. 1962 'S' has 956cc, convertible only. Manufactured by coachbuilders Brissoneau & Lotz. Big rust problems, some parts difficulties.

Engine 845cc 58x80mm/956cc 65x72mm, 4 cyl OHV, R/R
Max power 38bhp at 5000rpm/ 51bhp at 5500rpm
Max speed 79/84mph
Body styles coupé, convertible
Prod years 1959-62/1962-63
Prod numbers 177,122/n/a

PRICES
Coupé **A** 3000 **B** 1800 **C** 1000
Cabriolet **A** 4500 **B** 3000 **C** 2000

Renault Caravelle

The name Caravelle was used in America for the Floride and it was adopted in Europe from 1962, with the launch of the bigger-engined coupé. The 956cc five-bearing unit derived from the R8, as did the disc brakes and four-speed gearbox. The rear roof line was squared up to give better head room for rear passengers. From '63, the engine was expanded to 1.1 litres, the four-speed 'box became all-synchro and the fuel tank got bigger. At the same time, a new Caravelle Cabriolet was launched, which came as standard with a removable hardtop. All Florides and Caravelles are distinctly dodgy handlers.

Engine 956cc 65x72mm/1108cc 70x72mm, 4 cyl OHV, R/R
Max power 51bhp at 5500rpm/ 54bhp at 4500rpm
Max speed 84/88mph
Body styles coupé, convertible
Prod years 1962-68
Prod numbers n/a

PRICES
Coupé **A** 3500 **B** 2200 **C** 1300
Cabriolet **A** 5000 **B** 3500 **C** 2500

Renault

Renault R3/R4

Archetypal shed on wheels, sharing common elements with Citroën's 2CV. Renault's first front-wheel drive; engines transplanted from 4CV/Dauphine/R8. Torsion bar springing all round meant a bouncy ride and wayward handling, but it was amazing what you could fit into one. Gearbox ahead of engine, requiring an amazing push-pull gear linkage to operate it. Wheelbase different on each side! R3 was stripped-out export special with 603cc engine. 1108cc engine not offered until 1980 (on the GTL), while the 782cc engine was reserved for the French market. 1968-'71 Plein Air is doorless beach car, 4x4 also available.

Engine 603cc/747cc 54.5x80mm/782cc 55.8x80mm/845cc 58x80mm/956cc 65x72mm/1108cc 70x72mm, 4 cyl OHV, F/F & F/4x4
Max power 23bhp at 4800rpm to 34bhp at 4000rpm
Max speed 59-78mph
Body styles hatchback
Prod years 1961-92
Prod numbers c 8,000,000
PRICES
R4 **A** 1000 **B** 500 **C** 100
Plein Air **A** 4000 **B** 2800 **C** 1500

Renault 8

Successor of the Dauphine had angular styling but still the same rear-engined layout. New and larger 956cc engine, all-independent suspension, surprisingly four-wheel disc brakes. Even an automatic 'box was on offer, changed by pushing buttons on the dashboard. Hot Gordini version from '64 had quad headlamps, stripes, hemi cylinder head, twin carbs, lower suspension, brake servo and, from '67 1300, up to 103bhp (gulp!). '68 8S also had four headlamps but only 53bhp. '64 8-1100 (aka 8 Major) had bigger engine, synchromesh on all four gears and, from '65, longer nose and tail.

Engine 956cc 65x72mm/1108cc 70x72mm/1255cc 74.5x72mm, 4 cyl OHV, R/R
Max power 40bhp at 5200rpm to 103bhp at 6750rpm
Max speed 78-108mph
Body styles saloon
Prod years 1962-72
Prod numbers 1,329,372

PRICES
R8 **A** 1800 **B** 1000 **C** 600
R8 Gordini **A** 3500 **B** 2500 **C** 1200

Renault 10

From 1965, the 8 had also been sold with a long nose and rear end. Confusingly, it was sold in some markets as the 8-1100 or 8 Major, but in France it was called the 10! In other countries, including the UK, the 10 badge was reserved for the bored and stroked five-bearing 1.3-litre version of the long nose/tail 8 family, launched in 1969. Only badges and a black strip across the back identify it. Boring in the extreme, unless you reach the extremes of grip, when things get very interesting. Superseded by the 12. Difficult to think of a convincing reason to buy one these days.

Engine 1289cc 73x77mm, 4 cyl OHV, R/R
Max power 48bhp at 4800rpm
Max speed 82mph
Body styles saloon
Prod years 1969-71
Prod numbers n/a

PRICES
A 1200 **B** 600 **C** 300

Renault 16

A landmark car, often cited as the first of the hatchbacks. Modern in so many ways. Front-wheel drive, closed cooling system, front discs. Very spacious and flexible interior, with idiosyncratic Gallic touches. Column gear change is a demerit, but compensated by fine ride, sound roadholding and excellent performance. 1565cc engine introduced in '68 on 88bhp TS, in '69 for automatic cars, and from '71 for all. '73 TX (photo) is the ultimate 16, with its 1647cc engine, five-speed gearbox, quad headlamps and electric front windows. Some interest is now beginning to show, especially for the TX.

Engine 1470cc 76x81mm/1565cc 77x84mm/1647cc 79x84mm, 4 cyl OHV, F/F
Max power 55bhp at 5000rpm to 93bhp at 6000rpm
Max speed 88-105mph
Body styles hatchback
Prod years 1964-79
Prod numbers 1,846,000

PRICES
A 1500 **B** 1000 **C** 400

Renault 6

As the Citroën Dyane is to the 2CV, so Renault's 6 is to the 4. Basically the floorpan and running gear are identical to the 4, but the suit of clothes is a Euro-bland conventional hatchback, too upright to stand a chance of being pretty. Launched with the 845cc engine but also available from 1970 with the larger 1100 engine (which gives better all-round performance and comes with disc front brakes). When rust inevitably set in, no-one bothered to save them, so despite a run of almost two million, they are almost extinct. Leave them to poverty-stricken French customisers.

Engine 845cc 58x80mm/1108cc 70x72mm, 4 cyl OHV, F/F
Max power 38bhp at 5000rpm/48bhp at 5300rpm
Max speed 75/82mph
Body styles hatchback
Prod years 1968-79
Prod numbers 1,773,304

PRICES
A 700 **B** 300 **C** 100

Renault Rodeo

Some said Renault copied Citroën's 2CV when it developed the R4. There is no question that it copied Citroën's Mehari when it created the Rodeo. There were two distinct models: the 4 Rodeo, based on the R4, and the 6 Rodeo on the 6. The former had a raised bonnet and circular headlamps, the latter a flush bonnet and rectangular headlamps. The plastic bodies were made by Teilhol in various styles: Evasion (no roof), Chantier (covered cab) and Quatre Saisons (full folding roof and proper doors). A firm called Sinpar also offered an approved 4x4 conversion. Never sold in the UK.

Engine 845cc 58x80mm/1108cc 70x72mm, 4 cyl OHV, F/F
Max power 34bhp at 5000rpm/ 47bhp at 5300rpm
Max speed 62/80mph
Body styles utility
Prod years 1973-81
Prod numbers n/a

PRICES
A 2000 **B** 1200 **C** 700

Renault 12

The 12 finally killed off the last of the rear-engined Renaults. Despite its front-wheel drive format, it was conventional in every respect: rather bland three-box styling, front discs, four-speed gearbox (thankfully with a floor-mounted lever). Servo brakes on estate and '72 TS (60bhp). Automatic transmission optional from '74. Minor facelift in '75, dual-circuit brakes '77. Sadly, the most interesting 12 was not imported: the 1970-75 Gordini boasted a 110bhp 1565cc engine, stripes, and five speeds. The 12 America had a 1647cc engine and was for the US market only. Forgettable.

Engine 1289cc 73x77mm/1565cc 77x84mm/1647cc 79x84mm, 4 cyl OHV, F/F
Max power 54bhp at 5250rpm to 110bhp at 6250rpm
Max speed 90-115mph
Body styles saloon, estate
Prod years 1969-80
Prod numbers 2,865,079

PRICES
12 **A** 800 **B** 400 **C** 100
12 Gordini **A** 1500 **B** 1000 **C** 700

Renault 15/17

Playing the platforms game: this pair were based on the floorpan of the 12 and given distinctive coupé bodies. 15 had full rear three-quarter windows and single rectangular headlamps, while the 17 had quad lamps and reverse-angle C-pillars with louvred quarters. Range started at 1.3-litre 15TL (later GTL), through 1.6-litre 15TS, 1.6-litre 17TL, and fuel-injected 17TS. 1605cc engines for 17 in '73, renamed Gordini from '75 to '76, when 1647cc standard. Front discs on 15, all round discs and five speeds on 17, auto option for both. Convertible has full-length fabric sunroof. Not great cars, but more interesting than a Capri.

Engine 1289cc 73x77mm/1565cc 77x84mm/1605cc 78x84mm/1647cc 79x84mm, 4 cyl OHV, F/F **Max power** 60bhp at 5500rpm to 108bhp at 6000rpm **Max speed** 93-112mph
Body styles hatchback coupé, rolltop coupé
Prod years 1971-79
Prod numbers 207,854/92,589
PRICES
15 **A** 1200 **B** 700 **C** 400
17 **A** 1700 **B** 1000 **C** 500
17 Convertible **A** 2500 **B** 1600 **C** 1000

Renault 5

Chic styling, practical hatchback layout, great ride for a small car and high economy made the 5 Renault's best-selling car in the 1970s. Engines were initially the same as those in the 4 (including the France-only 782cc unit), but later TS and GTL models got the 1289cc engine and high gearing. 1108cc engine replaces 956cc in '79. Push-pull facia-mounted gear lever gave way to a floor shift in '77. Best one undoubtedly the '76 1.4-litre Gordini (not called Alpine in UK, as it was in France, because of Chrysler's objections): five speeds, 93bhp, alloys, spoiler and stripes.

Engine 782cc 55.8x80mm/845cc 58x80mm/956cc 65x72mm/1108cc 70x72mm/1289cc 73x77mm/1397cc 76x77mm, 4 cyl OHV, F/F
Max power 33bhp at 5200rpm to 93bhp at 6400rpm
Max speed 75-109mph
Body styles hatchback
Prod years 1972-84
Prod numbers 5,471,709
PRICES
5 **A** 900 **B** 600 **C** 200
5 Gordini **A** 2500 **B** 1500 **C** 700

Renault 20/30

Quite successful large front-drive hatchback Renaults. The 30 came first, fitted with the new joint Peugeot-Renault-Volvo V6 engine, providing good performance. All-independent suspension gave a very supple ride. 30TX from '79 had fuel injection and five speeds. The 20 shared its unremarkable styling and underpinnings, except for four-cylinder engines: 1.6 litres (TL) and 2.0 litres (TS), though new 2.2-litre in 1980 replaced both, and a new 2.1-litre diesel was launched at the same time. Suspect build quality and poor longevity means that most have now disappeared. Do they have any fans?

Engine 1647cc 79x84mm/1995cc 88x82mm/2165cc 88x89mm/2068cc 86x89mm/2664cc 88x73mm, 4/4/4/4 /V6 cyl OHV/OHC/OHC/diesel/OHC, F/F **Max power** 64bhp at 4500rpm to 142bhp at 5500rpm
Max speed 91-116mph
Body styles hatchback
Prod years 1975-84
Prod numbers 622,314/160,265

PRICES
A 1200 **B** 700 **C** 100

Riley 1½-Litre RMA/RME

Although now under Nuffield control, Riley's immediate post-war cars were true Rileys. Attractive timber-framed fabric-roofed sports saloon body, robust chassis, torsion bar IFS, semi-elliptics at the back, hydromech brakes, torque tube, plus of course the lovely 'hemi' 1.5-litre engine with its twin camshafts. Expect vice-free handling, firm ride, comfortable cruising, heavy steering. RMA replaced in '52 by the RME: hypoid axle, open propshaft, hydraulic brakes, bigger rear window. Wheel spats, new peaked front wings and no running boards from '53.

Engine 1496cc 69x100mm, 4 cyl OHV, F/R
Max power 54bhp at 4500rpm
Max speed 75mph
Body styles saloon
Prod years 1945-52/1952-54
Prod numbers 10,504/3446

PRICES
A 8000 **B** 5500 **C** 3500

Riley 2½-Litre RMB/RMC/RMD/RMF

Riley's 2.5-litre four had the longest stroke of any post-war British car and was a very flexible power unit. RMA chassis was lengthened by 6.5in and the bonnet was longer; the only other distinguishing mark was light, rather than dark, blue badge. Bigger inlet valves and 100bhp from '48. RMC was an export three-seater roadster (from '50 only two seats), RMD (photo) was four-seater four-light drophead. 1952 RMF replaced the RMB, boasting full hydraulic brakes, hypoid axle, better rear damping and a curved rear screen. All RM Rileys have rot-prone wood-framed bodies, and roofs can be porous.

Engine 2443cc 80.5x120mm, 4 cyl OHV, F/R **Max power** 90bhp at 4000rpm to 100bhp at 4500rpm
Max speed 85-100mph
Body styles saloon/sports/drophead coupé/saloon
Prod years 1946-52/1948-50/1948-51/1952-53
Prod numbers 6900/507/502/1050
PRICES
RMB/RMF **A** 9500 **B** 6000 **C** 3500
RMC **A** 16,000 **B** 11,000 **C** 8000
RMD **A** 14,000 **B** 9500 **C** 6500

Riley Pathfinder

The brave new era of Nuffield Rileys began here: the styling is drab Wolseley (soon to appear in the 6/90), but at least the old Riley 2.5-litre engine is retained. Suspension by torsion bars and wishbones at the front and torque arms, coil springs and Panhard rod at the rear. You got vacuum servo brakes, but rack-and-pinion steering was strangely discarded in favour of cam gear. Right-hand gear lever, plenty of leather and wood. Semi-automatic overdrive from '55, BMC C-type gearbox from the same time, very last ones had conventional semi-elliptic rear springs.

Engine 2443cc 80.5x120mm, 4 cyl OHV, F/R
Max power 110bhp at 4400rpm
Max speed 102mph
Body styles saloon
Prod years 1953-57
Prod numbers 5152

PRICES
A 3500 **B** 2500 **C** 1500

Riley 2.6

If the Pathfinder was unromantic, the 2.6 was even more so. BMC fitted the twin carb C-type six-cylinder engine, as in the near-identical Wolseley 6/90, but with a little extra power. Other differences from the Wolseley were the Riley grille, wider tyres, separate front seats and a rev counter. Spot a 2.6 from a Pathfinder by its bumper-mounted foglamps, separate side lights and indicators and larger rear window. As with the Pathfinder, both overdrive and automatic transmission were available as options. They rust in a big way, but at least there's a separate chassis to help restoration.

Engine 2639cc 79.4x88.9mm, 6 cyl OHV, F/R
Max power 102bhp at 4500rpm
Max speed 95mph
Body styles saloon
Prod years 1957-59
Prod numbers 2000

PRICES
A 3500 **B** 2500 **C** 1500

Riley 1.5

Mixed-parentage Riley, and blood-sister to the Wolseley 1500. Underneath, it's a Morris Minor floorpan and running gear but, unlike the Wolseley, its BMC B-series engine was given twin carbs to satisfy the sporting strain in the Riley driver's psyche (the engine was near-MGA spec). Its own unique front end with the traditional Riley chromework and wrap-around side grilles distinguished it externally. Modified combustion chambers and a new cam from '60, circular indicator/side lights from '61, new crank from '62. Notched up a good race record and is popular today in historic rallying.

Engine 1489cc 73x88.9mm, 4 cyl OHV, F/R
Max power 68bhp at 5400rpm
Max speed 85mph
Body styles saloon
Prod years 1957-65
Prod numbers 39,568

PRICES
A 3000 **B** 2000 **C** 1200

Riley 4/68 and 4/72

The ultimate in BMC Farinas (though that's not saying much). High performance thanks to MG Magnette spec twin carb engine, plus high level of trim (walnut, leather, etc). Unique Riley grille and a bonnet which follows the 'U' shape of its top edge. 4/72 introduced in September 1961 has the larger 1.6-litre engine, front anti-roll bar, rear stabilising bar, quieter exhaust, wider track and, for the first time, the option of automatic transmission. Duotone paint standard from October 1962, though most were finished that way before then. If you're into Farinas, this could be your dream car.

Engine 1489cc 73x88.9mm/1622cc 76.2x88.9mm, 4 cyl OHV, F/R
Max power 66.5bhp at 5200rpm/ 68bhp at 5000rpm
Max speed 91mph
Body styles saloon
Prod years 1959-61/1961-69
Prod numbers 10,940/14,151

PRICES
A 2500 **B** 1800 **C** 1000

Riley Elf

Issigonis would have nothing to do with this booted Mini – the marketing man's dream of a Mini with fins and chrome. Instead, Austin's style guru, Dick Burzi, did the design work. That meant off-the-shelf tradition: upright chrome grille (which lifted with the bonnet), wide side grilles and lashings of wood (the Elf was higher up in the hierarchy than the similar Wolseley Hornet, so it had a full-width veneer dash). Befinned boot was a little bigger than the Mini's. 1963 MkII has 998cc engine, better brakes. '66 MkIII has winding windows and concealed door hinges.

Engine 848cc 62.9x68.3mm/998cc 64.6x76.2mm, 4 cyl OHV, F/F
Max power 34bhp at 5500rpm/ 38bhp at 5250rpm
Max speed 73/77mph
Body styles saloon
Prod years 1961-69
Prod numbers 30,912

PRICES
A 1800 **B** 1200 **C** 700

Riley Kestrel/1300

BMC badge engineering reached its most pernicious height with the 1100/1300 range. Riley's version was basically an MG 1100 in its mechanical specification (twin carbs), with the inevitable grille forcing a unique bonnet. Walnut veneer facia, circular dials including a rev counter. 58bhp 1300 available from 1967, though 1100 MkII remains on sale for a few months more (reshaped rear lights, auto option). 1300 has all-synchro 'box, 65bhp twin carb engine from April '68, 70bhp from MkII of October '68 – known simply as Riley 1300 – which also had rocker switches and more chrome.

Engine 1098cc 64.6x83.7mm/1275cc 70.6x81.3mm, 4 cyl OHV, F/F
Max power 55bhp at 5500rpm to 70bhp at 6000rpm
Max speed 89-101mph
Body styles saloon
Prod years 1965-69
Prod numbers 21,529

PRICES
A 2500 **B** 1500 **C** 800

Rochdale Olympic

This glassfibre special body maker graduated to the more sophisticated Olympic in 1959. It was only the second true glassfibre monocoque road car ever made (after the Lotus Elite). Riley 1.5/Morris Minor running gear (but special coil spring and radius arm rear), choice of BMC A-series, B-series or Ford engines. Usually kit-built. Phase II of 1963 has rear hatch, bigger bonnet, Spitfire front suspension, front discs, Ford Anglia/Cortina GT engines. Its looks earned it the title 'British Porsche'. Strong club following.

Engine 948cc/997cc/1172cc/1489cc/ 1498cc/1558cc, 4 cyl OHV/OHV/SV/ OHV/OHV/OHV, F/R
Max power 37bhp at 5300rpm to 78bhp at 5200rpm
Max speed 80-114mph
Body styles coupé
Prod years 1959-68
Prod numbers c 400

PRICES
A 3500 **B** 1800 **C** 800

Rolls-Royce Silver Wraith

Initially for export only, Rolls-Royce re-entered production with the new Silver Wraith in 1946 – more accurately the chassis only, as all bodies were coachbuilt in aluminium. B60 straight-six inlet-over-exhaust-valve engine, initially 4.3 litres, from '51 4.6 litres. LWB chassis (133in versus 127in) from '51 on gains 4.9-litre engine from '54. Manual only until '52, when auto becomes optional. Character dependent on coachbuilder, though most are sedate chauffeur material. Mulliner's touring saloons are most numerous.

Engine 4257cc 89x114.3mm/4566cc 92x114.3mm/4887cc 95.25x114.3mm, 6 cyl IOE, F/R **Max power** 126bhp at 3750rpm/est 150bhp/est 175bhp
Max speed 85-94mph
Body styles various coachbuilt
Prod years 1946-59
Prod numbers 1783
PRICES
Saloon **A** 20,000-60,000 **B** 15,000-40,000 **C** 12,000-20,000
DHC **A** 45,000-70,000 **B** 30,000-50,000 **C** 22,500-35,000

Rolls-Royce Silver Dawn

A need was identified in export markets for a smaller owner-driver Rolls and so, in 1949, it adapted the Bentley MkVI to become the Silver Dawn. There are very few differences between the two: the Rolls has the flying lady, naturally, but also a less powerful engine and a different facia. Almost invariably it had the standard Pressed Steel body. The '52 Silver Dawn had the body of the Bentley R-Type and always the 4.6-litre engine. UK available from '53, by which time the Hydramatic auto 'box was available. Two wheelbase lengths: 127in and, from '51, 133in. Beware Bentley conversions.

Engine 4257cc 89x114mm/4566cc 92x114mm, 6 cyl IOE, F/R
Max power 126bhp at 3750rpm/est 150bhp
Max speed 90-95mph
Body styles saloon, some coachbuilt
Prod years 1949-55
Prod numbers 761

PRICES
Saloon **A** 24,000 **B** 16,000 **C** 11,000
DHC **A** 50,000 **B** 35,000 **C** 22,000

Rolls-Royce Phantom IV/V/VI

To very special order, the Phantom IV was launched in 1950. Its straight-eight B80 5.7-litre engine was widely used in military and commercial applications, so spares are not the problem they might be considering the production run was only 18, mostly for royalty with Hooper or Mulliner bodies. The Phantom V was a mass-produced bargain by comparison, had the same 145in wheelbase but four-speed auto transmission, PAS, dual-circuit brakes and a 6.2-litre V8. Phantom VI was equally gargantuan at 20ft long and nearly 3 tons, 6750cc from '78, drum brakes to the last. Strictly chauffeur-driven.

Engine 5675cc 88.9x114.3mm/6230cc 104.1x91.4mm/6750cc 104.1x99.1mm, 8/V8/V8 cyl IOE/OHV/HV, F/R
Max power 164bhp/est 200bhp/est 200bhp **Max speed** 90-105mph **Body styles** limousine, some coachbuilt
Prod years 1950-56/1959-68/1968-92
Prod numbers 18/793/373
PRICES
Phantom IV Too rare to value
Phantom V **A** 75,000 **B** 50,000 **C** 30,000. Phantom VI **A** 100,000 **B** 70,000 **C** 50,000

Rolls-Royce Silver Cloud I/II/III

The first Rolls-Royce to have its body manufactured by Rolls (as well as the chassis) was the 1955 Silver Cloud. Despite this, you could still buy a wide variety of coachbuilt bodies by James Young, Freestone & Webb, Mulliner, Hooper, etc. Only the grille sets it apart from the Bentley S-Type. 123in and 127in wheelbases, latter with Park Ward converted bodies. Six-port cylinder head and twin SUs, electrically-controlled dampers, standard auto. V8 from '59 Cloud II, when PAS became standard and a DHC was offered. Cloud III has double headlamps, better PAS and sorted V8.

Engine 4887cc 95.25x114.3mm/6230cc 104.1x91.4mm, 6/V8 cyl IOE/OHV, F/R **Max power** est 175bhp/est 200bhp **Max speed** 100-110mph **Body styles** saloon, limousine, drophead coupé
Prod years 1955-59/1959-62/1962-65
Prod numbers 2359/2716/2298
PRICES
Standard steel **A** 27,000 **B** 17,000 **C** 10,000 II Mulliner DHC **A** 100,000 **B** 75,000 **C** 55,000 III Mulliner DHC **A** 60,000 **B** 45,000 **C** 30,000

Rolls-Royce Silver Shadow I/II

Modern R-R with sharply styled unitary body, discs all round, standard auto and PAS, self-levelling suspension. Saloon standard, James Young two-door (straight-through waist-line) and Mulliner Park Ward two-door (kick in the rear wings) from launch, plus DHC from '67 – latter pair became Corniches in '71. Three-speed Hydra-matic from '68, LWB option with vinyl roof from '69. Bigger engine from '70, wider tyres and flared arches from '74. Bespoilered, rubber-bumpered Shadow II and LWB Silver Wraith from '77. Very durable, but avoid all-too-common cheap 'bangers' run by unsympathetic shoe-stringers.

Engine 6230cc 104.1x91.4mm/6750cc 104.1x99.1mm, V8 cyl OHV, F/R
Max power est 200bhp
Max speed 118/120mph
Body styles saloon, limousine, convertible **Prod years** 1965-77/1977-80 **Prod numbers** 19,493/10,566
PRICES
Shadow I **A** 12,500 **B** 9000 **C** 5500
MPW 2-door **A** 22,000 **B** 13,000 **C** 8000
Silver Shadow II **A** 15,000 **B** 10,000 **C** 7000

Rolls-Royce Corniche I

After Mulliner Park Ward (a part of Rolls-Royce by now) had built 504 convertibles and 571 Silver Shadow drophead coupés, they were effectively regularised on Rolls-Royce price lists in 1971 under the name Corniche. Apart from a deeper radiator grille and new facia, there were precious few changes. Mechanical history as Silver Shadow, except for ventilated front discs from '72, cruise control and tape player from '74, split-level air con from '76. Replaced by Corniche II in '77, which lasted until '94. Built to the most exacting standards, and it naturally shows.

Engine 6750cc 104.1x99.1mm, V8 cyl OHV, F/R
Max power est 200bhp
Max speed 120mph
Body styles saloon, convertible
Prod years 1971-77
Prod numbers 2013

PRICES
Saloon **A** 25,000 **B** 15,000 **C** 10,000
Convertible **A** 32,000 **B** 22,000 **C** 15,000

Rolls-Royce Camargue

This was the most expensive car listed in Britain when it was launched in 1975, equating to an averagely large house. Pininfarina did the styling, but most thought it a poor metamorphosis of the Fiat 130 Coupé. MPW built it on a Silver Shadow floorpan and you naturally had standard power steering, split-level air con and automatic transmission. More power thanks to four-barrel carbs, but no-one at Rolls told you how many horses there actually were. Camargues hit a value trough in the early 1990s, but are starting to rise now – a sure-fire future curiosity. Running costs should not be underestimated.

Engine 6750cc 104.1x99.1mm, V8 cyl OHV, F/R
Max power est 220bhp
Max speed 120mph
Body styles coupé
Prod years 1975-85
Prod numbers 531

PRICES
A 40,000 **B** 25,000 **C** 17,000

ROLLS-ROYCE & BENTLEY

Ghost Phantom Wraith Shadow Spirit....

...Visions of excellence brought to light

GARAGES

Since the company was founded in 1984, we at RR&B Garages have always believed that nothing but the best standard of service and craftsmanship was good enough for the cars in our care.

Coachwork, routine maintenance, repairs and manufacturing are all carried out to an uncompromising standard which enables us to guarantee all our workmanship.

Our dedicated and proficient staff offer an unrivalled service for pre-war to current models of Rolls-Royce and Bentley motor cars.

Our full colour company brochure is available upon request.

Shaw Lane Industrial Estate, Shaw Lane, Stoke Prior, Bromsgrove, Worcestershire, B60 4DT.
Telephone 01527 876 513, Facsimile 01527 876 513
·Routine Maintenance ·Coachwork ·Restoration ·Pre-Purchase Engineers Reports·

ROLLS-ROYCE & BENTLEY

Prescote Motor Carriages

RUN BY ENTHUSIASTS FOR ENTHUSIASTS

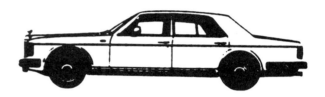

Rolls-Royce and Bentley Specialist

Restoration and servicing of engines, gearboxes, and all mechanical, electrical, chassis and coachwork.

Spare parts supplied and obsolete parts manufactured.

Veteran, vintage and classic cars restored.

Tel: (01703) 666682
Fax: (01703) 666882
out of hours:
Tel: (01590) 623352

Please contact Lawrie Smith or Peter Males to take care of your special car.

Mill House, Mill Road, Totton, Hampshire SO4 3ZQ

BEST CLASSIC

PETER BEST INSURANCE SERVICES LTD
Established 1985
INSURANCE BROKERS

Quality Collectors Car Insurance at Highly Competitive Premiums

© **THE Beaulieu POLICY**

Over 10 Years of Personal Service to the Classic Car Enthusiast

Endorsed by Beaulieu, home of the National Motor Museum.

Underwritten by

NORWICH UNION

Agreed Value, Limited Mileage for most collectable cars over 10 years old.
01621 840400 or Fax 01621 841401
The Farriers, Bull Lane, Maldon Essex CM9 4QB

Rover

Rover Ten/Twelve/Fourteen/Sixteen

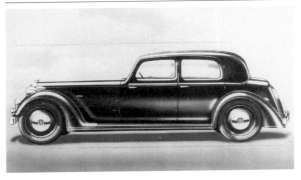

Immediately post-war, Rover's range of models was identical to those in production in 1939, except for disc wheels. The base model was the upright Ten, on a 105.5in (268cm) wheelbase. Underslung chassis, leaf springs all round, mechanical brakes, remote-control floor gear change with freewheel. Twelve had 112in (284cm) wheelbase, 1.5-litre engine and could be had with two-door tourer bodywork (some with Ten engines). Fourteen/Sixteen have 3in longer wheelbase, straight six engines. The three larger models could also be ordered with four-light Sports Saloon bodies. Well-built, stylish, archaic.

Engine 1389cc 66.5x100mm/1496cc 69x100mm/1901cc 63.5x100mm/ 2147cc 67.5x100mm, 4/4/6/6 cyl OHV, F/R
Max power 39bhp/48bhp/n/a/72bhp
Max speed 67-76mph
Body styles saloon, tourer
Prod years 1945-48
Prod numbers 2640/4840/1705/4150

PRICES
10/12 **A** 4200 **B** 2700 **C** 1500
12 DHC **A** 10,000 **B** 7500 **C** 5000
14/16 **A** 5000 **B** 3500 **C** 2200

Rover 60/75 (P3)

You'd be hard pushed to spot the differences between the new P3 and the old Twelve/Fourteen/Sixteen, but the wheelbase is shorter at 110.5in (281cm), overall length is less, and width greater. On the all-new chassis you will find independent front suspension by coil springs and wishbones, rear springs anchored directly to the body, hydromech brakes, new sloping-head four- and six-cylinder engines with side exhaust valves, synchromesh on third and top. Six-light and four-light sports saloon bodies, as before. Ultra-conservative in style and feel, and still rather insipid performers.

Engine 1595cc 69.5x105mm/2103cc 65.2x105mm, 4/6 cyl IOE, F/R
Max power 51bhp at 4000rpm/ 72bhp at 4000rpm
Max speed 70-75mph
Body styles saloon
Prod years 1948-49
Prod numbers 1274/7837

PRICES
A 5000 **B** 3500 **C** 2200

Rover 75/90/95/100/105/110 (P4)

The 1949 P4 was a revolution for Rover: bold, full-width, but still upright styling, featuring that curious 'Cyclops' central headlamp (which lasted till '52). Same P3 six engine on first 75 (expanded to 2.2 litres in '55), column shift (on the floor from '53), hydraulic brakes for '51. Wrap-around rear window from '54, by which time the 2.6-litre 90 was available. Non-freewheel overdrive option from '55. Twin carb 105 from '56 (S spec equals overdrive, servo brakes; R is iffy automatic – worth 20% less). New seven-bearing 2.6 engine in last 100, 95 and 110. Comfortable, durable but definitely not rust-free.

Engine 2103cc/2230cc/2638cc/ 2625cc, 6 cyl IOEV, F/R **Max power** 75: 75-80bhp/90: 90-93bhp/95: 102bhp/100: 104bhp/105: 108bhp/110: 123bhp **Max speed** 75-104mph **Body styles** saloon **Prod years** 1949-59/ 1953-59/1962-64/1960-62/1956-59/ 1962-64 **Prod numbers** 43,241/ 35,903/3680/16,521/10,781/4620

PRICES
75 'Cyclops' **A** 8000 **B** 6000 **C** 4000
75 '52-'54 **A** 6000 **B** 4500 **C** 2500
Other P4 **A** 5000 **B** 3500 **C** 1600

Rover 60/80 (P4)

While the six-cylinder P4s were the definitive items, Rover launched a poverty model in 1953 in the form of the 60. As such, it never had the Cyclops eye headlamp and always came with the ridged front wings common to post-'52 P4s. Four-cylinder engine derived from the Land-Rover. As a package the 60 was not very satisfactory. The later 80 is a better bet, though rarer. Engine source is still Land-Rover, and the 2.3-litre engine is the only OHV unit to go in a P4. Perkier and smoother than the 60. Unloved, so not very common. Check for rust in sills, chassis and body mounts.

Engine 1997cc 77.8x105mm/2286cc 90.5x88.9mm, 4 cyl IOE/OHV, F/R
Max power 60bhp at 4000rpm/77bhp at 4250rpm
Max speed 79/87mph
Body styles saloon
Prod years 1953-59/1959-62
Prod numbers 9666/5900

PRICES
60 **A** 5000 **B** 3800 **C** 1700
80 **A** 3500 **B** 2200 **C** 1000

Rover 3-Litre (P5)

Extremely successful in design terms, Rover's P5 was its first unitary effort, though there was still a separate front subframe. Traditional yet stylish, it was a favourite of government. Torsion bar IFS, semi-elliptic rear, optional auto or overdrive (standard from '60), servo front discs on all but earliest cars, optional PAS from '60. 1962 MkII has power boost, lowered suspension, close-ratio 'box. Coupé joins range at this time, though it's merely a saloon with a lower roof. 1965 MkIII has standard PAS, full-length steel side strip. Very durable but dreadful rust-traps.

Engine 2995cc 77.8x105mm, 6 cyl IOE, F/R
Max power 115bhp at 4250rpm to 134bhp at 5000rpm
Max speed 98-110mph
Body styles saloon, coupé
Prod years 1959-67
Prod numbers 48,541

PRICES
Saloon **A** 3500 **B** 2300 **C** 1300
Coupé **A** 4500 **B** 2600 **C** 1700

Rover 3.5-Litre (P5B)

Rover pulled off quite a coup when it secured the rights to make the light-alloy 3.5-litre V8 engine designed by Buick in 1961, but hardly used by them. In the P5 hull, its 161bhp permitted a 0-60mph time of 12.4 secs, which was excellent considering it was automatic only. But the V8's real strength was its flexibility and smoothness. Saloon and coupé styles again. Identify by Rostyle wheels, foglamps, twin exhausts, rubber-faced overriders and insignia. Used by British PMs into the 1980s. Check inner/outer wings, sills and floorpan for sadly rampant rust.

Engine 3528cc 88.9x71.1mm, V8 cyl OHV, F/R
Max power 161bhp at 5200rpm
Max speed 112mph
Body styles saloon, coupé
Prod years 1967-73
Prod numbers 20,600

PRICES
Saloon **A** 4500 **B** 3000 **C** 1600
Coupé **A** 5500 **B** 3500 **C** 2000

Rover 2000/2200 (P6)

Rover's P4-style 'auntie' image was well and truly swept aside by the P6 2000 in 1963. Winner of the inaugural Car of the Year Award, it was smoothly styled with double headlamps, had a new square five-bearing overhead cam engine, servo discs all round, all-synchro four-speed 'box, De Dion rear end. Excellent handling and decent performance. Automatic from October '66, as well as new TC (Twin Carb) with its 114-124bhp and rev counter. Plastic grille, twin bonnet bulges from '70. Bored-out 2200 replaces 2000 in '73, runs alongside SD1 1976-77. Mechanically sound, but hull is rust-prone.

Engine 1978cc 85.7x85.7mm/2205cc 90.5x85.7mm, 4 cyl OHC, F/R
Max power 90bhp at 5000rpm to 124bhp at 5500rpm/98bhp at 5000rpm to 115bhp at 5000rpm
Max speed 94-115mph
Body styles saloon
Prod years 1963-73/1973-77
Prod numbers 327,808/32,270

PRICES
SC **A** 2000 **B** 1200 **C** 700
TC **A** 2500 **B** 1500 **C** 800

Rover 3500 (P6)

Planting the ex-Buick alloy V8 in the 2000 produced a cracker of a car: fast, refined and flexible, but thirsty (no more than 20mpg). Standard auto 'box. Recognisable by extra grille below front bumper. 3500S is manual version launched in '71, but gearbox struggles to cope; recognise the S by its vinyl roof and, for the first two years, brushed steel spoked wheel trims. Auto 'box updated to 65BW unit in '73. Was the first car to be offered with Denovo run-flat tyres – but the idea itself was flat. A very cheap classic for what it is, and parts are no problem.

Engine 3528cc 88.9x71.1mm, V8 cyl OHV, F/R
Max power 140-161bhp at 5200rpm
Max speed 117mph
Body styles saloon
Prod years 1968-77
Prod numbers 79,057

PRICES
3500 **A** 3000 **B** 1700 **C** 1000
3500S **A** 3500 **B** 2000 **C** 1200

Rover 3500 (SD1)

Tradition went completely out of the window with the SD1, a BL/David Bache design with hardly a smidgen of chrome and no wood in sight. Car of the Year in 1977, it was built at a new plant at Solihull (from '82 at Cowley). V8 engine was the same, but there was a new five-speed 'box and PAS. Live rear axle and drum rear brakes seemed a retrograde step, though MacPherson struts at front made handling sharp. Novel interior treatment (instruments in a 'box') and practical hatchback. Suspect build quality at first gained it a reputation which was impossible to shake. 190bhp Vitesse from '82 is the interesting one (3897 made).

Engine 3528cc 88.9x71.1mm, V8 cyl OHV, F/R
Max power 155bhp at 5250rpm to 190bhp at 5280rpm
Max speed 125-135mph
Body styles hatchback
Prod years 1976-86
Prod numbers 113,966

PRICES
3500 **A** 1700 **B** 900 **C** 400
Vitesse **A** 2700 **B** 1400 **C** 700

Land-Rover SI

Rover's answer to the Jeep was the Land-Rover. It arrived in 1948 on an 80in wheelbase, with a steel box-section chassis and alloy body panels, 1.6-litre engine, and four-speed transfer box transmission. Tilt, pick-up, estate and innumerable special bodies. Headlamps moved from behind the grille in '51, when a 2-litre engine became standard. 86in and 107in wheelbases from '53, both stretched by 2in in '56. Four-door 10-seater from '55, diesel option from '57. Early SIs are becoming sought-after. Although collectors may prefer WW2 Jeeps, the Land-Rover *is* king.

Engine 1595cc 69.5x105mm/1997cc 77.8x105mm/2052cc 86x88.9mm, 4 cyl IOE/IOE/diesel, F/4x4
Max power 50-55bhp at 4000rpm/52-58bhp at 4000rpm/51bhp at 3500rpm
Max speed 60mph
Body styles utility, estate
Prod years 1948-58
Prod numbers 174,763

PRICES
'48-'51 **A** 5000 **B** 3500 **C** 1500
'51-'58 **A** 3500 **B** 2500 **C** 1000

Land-Rover SII/SIIA/SIII

The big news for the SII Landie was the fitment of a bigger 2¼-litre engine (apart from the very earliest examples, which retained the 2-litre). Track grew wider, the styling – such as it was – became less angular and the fuel filler moved from under the driver's seat to the offside of the body. Otherwise, same rock-hard leaf springs, unbreakable (but rust-prone) chassis, and alloy bodies. Diesel grew to 2.3 litres as well in '61, with the arrival of the SIIA. 2.6-litre 'six' option from '67, headlamps on wings from '69. SIII from '71 has all-synchro 'box, new grille, safety facia. V8 power option from '80.

Engine 2052cc/2286cc/2625cc/3528cc, 4/4/6/V8 cyl diesel & OHV/diesel & OHV/OHV/OHV, F/4x4
Max power 51-55bhp/62-77bhp/85bhp/91bhp
Max speed 60-91mph
Body styles utility, estate
Prod years 1958-61/1961-71/1971-85
Prod numbers SII/SIIA: c 450,000/SIII: c 805,000
PRICES
SII/SIIA **A** 3000 **B** 2000 **C** 800
SIII **A** 4000 **B** 2500 **C** 800

Range Rover

Stole a march on everyone: brilliant product planning exercise, mating Rover luxury with Land-Rover off-road ability. Always V8 power (or 2.4-litre turbodiesel 'four' from '86), permanent 4WD, four-wheel discs, coil springs all round, self-levelling rear suspension, separate chassis, smart alloy body panels, overdrive option. Better than anything off-road, yet impressively civilised and fairly rapid. Four-door option from '81, five speeds '83. Daunting running costs, though working on them is fairly simple and spares are everywhere and almost as cheap as Land-Rover parts.

Engine 3528cc 88.9x71.1mm/3947cc 94x71.1mm/4278cc 94x77mm/2393cc 92x90mm/2494cc 90.5x97mm, V8/V8/V8/4 cyl OHV/OHV/OHV/diesel, F/4x4
Max power 112bhp at 4200rpm to 200bhp at 4850rpm
Max speed 92-112mph
Body styles estate
Prod years 1970-95
Prod numbers c 310,000
PRICES
'70-'81 **A** 4000 **B** 2500 **C** 1100

Saab 92

Aircraft maker Saab's first car was conceived by Gunnar Ljungstrom and styled by Sixten Sason: an aerodynamic, lightweight, unitary saloon of extraordinary appearance. Transverse two-stroke twin drove the front wheels through a three-speed-and-freewheel 'box, and there was all-independent suspension and rack-and-pinion steering. Split windscreen and small rear screen identify it, though '53 92B has larger rear window (and a boot lid). Extra 3bhp from '54. Very rare outside Scandinavia. Prototype is pictured.

Engine 764cc 80x76mm, 2 cyl TS, F/F
Max power 25bhp at 3800rpm to 28bhp at 4000rpm
Max speed 62mph
Body styles saloon
Prod years 1950-56
Prod numbers 20,128

PRICES
A 6500 **B** 4000 **C** 2500

Saab 93/GT750

Though the basic shape of the Saab changed very little, the 93 was recognisable by its new nose and tall central grille. Three-cylinder two-stroke engine now mounted longitudinally (but still boasting hemispherical combustion chambers), plus a new gearbox. 92's oversteer cured by coil springs instead of torsion bars and wider track, and rear axle became rigid. One-piece windscreen for 93B of '58, when GT750 launched to exploit the Saab's popularity in rallying: 45bhp (or 55bhp with twin carbs), better seats, rev counter and twin chrome stripes. 1959 93F has forward-hinging doors.

Engine 748cc 66x72.9mm, 3 cyl TS, F/F
Max power 33bhp at 4200rpm to 55bhp at 5000rpm
Max speed 68-100mph
Body styles saloon
Prod years 1956-60/1958-62
Prod numbers 52,731

PRICES
93 **A** 6000 **B** 4000 **C** 2500
GT750 **A** 7500 **B** 5000 **C** 3000

Saab 95

The 95 estate was launched in June 1959 – before the 96 saloon – boasting an expanded 841cc two-stroke engine and four speeds. Estate rear body necessitated mini-tailfins, and there was a two-piece tailgate. A small extra bench seat could be raised from the floor so that seven people could be carried. Suicide doors for first six months only. Extended front from '64, triple carbs from '66, Ford V4 option from '67 (1.7 for USA) and no two-strokes after '68. Larger grille and revised tail-lights from '69, impact bumpers from '75. 95s have their own appreciative following. They'll last and last.

Engine 841cc 70x72.9mm/1498cc 89x58mm/1698cc 90x66.8mm, 3/V4/V4 cyl TS/OHV/OHV, F/F **Max power** 38bhp at 4250rpm to 42bhp at 5000rpm/65bhp at 4700rpm/65bhp at 5000rpm **Max speed** 79-90mph
Body styles estate **Prod years** 1959-78 **Prod numbers** c 110,500

PRICES
'59-'64 **A** 4000 **B** 2500 **C** 1200
'64-'68 **A** 3500 **B** 2200 **C** 1100
V4 **A** 3500 **B** 2000 **C** 1000

Saab 96/GT850

The definitive Saab arrived in 1960, sharing its mechanical spec with the 95, except for a three-speed 'box on two-stroke cars (although four speeds was optional). Larger wrap-around rear screen, bigger boot, revised rear lights. GT850 arrives '62 (named Sport in UK) and has triple carb two-stroke engine and front discs. German Ford V4 engine from '67, new grille and rectangular headlamps in '69. 1698cc V4 is US-only emissions engine from '71. Rally-derived gearbox casing from '75. UK imports end in '76 but production goes on to 1980, the last 300 finished in metallic blue.

Engine 841cc 70x72.9mm/1498cc 89x58mm/1698cc 90x66.8mm, 3/V4/V4 cyl TS/OHV/OHV, F/F
Max power 38bhp at 4250rpm to 55bhp at 5000rpm/65bhp at 4700rpm/65bhp at 5000rpm **Max speed** 79-90mph **Body styles** saloon
Prod years 1960-80/1962-66
Prod numbers 547,221 (incl 95)
PRICES
96 '60-'68 **A** 5000 **B** 2800 **C** 1500
GT850/Sport **A** 5500 **B** 3000 **C** 1600
96 '67-'80 **A** 3500 **B** 2000 **C** 800

Saab Sonett II/III

In the mid-1950s Rolf Mellde styled a sports car for Saab which never passed the prototype stage. Several years later, Mellde showed Saab sketches for a Sonett II and that, with design work by Bjorn Karlstrom, went into production. GT850 floorpan and mechanicals (including a four-speed transaxle), aerodynamic glassfibre body, non-opening wrap-around rear window. V4 engine after only 258 two-strokes built. US dealers requested a redesign, so Coggiola restyled the front and rear ends to make the Sonett III: pop-up headlamps, flush rear hatchback screen, revised dash, floor-mounted gear lever. LHD only.

Engine 841cc 70x72.9mm/1498cc 89x58mm/1698cc 90x66.8mm, 3/V4/V4 cyl TS/OHV/OHV, F/F
Max power 55bhp at 5000rpm/65bhp at 4700rpm/65bhp at 5000rpm
Max speed 96-100mph
Body styles coupé
Prod years 1966-70/1970-74
Prod numbers 1768/8368

PRICES
II 841cc **A** 9000 **B** 6000 **C** 4000
II/III V4 **A** 6500 **B** 4500 **C** 2700

Saab 99/90

Saab grows up: distinctively-styled bigger saloon with the same basic layout as the 96 (front-drive, front double wishbones, coil sprung rigid rear axle). Engine range is all-new, developed jointly by Saab and Triumph and initially manufactured by the British firm. Two doors only to start, four from '70, Combi fastback hatch from '74. Optional fuel injection and automatic from '69 (though not for UK), 1854cc engine from '71, 1985cc in '72 (fuel-injected option from '73, standard from '75). Renamed 90 in '84. Turbo too young for inclusion. Great handling, solid character and quality engineering.

Engine 1709cc 83.5x78mm/1854cc 87x78mm/1985cc 90x78mm, 4 cyl OHC, F/F
Max power 80bhp at 5300rpm to 115bhp at 5500rpm
Max speed 92-110mph
Body styles saloon, hatchback
Prod years 1967-84/1984-87
Prod numbers 588,643/25,378

PRICES
A 1600 **B** 1000 **C** 500

Salmson S4

The French luxury car maker had introduced its new S4 model just before the war as a thoroughly modern and attractively styled saloon and drophead coupé. It returned in 1946 with headlamps mounted in the front wings, but everything else was the same: transverse leaf IFS, cantilever rear springs, servo brakes, four speeds and a choice of 14HP or 18HP twin cam engines. The larger version had a 6in longer wheelbase. Hydraulic brakes from 1950. A reasonable seller for the few years it was on offer.

Engine 1730cc 75x98mm/2320cc 84x105mm, 4 cyl DOHC, F/R
Max power 50bhp at 4200rpm/ 70bhp at 3700rpm
Max speed 75/85mph
Body styles saloon, drophead coupé
Prod years 1946-52
Prod numbers 2983

PRICES
Saloon **A** 10,000 **B** 7000 **C** 3500
DHC **A** 20,000 **B** 12,500 **C** 6000

Salmson Randonnée

In 1951, Salmson showed a fresh design called the Randonnée ('Touring'). The aluminium bodywork was both handsome and classic, and was constructed by Tolerie. The bore of the 18HP engine was reduced and the entire engine (still a twin cam) was now made of light alloy. As with other Salmsons, Cotal electromagnetic four-speed pre-selector transmission was standard. This was a large car (180in/457cm long), so its performance was more 'touring' than 'sports'. Like most other big French cars, it struggled against monumental government tariffs, and few were made.

Engine 2248cc 82x105mm, 4 cyl DOHC, F/R
Max power 71bhp at 4000rpm
Max speed 90mph
Body styles saloon, drophead coupé
Prod years 1951-52
Prod numbers 545

PRICES
Saloon **A** 10,000 **B** 7000 **C** 3500
DHC **A** 20,000 **B** 12,000 **C** 6000

Salmson 2300 Sport

Salmson's swansong model was also its most attractive one. Presented at the 1953 Paris Salon and billed as 'the first French sports car', it was a pretty shark-nosed 2+2 coupé using a tuned (twin choke) version of the Randonnée engine, plus either Cotal electric or conventional ZF manual transmission. There was also a two-seater convertible and, very briefly, a four-door saloon (only three built). Some racing success was scored, but was never convincing. Salmson was already in trouble and Renault bought the company in 1956, killing it off within a year.

Engine 2248cc 82x105mm, 4 cyl DOHC, F/R
Max power 105bhp at 5000rpm
Max speed 110mph
Body styles coupé, convertible, saloon
Prod years 1953-57
Prod numbers 209

PRICES
Coupé **A** 15,000 **B** 11,000 **C** 6000
Convertible **A** 25,000 **B** 17,000 **C** 10,000

Seat 600/800

The Spanish branch of Fiat opened its factory gates in 1953. In 1957, it too began production of the 600 and, after 1969, it was the sole manufacturing source for export 600s. It made the 600E (basic) and 600L (luxury) from 1970. One interesting departure was the Seat 800, essentially a 600D stretched by 7in and given an extra pair of front-hinged doors. This put the weight up, but the drive train was untouched, so its was quite a sluggard. Seat also made a four-door 850, plus the 850/850 Coupé well after Italian production ceased.

Engine 767cc 62x63.5mm, 4 cyl OHV, R/R
Max power 29bhp at 4800rpm
Max speed 64-68mph
Body styles saloon
Prod years 1957-73/1967-73
Prod numbers n/a

PRICES
A 2000 **B** 1200 **C** 700

Seat 133

Seat made several modified Fiats, like four-door 127s, but the 133 was its own product. Underneath the shrunken-127 style body lay the floorpan and running gear of a Fiat 850, and even the 850 facia. That meant a rear-mounted 843cc engine, drum brakes, and cramped cabin. Rather feeble performance and desperately unrefined compared to mid-1970s superminis, but at least you got 40mpg. Fiat sold the 133 in the UK with 'Fiat costruzione Seat' badges from 1975, but it lasted here less than a year and was a total flop. Undeservedly popular, though, in import-phobic Spain. Zero-rated classic credentials.

Engine 843cc 65x63.5mm, 4 cyl OHV, R/R
Max power 34bhp at 4800rpm to 44bhp at 6400rpm
Max speed 75-84mph
Body styles saloon
Prod years 1974-80
Prod numbers n/a

PRICES
A 600 **B** 300 **C** 100

Seat 1200 Sport/1430 Sport

The only truly individual Seat of the 1970s was the 1200 Sport. This car grew out of the Barcelona factory's refusal to build the 128 3P under licence: it wanted something that would appeal to *Spanish* tastes. Floorpan is Fiat 127, but the engine is the old 1.2-litre 124, which was still being made in Spain (later a bored out '1430' unit). 2+2 coupé body is very compact and pretty modern – the matt-black polymer bumpers were ahead of their time. Fairly rapid (0-60mph in 13 secs) and excellent handling, but underdeveloped compared to the 3P. Never imported to the UK, so LHD only. Pick one up in Torremolinos.

Engine 1197cc 73x71.5mm/1438cc 80x71.5mm, 4 cyl OHV, F/R
Max power 67bhp at 5600rpm/ 77bhp at 5600rpm
Max speed 100mph
Body styles coupé
Prod years 1975-80
Prod numbers n/a

PRICES
A 2000 **B** 1200 **C** 600

Siata Daina/Rallye

Siata started building special bodies on Fiats as early as 1926. Post-war, it made Amica bodies for the Topolino, but the sporting Daina of 1950 was based on the Fiat 1400. The engine was tuned and fitted with twin Webers, or there were optional 1.5-litre (Sport) and 1.8-litre (GT) engines. Siata would also add a fifth gear on request. Bodywork was by Farina (photo) or Bertone. The Rallye was another 1400-based project, having MG TD inspired touring bodywork. This particular model appears to be very rare.

Engine 1395cc 82x66mm/1479cc 82x70mm/1795cc 84x82mm, 4 cyl OHV, F/R
Max power 65bhp at 5000rpm/70bhp at 4500rpm/80bhp at 4500rpm
Max speed 90-105mph
Body styles coupé, convertible
Prod years 1950-58/1951-58
Prod numbers n/a

PRICES
A 12,000 **B** 8000 **C** 3500

Siata 208S

As a close ally of Fiat, Siata was keener than anyone to get hold of Fiat's new V8 engine, as seen in the 8V sports coupé. It made the 208S, using the Fiat V8 power unit as its basis. Coachwork was by Vignale in coupé or convertible forms. The 208S featured pop-up headlamps and low-set foglamps in the vertically-slatted grille. Gearbox was invariably a five-speeder. Some cars destined for America were fitted with tuned Chrysler or Cadillac V8 engines, and were sold with the suffix CS. If you find one, it won't be cheap.

Engine 1997cc 72x61.3mm/5426cc 97x92mm, V8 cyl OHV, F/R
Max power 105-125bhp at 6500rpm/300bhp at 5000rpm
Max speed 112-140mph
Body styles coupé, convertible
Prod years 1952-54
Prod numbers 56

PRICES
Rarely offered for sale, though good ones should fetch a six-figure sum

Siata Amica

The first Amica appeared in 1949, and was based on the 500 Topolino, with an optional 750cc engine or even American Crosley power. A new Amica arrived in 1955, based on the Fiat 600 platform and running gear. There was one basic body style (a two-seater offered in open and coupé forms), but styling details changed on a regular basis. Siata quoted 59mph and 47mpg from the standard 21bhp 600 engine; tuning packages from outside firms were available. Exact production figures are unknown, but quite a few were made. Siata also made a 750 Coupé and various bodies on the Nuova 500.

Engine 633cc 60x56mm, 4 cyl OHV, R/R
Max power 21bhp at 4600rpm
Max speed 59mph
Body styles coupé, convertible
Prod years 1955-60
Prod numbers n/a

PRICES
A 8000 **B** 5000 **C** 3000

Siata Spring/Orsa Spring

Siata went out with a whimper, not a bang, as the Spring was a pale pastiche of the MG TD based on a Fiat 850S floorpan. The 2+2 convertible even had a 'traditional' chrome grille, despite the fact that the engine was in the boot! The spare wheel was mounted on the tail, there were 'bug eye' headlamps and an optional hardtop. Sadly this was probably Siata's best-selling car: it is believed that several hundred of these joke cars were sold in America before Siata went bust in 1970. A firm called Orsa revived it in 1971 and made a handful more. A bit too much like a kit car for comfort.

Engine 843cc 65x63.5mm, 4 cyl OHV, R/R
Max power 37bhp at 5200rpm to 47bhp at 6400rpm
Max speed 80-87mph
Body styles convertible
Prod years 1967-70/1971-75
Prod numbers n/a

PRICES
A 5000 **B** 3500 **C** 2000

Simca 5/6

Simca started building Fiats under licence from 1934 and began constructing the 5, a French-made 500B Topolino, in 1936. Unusually, production continued in a small way throughout the war and it was made up until 1949. In 1948 came the 6, essentially a licence-built Fiat 500C (the 6 is pictured). In all respects the Simcas duplicated the Topolino, except for their badging. They were produced in rather smaller numbers than in Turin, but are not worth any more than the equivalent Topolino. The model was axed in readiness for Simca's own Aronde.

Engine 569cc 52x67mm, 4 cyl OHV, F/R
Max power 13bhp at 4000rpm to 16.5bhp at 4400rpm
Max speed 52-60mph
Body styles coupé, convertible, estate
Prod years 1936-49/1948-50
Prod numbers 58,694/16,512

PRICES
5 **A** 5000 **B** 2500 **C** 1700
6 **A** 5000 **B** 2000 **C** 1500

Simca 8

The other pre-war licence-built Fiat was the 8, a local version of the 508C Millecento (1100). Production began in 1937 and continued virtually unchanged after 1945, although again a handful had been made during the war years. Like the Fiat, it was a pillarless 'clap-hands' four-door saloon, but there were also Découvrable (open), Coupé, Break (estate) and Fourgonette/Camionette commercial models. A 1200 engine replaced the old 1100 in 1949, which also powered the first Arondes. 8 production overlapped slightly with the new Aronde, but the latter's success drew a veil over Simca's links with Fiat.

Engine 1089cc 68x75mm/1221cc 72x75mm, 4 cyl OHV, F/R
Max power 32-35bhp at 4400rpm/40bhp at 4000rpn
Max speed 62-75mph
Body styles saloon, estate, coupé, convertible
Prod years 1937-52
Prod numbers 89,457
PRICES
Saloon **A** 4000 **B** 2500 **C** 1500
Coupé/Convertible **A** 6000 **B** 4000 **C** 2200

Simca

Simca 8/9 Sport Cabriolet/Coupé

Although it had made a coupé version of the 8, the first true car of Simca's own design was the 8 Sport Cabriolet, first shown in 1948 but not produced until the following year. Pinin Farina did the styling and Facel Metallon (who would later make the Facel Vega) built it. Underneath it lay the mechanicals of the 8, featuring a mildly tuned 1200 engine. Thus it was something of a sheep in wolf's clothing. A coupé arrived in 1950 and the 9 had the Aronde's coil spring/wishbone front suspension, cleaner styling and a new grille. There were class wins in the Alpine and Monte Carlo rallies.

Engine 1221cc 72x75mm, 4 cyl OHV, F/R
Max power 50bhp at 4800rpm
Max speed 84mph
Body styles convertible, coupé
Prod years 1949-52/1950-54
Prod numbers 4822

PRICES
Coupé **A** 12,000 **B** 8000 **C** 4000
Cabriolet **A** 18,000 **B** 12,000 **C** 8000

Simca Aronde

Amazingly successful car which established Simca as one of France's leading independent makes. Despite the fact that the same Fiat 1.2-litre engine lay under the bonnet, the Aronde ('Swallow') looked modern. Better handling than the Fiat thanks to coil spring/wishbone/anti-roll bar front suspension. Four-speed 'box, column shift, Simcamatic clutch option, hydraulic brakes, worm-and-roller steering. Chatelaine was the estate, Grand Large hardtop coupé for '52, 1.3-litre 'Flash' engine from '55, big valve engine from '57. Restyled for '56 model year with hooded headlamps and rear fins.

Engine 1221cc 72x75mm/1290cc 74x75mm, 4 cyl OHV, F/R
Max power 45bhp at 4500rpm to 57bhp at 5200rpm
Max speed 75-90mph
Body styles saloon, estate, coupé
Prod years 1951-60
Prod numbers 1,014,355

PRICES
Saloon/Estate **A** 2000 **B** 1200 **C** 700
Coupé **A** 3000 **B** 1800 **C** 1000

Simca Coupé de Ville/Week-End

On the basis of the Aronde, Simca launched a low-bodied coupé with the grand name of Coupé de Ville. This was effectively the replacement for the 9 Sport and had slim pillars, a rather American chrome grille and the mechanicals of the Aronde 1300, the 'Flash' engine being tuned to produce 55-60bhp. The convertible version was launched late in the day, in 1956, and was named Week-End. These models are little known in Britain. Most were exported to America and, if you want one, that's the place to start looking – or go for the much more common Plein Ciel/Océane (below).

Engine 1290cc 74x75mm, 4 cyl OHV, F/R
Max power 55-60bhp at 5200rpm
Max speed 87mph
Body styles coupé/convertible
Prod years 1954-57/1956-57
Prod numbers 3136/678

PRICES
Coupé de Ville **A** 5000 **B** 3500 **C** 2000
Week-End **A** 8000 **B** 5000 **C** 3000

Simca Océane/Plein Ciel

The previous Coupé/Week-End models were much restyled by Simca in 1957, creating the Océane (convertible) and Plein Ciel (coupé). There were now wrap-around screens front and rear, a conventional mesh grille, recessed front indicators, hooded headlamps and a sharper tail. Whether these changes actually marked an improvement is debatable. Still an Aronde underneath, though with the Flash Special engine (last ones have the 70bhp Rush Super unit). Cowhorn front bumper and extra chrome from '58, stripped-out models from '59, five-bearing crank from '60. Some RHD examples still exist.

Engine 1290cc 74x75mm, 4 cyl OHV, F/R
Max power 57-70bhp at 5200rpm
Max speed 87-93mph
Body styles convertible/coupé
Prod years 1957-61
Prod numbers 11,560

PRICES
Océane **A** 4500 **B** 3000 **C** 2500
Plein Ciel **A** 3200 **B** 2200 **C** 1500

Simca Aronde P60

The last and best incarnation of the Aronde series was the 1958 P60. The 'P' stood for *personnalisation*, suggesting that there was a model to suit every customer. As there were 12 in total, that was not an idle claim. Top-of-the-range luxury Montlhéry has Flash Special engine, as does Monaco two-door coupé. All P60s had a new concave oval grille, lower bonnet-line, slimmer pillars, extended roof, bigger glasshouse, wrap-around rear screen. Étoile was the entry-level model, also offered with the ancient 1089cc engine, though not in UK. Reliable, well-made and very cheap.

Engine 1089cc 68x75mm/1290cc 74x75mm, 4 cyl OHV, F/R
Max power 48bhp at 5200rpm to 70bhp at 5200rpm
Max speed 80-90mph
Body styles saloon, estate, coupé
Prod years 1958-64
Prod numbers 260,504

PRICES
Saloon/Estate **A** 1800 **B** 1000 **C** 500
Monaco/Grand Large coupé **A** 2700 **B** 1700 **C** 1000

Simca

Simca Vedette/Ariane

In 1954 Simca bought Ford of France and hastily restyled the Detroit-designed Ford Vedette: all-new unitary body, MacPherson strut front suspension, but leaf sprung rear and antique flat-head V8. Range spanned chrome-less Trianon, mid-range Versailles, luxury Régence, special-order Présidence. Ariane was Suez Crisis special: a Vedette with the 1290cc Aronde engine (the word 'underpowered' comes to mind). Range restyled in '58 with fins and, apart from the Ariane and Marly estate, a wrap-around screen. Ferlec clutch option '58, overdrive '59. Vedettes were popular as taxis, which sums up their appeal.

Engine 2353cc 66x85.7mm/1290cc 74x75mm, V8/4 cyl SV/OHV, F/R
Max power 80bhp at 4400rpm to 84bhp at 4800rpm/48bhp at 5200rpm
Max speed 87/75mph
Body styles saloon, estate
Prod years 1954-61/1957-63
Prod numbers 166,895/159,418

PRICES
Vedette **A** 3500 **B** 2200 **C** 1000
Ariane **A** 2700 **B** 1800 **C** 900

Simca 1000

Following continental small car fashion, the 1000 had its engine in the tail, mounted longitudinally. Mario Revelli-Beaumont designed its uncompromisingly square shape. All-independent suspension, drum brakes, cam-and-roller steering. More power from '63, semi-automatic offered from '65. '67 Sim 4 is 777cc France-only base model. Minor restyle in '68, plus better suspension and rack-and-pinion steering and 1118cc engine on GLS/Special. Special gains front discs in '69, 86bhp 1294cc engine from '72-'77. An unremarkable and poor-handling little car: buy a Fiat 850 instead.

Engine 777cc 68x53.5mm/944cc 68x65mm/1118cc 74x65mm/1294cc 76.7x70mm, 4 cyl OHV, R/R
Max power 31bhp at 6100rpm to 60bhp at 5800rpm
Max speed 72-93mph
Body styles saloon
Prod years 1961-78
Prod numbers 1,642,091

PRICES
A 900 **B** 500 **C** 100

Simca 1000 Rallye 1/2

In 1970, Simca cashed in on the go-faster market with the Rallye. At first it had the 1118cc engine, but from '72 gained the 1294cc unit, developing 60bhp. You got special sports seats, leather steering wheel, different dash with rev counter, foglamps, matt black bonnet and stripes. The Rallye 2 arrived in 1973 and was sold alongside the first edition. It had a tuned twin carb 82bhp engine, front-mounted radiator, wider tyres, four-wheel discs and a larger fuel tank, plus special decals. Though they were quite fun, the handling was always on the eccentric side.

Engine 1118cc 74x65mm/1294cc 76.7x70mm, 4 cyl OHV, R/R
Max power 49bhp at 5400rpm to 60bhp at 5800rpm/82bhp at 6000rpm
Max speed 90-95/106mph
Body styles saloon
Prod years 1970-76/1973-76
Prod numbers see 1000

PRICES
Rallye 1 **A** 1800 **B** 1300 **C** 900
Rallye 2 **A** 3000 **B** 2000 **C** 1200

Simca 1000 Coupé/1200S Coupé

If the 1000 saloon looked like a brick, the Coupé had been transformed into an elegant coupé. The thanks went to Bertone, who did the styling and also built the bodies, before despatching them to France on special trains for final assembly. Four-wheel discs, but 52bhp not really enough to call this a sports car. If you wanted extra performance, Abarth did a DOHC 1.3-litre conversion. The '67 1200S was a better performer, thanks to its 80bhp 1.2-litre engine, also brake servo and, from '68, rack-and-pinion steering. The 1200S also had an all-new front end with a dummy grille and recessed headlamps.

Engine 944cc 68x65mm/1204cc 74x70mm, 4 cyl OHV, F/R
Max power 52bhp at 5400rpm/80-85bhp at 6000rpm
Max speed 94/105mph
Body styles coupé
Prod years 1962-67/1967-71
Prod numbers 10,011/14,741

PRICES
1000 **A** 3000 **B** 1800 **C** 900
1200S **A** 3500 **B** 2000 **C** 1200

Simca 1300/1500 and 1301/1501

Aronde replacement, which looked sharp for 1963 and suspiciously like a Fiat. 1300 and 1500 engines, the latter with horizontal grille bars and front discs. Column shift, with optional automatic or floor-mounted lever, which had an odd reverse gate prior to '66. Estates arrive in '65, front discs for 1300 in '66. 1301/1501 have longer boots and bonnets and new nose treatment, otherwise untouched. Servo brakes on 1501 from '69, 1301 Special (GL/S in UK) has twin-choke carb. Not very exciting, but fairly high survival rate despite a reputation for rusting.

Engine 1290cc 74x75mm/1475cc 75.2x83mm, 4 cyl OHV, F/R
Max power 60bhp at 5200rpm to 67bhp at 5400rpm/73-80bhp at 5400rpm
Max speed 84-100mph
Body styles saloon, estate
Prod years 1963-67/1967-76
Prod numbers 275,626/162,183/637,263/267,835

PRICES
A 800 **B** 400 **C** 100

Simca · Singer

Simca 1100/1204

Really quite an advanced car for 1967: transverse engine, front-wheel drive, front disc brakes, independent suspension by torsion bars all round and practical hatchback bodywork. Lots of good points, such as supple ride, comfortable and adaptable interiors and good handling. But they rusted badly, suffered from rattly engines and remained sorely underdeveloped in the 1970s, which meant that no-one cried at the crusher. Generally called 1100, despite range of engines (944cc unit did not come to UK). Ultimate version was the 82bhp twin carb TI.

Engine 944cc 68x65mm/1118cc 74x65mm/1204cc 74x70mm/1294cc 76.7x70mm, 4 cyl OHV, F/F
Max power 45bhp at 6000rpm to 82bhp at 6000rpm
Max speed 80-103mph
Body styles hatchback, estate
Prod years 1967-82
Prod numbers c 2,000,000

PRICES
A 750 **B** 300 **C** 100

Singer Nine Roadster

Singer's smallest post-war model was sold as a Roadster only. The beam axle/leaf spring chassis is basically the same as the Ten/Twelve saloons (below), although it has a shorter wheelbase. The timber-framed bodywork is exclusively a touring convertible with cutaway doors and a fold-flat windscreen. Overhead cam engine is advanced, but mechanical brakes and three-speed gearbox feel archaic. Four speeds and quieter rear axle from 1949 4A. The 4AB of 1950 had coil-over-shock IFS, larger hydromech brakes, shorter grille, longer unvalanced front wings, plain disc wheels.

Engine 1074cc 60x95mm, 4 cyl OHC, F/R
Max power 35-36bhp at 5000rpm
Max speed 65mph
Body styles tourer
Prod years 1946-52
Prod numbers 6890

PRICES
A 7000 **B** 5000 **C** 3000

Singer SM Roadster

After an experimental batch of the so-called 4AC (fitted with a sleeved-down 1194cc SM engine), the SM Roadster (also known as the 4AD) arrived in production in 1951, though for export only to begin with. British sales began in late 1953. In order to make it eligible for under-1500cc classes in competition, the SM1500 engine was short-stroked down to 1497cc. Twin carbs could be ordered as an option from September 1952, in which form the engine developed 58bhp and offered far superior top-gear acceleration – not a strong point on the 48bhp version. Replacement plastic-bodied SMX made as prototypes only.

Engine 1497cc 73x89.4mm, 4 cyl OHC, F/R
Max power 48bhp at 4200rpm to 58bhp at 4800rpm
Max speed 75-80mph
Body styles tourer
Prod years 1951-55
Prod numbers 3440

PRICES
A 8000 **B** 6000 **C** 4000

Singer Super Ten/Super Twelve

First seen in 1938, the 10HP Singer returned to production in December 1945. A functional-looking six-light saloon, it had four suicide doors, standard sliding roof, opening windscreen and a boot lid which could fold flat. Larger bore version of the Nine engine, hydraulic brakes and a four-speed 'box. Super Twelve was similar mechanically and visually, though it had an 8in longer wheelbase, 1.5-litre engine, larger plain disc wheels and a protruding boot; there was also a very rare drophead body. Essentially staid and dull machines, but technically more engaging than many other immediately post-war 10s and 12s.

Engine 1193cc 63.25x95mm/1525cc 68x105mm, 4 cyl OHC, F/R
Max power 37/43bhp at 5000rpm
Max speed 62/70mph
Body styles saloon, drophead coupé
Prod years 1945-49/1946-49
Prod numbers 10,497/1098

PRICES
Ten **A** 3000 **B** 2000 **C** 1000
Twelve **A** 3500 **B** 2200 **C** 1200

Singer SM1500

Although first seen in 1947, production of Singer's bold new saloon did not begin until June 1949. Its full-width style was hardly the last word in elegance and the upright Singer grille had been discarded for rows of horizontal chrome slats, but it was a very spacious 5/6 seater. Robust chassis, coil-sprung IFS, hydraulic brakes, larger bore/shorter stroke version of the Twelve engine. Short-stroke (4AD-type) engine from '51, higher headlamps, larger rectangular rear window and better interiors from '52, twin carb option from '53. Last ones had chrome flashes on their wings.

Engine 1506cc 73x90mm/1497cc 73x89.4mm, 4 cyl OHC, F/R
Max power 48bhp at 4200rpm/48-58bhp at 4800rpm
Max speed 72-78mph
Body styles saloon
Prod years 1949-54
Prod numbers 19,208

PRICES
A 2300 **B** 1500 **C** 900

Singer Hunter

Updating the SM1500, notably having the upright grille and glassfibre bonnet/front wing valances of the aborted SMX Roadster. Up-market spec included horse's head bonnet mascot, heater, rev counter, clock and foglamps. Floor gearchange was a novel option – Singer was the first British maker to offer this – while a column lever remained standard. The twin carb engine was still available. Stripped-out Hunter S of '55 had no spare or mascot, less chrome and a painted grille. Sadly, the 75bhp DOHC alloy-cam Hunter 75 announced in 1955 never became a true production model.

Engine 1497cc 73x89.4mm, 4 cyl OHC, F/R
Max power 50bhp at 4200rpm to 58bhp at 4800rpm
Max speed 72-78mph
Body styles saloon
Prod years 1954-56
Prod numbers 4750

PRICES
A 2700 **B** 1700 **C** 1000

Singer Gazelle

Singer was absorbed into the Rootes Group in 1956. Although existing Singer models were axed, Sir William Rootes' apprenticeship with Singer persuaded him to keep the overhead cam engine in production. This engine was therefore found in the Gazelle, an up-market Hillman Minx. Upright grille, leather and walnut inside and fancy paint schemes outside distinguished it. Hillman pushrod engine from '58 MkIIA and thereafter mechanical/styling changes as per the Minx (see page 103), except for overdrive option on MkV/VI ('63 on). No convertibles after mid-1962.

Engine 1497cc 73x89.4mm/1494cc 79x76.2mm/1592cc 81.5x76.2mm, 4 cyl OHC/OHV/OHV, F/R **Max power** 52.5bhp at 4500rpm/53-57bhp at 4600rpm/57bhp at 4100rpm **Max speed** 80-84mph **Body styles** saloon, estate, convertible **Prod years** 1956-67 **Prod numbers** 86,068

PRICES
MkI/II **A** 2200 **B** 1500 **C** 1000
MkIIA-MkVI **A** 1900 **B** 1200 **C** 800
Convertible **A** 3500 **B** 2500 **C** 1500

Singer Vogue

Badge engineering transfers to the Hillman Super Minx: oval grille with upright slats, double headlamps in bigger chrome-edged cowls with air intakes below, 58bhp OHV engine. However, there were never any Vogue convertibles, and estates arrived in 1962. All major changes as per the Super Minx, including front discs from '62 MkII, six-light styling and squarer roof-line from '64 MkIII. The 1725cc engine on the latter is the same spec as the Humber Sceptre/Sunbeam Rapier, which means 78-85bhp, some 13-20bhp more than the equivalent Super Minx. Saloon discontinued in 1966, estate follows it in '67.

Engine 1592cc 81.5x76.2mm/1725cc 81.5x82.5mm, 4 cyl OHV, F/R
Max power 58bhp at 4400rpm/78bhp at 5000rpm to 85bhp at 5500rpm
Max speed 80-95mph
Body styles saloon, estate
Prod years 1961-67
Prod numbers c 48,000

PRICES
A 1300 **B** 900 **C** 500

Singer Chamois/Chamois Sport/Chamois Coupé

Plush Hillman Imp, sporting dummy plated grille, double chrome side mouldings, chrome strips on engine lid, bumper overriders, slotted trims on wider-rim wheels, extra gauges, padded facia and door cappings with a touch of walnut veneer beneath, plus a standard heater. Rapid 51bhp twin carb Sport launched '66, also has servo brakes. Coupé arrives in '67: similar to Imp but with luxury trim; sadly, not the twin carb engine, though. All Chamois get double headlamps and a narrower grille from '68. Usual Imp faults (quirky pneumatic throttle pre-'65, fragile engines, rust), but more desirable.

Engine 875cc 68x60.4mm, 4 cyl OHC, R/R
Max power 39bhp at 5000rpm/51bhp at 6100rpm
Max speed 81-90mph
Body styles saloon, coupé
Prod years 1963-70/1966-70/1967-70
Prod numbers 40,678/4149/4971

PRICES
Chamois **A** 1300 **B** 900 **C** 400
Chamois Sport **A** 1750 **B** 1200 **C** 600
Chamois Coupé **A** 1600 **B** 1200 **C** 600

Singer Gazelle/Vogue

One final flutter of badge-engineering sees Singer to a sad grave. Hillman Hunter gets the special grille treatment in the '66 Vogue. Up-market, so 1725cc and overdrive/auto optional. Estate arrives in April '67 with 68bhp but rises to saloon's 74bhp after first six months. Gazelle slots in below Vogue as saloon only: manual versions have 1496cc engines, automatics have 1725cc. Servo option from May '68, standard from September. Chrysler pulled the plug on Singer in April 1970, but the Vogue lived on (very) briefly with Sunbeam badges (see page 205).

Engine 1496cc 81.5x71.6mm/1725cc 81.5x82.55mm, 4 cyl OHV, F/R
Max power 60bhp at 4800rpm/68-74bhp at 5000rpm
Max speed 86-92mph
Body styles saloon, estate
Prod years 1966-70
Prod numbers 31,482/47,655

PRICES
A 1200 **B** 900 **C** 400

Skoda

Skoda 1101/1102

The 1946 Czech-made Skoda was very much like its pre-war forebear, the Popular. That meant a steel backbone chassis, steel-over-wood-frame bodywork, swing axles at the rear, and all-independent suspension by transverse leaf springs to cope with bad Czech roads. OHV engine, floor-shift four-speed 'box, hydraulic brakes, central chassis lubrication, automatic ignition. Bodies more modern than pre-war, having a touch of Chevy about them. Not very romantic, but pretty tough. Quite a few still in use in Eastern Europe.

Engine 1089cc 68x75mm, 4 cyl OHV, F/R
Max power 32bhp at 3800rpm
Max speed 65mph
Body styles saloon, rolltop saloon, estate, convertible
Prod years 1945-52
Prod numbers 81,140

PRICES
Saloon **A** 2000 **B** 1200 **C** 500
Convertible **A** 3500 **B** 2000 **C** 800

Skoda 1200/1201/1202

Modern full-width appearance and all-steel construction for Skoda's new 1200 of 1952. Apart from the split windscreen pillar, there was hardly a straight line to be seen. Longer wheelbase (106in/269cm versus 97in/248cm), but similar chassis construction and suspension. Bored-out engine produced more power, especially in the high-compression 1201 which replaced the 1200 in 1956. Post-'55 models had the unusual feature of amber tail-lights which glowed under deceleration. 1202 estate lasted until 1970. Sales in western Europe were very limited, so if you want one, prepare to take a trip to Prague.

Engine 1221cc 72x75mm, 4 cyl OHV, F/R
Max power 36bhp at 4000rpm/ 45bhp at 4200rpm
Max speed 72-78mph
Body styles saloon, estate
Prod years 1952-56/1956-58/ 1958-70
Prod numbers 93,741

PRICES
A 1500 **B** 900 **C** 400

Skoda 440/445/Octavia

The 440 was the 'Spartak' people's car of Czechoslovakia, known in some export markets as the Orlik or Rapid. The backbone chassis remained, though wheelbase was shorter. Front suspension gains lever-type dampers, and rear dampers except on a stripped-out home-market special. Usually column-shift and the front and rear screens are interchangeable! '57 445 had 1201's bigger engine and higher gearing. Both replaced by Octavia and 1.2-litre Octavia Super in '59: one-piece grille, coil spring and wishbone IFS. TS has twin carb Felicia 1.1/1.2 engines, and is rare (only 2273 built). No saloons after '64.

Engine 1089cc 68x75mm/1221cc 72x75mm, 4 cyl OHV, F/R
Max power 40bhp at 4200rpm to 50bhp at 5600rpm/45bhp at 4400rpm to 53bhp at 5100rpm
Max speed 75-95mph
Body styles saloon, estate
Prod years 1955-59/1957-59/1959-71
Prod numbers 440/445: 84,792/Octavia: 279,724

PRICES
A 1500 **B** 1000 **C** 500

Skoda 450/Felicia

The convertible version of the 440 was called the 450, simply because the engine developed 50bhp, thanks to a higher compression ratio and twin carbs. The front end was different: it had an oval grille instead of the saloon's split type. Rag top could be supplemented by a removable glassfibre hardtop, designed by Ghia. Renamed Felicia in '59, concurrent with the Octavia, and given a choice of the same 1089cc engine or, from '61, the 1221cc unit, in which case the car was called Felicia Super. Some were imported to Britain, but you can still find them cheaply and plentifully on the continent.

Engine 1089cc 68x75mm/1221cc 72x75mm, 4 cyl OHV, F/R
Max power 50bhp at 5600rpm/ 53bhp at 5100rpm
Max speed 90-95mph
Body styles convertible
Prod years 1958-59/1959-64
Prod numbers 15,864

PRICES
A 2500 **B** 1800 **C** 1000

Skoda 1000MB/1100MB

Just as most manufacturers were beginning to realise the benefits of front-wheel drive, Skoda turned to rear engines and stuck it out for over 25 years. New features for Skoda: unitary construction, new engines, all-synchro 'box. It wasn't a bad-looking car, the interior was spacious (you could even fold the seats to make a bed) and mild excitement attended the twin carb MBG and two-door hardtop saloon MBX versions. The problem was those swing axles at the back, which gave every owner a fright going round corners at some time or another. 1100MB has a gas-flowed 1.1-litre engine.

Engine 988cc 68x68mm/1107cc 72x68mm, 4 cyl OHV, R/R
Max power 45bhp at 4650rpm to 52bhp at 4800rpm/54bhp at 4600rpm
Max speed 75-82mph
Body styles saloon
Prod years 1964-69
Prod numbers 1,239,327

PRICES
Saloon **A** 1000 **B** 500 **C** 200

Skoda S100/S110/S110R

In the last year of the 1000MB's life, the roof-line had been altered and the rear window made larger. In 1969, further changes were made and the model was renamed S100. The front and rear ends were restyled, lower-set recessed headlamps being the most notable difference. You also got standard front discs. S110 has bigger engine, while top-spec S110L has twin carbs and 52bhp. The same engine went into the S110R Coupé (photo), a cheeky-looking 2+2, usually supplied with built-in foglamps and even a wood facia. Could be an interesting tail-slide special, if you can find one that hasn't rusted.

Engine 988cc 68x68mm/1107cc 72x68mm, 4 cyl OHV, R/R
Max power 43bhp at 4650rpm/48bhp at 5000rpm to 52bhp at 5500rpm
Max speed 78-90mph
Body styles saloon, coupé
Prod years 1969-77/1969-77/1970-81
Prod numbers Saloon: 323,848/Coupé: 56,902

PRICES
S100/S110 **A** 700 **B** 300 **C** 100
S110R **A** 1100 **B** 600 **C** 300

Spatz

Egon Brütsch was a prodigious German microcar designer and to him goes the credit for making the world's first glassfibre monocoque, the 1954 Spatz. Sadly, this three-wheeler simply had all its components bolted to the plastic body, which proceeded to crack all over the place, leading a German court to ban it from the road. Hans Ledwinka (ex-Tatra) redesigned it with four wheels and a chassis, plus a Sachs engine. 'Bike maker Victoria got involved in '56. One of the better micros, but blighted by a chequered production career. Ledwinka's connection ensures a strong following.

Engine 191cc 65x58mm/248cc 67x70mm, 1 cyl TS, R/R
Max power 10bhp at 5250rpm/14bhp at 5200rpm
Max speed 47/56mph
Body styles sports
Prod years 1955-58
Prod numbers 1588

PRICES
A 6000 **B** 4000 **C** 2500

Standard 8

Warmed-over pre-war material was Standard's enforced output from 1945. The 'Flying Eight' had an underslung chassis, IFS by transverse leaf spring and wishbones, semi-elliptic rear, four speeds post-war, sidevalve engine and a choice of bodies: two-door four-light saloon, drophead coupé and two-door tourer with cutaway doors. Estate offered from '48 is rare. More sprightly than most 8HP cars, and fairly spacious despite its compact size (11ft 7in/353cm long). Tourer is a cheap route into vintage-style open motoring.

Engine 1009cc 56.7x100mm, 4 cyl SV, F/R
Max power 28bhp at 4000rpm
Max speed 60mph
Body styles saloon, estate, drophead coupé, tourer
Prod years 1945-48
Prod numbers 53,099

PRICES
Saloon **A** 2000 **B** 1200 **C** 600
DHC/Tourer **A** 3000 **B** 2000 **C** 1300

Standard 12/14

The bodywork of Standard's larger cars was 3in wider than it was pre-war, providing these four-door saloons and estates with a large cabin. A two-door drophead coupé was also available, having an external luggage carrier and spare wheel on the back. Both used the same chassis, the 12 with a 1.6-litre engine, the 14 1.8 litres. The latter was reserved for export until June '46. Improved steering and suspension in '46, but both models were swept aside by the Vanguard, a result of a government-encouraged 'one-model' policy. Sturdy, reliable and smooth-running, but deadly slow.

Engine 1609cc 69.5x106mm/1776cc 73x106mm, 4 cyl SV, F/R
Max power 44bhp/50bhp at 3800rpm
Max speed 65-70mph
Body styles saloon, estate, drophead coupé
Prod years 1945-48
Prod numbers 9959/22,229

PRICES
Saloon **A** 2500 **B** 1500 **C** 800
DHC **A** 3600 **B** 2300 **C** 1500

Standard Vanguard I

An important model for Standard: one of Britain's first fresh post-war designs, still having a separate chassis, but now coil spring IFS, hydraulic brakes, new overhead valve engine, three-speed all-synchro gearbox with right-hand column shift and, for the first time, a heater. Detroit full-width styling in a choice of four-door saloon and van-like all-steel estate bodies. Rear wheel spats from '49, when gear lever transfers to left of column. Laycock-de-Normanville overdrive option from '50, new grille and bigger rear window from '51. Incredibly strong but does not like corners at all.

Engine 2088cc 85x92mm, 4 cyl OHV, F/R
Max power 68bhp at 4200rpm
Max speed 80mph
Body styles saloon, estate
Prod years 1948-52
Prod numbers 184,799

PRICES
A 3500 **B** 2500 **C** 1400

Standard Vanguard II

Phase II arrived in January 1953. Apart from a new bonnet ornament and one less grille bar, it looked exactly the same from the front, but the B-pillars moved back 5in, wider doors were fitted and the rear end became a notchback, therefore more boot space and head room. Mechanically, things were much as before, except for a hydraulic clutch. Rare two-door estate version sold 1953-54 (four-door more common), and convertibles made in Standard's Belgian plant. Depressing 2092cc diesel from '54 was Britain's first passenger diesel, boasted standard double overdrive (1973 made). Petrol saloons replaced in '55.

Engine 2088cc 85x92mm/2092cc 81x101.6mm, 4 cyl OHV/diesel, F/R
Max power 68bhp at 4200rpm/40bhp at 3000rpm
Max speed 75/65mph
Body styles saloon, estate, convertible
Prod years 1953-56
Prod numbers 83,067

PRICES
A 2500 B 1200 C 700

Standard Vanguard III/Vignale Vanguard/Sportsman/Ensign/Ensign De Luxe

Other than its engine and gearbox, the Phase III Vanguard was all-new: unitary build, longer, lower and wider shape. Final drive higher and weight reduced, so top speed and mpg are up. Phase III Estate and Sportsman arrive '56, latter having twin carb 90bhp TR3-type engine, upright grille, dual overdrive, two-tone paint. Vanguard gets overdrive option from '57, also auto. '58 restyle by Vignale (new grille, bigger glasshouse). Ensign has 1.7-litre engine and dressed-down trim (though De Luxe 1962-63 has 2138cc engine and front disc option). Late ones have TR3 'box option. All are robust, but heavy to handle.

Engine 2088cc/1670cc/2136cc, 4 cyl OHV, F/R **Max power** 68-90bhp/60bhp /75bhp **Max speed** 80-95mph
Body styles saloon, estate
Prod years 1955-58/1958-61/1956-57/1957-61/1962-63
Prod numbers 37,194/26,276/901/18,852/2318

PRICES
Vanguard/Vignale/Ensign De Luxe
A 2400 B 1300 C 500
Sportsman A 3000 B 1800 C 1000
Ensign A 2200 B 1000 C 500

Standard Vanguard Luxury Six

Styling is the same as Michelotti's Vignale Vanguard, but the engine is a new 80bhp twin carb 2-litre straight six, later to be fitted to the Triumph 2000. The Luxury Six's longer rear springs and higher gearing also found their way on to late four-cylinder Vanguards and Ensigns. Standard gearbox is the three-speed column-change unit, but overdrive is optional, as is the TR3-type floor-shift four-speeder and, from '61, automatic. Front discs are a desirable optional extra, available from July 1961. Only badges, some extra chrome and luxury fittings identify it.

Engine 1998cc 74.7x76mm, 6 cyl OHV, F/R
Max power 80bhp at 4400rpm
Max speed 92mph
Body styles saloon, estate
Prod years 1960-63
Prod numbers 9953

PRICES
A 2500 B 1300 C 600

Standard 8/10/Pennant

The 8 marked a return to the small car market abandoned for five years. Unitary construction, unburstable 803cc OHV engine, coil spring IFS, hypoid rear axle. First ones were very cheap and had very basic trim, sliding windows and fold-down rear seats (there was no boot lid until 1957!). Optional overdrive/auto clutch from '57. 10 has 948cc engine, chrome grille bars, winding windows and boot lid. Tuning kits available for up to 42bhp and two-pedal option from '56. Companion is estate, Family 10 is 8 bodyshell with 948cc engine. Pennant is 10 with extended front and rear wings, two-toning and remote shift.

Engine 803cc 58x76mm/948cc 63x76mm, 4 cyl OHV, F/R
Max power 26bhp at 4500rpm to 33bhp at 5000rpm/33bhp at 4500rpm to 37bhp at 5000rpm
Max speed 63-70mph
Body styles saloon, estate
Prod years 1953-59/1954-60/1957-60
Prod numbers 136,317/172,500/42,910

PRICES
A 1800 B 1000 C 500

Stanguellini

Vittorio Stanguellini cut his automotive teeth tuning Fiat engines. He began making sports cars in 1947, with a Bertone-bodied four-seater based on the Fiat 1100. Stanguellini also made twin cam (*bialbero*) versions of Fiat 750 and 1100 engines. Each car was bespoke built, though a typical specification included a Fiat engine, transverse leaf and wishbone front suspension and coil sprung live rear axle. Racing cars were the main output, however, and production ended in '66. A Bertone body is pictured.

Engine 741cc 58x70mm/1089cc 68x75mm, 4 cyl DOHC, F/R
Max power 70-80bhp at 7500rpm/90bhp at 7000rpm
Max speed 110/125mph
Body styles sports, coupé
Prod years 1947-66
Prod numbers n/a

PRICES
Too rare for accurate valuation

Steyr-Puch · Studebaker · Stutz · Subaru

Steyr-Puch 500/650/700

Austria's motor industry *is* Steyr-Puch. After the war, it made various Fiats under licence, but the little 500 was different: although it was a Nuova 500, it had its own air-cooled OHV flat-twin engine, its own swing axle rear end, a synchro 'box, and bigger brakes. It gained a reputation as a good hillclimb and circuit performer and soon 643cc (650) and 660cc (650TR) versions were available. Special grille identifies it, and most have a full-length fabric sunroof. 700 was the estate version. Twin fuel tanks and a different dash featured on the 650TR.

Engine 493cc/643cc/660cc, 2 cyl OHV, R/R
Max power 16-20bhp at 4600rpm/20-25bhp at 4800rpm/27-40bhp at 5800rpm
Max speed 60-90mph
Body styles rolltop saloon, estate
Prod years 1957-69
Prod numbers 54,300

PRICES
500/700 **A** 3000 **B** 1800 **C** 1000
650TR **A** 4500 **B** 3000 **C** 2200

Studebaker 'Loewy' Coupe

To celebrate its first century, Studebaker launched a range of coupes, now nicknamed 'Loewy', though they were actually styled by Bob Bourke. Based on the Land Cruiser chassis, they were perhaps America's prettiest car of the 1950s. Range began with pillarless Regal Starliner and pillared Starlight, Commander and Champion, and later a President. Facelifted for '56 with square grilles and fins, and named Flight Hawk, Power Hawk, Sky Hawk and Golden Hawk, the latter being available with supercharging. Pillared Silver Hawk followed. Attractive and sought-after.

Engine 2779cc/3042cc 6 cyl SV, 3675cc/3811cc/4248cc/4737cc/5768cc V8 cyl OHV, F/R
Max power 85bhp at 4000rpm to 275bhp at 4800rpm **Max speed** 85-125mph **Body styles** coupe, hardtop coupé **Prod years** 1953-61
Prod numbers n/a

PRICES
'53-'55 **A** 7500 **B** 5000 **C** 2500
'56-'61 **A** 9000-12,000 **B** 6500-9000 **C** 2500-4500

Studebaker Avanti

Raymond Loewy designed the Avanti in a remarkably short space of time, and it was a remarkable car. The swoopy glassfibre body was as untypically Detroit as you could get. Under it lay a mere beefed-up Lark chassis with anti-roll bars, rear radius arms and America's first front disc set-up. Jet Thrust V8 engine available in various forms: base R1, supercharged R2, bored-out and blown R3, bored-out twin carb R4 and experimental fuel-injected twin-supercharged R5 (with 575bhp!). Early quality control problems and mounting losses effectively killed it, though it was revived and is still in production.

Engine 4737cc 90.5x92.1mm/4973cc 92.8x92.1mm, V8 cyl OHV, F/R
Max power 240-335bhp at 4500rpm
Max speed 120-160mph
Body styles coupé
Prod years 1963-64
Prod numbers 4643

PRICES
A 18,000 **B** 12,000 **C** 6000

Stutz Blackhawk

Americans have a fascination with reviving the past: there have been 'new' Cords, Duesenbergs and even Stutzes. An industrialist bought the Stutz name and launched his own Virgil Exner designed coupe in 1969. Underneath the frankly vulgar body (made by Carrozzeria Padama in Italy) lay a Pontiac Grand Prix chassis. Massive power options initially, dropping rapidly in the 1970s. Models offered included the Blackhawk two-door hardtop coupe, Bearcat two-door convertible, Victoria four-door saloon (pictured), and mammoth (24ft 7in/750cm) Royale state limo. Strictly for Hollywood and Riyadh.

Engine 5736cc 103.8x86mm/6605cc 110.5x85.8mm/6964cc 104x103mm/7456cc 105.3x106.9mm, V8 cyl OHV, F/R
Max power 160-430bhp
Max speed 106-145mph
Body styles saloon, limousine, coupe, convertible
Prod years 1969-c93
Prod numbers n/a

PRICES
A 30,000 **B** 15,000 **C** 5000

Subaru 360

Completely bizarre car spawned by Japan's microcar regulations. Subaru's first passenger car, it sported a 356cc two-stroke two-cylinder engine in the rear of what was a cross between a Dinky toy and a jelly bean. Hooded headlamps, optional roll-top, all-independent suspension and rear swing axles. Early ones had three speeds and mechanical brakes. Custom (estate) even odder, and commercial version with flop-down rear windows odder still. 422cc option from '60, in which form it was called the 450 or Maia. No UK imports, but a growing following in Japan and the USA.

Engine 356cc 61.5x60mm/422cc, 2 cyl TS, R/R
Max power 16-22bhp at 5000rpm
Max speed 56-62mph
Body styles saloon, rolltop saloon, estate
Prod years 1958-70
Prod numbers n/a

PRICES
A 3000 **B** 1500 **C** 800

Subaru Leone

Leone was the home-market name of the first Subaru to be imported into the UK (where it was known simply as the 1600). This event happened in 1977, although the Leone had been on sale in Japan since 1971. Flat four engines, all-independent springing (torsion bars at the back), front discs. Front-wheel drive initially, but Subaru's pièce de resistance – four-wheel drive – from '74 made it the world's first popular 4WD passenger car. We got all body types, but only the estate was offered here with 4WD. Welsh hill farmers lapped 'em up but unfortunately mountain weather ate them up.

Engine 1176cc 79x60mm/1361cc 85x60mm/1595cc 92x60mm, 4 cyl OHV, F/F & F/4x4
Max power 68bhp at 6000rpm to 95bhp at 6400rpm
Max speed 90-105mph
Body styles saloon, estate, coupé
Prod years 1971-79
Prod numbers 1,269,000

PRICES
A 900 B 500 C 200

Sunbeam-Talbot Ten/2-Litre

The pre-war Talbot 10 was revived in 1945 as the Sunbeam-Talbot Ten. Effectively a rebodied 1935 Hillman Minx, the styling was the car's most distinctive feature: flowing wings, reverse-angle C-pillar, frameless rear glass. Beam axles, mechanical brakes. Minx engine was tuned, which was necessary to counteract the greater weight. 2-Litre had a 3in longer wheelbase and stiffer chassis, Humber Hawk sidevalve engine, hydraulic brakes, and an awkward four-speed 'box. Needed to be faster, though viewed as desirable machines in their day.

Engine 1185cc 63x95mm/1944cc 75x110mm, 4 cyl SV, F/R
Max power 41bhp at 4400rpm/56bhp at 3800rpm
Max speed 67/71mph
Body styles saloon, sports tourer, drophead coupé
Prod years 1938-48/1939-48
Prod numbers 7673/1306

PRICES
Saloon A 2700 B 2000 C 1500
DHC/Tourer A 5000 B 3200 C 2500

Sunbeam-Talbot 80/90

Take a 2-Litre chassis, update the bodywork and graft on an overhead valve layout on the 1.2 and 2.0 engines and you have the 80 and 90. Bodywork by Thrupp & Maberly has curved windscreen, near-full-width styling, rear wheel spats. Also column-change 'box, softer springing, twin leading shoe hydraulic brakes. 1950 90 MkII has new cruciform chassis, coil spring IFS, hypoid axle, 2.3-litre engine (spot it by its extra air intake grilles). 1952 MkIIA has bigger brakes, better steering and suspension, perforated disc wheels and no wheel spats. Higher compression from October '53. 90s are rugged and thrashable.

Engine 1185cc 63x95mm/1944cc 75x110mm/2267cc 81x110mm, 4 cyl OHV, F/R
Max power 47bhp at 4800rpm/64bhp at 4100rpm/70-77bhp at 4100rpm
Max speed 72-87mph
Body styles saloon, drophead coupé
Prod years 1948-50/1948-54
Prod numbers 3500/c 17,500

PRICES
Saloon A 4500 B 2800 C 1800
DHC A 6500 B 4500 C 3000

Sunbeam MkIII/MkIIIS

'Talbot' dropped out of Sunbeam in 1954, while the 90 tag was also discarded in favour of just MkIII. High compression Humber Hawk engine, slotted chrome-plated wheel trims, chromed side grilles, three portholes below each side of the bonnet, seats from the Alpine. Overdrive and a rev counter were optional. Dropheads discontinued in October '55 and saloons in January 1957, but a batch of unsold MkIIIs was converted in 1957 by Castles Motors with Rootes approval: these had a high compression head, bigger carb, remote gear-change, overdrive, bonnet air intake and upward-opening boot lid, and were known as the MkIIIS.

Engine 2267cc 81x110mm, 4 cyl OHV, F/R
Max power 80bhp at 4200rpm/92bhp at 4400rpm
Max speed 95-100mph
Body styles saloon, drophead coupé
Prod years 1954-57/1957
Prod numbers c 2250/40

PRICES
Saloon A 4500 B 2800 C 1800
DHC A 7000 C 4500 C 3000

Sunbeam Alpine

Stirling Moss won the Alpine Rally in a 90 and so this name was a natural for the two-seater sports version. A stiffened 90 chassis, tuned engine and higher-geared steering made it an accomplished if somewhat heavy cruising car. You got detachable side screens with plastic sliding panels and a removable hood, so that completely open-air motoring was possible. Overdrive and rev counter standard from '54. Most exported to America, helped by a film appearance with Grace Kelly and Cary Grant in one (*Over the Roofs of Nizza*). Rather attractive cars, but scarce in UK.

Engine 2267cc 81x110mm, 4 cyl OHV, F/R
Max power 80bhp at 4200rpm
Max speed 100mph
Body styles sports
Prod years 1953-55
Prod numbers c 3000

PRICES
A 12,500 B 8000 C 6000

Sunbeam Rapier I/II/III/IIIA/IV/V

The vague resemblance of the Rapier to the Studebaker Hawk is no coincidence, since Raymond Loewy was in charge of styling both. The first of the unitary construction Rootes cars which sired the Hillman Minx and Singer Gazelle. Always a two-door, with tuned Minx engines, standard overdrive, twin carbs from '56. MkII has upright grille, fins, 1.5-litre engine, floor change, no overdrive. MkIII has alloy head and front discs, MkIIIA gains 1.6-litre engine, MkIV has smaller wheels, more power, brake servo, unhooded headlamps. Final MkV has 1725cc engine. Last convertibles made October '63.

Engine 1390cc/1494cc/1592cc/1725cc, 4 cyl OHV, F/R **Max power** 62bhp at 5000rpm to 94bhp at 5200rpm **Max speed** 85-100mph **Body styles** coupé, convertible **Prod years** 1955-57/1958-59/1959-61/1961-63/1963-65/1965-67 **Prod numbers** 68,809 (7477/15,151/15,368/17,354/9700/3759)
PRICES
I Coupé **A** 2400 **B** 1500 **C** 1000
II-V Coupé **A** 2700 **B** 1700 **C** 1000
II-IIIA Convertible **A** 5000 **B** 3500 **C** 2000

Sunbeam Alpine I/II/III/IV/V

Nothing like the old Alpine: unitary construction (based on the Hillman Husky, of all things), sharp styling including tail fins, and a rorty twin carb version of the Hillman Minx engine. As an MGA rival it was struggling, but it offered nearly 100mph and marginal 2+2 seating. Overdrive was optional, as was a hardtop. MkII has 1.6 litres, MkIII has single twin-choke carb, improved suspension, quarter lights (also available as a cleanly-styled GT coupé with less power). MkIV has cut-down fins, new grille, auto option; final MkV has five-bearing 1725cc twin carb engine but no auto option. GTs worth 10% less.

Engine 1494cc 79x76.2mm/1592cc 81.5x76.2mm/1725cc 81.5x82.55m, 4 cyl OHV, F/R
Max power 78bhp at 5300rpm to 93bhp at 5500rpm
Max speed 95-102mph
Body styles sports, hardtop coupé
Prod years 1959-60/1960-63/1963-64/1964-65/1965-68
Prod numbers 69,251 (11,904/19,956/5863/12,406/19,122)
PRICES
A 7000 **B** 4500 **C** 2500

SUNBEAM SUPREME

The Largest Specialists of New and Used Parts for Sunbeam Alpines and Tigers

- Service and restoration ● Reproduction rare items a speciality ● Membership club discounts ● All major credit cards welcome ● Fast next day Mail Order ● Worldwide exports ● Free advice and technical help ● Free brochure and price list ● Cars for sale, complete or shells

BODY PANELS AS GOOD AS ORIGINAL FACTORY PARTS
Our supplier has been manufacturing for 10 years and has vast experience in the marque

Vast quantity of new spares in stock plus and a full range of second-hand parts

Contact: Simon or Chris on
Leicester (0116) 2536214. Fax (0116) 2513231
Open 9.00am - 10.00pm seven days

WANTED: ALPINES, TIGERS – PARTS OR UNFINISHED PROJECTS

The Definitive Rally Cars Reference Book

A-Z OF WORKS RALLY CARS
FROM THE 1940s TO THE 1990s
by GRAHAM ROBSON

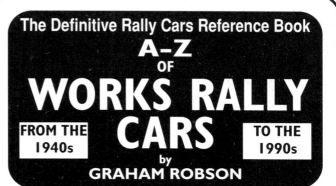

✳ From AC to Wartburg, an exhaustive guide to works rally cars built for international rallying

✳ Features all the significant cars in the history of rallying

✳ More than 250 photographs including rare archive shots

✳ 176 pages, hardcover, 260 × 190mm

✳ Price £19.95

'Fascinating and comprehensive – you'll pick this up time and time again.' *Classic and Sportscar*

Available from good bookshops, but in case of difficulty call
Bay View Books
Tel: 01237 429225/421285

Sunbeam Harrington Alpine A/B

Conversions are outside the frames of reference of this guide, but the Harrington is different. It was officially approved by Rootes and, after a good showing at Le Mans in 1962, adopted as an official Sunbeam. Harrington put a glassfibre roof on a roadster, opening up room for heads and luggage. George Hartwell offered tuning packages (up to 104bhp). Type B (photo) has more radical roof/tail chop and a finless fastback rear incorporating a hatchback and chrome body side strips. Spec as Alpine II. Very rare and consequently sought-after by Alpine fanatics.

Engine 1592cc 81.5x76.2mm, 4 cyl OHV, F/R
Max power 88-104bhp at 6000rpm
Max speed 100-110mph
Body styles coupé
Prod years 1961-62/1962-63
Prod numbers 150/250

PRICES
Type A **A** 8000 **B** 5000 **C** 3000
Type B **A** 9000 **B** 6000 **C** 3500

Sunbeam Tiger I/II

Just like the AC Cobra, Carroll Shelby fitted Ford's 260 V8 engine into the shell of the Alpine – shareholder Chrysler's V8 wouldn't fit. It looked almost identical from the outside (there was a full-length body strip), but the whole structure was stiffened up, including the springs. As well as the V8, the four-speed 'box came from Ford of Detroit, and you got rack-and-pinion steering and optional LSD. US-only Tiger II got the 289 V8 from the Mustang, plus wider gearing, a cross-hatch grille and body stripes. UK sales 1965-66 only. Not as dramatic or fast as a Cobra, and they overheat, but much cheaper!

Engine 4261cc 96.5x73mm/4737cc 101.6x73mm, V8 cyl OHV, F/R
Max power 164bhp at 4400rpm/200bhp at 4400rpm
Max speed 120/125mph
Body styles sports
Prod years 1964-66/1967-68
Prod numbers 6495/571

PRICES
Tiger I **A** 12,000 **B** 8000 **C** 5000
Tiger II **A** 15,000 **B** 10,000 **C** 6000

Sunbeam Venezia

By no stretch of the imagination is this a sports car, though Carrozzeria Touring built it according to its *superleggera* method of aluminium panels over a tubular frame. Underneath, it's a Humber Sceptre floorpan. The engine is tuned to yield 88bhp and you get standard front discs and overdrive. Pretty styling somewhat compromised by the use of a Sceptre windscreen, but four hooded circular headlamps and upright chrome grille give it some presence. Extremely rare and never sold in Britain, though many have filtered back here. They're not expensive, either.

Engine 1592cc 81.5x76.2mm, 4 cyl OHV, F/R
Max power 88bhp at 5800rpm
Max speed 100mph
Body styles coupé
Prod years 1963-64
Prod numbers 143

PRICES
A 5000 **B** 3500 **C** 2200

Sunbeam Imp Sport/Stiletto

The Imp Sport, officially born to supply rally driving Imp-men, was all but identical to the Singer Chamois Sport: luxurious trim, extra instruments and uprated 51bhp engine. The extra power was from twin carbs and high-lift cam, and you got an oil cooler and servo brakes as well. Differentiated from Singer by its single horizontal chrome bar across the front, wheel trims and 'Imp Sport' badges on the doors. Simplified facia for '69, Chamois-style grille for brief period in 1970, four headlamps and horizontal bar thereafter. Stiletto has Imp Californian's dashing coupé looks, vinyl roof *and* the 51bhp engine.

Engine 875cc 68x60.4mm, 4 cyl OHC, R/R
Max power 51bhp at 6100rpm
Max speed 90mph
Body styles saloon, coupé
Prod years 1966-76/1967-72
Prod numbers c 10,000/c 10,000

PRICES
Imp Sport **A** 1750 **B** 1200 **C** 600
Stiletto **A** 1900 **B** 1300 **C** 800

Sunbeam Alpine/Rapier

The pillarless coupé shape may have exotic suggestions of the Plymouth Barracuda – plagiarism was rife back then – but the Rapier was not a sports coupé. The floorpan is a simple Hillman Hunter, sharing its 1725cc engine, servo front discs and recirculating ball steering. You got twin carbs and 88bhp, a close-ratio dual-overdrive 'box (auto option) and fancy wheel trims. H120 Holbay-tuned 105bhp version arrives October '68: higher final drive, Rostyles, matt black grille, tail spoiler, stripes. Alpine is single carb 74bhp poverty model with painted sills, plain wheels, less trim and no overdrive.

Engine 1725cc 81.5x82.55mm, 4 cyl OHV, F/R
Max power 74bhp at 5000rpm to 105bhp at 5200rpm
Max speed 92-106mph
Body styles coupé
Prod years 1969-76/1967-76
Prod numbers 46,206

PRICES
Alpine **A** 1200 **B** 800 **C** 400
Rapier **A** 1500 **B** 1000 **C** 500
Rapier H120 **A** 2000 **B** 1250 **C** 600

Sunbeam · Swallow · Talbot · Tatra

Sunbeam Vogue

Rootes squashed the Singer marque in 1970 but curiously decided to continue the Vogue (an up-market Hillman Hunter) with Sunbeam badges. The exercise basically showed up badge-engineering for the vapid smokescreen it was: Rootes themselves were confused by this, and the Sunbeam Vogue lasted just six months, from April to October 1970. Only the badges differed over the Singer, so you got the single carb (Sunbeam Alpine spec) 1725cc five-bearing engine and servo disc front brakes. Saloon and estate bodies available, but the number built was barely into the hundreds.

Engine 1725cc 81.5x82.55mm, 4 cyl OHV, F/R
Max power 74bhp at 5000rpm
Max speed 92mph
Body styles saloon, estate
Prod years 1970
Prod numbers n/a

PRICES
A 1200 **B** 900 **C** 400

Swallow Doretti

William Lyons sold his Swallow sidecar business during the war to Tube Investments, who tried their luck with a sports car in 1954. The Doretti used Triumph TR2 parts fitted into Swallow's own tubular frame, using an inner skin of steel and a hand-built aluminium body. That double skinning meant weight was up compared to the TR2, so less performance, and it was £230 more expensive than a TR2. Charms are leather-trimmed cockpit and extra refinement. 2+2 Sabre version not productionised. Short life because of objections from Lyons himself.

Engine 1991cc 85x92mm, 4 cyl OHV, F/R
Max power 90bhp at 4800rpm
Max speed 101mph
Body styles convertible
Prod years 1954-55
Prod numbers 276

PRICES
A 12,000 **B** 9000 **C** 6500

Talbot Lago Record/Baby

Although a 4½-litre Talbot Lago won Le Mans in 1950, road car versions were almost invariably overblown and overweight. However, they were magnificent *grandes routières*, bodied by firms like Graber, Figoni & Falaschi and Saoutchik. The heart of the machine was its triple carb straight six engine, and there was IFS, hydraulic braking and a pre-selector 'box. Wheelbase grew over the years, and the last flagship model (1953) was a pretty in-house coupé. 'Baby' had a short-stroke four and optional four-speed 'box.

Engine 4482cc 93x110mm/2690cc 93x99mm, 6/4 cyl OHV, F/R
Max power 170-210bhp at 4200rpm/120bhp at 4500rpm
Max speed 108/98mph
Body styles saloon, coupé, convertible
Prod years 1946-55/1949-55
Prod numbers c 750

PRICES
Vary according to coachbuilder; Baby starts at £10,000

Talbot Lago Sport/America/Simca

The 1953 Grand Sport coupé was shrunken on to a short-wheelbase version of the Record chassis to create the attractive Sport 2500, which had a 120bhp cross-pushrod alloy-head 'four' developed from the Baby. Choice of ZF and Pont-à-Mousson 'boxes. Not as fast (or cheap) as an XK120, so optional BMW V8s in three sizes were offered, in which form it was known as the America. In 1959, Simca absorbed Talbot and installed the old and rather unsuitable Ford/Simca flathead V8, but this did not last long.

Engine 2491cc 89.5x99mm/2476cc 72.5x75mm/2580cc 74x75mm/3168cc 82x75mm/2351cc 66.1x85.7mm, 4/V8/V8/V8/V8 cyl OHV/OHV/OHV/OHV/SV, F/R **Max power** 84bhp at 4800rpm to 150bhp at 5000rpm
Max speed 93-121mph **Body styles** coupé **Prod years** 1954-57/1957-59/1959-60 **Prod numbers** 54/12/10

PRICES
Too rare for accurate valuation, but probably in £15-40,000 range

Tatra 57B

Immediately after the war, the leading Czech car maker returned with the pre-war 57, the smaller of its remarkable pair of models, which had first appeared as early as 1932. It was also known as the Hadimrska and was licence-built in Germany as the Rohr Junior and in Austria as the Austro-Tatra. The lightweight two-door 57B had a front-mounted air-cooled flat four engine. Coachbuilder Sodomka built a number of special bodies on the antiquated chassis, such as the 1947 Prague Motor Show car pictured.

Engine 1256cc, 4 cyl OHV, F/R
Max power 25bhp at 2800rpm
Max speed 56mph
Body styles saloon
Prod years 1938-49
Prod numbers n/a

PRICES
A 5000 **B** 3000 **C** 1800

Tatra 87

The 1934 Tatra 77 had the world's first all-enveloping streamlined saloon body. The 87 which replaced it in 1937 was a genuine 100mph saloon – considering it measured 15ft 6½in (474cm) long, this was an achievement. Remarkable in so many ways, the Tatra was designed by Hans Ledwinka. Under its extraordinary lightweight saloon body lay an independently sprung box section chassis powered by an air-cooled OHC V8 mounted behind the rear axle. This is probably the most sought-after Tatra today, simply because of its unique landmark design and surprising all-round competence.

Engine 2968cc 75x84mm, V8 cyl OHC, R/R
Max power 73bhp at 3600rpm
Max speed 100mph
Body styles saloon
Prod years 1937-50
Prod numbers 3056

PRICES
A 12,000 **B** 7500 **C** 4000

Tatra T600 Tatraplan

This was a smaller brother for the 87, shrunk down in every direction but self-evidently related and still modern-looking, and unitary-built. The extra central headlamp was absent, as was the rear fin, and the windscreen was split into two sections, not three as before. In line with its intended role as a cheaper Tatra, the rear-mounted engine was a flat-four unit (still air-cooled) and had no overhead cam gear. The gearbox and differential were integral with the engine and the four-speed 'box was operated by a column lever. Swing axles at the back, so watch that tail-spin.

Engine 1950cc 85x86mm, 4 cyl OHV, R/R
Max power 52bhp at 4000rpm
Max speed 80mph
Body styles saloon
Prod years 1947-52
Prod numbers 6350

PRICES
A 7500 **B** 5000 **C** 2500

Tatra T603

After a gap of three years during which no cars were built (only trucks), the T603 arrived in 1955. Aerodynamics were still high on the agenda, but there was more of a saloon look to the car. The rear-mounted air-cooled V8 returned in a steel unibody with a flat floor, but the camshaft was now mounted in the centre of the 'V'. Early ones had three headlamps in a single panel, later ones (T2-603) four headlamps in a split 'grille'. New coil spring suspension replaced the old leafs, providing a superior ride. The last models boasted a twin carb engine, servo brakes and power steering.

Engine 2472cc 74.9x70mm/2544cc 74.9x72mm, V8 cyl OHC, R/R
Max power 95-105bhp at 4800rpm
Max speed 99-102mph
Body styles saloon
Prod years 1955-75
Prod numbers 22,422

PRICES
A 5000 **B** 2800 **C** 1200

Tatra T613

The T613 shared the basic layout of the T603, but had an all-new body designed by Vignale, which abandoned any disposition towards aerodynamics. The expanded 3.5-litre air-cooled V8 was required to power a car that weighed well over 1.5 tons, and it was mounted ahead of the rear axle for the first time. The V8 was a marvel: light alloy heads, four overhead camshafts, two twin-barrel carbs. While Russia relied on Fiat technology, Tatra pressed ahead on its own: you got powered rack-and-pinion steering and servo discs all round. Much revised, it is still being made today by a fiercely independent Tatra.

Engine 3495cc 85x77mm, V8 cyl QOHC, R/R
Max power 165bhp at 5200rpm
Max speed 118mph
Body styles saloon, limousine
Prod years 1974-date
Prod numbers n/a

PRICES
A 5000 **B** 2800 **C** 1000

Thurner RS

German specialist manufacturers are not encouraged by strict construction laws, but Rudolf Thurner succeeded in making some impact with his RS. The glassfibre gull-winged body looked dramatic, even though, underneath, the floorpan was merely a shortened NSU 1200C. Engines derived from the NSU range, although a 135bhp racing version was offered with fuel injection. Another project to sell cars with VW Beetle floorpans and Porsche 914 power came to nothing. The ultimate car for the NSU collector?

Engine 996cc 69x66.6mm/1085cc 72x66.6mm/1177cc 75x66.6mm, 4 cyl OHC, R/R
Max power 65bhp at 5500rpm to 135bhp at 6200rpm
Max speed 110-125mph
Body styles coupé
Prod years 1969-74
Prod numbers 121

PRICES
A 12,500 **B** 9000 **C** 6000

Toyota Sports 800

Toyota was a virtually unknown quantity in the west until the 1970s, so we missed out on the interesting little S800. This very compact two-seater featured a fixed rear roof and twin removable targa panels. The engine was an air-cooled twin carb two-cylinder unit which was nevertheless powerful enough to give MG Midget levels of performance. Based on the Publica saloon, it was technically conservative: IFS by torsion bars, rigid rear axle with semi-elliptics, drum brakes and worm and sector steering. Very few outside Japan, where all the interest is.

Engine 790cc 83x73mm, 2 cyl OHV, F/R
Max power 45bhp at 5400rpm
Max speed 96mph
Body styles targa coupé
Prod years 1965-70
Prod numbers n/a

PRICES
A 12,000 **B** 9500 **C** 5500

Toyota 2000GT

Immortalised in the Bond film *You Only Live Twice*, although that car was a unique convertible. The 2000GT coupé was a quite stunning machine, masterfully styled by Count Albrecht Goertz. Pop-up headlamps, aggressive driving lamps behind glass cowls and subtle curves make it a true design classic. Yamaha designed and built the glorious twin cam engine (indeed they built the whole car). Optional Competition tuning could raise power to 200bhp. You got an all-synchro five-speed 'box, four-wheel discs, A-arm and coil spring independent suspension all round. Japanese go mad for them, and pay the earth.

Engine 1988cc 75x75mm, 6 cyl DOHC, F/R
Max power 150bhp at 6600rpm
Max speed 137mph
Body styles coupé
Prod years 1967-70
Prod numbers 337

PRICES
A 80,000 **B** 50,000 **C** 30,000

Toyota Corona/Corona MkII

The Corona deserves a mention because it was the first Toyota to be sold in Britain, arriving here in October 1965. Its inclined grille and double headlamps are obvious ID features. Utterly conventional in every way: coil sprung front, leaf sprung rear, hypoid final drive, worm and sector steering, drum brakes, floor-change four-speed synchro gearbox (or automatic option). 1200/1350 options in Japan, but 1500 or 1600S engines in UK, the latter with front discs and twin carbs. MkII has five-bearing 1900 OHC engine in UK and servo front discs. Twin carbs on SL Coupé, which is the only one of any interest.

Engine 1490cc 78x78mm/1587cc 80.5x75mm/1858cc 86x80mm/1879cc 88x78mm, 4 cyl OHV/OHV/OHC/OHV, F/R
Max power 74bhp at 5000rpm to 108bhp at 5500rpm
Max speed 87-100mph
Body styles saloon, estate, coupé
Prod years 1964-72/1968-72
Prod numbers c 1,788,000
PRICES
Saloon/Estate **A** 1500 **B** 1000 **C** 400
Coupé **A** 2000 **B** 1200 **C** 600

Toyota Crown

Although the Crown name dates back to 1955, the first one to make it here was the '67-'71 series, first imported in late 1968. The version sold here had a seven-bearing 2.3-litre overhead cam straight six engine. Dull specification included a live rear axle, coil spring/damper suspension, recirculating ball steering and drum brakes. UK cars had standard Toyoglide automatic and power front disc brakes, and we only got the saloon and estate. There was an S model as well as a hardtop coupé which had 2 litres, twin carbs and 105-125bhp, but that was reserved for other markets. Do any still exist here?

Engine 1998cc 75x75mm/2253cc 75x85mm, 6 cyl OHC, F/R
Max power 105-125bhp at 5800rpm/115bhp at 5200rpm
Max speed 96-103mph
Body styles saloon, estate, coupé
Prod years 1967-71
Prod numbers 352,882

PRICES
A 1500 **B** 1000 **C** 500

Toyota Crown

The next Crown series became a much more familiar sight on British roads. Its styling was nothing if not distinctive: slab sides, angular C-pillar treatment, cowhorn bumpers and a curious extra air intake above the quad headlamps. A huge, vulgar, bumbling contraption, weighing nearly 1.5 tons. Still a rigid rear axle, so handling is ponderous in the extreme. Estate has an extra row of rear-facing seats in the boot, while the fastback pillarless coupé (which did come to the UK) vaguely resembled a Mustang. 2-litre engine for Japan, 2300 for some markets, but exclusively 2600 in UK.

Engine 1998cc 75x75mm/2253cc 75x85mm/2563cc 80x85mm, 6 cyl OHC, F/R
Max power 115bhp at 5800rpm to 130bhp at 5200rpm
Max speed 100-106mph
Body styles saloon, estate, coupé
Prod years 1971-74
Prod numbers n/a

PRICES
A 1600 **B** 1000 **C** 500

Toyota Celica

Virtually every boring Japanese saloon had its coupé equivalent, so it was pleasing to see a specially-conceived coupé from Toyota – even though, underneath, it was only a Carina. Pretty styling, initially in notchback coupé form, but from '73 (UK '76) also as the Liftback (fastback hatch). UK imports consisted of the 1600ST (OHV), 1600GT (OHC), 2000ST (OHC) and 2000GT (DOHC), the 2-litre ones in Liftback form only. Japan also had a 1400 model. Lots of equipment (dash was chock full of gauges) *and* some character. They are now gaining a following in Britain.

Engine 1407cc 80x70mm/1588cc 85x70mm/1968cc 88.5x80mm, 4 cyl OHV/OHV & OHC/OHC & DOHC, F/R **Max power** 86bhp at 6000rpm/100bhp at 6000rpm to 115bhp at 6400rpm/100bhp at 5500rpm to 130bhp at 6000rpm **Max speed** 102-112mph **Body styles** coupé, hatchback coupé **Prod years** 1970-77 **Prod numbers** c 1,500,000
PRICES
1600/2000ST **A** 2500 **B** 1300 **C** 500
1600/2000GT **A** 3500 **B** 1600 **C** 600

Trabant P50/500/600

A variety of 'people's cars' was built at East Germany's Sachsenring plant at Zwickau, starting with the DKW-derived IFA F8 and culminating with the Trabant, which began life as the P50. This unpleasant little car had a 500cc two-stroke twin engine, Duroplast body (a mix of textile and plastic fibres) over a steel tube skeleton, crash 'box with freewheel and hydraulic drum brakes. Estate car was called Universal. More power for the 1960 500, and bored-out engine for the 600 from 1963. Marginally more interesting than the later 601.

Engine 500cc 66x73mm/594cc 72x73mm, 2 cyl TS, F/F **Max power** 18bhp at 3750rpm/20bhp at 3900rpm/23bhp at 3900rpm **Max speed** 56/59/62mph **Body styles** saloon, estate **Prod years** 1958-60/1960-63/1963-65 **Prod numbers** P50/500: 131,495/600: 106,117

PRICES
A 700 **B** 500 **C** 300

Trabant 601/1.1

Significant as a symbol of crumbling communist regimes, but little else. East Germans had to wait years to own one, but now they're giving them away (literally). Same choking two-stroke engine, therefore same painful progress, but new squared-up Duroplast body. Taken over by VW after reunification, so post-'90 ones have VW Polo 1.1-litre engines. When the Berlin wall fell, a flood of them came west and it was (briefly) a 'fashion' statement to own one, but the novelty quickly wore off. If you must have a Trabi, make it a military-derived open jeep-style Tramp, offered from 1978.

Engine 594cc 72x73mm/1043cc 75x79mm, 2/4 cyl TS/OHV, F/F **Max power** 23bhp at 3900rpm to 26bhp at 4200rpm/40bhp at 5300rpm **Max speed** 59-67mph **Body styles** saloon, estate, utility **Prod years** 1964-90/1990-91 **Prod numbers** 2,860,214 (of which c 55,000 1.1)

PRICES
A 600 **B** 400 **C** 100

Trident Clipper

This should have been a TVR: the Blackpool maker commissioned Trevor Fiore to design it and Fissore to build it on a Grantura chassis but TVR went bust in 1965. It fell to TVR dealer Bill Last to make them in series, albeit modified with glassfibre bodies and Austin-Healey 3000 chassis, into which he plonked a 390bhp Ford V8. Gulp. The Healey left production in '68, so a lengthened TR6 chassis was then used. From '71, a Chrysler 5.6-litre V8 became standard. 1976 revival had impact bumpers and a Ford 5-litre V8. Something of a brute.

Engine 4727cc 101.6x73mm/5572cc 102.6x84.1mm/4950cc 101.6x76.2mm, V8 cyl OHV, F/R **Max power** 140bhp at 3600rpm to 390bhp at 6000rpm **Max speed** 120-150mph **Body styles** coupé, convertible **Prod years** 1966-77 **Prod numbers** c 135 (incl V6)

PRICES
A 13,000 **B** 9000 **C** 5000

Trident Venturer V6/Tycoon

Although the only way to tell a Venturer apart from a Clipper was its square grille-mounted headlamps, it was a very different car. Under the plastic bonnet sat a Ford 3-litre V6. Apart from the very first Healey-based cars, all had the lengthened TR6 platform, which means all-independent suspension (though '76 revival cars have rigid rear axles). Tycoon has Triumph 2.5 Lucas injection engine and an automatic option, but only six were made. Died in '74, only to be revived in '76 with US Federal bumpers and quad headlamps. No convertibles, though most have sliding canvas sunroofs.

Engine 2994cc 93.7x72.4mm/2498cc 74.7x95mm, V6/6cyl OHV, F/R **Max power** 136bhp at 4750rpm/143bhp at 5500rpm **Max speed** 125mph **Body styles** coupé **Prod years** 1969-77/1971-74 **Prod numbers** see Clipper

PRICES
A 7500 **B** 5000 **C** 3000

Triumph

Triumph 1800/2000/Renown

Joined with Standard in 1945, Triumph launched its first post-war cars in '46. Roadster (below) and saloon share the same twin-rail chassis. Razor-edge styling, aluminium-over-ash construction, transverse leaf and wishbone IFS, leafs out back, engine derived from Standard 14, obstructive column-change with four-speed 'box, hydraulic brakes. Interim 2000 has Vanguard engine and three speeds. Renown also has Vanguard's box section chassis and coil spring IFS, though the same body. Limo '51-'52 has 3in longer wheelbase, standard on LWB MkII from '52.

Engine 1776cc 73x106mm/2088cc 85x92mm, 4 cyl OHV, F/R
Max power 65bhp at 4400rpm/68bhp at 4200rpm
Max speed 72mph
Body styles saloon, limousine
Prod years 1946-49/1949/1949-54
Prod numbers 4000/2000/9491

PRICES
A 6500 **B** 3500 **C** 2000

Triumph 1800/2000 Roadster

Quite unique post-war sports car: very wide body (three abreast bench seat – courtesy of column change – and triple wipers!), plus a dickey seat and cover which doubles up as a second glass screen. Chassis identical to the razor-edge saloon, so suspension is firm but overall feel is heavy and performance lacking. Lots of luxury, though. Vanguard engine, three-speed 'box and back axle from '48 – so even less sporting appeal. For all that, an immensely appealing package – the style and character are friendly and the dickey seat is great fun ('invigorating' said one journal). Values remain high.

Engine 1776cc 73x106mm/2088cc 85x92mm, 4 cyl OHV, F/R
Max power 65bhp at 4400rpm/68bhp at 4200rpm
Max speed 75mph
Body styles drophead coupé
Prod years 1946-48/1948-49
Prod numbers 2501/2000

PRICES
A 15,000 **B** 10,000 **C** 5000

Triumph Mayflower

The razor-edge approach falls a bit flat on a car of the Mayflower's size, especially with its straight-through wing-line. At least it's spacious and airy inside. Unitary construction, pre-war Standard 10 sidevalve engine, Vanguard three-speed 'box and back axle, coil spring IFS. Sluggish and a dreadful handler but high level of build quality and durability. In '51, improved rear suspension, wider track, press-button door handles. Only 11 drophead coupés made 1950-51. Will never be a desirable classic, but it has its fans who care not a jot for perceived fashionableness.

Engine 1247cc 63x100mm, 4 cyl SV, F/R
Max power 38bhp at 4200rpm
Max speed 60mph
Body styles saloon, drophead coupé
Prod years 1949-53
Prod numbers 35,000

PRICES
A 2500 **B** 1200 **C** 600

Triumph TR2/TR3/TR3A/TR3B

Successful sports cars aimed at America. Separate chassis, front suspension developed from the Mayflower, 2-litre engine based on the Vanguard. Although handling was not too sharp, its top speed certainly was. That, combined with a bargain price and a win in the 1954 RAC Rally, brought it instant success. Shorter doors from '55. TR3 has egg crate grille, 95-100bhp, the novelty of front discs from '56; optional triple overdrive and rear seats. TR3A has full-width grille, door handles, bigger engine option from '59. 1962 'TR3B' is US-only overlap car with TR4 all-synchro 'box. Raw and rugged motoring for spartans.

Engine 1991cc 83x92mm/2138cc 86x92mm, 4 cyl OHV, F/R
Max power 90bhp at 4800rpm to 100bhp at 5000rpm/100bhp at 4600rpm **Max speed** 103-110mph
Body styles sports **Prod years** 1953-55/1955-57/1957-62/1962
Prod numbers 8628/13,377/58,236/3331
PRICES
TR2 **A** 10,000 **B** 7500 **C** 5000
TR3/TR3A/TR3B **A** 9000 **B** 7000 **C** 5000

Triumph Italia

Triumph's styling consultant, Giovanni Michelotti, penned this typically Italian fixed-head design on the TR3A chassis in 1958. It went into production at the Vignale works, Triumph delivering the unmodified rolling chassis for the Italian coachbuilder to fit its all-steel bodies. All Italias had chrome wire wheels. Elements of the later TR4 can be seen in the bodywork. No Italias were officially sold in the UK, most of them being supplied to well-heeled Italian Triumph enthusiasts, but despite their rarity, you do come across them. Vignale's metalwork was not always very wonderful, though.

Engine 1991cc 83x92mm, 4 cyl OHV, F/R
Max power 100bhp at 5000rpm
Max speed 108mph
Body styles coupé
Prod years 1959-63
Prod numbers 329

PRICES
A 20,000 **B** 16,000 **C** 10,000

Triumph

Triumph TR4/TR4A

While the TR2/3/3A were confirmed old school sports cars, the TR4 brought Triumph right up to date. Naturally, Michelotti did the styling on essentially the same chassis, though rack-and-pinion steering and servo brakes were now standard. Longer, lower, wider and heavier, the TR4 was more comfortable and practical, if still hard-riding. The 2-litre engine was usually requested for sub-2000cc competition use. 'Surrey' steel hardtop is a precursor of the targa roof. Coil spring IRS on TR4A added weight, though a little extra power compensated. Recognise a TR4A by its wing-mounted sidelights.

Engine 1991cc 83x92mm/2138cc 86x92mm, 4 cyl OHV, F/R
Max power 100bhp at 5000rpm/ 100bhp at 4600rpm to 104bhp at 4700rpm
Max speed 110mph
Body styles sports
Prod years 1961-65/1964-67
Prod numbers 40,253/28,465

PRICES
A 8500 **B** 6000 **C** 4000

Triumph TR5/TR250

The first British production car to gain fuel injection was the TR5 PI, at least on the home market. Lucas supplied the technology, which was fitted to a Triumph 2500 six-cylinder engine. Body is the same as the TR4A, but you have bigger brakes, stiffer rear suspension (ouch) and a lot more performance. Unless, that is, you buy an American market TR250, which has twin Strombergs and only 104bhp, and is recognised by its racing stripes on the bonnet – best avoid this watered-down TR5. Sought-after, but pay attention to corrosion in separate chassis and trouble with fuel injection.

Engine 2498cc 74.7x95mm, 6 cyl OHV, F/R
Max power 150bhp at 5500rpm/ 104bhp at 4500rpm
Max speed 125/107mph
Body styles sports
Prod years 1967-68
Prod numbers 2947/8484

PRICES
TR5 **A** 10,000 **B** 7500 **C** 5000
TR250 **A** 7000 **B** 4500 **C** 3000

Triumph TR6

Simply by redesigning the front and rear ends (German coachbuilder Karmann did the work), the TR5 suddenly looked more modern. Seats are new and you get a front anti-roll bar, but underneath it's beginning to look a bit antique. Optional hardtop is a squarer one-piece steel affair, no longer a 'Surrey'. 150bhp on fuel-injected European cars (but only 124bhp from the end of '72), and for the USA, the same twin carb set-up as TR250. Gear ratios altered in '71, lip spoiler and standard overdrive added in '73. PI production ends in '75, though twin carb runs until mid-'76. RIP the great macho Triumph sports car.

Engine 2498cc 74.7x95mm, 6 cyl OHV, F/R
Max power 104bhp at 4500rpm to 150bhp at 5500rpm
Max speed 107-125mph
Body styles sports
Prod years 1969-76
Prod numbers 94,619

PRICES
TR6 PI **A** 9500 **B** 6500 **C** 4500
TR6 Carb **A** 7000 **B** 4500 **C** 3000

Triumph Herald/Herald 1200/Herald 12/50/Herald 13/60

Michelotti made the Standard 10 replacement look sharp, but Triumph stayed conservative under it all: separate chassis, tuned Standard engine, drum brakes – though rack-and-pinion steering and all-independent suspension. Twin carb option has 45bhp, standard on coupé and convertible. Optional front discs from '61. 1200 has bigger engine, estate body option; joined by up-market 12/50 (standard sunroof, front discs, heater and 51bhp). 13/60 has more powerful 1296cc engine and slanted Vitesse-style bonnet. Both chassis and body will rust merrily if neglected; but at least most body panels simply bolt on.

Engine 948cc 63x76mm/1147cc 69.3x76mm/1296cc 73.7x76mm, 4 cyl OHV, F/R **Max power** 35bhp at 4500rpm to 61bhp at 5000rpm
Max speed 72-87mph **Body styles** saloon, estate, coupé, convertible
Prod years 1959-64/1961-70/1963-67/1967-71 **Prod numbers** 525,767 (100,275/289,575/53,267/82,650)
PRICES
Saloon/Estate **A** 1600 **B** 1000 **C** 500
Coupé **A** 2200 **B** 1300 **C** 800
Convertible **A** 3000 **B** 2000 **C** 1000

Triumph Vitesse/Vitesse 2-Litre

One of the smallest six-cylinder engines ever seen went into the Vitesse. It was in most respects an uprated Herald, having a stronger chassis, close-ratio 'box, standard front discs and optional overdrive. Easily spotted thanks to its slanted four-headlamp treatment. 95bhp 2-litre (GT6) engine and all-synchro 'box from '66, but still swing rear axles which can achieve extraordinary attitudes in corners. 1968 MkII has rear wishbones and lever-arm dampers, plus more power and a grille of thick bars. Can be a lot of fun, especially in convertible form, and Triumph parts are cheap and plentiful.

Engine 1596cc 66.75x76mm/1998cc 74.7x76mm, 6 cyl OHV, F/R
Max power 70bhp at 5000rpm/95bhp at 4700rpm to 104bhp at 5300rpm
Max speed 88-102mph
Body styles saloon, convertible
Prod years 1962-66/1966-71
Prod numbers 31,261/ 19,951
PRICES
1600 Saloon **A** 2500 **B** 1500 **C** 800
1600 Convertible **A** 3800 **B** 2500 **C** 1400
2-litre worth 15% extra

421 Aldermans Green Road, Coventry CV2 1NP

HERALD — VITESSE — SPITFIRE — GT6

Quality parts at reasonable prices
Friendly service ● Worldwide Mail Order Service ● Fully illustrated Catalogue (free on request)

The Specialists for Engines, Gearboxes, Overdrives and Differentials

Tel: (01203) 645333
Fax: (01203) 645030

Shop open Monday to Friday 9.30 to 5.30;
Saturday 9.00 to 1.00

TRIUMPH!
TRIUMPH!

1953-1982 - WE CAN SUPPLY PARTS FOR ALL
COMPARE OUR PRICES
TR7 PRICE LIST AVAILABLE.
ORDER YOUR FREE COPY NOW
TR8 PRICE LIST ALSO AVAILABLE

Full workshop facilities available from £15 per hour + VAT
Small jobs a pleasure - or a full rebuild to suit your pocket

A large selection of cars for sale always available

TR7 convertibles   £600-£5,000
TR7 FHCs   £350-£2,500
TR7V8   £1,000-£7,500
TR2-6   £1,500-£15,000

Please call for details of current stock

TRs ALWAYS WANTED

TR7, TR7, TR7 .....
We stock the lot, new and guaranteed secondhand spares

ROBSPORT INTERNATIONAL
Cokenach, Barkway, Royston, Herts

Call Rob or Simon: 01763 848673
or Fax 01763 848167

British Sports Car Spares

SPITFIRE...GT6...VITESSE...HERALD... MGB...
MG MIDGET...AUSTIN HEALEY SPRITE

Cut the cost of your Repairs
New & Second-hand Parts Suppliers...

PHONE/FAX FOR OUR PRICE LIST
FRIENDLY ADVICE ALWAYS GIVEN

WORLDWIDE MAIL ORDER SERVICE AVAILABLE

303 GOLDHAWK ROAD, LONDON W12 8EU   TEL: 0181-748 7823 FAX: 0181-563 0101

Triumph

Triumph Spitfire 4/MkII

Considering that it's essentially a Herald underneath, the Spitfire 4 did a good job of extracting sporting character. Shortened backbone chassis, standard front discs, twin carbs, high compression ratio, and welded body panels styled, as ever, by Michelotti. Improbably tight turning circle, front-mounted bonnet hinges and tip-toe cornering inherited from Herald. Optional overdrive and hardtop from '63. MkII gained 4bhp, a new grille, and better upholstery. Extra stages of tune available. Rust is the main enemy of the Spitfire: chassis and most body panels are susceptible.

Engine 1147cc 69.3x76mm, 4 cyl OHV, F/R
Max power 63bhp at 5750rpm/67bhp at 6000rpm
Max speed 90/93mph
Body styles sports
Prod years 1962-65/1965-67
Prod numbers 45,753/37,409

PRICES
A 4000 **B** 2300 **C** 1200

Triumph Spitfire MkIII/IV

Higher bumpers to satisfy American safety laws identified the MkIII Spitfire, though the main change was the fitment of the larger-bore 1.3-litre engine, which upped performance (0-60mph in 13.6 secs). Also larger front disc calipers, wood veneer facia and permanently fixed hood. For the MkIV, Michelotti was again called in to change the front and rear end styling; one-piece front wings and a Kamm tail are obvious changes. Other MkIV alterations were an all-synchro 'box, higher final drive and improved transverse leaf-to-pivot rear suspension, but power fell back and performance suffered.

Engine 1296cc 73.7x76mm, 4 cyl OHV, F/R
Max power 75bhp at 6000rpm/63bhp at 6000rpm
Max speed 90-95mph
Body styles convertible
Prod years 1967-70/1970-74
Prod numbers 65,320/70,021

PRICES
MkIII **A** 3700 **B** 2100 **C** 1000
MkIV **A** 3500 **B** 1900 **C** 800

Triumph Spitfire 1500

Some of the later MkIVs had been fitted with 1500 engines for the US market from 1973, but European buyers did not get the engine until December 1974, at the same time as the MG Midget. Chassis is unchanged over MkIV, so that means the superior handling set-up of the swing-spring rear end, helped by a wider rear track. Power was up, but still not as high as the MkIII days. Transfers on the bonnet and boot replace badges and there's a black rear panel, but otherwise it's very hard to spot the changes. No wire wheel option, either. The last separate-chassis Triumph.

Engine 1493cc 73.7x87.5mm, 4 cyl OHV, F/R
Max power 71bhp at 5250rpm
Max speed 98mph
Body styles convertible
Prod years 1974-80
Prod numbers 95,829

PRICES
A 3600 **B** 2000 **C** 1000

Triumph GT6 MkI/II/III

Clever MGB GT rival formed by mating the Spitfire chassis to the Vitesse 2-litre twin carb engine and adding a fixed steel roof and hinged rear window. Optional overdrive improves character immensely. Running gear is Spitfire/Herald-derived, so handling interesting on the limit. That's solved in the MkII, which has the same double wishbone rear suspension as the Vitesse MkII, Spitfire-type raised bumpers, 104bhp, Rostyle wheels and better cabin. MkIII shares Michelotti's Spitfire MkIV restyle, plus perforated disc wheels, reshaped rear quarter windows. Last ones had Spitfire MkIV rear suspension.

Engine 1998cc 74.7x76mm, 6 cyl OHV, F/R
Max power 95bhp at 5000rpm to 104bhp at 5300rpm
Max speed 108-112mph
Body styles coupé
Prod years 1966-68/1968-70/1970-73
Prod numbers 15,818/12,066/13,042

PRICES
Mk I/II **A** 3600 **B** 2500 **C** 1300
MkIII **A** 4500 **B** 3000 **B** 1500

Triumph 2000 MkI/II

Arch-rival Rover's 2000 had a strong competitor from Triumph. Michelotti styled the unitary body, the Standard Vanguard donated most of the mechanical side. Twin carbs, synchromesh on all four gears, servo front discs, all-independent suspension, optional Laycock overdrive or automatic. Estates arrive for the '66 model year. MkII gains attractive restyle front and rear, which lengthens both ends. Also new wheel trims, full width wood dash, power steering option, bigger brake servo. Power up and higher gearing in '75, when estate is pensioned off. Not expensive yet, but there's a strong following.

Engine 1998cc 74.7x76mm, 6 cyl OHV, F/R
Max power 90bhp at 5000rpm to 93bhp at 4750rpm
Max speed 100mph
Body styles saloon, estate
Prod years 1963-69/1969-77
Prod numbers 120,645/99,171

PRICES
A 2200 **B** 1000 **C** 500

Triumph

Triumph 2.5 PI MkI/2.5 PI MkII/2500TC/2500S

Fitting a detuned TR5 fuel-injected 2.5-litre engine in the 2000 hull produced a real executive express. But it never really shook off the reputation of Lucas's unreliable injection system. Outward differences were black vinyl C-pillar and Rostyle wheel trims. MkII (photo) had the same restyle as the 2000, except for a matt black grille. Renamed 2500PI in '74 when 99bhp twin carb 2500TC launched (fuel injection was quietly forgotten in '75). 2500S was an up-market non-injection model with standard PAS, overdrive and alloy wheels – it's easily the best (though not the fastest) of the bunch.

Engine 2498cc 74.7x95mm, 6 cyl OHV, F/R
Max power 132bhp at 5450rpm/99bhp at 4700rpm/106bhp at 4700rpm
Max speed 105-110mph
Body styles saloon, estate
Prod years 1968-69/1969-75/1974-77/1975-77
Prod numbers 97,140 (9029/47,455/32,492/8164)

PRICES
A 2500 **B** 1250 **C** 700

We have the largest stock of used parts anywhere in the country, so why not buy where the trade buys **AT TRADE PRICES**. We guarantee if you can buy cheaper from any other trader, then please let us know and we will refund the difference.

NOW NO VAT! NOW NO VAT! NOW NO VAT! NOW NO VAT!

Water pumps, all models and Herald	£10
Fans, all models and Herald, from	£5
Carbs, all models per pair	£35
Manifolds, all Spitfires, inlet and outlet, from	£10
Speedos, rev counters, Jaeger & Smiths, each	£8
Fuel & temp, Jaeger & Smith, each	£5
Dash tops, MkIV, 1500, GT6 III	£25
Grill, MkIV, 1500, GT6 III, TO CLEAR	£5
Steering wheel, MkIV, 1500, GT6 III	£10
Driveshafts, MkI, II, III, Herald & early MkIV complete	£35
Doors, MkIV, 1500, complete from	£25
Wheels, all models and Herald, ONLY	£5
Wheel trims, MkIV 1500, black or silver	£3
Starter motor, all models, TO CLEAR	£5
Alternator, MkIV, 1500, GT6 III	£12
Dynamo, MkI, II, III, Herald & early GT6, TO CLEAR	£5
Wiring looms, all models, complete	£25
Dash support bracket, all models	£5
Bumpers, MkIV, 1500, GT6 III, from	£35
Propshafts, all models, non overdrive	£15
Hood frame, MkIV, 1500	£45
Steering rack, all models, TO CLEAR	£10
Springs, all models, from	£10
GT6/Vitesse 2000cc & 2500cc engines only	£95
Chassis – all models, all good	£85
Bonnet, side catch, all models	£5
Headlight cowl, MkIV, 1500, GT6 III	£15
Anti roll bar, all models	£5

NEW PARTS AVAILABLE

Boot racks, polished alloy, all chrome or wood slat type, all at only £22 EACH

VARIOUS PANELS AVAILABLE

Front corner valances, each £19
Soft top, MkI, II, III, IV, 1500, good quality, rear zip window, steel poppers, and free popper fixing tool, ONLY £68

If the part you require is not listed here, please ring, I am sure we will have it somewhere amongst our huge stock of used parts. We can deliver next day if required and we accept Access and Visa, or if you prefer to call in, we are just 10 minutes from Junction 11 of the M4, or 10 minutes Junction 4A M3. PLEASE RING FIRST.

JINGLES FARM, NEW MILL ROAD, FINCHAMPSTEAD, BERKS, RG40 4QT. Tel or Fax:
01734 732648
'SPITZBITZ' THE ONLY NAME TO REMEMBER

★ HERALD ★ SPITFIRE ★ GT6 ★ VITESSE ★
★ TR2 ★ TR3 ★ TR4 ★ TR5 ★ TR6 ★ TR7 ★

TRGB

CHEAPER TRIUMPH SPARES FROM TRGB

WE HAVE A HUGE SELECTION OF STOCK
WE MAIL ORDER AND ACCEPT ACCESS,
VISA AND SWITCH.
3 DAY SERVICE FROM £5.50
OVERNIGHT FROM £9.50

Unit 1, Sycamore Farm Industrial Estate, Long Drove,
Somersham, Huntingdon, Cambridgeshire, PE17 3HJ
Tel: 01487 842168 Fax: 01487 740274

TRIUMPH WALES

SALES & SERVICE & SPARES & RESTORATION

BREAKING AND RESTORING
HERALDS   SPITFIRES
DOLOMITES
1850/1500/1300
1500 FWD, 1300 FWD
2000, 2500, TR7, VITESSE, GT6
2 x TR6s
MOST PARTS AVAILABLE • FAST MAIL ORDER

STEVE GILL SPORTSCARS
STEDMANS YARD, TALYWAIN, PONTYPOOL, GWENT
01495 774963

The World's Softest Car Covers

Tailor-made from over 25,000 individual patterns
Covercraft offer car covers tested and approved by
the world's leading manufacturers to suit your requirements.

For free Car Care Guide,
brochures and samples contact:

60 Maltings Place, Bagleys Lane
London SW6 2BX
Tel: 0171 736 3214 • Fax: 0171 384 2384

Triumph 1300/1500/Toledo/Dolomite

The 1300/1500 was Triumph's first and only front-wheel drive car, though perversely the later 1500TC, Toledo and Dolomite had rear-wheel drive under the same exterior! Michelotti styling, Herald engine (twin carbs on TC), all-independent suspension, front discs (servo on TC). 1500 engine and longer bodywork from '70. Toledo had beam rear axle and gearbox resited from under the engine to in-line. Attractive Dolomite 1850 model arrived '72, fitted with a new slant-four OHC engine. Toledo renamed Dolomite in '76. All sold as quality small saloons (*à la* BMW), but have far less cachet today than in the 1960s.

Engine 1296cc 73.7x76mm/1493cc 73.7x87.5mm/1854cc 87x78mm, 4 cyl OHV/OHV/OHC, F/F and F/R
Max power 58-76bhp/61-65bhp/92bhp **Max speed** 85-100mph
Body styles saloon **Prod years** 1965-70/1970-76/1970-76/1972-80
Prod numbers 148,350/91,902/119,182/154,296

PRICES
1300/1500 **A** 1200 **B** 700 **C** 200
1850 **A** 1600 **B** 900 **C** 400

Triumph Dolomite Sprint

It's the engine which sets the Sprint apart from most other British '70s saloons. 16-valve technology was unusual for 1973, and Triumph overhead cam unit was both simple and effective (0-60mph in under 9 secs was blistering for '73). TR6/Stag-type four-speed 'box could be had with optional overdrive, or there was an automatic (only 1634 built). Also larger brakes, and alloy wheels (the first on any British mass-produced car); otherwise similar to Dolomite 1850. Overheating and head gasket problems mar a brilliant car; also engine rebuild costs high. Prices rising.

Engine 1998cc 90.3x78mm, 4 cyl OHC, F/R
Max power 127bhp at 5700rpm
Max speed 115mph
Body styles saloon
Prod years 1973-80
Prod numbers 22,941

PRICES
A 3500 **B** 2300 **C** 1000

Triumph Stag

The convertible four-seater idea was Michelotti's and a trend-setter. Unitary construction was a first for a sporting Triumph, and there were MacPherson struts up front and trailing arms and coil springs at the rear. Dual circuit braking system featured servo front discs. The 3-litre V8 engine was a development of the Saab/Triumph 1854cc 'four'. Well-behaved if not outrageously fast. Standard PAS and most had auto 'boxes (manual and overdrive units were also offered). T-bar for safety and soft and hard tops available. Build quality suspect, and the V8 needs looking after, but the Stag has a large following.

Engine 2997cc 86x64.5mm, V8 cyl OHC, F/R
Max power 145bhp at 5500rpm
Max speed 118mph
Body styles targa convertible
Prod years 1970-77
Prod numbers 25,939

PRICES
A 10,000 **B** 6500 **C** 4000

Triumph TR7

Purists moaned, the press attacked Harris Mann's quirky styling and customers became disillusioned by the sad build quality. The TR7 was, in short, a lemon. Yet that's all changing now. The TR7 matured into a respectable sports car from compromised beginnings as a 'US safety' coupé only. Four speeds graduated to five (optional '76, standard from '79), far more desirable convertible arrived '79. 2-litre Dolomite alloy-head engine (but only 8 valves) is fairly perky but prone to overheating. Brakes were never really up to it; many have been uprated. Avoid cheap bangers, but good ones certain to appreciate.

Engine 1998cc 90.3x78mm, 4 cyl OHC, F/R
Max power 105bhp at 5500rpm
Max speed 110mph
Body styles coupé, convertible
Prod years 1975-81
Prod numbers 112,368

PRICES
Coupé **A** 2500 **B** 1300 **C** 600
Convertible **A** 4000 **B** 2500 **C** 1500

Tucker

Preston Thomas Tucker's dream to make 'the most completely new car in 50 years' almost worked. Alex Tremulis styled its long, low shape with such features as a turning 'Cyclops' headlamp, doors which cut into the roof, and myriad safety items. Its engineering was advanced: all-independent suspension, helicopter-derived flat-six alloy engine with the world's first sealed water cooling. It was fast, aerodynamic and good-handling, despite tail-mounted engine. Over-ambitious finance deals killed it after only 50 had been made. Interest became higher after Tucker owner Francis Coppola made a Hollywood blockbuster.

Engine 5540cc, 6 cyl OHV, R/R
Max power 166bhp
Max speed 120mph
Body styles saloon
Prod years 1948
Prod numbers 50

PRICES
Too rare for accurate valuation

Turner A30 Sports

Jack Turner began making bespoke specials in 1950. His first standard offering was the A30 Sports, an Austin A30 powered sports car using a tubular chassis. The Austin also provided its coil spring front suspension, gearbox and hydromech brakes, anticipating the Austin-Healey Sprite. Rear suspension was by trailing arms and torsion bars. Glassfibre bodywork kept weight low and potential speed high. Sold in kit form as well as with new parts (which Turner had to buy from dealers, since Austin refused to supply him).

Engine 803cc 58x76mm, 4 cyl OHV, F/R
Max power 30bhp at 4800rpm
Max speed 80mph
Body styles sports
Prod years 1955-57
Prod numbers 90

PRICES
A 5000 B 3500 C 2500

Turner 950/Sports MkI/II/III

The 950 had the same A30 Sports chassis and body (although now with tiny tail fins), but the A35's hydraulic brakes and engine (which could be tuned as high as 60bhp). The 1959 Sports was the definitive item: revised front and rear bodywork, optional front discs. Sports has cut-down fins, same wide spread of power options (including the Alexander crossflow head). MkII has Ford Anglia/Classic/Cortina power options, in which case Triumph Herald front suspension was fitted. Some Turners had 75-90bhp Coventry-Climax 1100/1200 engines. Most went to the USA. Makes an interesting historic racer.

Engine 948cc/948cc/997cc/1340cc/1498cc/1098cc/1216cc, 4 cyl OHV/OHV/OHV/OHV/OHV/OHC/OHC, F/R **Max power** 43bhp at 4000rpm to 90bhp at 6900rpm **Max speed** 80-100mph **Body styles** sports
Prod years 1957-59/1959-60/1960-63/1963-66
Prod numbers 170/160/150/100
PRICES
950 A 5000 B 3500 C 2500
Sports I/II/III A 6000 B 4000 C 2800
Turner-Climax A 8000 B 5500 C 3800

TVR Grantura I/II/IIA

Trevor Wilkinson produced about 30 cars and chassis before the first definitive TVR: the Grantura. Dumpy glassfibre coupé with tubular backbone chassis, VW torsion bar and trailing arm suspension, wire wheels, Austin-Healey drums, Ford Consul windscreen. Large choice of engines: Ford sidevalve (with optional 56bhp supercharging), Coventry-Climax 1100/1200, BMC B-series (usually MGA) and Ford 105E. MkII has Ford Classic engine option, little rear fins and a blister over each rear wheelarch, plus front disc option, which becomes standard on MkIIA.

Engine 1172cc/997cc/1340cc/1489cc/1588cc/1622cc/1098cc/1216cc, 4 cyl SV/OHV/OHV/OHV/OHV/OHV/OHC/OHC, F/R
Max power 35bhp at 4500rpm to 108bhp at 6000rpm
Max speed 78-110mph
Body styles coupé
Prod years 1958-60/1960-61/1961-62
Prod numbers I: 100/II & IIA: 400

PRICES
A 4500 B 3300 C 2200

TVR Grantura III/1800S/MkIV

Development of TVRs had progressed in a very piecemeal way, but Granturas all looked alike from the MkIII. The chassis was stiffer and the wheelbase longer, though the overall length remained the same. All-independent suspension was by coil springs and wishbones, and you got front discs and rack-and-pinion steering. Some early ones still had 109E and Climax engines, but the spec solidified around MGA or, from '63, MGB engines; overdrive option with the latter. 1800S has fin-less cut-off tail and MGB power. Confusingly-named MkIV has 115bhp option, steel wheels and better trim.

Engine 1216cc 76.2x66.7mm/1340cc 81x65.1mm/1622cc 76.2x88.9mm/1798cc 80.3x88.9mm, 4 cyl OHC/OHV/OHV/OHV, F/R
Max power 83bhp at 6000rpm/57bhp at 5000rpm/90bhp at 5500rpm/95bhp at 5400rpm
Max speed 101-110mph
Body styles coupé
Prod years 1962-64/1964-66/1966-67
Prod numbers 90/128/78
PRICES
A 4700 B 3500 C 2300

TVR Vixen S1/S2/S3/S4

Grantura replacement standardises around the Ford Cortina GT engine (though 12 early cars had MGB power). Externally, it all looks very familiar, apart from a new bonnet scoop. Steel wheels were standard and the option of wires was supplemented by alloys. Less power but also less weight, so performance still excellent. S2 has longer Tuscan SE-type 90in wheelbase and bolted-on bodywork instead of bonding (which helps restorers). S3 has standard alloys, 92bhp Capri engine and Zodiac MkIV square side vents. S4 is interim car with 2500M chassis. Hard-riding as ever, but now better built and more comfortable.

Engine 1599cc 81x77.6mm/1798cc 80.3x88.9mm, 4 cyl OHV, F/R
Max power 88bhp at 5400rpm to 92bhp at 6000rpm/95bhp at 5400rpm
Max speed 106-112mph
Body styles coupé
Prod years 1967-68/1968-70/1970-72/1972-73
Prod numbers 117/438/168/23

PRICES
A 5000 B 3500 C 2000

TVR Griffith 200/400

America had been virtually TVR's first market. Importer Jack Griffith fitted Ford 289 V8s into Grantura IIIs shipped over from Blackpool, together with Ford four-speed manual or automatic 'boxes. 200 is the early car with BMC final drive; 400 has the cut-off tail, plus a larger radiator and electric fans to cure overheating. Handling was brutish, performance was terrifying (0-60mph in less than 6 secs). All but 20 were sold in the USA, where its shortcomings (zero ride quality, cramped interior, patchy build quality) were kindly overlooked. Extremely collectable, especially in RHD.

Engine 4727cc 101.6x72.9mm, V8 cyl OHV, F/R
Max power 195bhp at 4400rpm/ 271bhp at 6500rpm
Max speed 140-155mph
Body styles coupé
Prod years 1963-64/1964-66
Prod numbers 310

PRICES
A 20,000 **B** 16,000 **C** 11,000

TVR Tuscan V8/V8 SE/V8 SE 'Wide'/V6

Griffith revived by new management after a short gap. Identical spec to old Griffith, the V8 having the 195bhp engine, the SE boasting the 271bhp unit, though some late ones came with 220bhp Ford 302 V8s (outputs are gross and probably to be taken with a pinch of salt). From 1968, the Tuscan got a 4.5in longer wheelbase and a new 4in wider body and chunky 6in rim wire wheels. Comparatively good value for the performance, but bombed in increasingly suspicious US market. Only two RHD wide-bodied cars were made. V6 has Vixen body, 3-litre Capri engine, optional dual overdrive.

Engine 4727cc 101.6x72.9mm/4950cc 101.6x76.2mm/2994cc 93.7x72.4mm, V8/V8/V6 cyl OHV, F/R **Max power** 195bhp at 4400rpm to 271bhp at 6500rpm/220bhp at 4600rpm/128bhp at 4750rpm **Max speed** 125-155mph
Body styles coupé **Prod years** 1967/1967-68/1968-70/1969-71
Prod numbers 28/24/21/101

PRICES
V8 **A** 18,000 **B** 15,000 **C** 10,000
V6 **A** 6500 **B** 4500 **C** 3000

TVR 1300/2500

Under Martin Lilley, TVR supplemented its traditional Ford-powered Vixen by courting Triumph. 1300 was an economy special with a Spitfire engine, while the 2500 had a TR6 engine and gearbox (a move to get around US emissions regulations). 1300 virtually non-existent (have any not been re-engined?), but the 2500 was successful. Structure and body is Vixen, though the last examples of both types had the M-type chassis (just like the Vixen S4 which arrived just after these models). If you want a TR-engined TVR but like the earlier body style, this is for you.

Engine 1296cc 73.7x76xmm/2498cc 74.7x95mm, 4/6 cyl OHV, F/R
Max power 63bhp at 6000rpm/ 106bhp at 4950rpm to 150bhp at 5500rpm
Max speed 100/110-115mph
Body styles coupé
Prod years 1971-72
Prod numbers 15/385
PRICES
1300 Impossible to value (probably none exists)
2500 **A** 6000 **B** 4000 **C** 2700

TVR 1600M/2500M/3000M

Although the new TVR M series chassis had already made its appearance in the last 1300/2500s and Vixen S4, the first true 'M' car was the 2500M. Over the stiffer 90in chassis sat a new, longer body (still in glassfibre and with only two seats). 2500M had a smog-restricted TR6 engine/gearbox, front discs, standard alloys and optional overdrive, but was intended mainly for America. Britain got the 1600M (Cortina GT power train) and 3000M (Capri V6), the latter being available with optional overdrive. 1600M briefly left production 1973-75. Far fewer Ms were supplied in kit form than previous TVRs.

Engine 1599cc 81x77.6mm/2498cc 74.7x95mm/2994cc 93.7x72.4mm, 4/6/V6 cyl OHV, F/R **Max power** 86bhp at 5500rpm/106bhp at 4950rpm/138bhp at 5000rpm
Max speed 105/110/120mph
Body styles coupé **Prod years** 1972-73 & 1975-77/1972-77/1972-79
Prod numbers 148/947/654
PRICES
1600M **A** 6000 **B** 4000 **C** 2800
2500M **A** 5000 **B** 3500 **C** 2500
3000M **A** 6500 **B** 4500 **C** 3000

TVR 3000M Turbo/Taimar Turbo/3000S Turbo

TVR really bit the bullet with the Turbo. Virtually no-one was into turbocharging in 1975 when TVR listed its first Turbo model, compressor courtesy of Broadspeed. First model out was the 3000M version, followed by Taimar (hatch) and S (convertible) interpretations. Only four high-spec SE models were made – three Taimars and one S. The changes were minimal: fatter tyres, higher gearing and some stripes, but that was it. Therefore you could expect a very interesting 0-60mph in 5.8 secs, surprisingly little turbo lag, 19mpg and severe under-braking. Massively desirable but extremely rare.

Engine 2994cc 93.7x72.4mm, V6 cyl OHV, F/R
Max power 230bhp at 5500rpm
Max speed 139mph
Body styles coupé, hatchback coupé, convertible
Prod years 1975-79/1976-79/1978-79
Prod numbers 20/30/13
PRICES
3000M Turbo **A** 9500 **B** 7500 **C** 5500
Taimar Turbo **A** 11,000 **B** 8500 **C** 6000
3000S Turbo **A** 17,000 **B** 12,500 **C** 10,000

TVR Taimar

Since the very first TVR, not one had had an opening boot. Owners were expected to manhandle their suitcases over the seats into a cramped luggage compartment. In 1976, the rear end was restyled around a large opening tailgate, which made the whole car very much more practical, and the Taimar immediately became TVR's best-selling model. The weight penalty was negligible, so performance remained exactly as per the 3000M (only 3-litre Taimars were made). Some 30 Taimars received turbocharged engines (see page 216). Today, the Taimar is one of the most sought-after 1970s TVRs.

Engine 2994cc 93.7x72.4mm, V6 cyl OHV, F/R
Max power 138bhp at 5000rpm
Max speed 120mph
Body styles hatchback coupé
Prod years 1976-79
Prod numbers 395

PRICES
A 7500 **B** 5000 **C** 3500

Unipower GT

Undoubtedly one of the best of the many Mini-based sports cars of the 1960s was the Unipower. Designed by BMC works driver Andrew Hedges and Tim Powell, it was made by fork-lift manufacturer Universal Power Drives (hence Unipower). Mini-Cooper/Cooper 'S' engine mounted just ahead of rear axle in tubular space frame, to which the glassfibre body was bonded. Right-hand gearchange, front discs. Only 40in high and tiny doors, so ingress was an art-form. Fabulous handling and high performance due to light weight (½ ton/570kg). Japan has sucked most of them away.

Engine 998cc 64.6x76.2mm/1275cc 70.6x81.3mm, 4 cyl OHV, M/R
Max power 55bhp at 5800rpm/ 76bhp at 5800rpm
Max speed 100/120mph
Body styles coupé
Prod years 1966-70
Prod numbers 75

PRICES
A 8000 **B** 5500 **C** 3000

Vanden Plas Princess 4-Litre

This began life as the long-wheelbase Austin A135 Princess in 1952, but was badged merely Princess from 1957. The wheelbase was a sizeable 11ft (335cm) and you could choose between a six-light limousine with division or, from '53, a saloon version (usually four-light). Tubeless tyres and auto option from late '55, PAS is a desirable option (from '60). Servo brakes in '62, Selectaride dampers optional from '62. Saloon dies in '67, limo in '68. Many chassis-only deliveries, plus a few landaulettes.

Engine 3995cc 87x111mm, 6 cyl OHV, F/R
Max power 135bhp at 3700rpm
Max speed 89mph
Body styles saloon, limousine, landaulette
Prod years 1952-68
Prod numbers 3344

PRICES
A 4500 **B** 3000 **C** 1200

Vanden Plas Princess IV

Although announced as an Austin, the Princess IV went into production badged as a Princess, but since it was made by Vanden Plas, it is listed here. New long-wheelbase chassis, Hydramatic automatic 'box, power steering, tubeless tyres, and a tweaked 4-litre engine. The bodywork was styled by Vanden Plas in an inelegantly slab-sided fashion: a two-ton 18½ft (499cm) barge. As it cost more than a Jaguar MkVIII, it's no surprise it didn't sell – Austins aren't supposed to cost that much. Rare touring limousine version had a glass division (22 built). Few survivors, and beware of rust and thirst.

Engine 3995cc 87x111mm, 6 cyl OHV, F/R
Max power 150bhp at 4100rpm
Max speed 99mph
Body styles saloon, touring limousine
Prod years 1956-59
Prod numbers 200

PRICES
A 6000 **B** 3500 **C** 2000

Vanden Plas 3-Litre MkI/II

Super-luxury version of the separate-chassis Austin A99 Westminster and a much better bet than the Princess IV it replaced. Vanden Plas grille and hub caps, twin foglamps, no bonnet scoop, extra sound deadening and oodles of leather and wood marked this out as the top-notch A99 derivative. Overdrive or auto transmission and Touring Limousine option. MkII follows Austin A110 update: more power, twin exhausts, floor-mounted gear lever, rearranged facia. Desirable PAS option from July '62. A better buy than the dowdy old Austin (and at least *some* following), but watch that rust.

Engine 2912cc 83.3x88.9mm, 6 cyl OHV, F/R
Max power 103bhp at 4500rpm/ 120bhp at 4750rpm
Max speed 98/106mph
Body styles saloon, touring limousine
Prod years 1959-61/1961-64
Prod numbers 4715/7900

PRICES
A 2700 **B** 1700 **C** 1000

Vanden Plas 4-Litre R

'R' is BMC's understated way of saying that this car has a Rolls-Royce engine under the bonnet. It's a twin carb alloy straight six, with Rolls' favoured inlet-over-exhaust valve layout. Body is identical to the 3-Litre except for the cut-down tail fins, longer and flatter roof-line, horizontal tail lights and built-in foglamps. Standard Borg Warner automatic, power steering, servo front discs and optional electrically adjustable dampers. Rolls even considered making their own version. Expensive on fuel and repair bills and just as rusty as any A110 family member. Overlight steering is vague, but this model has its followers.

Engine 3909cc 95.25x91.4mm, 6 cyl IOE, F/R
Max power 175bhp at 4800rpm
Max speed 100mph
Body styles saloon
Prod years 1964-68
Prod numbers 6555

PRICES
A 4000 **B** 2500 **C** 1000

Vanden Plas Princess 1100/1300

Top-of-the-range version of the most badge-engineered car of all time. At least this one's slightly more than a fancy grille job: the trimming was done by Vanden Plas in London, and consisted of leather, wood, picnic tables in the backs of the front seats, veneer dash and extra sound deadening. 1100 always has twin carbs, plus sliding sunroof option from '65. 1100 MkII can also come with an auto 'box. Single carb 1300 from '67, but post-'68 manuals have twin carbs and 65bhp. Not really what you'd call a classy car, but quaint nonetheless. Japanese BMC fans love them.

Engine 1098cc 64.6x83.7mm/1275cc 70.6x81.3mm, 4 cyl OHV, F/F
Max power 55bhp at 5500rpm to 65bhp at 5750rpm
Max speed 80-90mph
Body styles saloon
Prod years 1963-68/1967-74
Prod numbers 16,007/27,734

PRICES
1100 **A** 3800 **B** 1600 **C** 400
1300 **A** 5000 **B** 2300 **C** 800

Vanden Plas 1500/1.7

There was still a band of pensioners who remained loyal to the idea of a BL small car with picnic tables. The Vanden Plas 1500 was the answer to their dreams. An Austin Allegro was spirited down to Vanden Plas in London and fitted with a snout-like grille, foglamps, walnut veneer for the dash and door cappings, plus leather and, naturally, picnic tables. Five speeds or automatic could be chosen. Assembly moved to Abingdon in late '79, when a 1.7-litre engine became standard, but only if you ordered automatic. The press regarded it as a joke but it does have its following. Look out for pampered one-owner examples.

Engine 1485cc 76.2x81.2mm/1748cc 76.2x95.8mm, 4 cyl OHC, F/F
Max power 68bhp at 5500rpm/90bhp at 5500rpm
Max speed 91/94mph
Body styles saloon
Prod years 1974-80/1979-80
Prod numbers 11,842

PRICES
A 1400 **B** 800 **C** 100

Vauxhall 10/12/14

Before Vauxhall launched its definitive 1946 line-up, it made half a dozen pre-war 1.4-litre six-light 12s. The same engine went in the new four-light HIX 12, which was also sold with a 1.2-litre engine as the 10. Unitary construction, with engine on a subframe, Dubonnet IFS (which made steering feel odd), three-speed 'box. The 14 had a longer wheelbase, six-cylinder engine, six-light styling, projecting boot and adjustable steering column. Rust is a killer, so very rare today. All are rather slow, although torque makes up for it.

Engine 1203cc 63.5x95mm/1442cc 69.5x95mm/1781cc 61.5x100mm, 4/4/6 cyl OHV, F/R
Max power 31.5bhp at 3600rpm/35bhp at 3600rpm/47.5bhp at 3600rpm
Max speed 58/65/70mph
Body styles saloon
Prod years 1946-47/1946-48/1946-48
Prod numbers 10 & 12: 44,047/14: 30,511

PRICES
A 2200 **B** 1400 **C** 700

Vauxhall Wyvern/Velox

Although the central hull was the same as the old 12/14, the new Wyvern and Velox looked very different, thanks to long front wings incorporating integral headlamps, plain disc wheels, alligator bonnet, low grille of horizontal bars and lengthened tail. Inside, there was a new column change for the three-speed 'box. Wyvern had a modified 12 engine and body-coloured wheels, whereas the Velox had a wider track, cream-coloured wheels and a new 2.3-litre six-cylinder engine. In 1949, both gained separate sidelamps, worm-and-peg steering and leather upholstery. Undistinguished.

Engine 1442cc 69.5x95mm/2275cc 69.5x100mm, 4/6 cyl OHV, F/R
Max power 33bhp at 3600rpm/54bhp at 3500rpm
Max speed 63/75mph
Body styles saloon
Prod years 1948-51
Prod numbers 55,409/76,919

PRICES
Wyvern **A** 2000 **B** 1200 **C** 500
Velox **A** 2200 **B** 1500 **C** 700

Vauxhall Wyvern/Velox/Cresta

Series EIX Wyvern and EIP Velox arrived in 1951 as the first all-new post-war Vauxhalls. The engines remained the same, but new full-width bodywork sat on a longer wheelbase and wider track. Novelties were a curved windscreen, thick chrome grille bars, two-way opening bonnet, hypoid rear axle and coil and wishbone IFS. Short-stroke engines from April '52, rear wing stone-guards from October, high compression engines and recirculating ball steering from '53, slatted grille from '54, wrap-around rear screen from '55. Estates on Velox only. Velox-based Cresta has two-toning, chrome wheel trims, mascot.

Engine 1442cc 69.5x95mm/1507cc 79.4x76.2mm/2275cc 69.5x100mm/2262cc 79.4x76.2mm, 4/4/6/6 cyl OHV, F/R **Max power** 35bhp/ 40-48bhp/ 55bhp/64-68bhp
Max speed 65-84mph **Body styles** saloon, estate **Prod years** 1951-57/1951-57/1954-57 **Prod numbers** 110,588/235,296/166,504
PRICES
Wyvern **A** 2500 **B** 1400 **C** 600
Velox **A** 2700 **B** 1600 **C** 800
Cresta **A** 3000 **B** 1800 **C** 1000

Vauxhall Victor F

One of the most blatant attempts to impose Detroit fads on the leafy suburbs of middle-class England: reverse-angle wrap-around windscreen, 'bullet' bumpers and gashes down the sides. On a car measuring 14ft long, the effects were gruesome. All-synchro 'box, IFS, semi-elliptic live rear. Estates from '58, when Newtondrive two-pedal control also offered. Exhaust moved from through rear bumper to underneath in late '58. Series 2 has full-width grille, protruding sidelamps, wrap-around bumpers, no flash in the rear door. Bigger rear window from '61. One of the all-time worst rust traps.

Engine 1507cc 79.4x76.2mm, 4 cyl OHV, F/R
Max power 55bhp at 4200rpm
Max speed 75mph
Body styles saloon, estate
Prod years 1957-61
Prod numbers 390,745

PRICES
A 2000 **B** 1250 **C** 500

Vauxhall Velox/Cresta PA/PADX & PASX

The Victor became Vauxhall's four-cylinder model from '57, and these were its sixes. Size had grown significantly over the last Velox and GM's American influence was plainly visible in the wrap-around screens and tail fins. Mechanically similar to last sixes except for automatic choke. Bizarre-looking Friary estates from April '59. One-piece wrap-around rear screen, curved chrome grille top and smoother sides from August '59. PADX Cresta and PASX Velox arrive in 1960: square 2.6 engine, larger wheels, bigger fins, Hydramatic and power front disc brake options (latter standard from '61). Ageing rock 'n' rollers' favourite motors.

Engine 2262cc 79.4x76.2mm/2651cc 82.55x82.55mm, 6 cyl OHV, F/R
Max power 78bhp at 4400rpm/ 95bhp at 4600rpm
Max speed 90/97mph
Body styles saloon, estate
Prod years 1957-60/1960-62
Prod numbers 81,841/91,923

PRICES
A 4500 **B** 3000 **C** 1200

Vauxhall Victor/ VX 4/90 FB

Gone were the wrap-arounds and break-your-knees dog-leg of the old Victor F. In its place came a much more soberly styled machine. Same durable engine but with pressurised cooling; choice of column three-speed or floor-change four-speed 'box. De Luxe has separate front seats, leather and side chrome. VX 4/90 (pictured) is saloon-only twin carb model, with servo front discs, floor change, vertical-slatted grille and coloured body flashes. 1595cc engine from September 1963, when Victor gains front discs, VX 4/90 gains walnut dash. Not as rust-prone as previous Victor but styling is bland and generally lacks character.

Engine 1507cc 79.4x76.2mm/1595cc 81.6x76.2mm, 4 cyl OHV, F/R
Max power 50bhp at 4600rpm to 71bhp at 5200rpm/58.5bhp at 4800rpm to 74bhp at 5200rpm
Max speed 78-94mph
Body styles saloon, estate
Prod years 1961-64
Prod numbers 328,640/n/a

PRICES
Victor **A** 1400 **B** 800 **C** 400
VX 4/90 **A** 1700 **B** 1000 **C** 500

Vauxhall Velox/Cresta PB

Expanded Victor FB styling looks even more bland on the Velox and Cresta. Wheelbase up by 2.5in over PA and overall length now exceeds 15ft. Standard servo front discs, three speeds on the column (though still Hydramatic and overdrive options). Higher axle ratio from '63 and Martin Walter-converted estates from the end of '63. 3.3-litre engine from October '64, plus optional four-on-the-floor, wider grille incorporating headlamps. For last six months, power steering becomes an option and the automatic is two-speed Powerglide, not Hydramatic. Utterly lacking in romance, but 3.3 was a formidable tool.

Engine 2651cc 82.55x82.55mm/ 3294cc 92.1x82.55mm, 6 cyl OHV, F/R
Max power 95bhp at 4600rpm/ 113bhp at 4600rpm
Max speed 94-102mph
Body styles saloon, estate
Prod years 1962-65
Prod numbers 87,047

PRICES
A 1500 **B** 800 **C** 300

Vauxhall

Vauxhall Victor/VX 4/90 FC

Roomier body style for the so-called 101 Victor. Higher compression gave more power, but inappropriate two-speed Powerglide auto option (from '65) crippled performance. VX 4/90 remains the sporty twin carb member of the family, distinguished by its thick central grille bar and uprights, plus contrasting body flash and different facia with four hooded instruments. VX 4/90 gets limited-slip diff as standard from late '65, plus a wood facia and chrome side strip from '66. Victor gets cross-hatch grille from '66. Less rust again, but hardly outstanding.

Engine 1595cc 81.6x76.2mm, 4 cyl OHV, F/R
Max power 60-66bhp at 4600rpm/74bhp at 5200rpm
Max speed 86-96mph
Body styles saloon, estate
Prod years 1964-67
Prod numbers 219,814/13,449

PRICES
Victor **A** 1250 **B** 750 **C** 300
VX 4/90 **A** 1500 **B** 900 **C** 500

Vauxhall Cresta/Viscount PC

Velox name is dropped for the '66 model year and the Cresta continues as Vauxhall's top-of-the-range car. The styling's crisper, if still Americanised, but underneath it's basically the same story as the PB (but 3.3-litre engine only). De Luxe has four headlamps, more chrome, Ambla upholstery. Martin Walter estate for just one year ('67-'68). Luxury Viscount (saloon only) arrives '66, with vinyl roof, mesh grille, PAS, electric windows, leather, wood, Powerglide auto (though four-speed manual option). GM automatic replaces Powerglide from '70. Viscount is now quite rare.

Engine 3294cc 92.1x82.55mm, 6 cyl OHV, F/R
Max power 123bhp at 4600rpm
Max speed 102mph
Body styles saloon, estate
Prod years 1965-72/1966-72
Prod numbers 53,912/7025

PRICES
Cresta **A** 1300 **B** 700 **C** 300
Viscount **A** 1500 **B** 800 **C** 300

Vauxhall Victor/Ventora/VX4/90 FD

Coke bottle styling arrives for the Victor FD. American influence persists in three-speed 'box with column-change (though four-on-the-floor/overdrive/automatic are options). Two new engines: hemi-head overhead cam units in 1600 and 2000 sizes. Servo front discs on the latter (optional on 1600), rack-and-pinion steering on all. Ventora is FD shell with the torquey old 3.3-litre six fitted, plus vinyl roof and more fittings. Confusingly, 3.3 estate version badged Victor 3300. GM 'Strasbourg' auto 'box replaces Powerglide in late '69. 1969 VX 4/90 has 2-litre twin carb, overdrive and Rostyle wheels.

Engine 1599cc 85.7x69.2mm/1975cc 92.25x69.2mm/3294cc 92.1x82.55mm, 4/4/6 cyl OHC/OHC/OHV, F/R
Max power 72bhp at 5600rpm/88bhp at 5500rpm to 100bhp at 5400rpm/123bhp at 4600rpm **Max speed** 90-107mph **Body styles** saloon, estate
Prod years 1967-72/1968-72/1969-72
Prod numbers 212,362
PRICES
Victor **A** 1200 **B** 700 **C** 250
VX 4/90 **A** 1300 **B** 800 **C** 400
Ventora **A** 1500 **B** 800 **C** 400

Vauxhall Victor/Ventora/VX4/90/VX FE

Vauxhall and its GM European partner, Opel, began to move closer together. The FE Victor shared its floorpan with the Opel Rekord, but the running gear and engines were different. New 1800 and 2300 engines, standard four-speed 'box and floor change (overdrive and auto options remained), plus front discs. Ventora continued as the 3300 version (estate called Victor 3300 until '73). VX 4/90 had twin carb 2300 engine and Rostyles as before, but overdrive till '73 only, five speeds from '77. Victor renamed VX1800/VX2300 from '76. Better than a Cortina, but rusts badly.

Engine 1759cc 85.7x76.2mm/2279cc 97.5x76.2mm/3294cc 92.1x82.55mm, 4/4/6 cyl OHC/OHC/OHV, F/R
Max power 77bhp/100-116bhp/123bhp **Max speed** 96-104mph
Body styles saloon, estate
Prod years 1972-78
Prod numbers c 90,000

PRICES
Victor/VX **A** 1000 **B** 600 **C** 200
VX 4/90 **A** 1250 **B** 750 **C** 400
Ventora **A** 1100 **B** 700 **C** 300

Vauxhall Viva HA

Moderately successful but unloved new small Vauxhall, a new market area for the Luton firm. Unbecoming boxy styling hid a specification which was quite advanced for a small car in 1963: all-synchro four-speed floor-change 'box, rack-and-pinion steering, IFS and optional front discs. First British car to have an acrylic lacquer finish. Basic ones have fixed rear side windows. De Luxe 90 and SL90 launched in October '65 have high compression (54bhp) engine, reinforced propshaft and standard servo front discs, and are the HAs to have. Almost no collector value at all.

Engine 1057cc 74.3x61mm, 4 cyl OHV, F/R
Max power 44bhp at 5200rpm to 54bhp at 5600rpm
Max speed 80-85mph
Body styles saloon
Prod years 1963-66
Prod numbers 321,332

PRICES
A 900 **B** 600 **C** 300

Vauxhall Viva HB/Viva GT

Shrunken coke bottle styling, larger dimensions, coil springs all round (though live axle rear), bored-out engine (and OHC 1600 from '68), Borg Warner automatic option from '67 on De Luxe and SL. Viva 90 is high compression/Stromberg carb variant with front discs. Three-door estate arrives mid-'67. 1967 Brabham Viva has twin Strombergs and 69bhp, identified by stripes around front grille. True performance model is the '68 GT (photo), which has an inclined Victor 2-litre twin carb engine, close-ratio 'box, bonnet scoops, black grille (and black bonnet till '69).

Engine 1159cc 77.7x61mm/1599cc 85.7x69.2mm/1975cc 95.25x69.2mm, 4 cyl OHV/OHC/OHC, F/R **Max power** 47bhp at 5200rpm to 69bhp at 5600rpm/72bhp at 5600rpm/104bhp at 5600rpm **Max speed** 82-100mph **Body styles** saloon, estate **Prod years** 1966-70/1968-70 **Prod numbers** 91,813/4606
PRICES
Viva **A** 800 **B** 500 **C** 200
Brabham Viva **A** 3500 **B** 2500 **C** 1800
Viva GT **A** 1600 **B** 1100 **C** 800

Vauxhall Viva HC/Magnum/Firenza

Evolutionary style for new Viva, based on old HB platform. Launched with saloon and fastback estate bodies, 1200 and 1600 engines (latter boasting servo front discs and auto option). 1256cc engine replaces 1159cc in '71, 1800 and 2300 arrive '72, though 2300 badged Magnum after '73 (and also offered with 1800 engine). Coupé called Firenza, initially with 1200/1600/2000 engines, from '72 with 1300/1800/2300. After '73, coupés marketed as Magnums but for 16 months only. Bewildering range of models, but none can touch Ford Escort/Capri for 'cool' quotient.

Engine 1159/1256cc 4 cyl OHV, 1599/1759/1975/2279cc 4 cyl OHC, F/R **Max power** 49bhp at 5300rpm to 110bhp at 5200rpm **Max speed** 83-103mph **Body styles** saloon, estate, coupé **Prod years** 1970-79/1973-77/1971-73 **Prod numbers** 640,863/20,300/18,352
PRICES
Viva **A** 800 **B** 400 **C** 100
Magnum Saloon/Estate **A** 1500 **B** 1200 **C** 500 Magnum/Firenza Coupé **A** 1800 **B** 1500 **C** 700

Vauxhall Firenza Droop-Snoot/Magnum Sports Hatch

The 'standard' Firenza died in September '73 and was replaced by a rather different coupé: the Droop-Snoot, so called because of its distinctive slanted glassfibre nose and spoiler, incorporating double rectangular headlamps behind glass covers. There was also a boot spoiler and a set of meaty alloy wheels. Tuned engine (0-60mph in 8.5 secs), ZF rally-derived five-speed 'box, black facia. Magnum Sports Hatch had the same nose treatment and alloy wheels, but the fastback estate body, complete with rear spoiler. Both are very rare but the Firenza is the more desirable. Only 100 are known to remain.

Engine 2279cc 97.5x76.2mm, 4 cyl OHC, F/R
Max power 131bhp at 5500rpm
Max speed 116mph
Body styles coupé/estate
Prod years 1973-75/1976
Prod numbers 204/n/a

PRICES
Droop-Snoot **A** 3500 **B** 2300 **C** 1500
Sports Hatch **A** 2200 **B** 1700 **C** 1000

Vauxhall Chevette

Opel had introduced its Kadett as early as August 1973 and the Chevette was a direct derivative, launched almost two years later. Based on the three-door Kadett City hatch launched at the same time, it had a unique 'droop-snoot' nose and Viva-derived 1256cc engine and four-speed gearbox. Saloon and estate versions arrived in April '76, but not any of the other Kadett variations. Flush headlamps from '79. All had the same old engine, so unexciting performance. Outlived the Kadett for five years, and a successful car for Vauxhall. Still regarded as banger material.

Engine 1256cc 81x61mm, 4 cyl OHV, F/R
Max power 58bhp at 5600rpm
Max speed 92mph
Body styles saloon, hatchback, estate
Prod years 1975-84
Prod numbers 415,608

PRICES
A 700 **B** 400 **C** 100

Vauxhall Chevette HS

Rally homologation special. DTV raced them from '76 but the road version did not arrive until mid-'78. Racing Lotus heads and ZF 'box swapped in road car for DTV-developed 16-valve twin cam head and Getrag five-speed 'box. Suspension based on Opel Kadett GT/E (double front wishbones, rear twin trailing arms, torque tube, coil springs and Panhard rod). Bodywork received a hefty front air dam, wide wings, understated rear spoiler, metallic silver paint with red stripes and fat alloy wheels. 0-60mph in 8.5 secs wasn't bad, but it was crude and cost £1000 more than an RS2000, so sales slow. Desirable.

Engine 2279cc 97.5x76.2mm, 4 cyl OHC, F/R
Max power 135bhp at 5500rpm
Max speed 115mph
Body styles hatchback
Prod years 1978-79
Prod numbers 400

PRICES
A 3500 **B** 2500 **C** 1800

Vauxhall Cavalier

British version of the Opel Ascona, distinguished by its reworked Manta-style nose. Smaller range of engines than the Ascona (only 1600/1900 initially, 2000 from '78 and unique-to-Britain Viva-engined 1300, with manual gearbox only, from '77). Built in Antwerp and Luton, it gave Vauxhall a credible alternative to the Cortina, though it was equally dull in the final reckoning. Coupé launched in '75 and hatchback in '78 were straight duplications of the Manta, sharing their engines (there were no Viva-powered 1300 versions). Coupé sold only with 1900/2000 engines.

Engine 1256cc 81x61mm/1584cc 85x69.8mm/1897cc 93x69.8mm/1979cc 95x69.8mm, 4 cyl OHV/OHC/OHC/OHC, F/R
Max power 58bhp at 5600rpm to 100bhp at 5400rpm **Max speed** 91-109mph **Body styles** saloon, hatchback coupé, coupé
Prod years 1975-81
Prod numbers 238,980
PRICES
Saloon **A** 800 **B** 400 **C** 100
Coupé/Hatch **A** 1700 **B** 1000 **C** 500

Veritas Comet/Saturn/Scorpion/Nürburgring

Ex-BMW engineer Ernst Loof created the Veritas marque to build racing and sports cars. The first was the BMW 328-chassised Meteor (though this was really a race car). The 2-litre series used 80bhp BMW 328 engines in a space frame, while there was also a short-stroke 1.5-litre option. Post-1950 cars had Loof's own Heinkel-built 1988cc 100-147bhp OHC dry sump unit and a five-speed 'box. Comet was SWB sports, Saturn was 2+2 coupé, Scorpion 2+2 convertible. The 1951 Nürburgring was the Spohn-bodied successor, offered in two wheelbase lengths.

Engine 1500cc/1971cc 66x96mm/1988cc 75x75mm, 6 cyl OHV/OHV/OHC, F/R
Max power 55bhp at 4200rpm to 147bhp at 7500rpm
Max speed 93-142mph
Body styles sports, coupé, convertible
Prod years 1947-53
Prod numbers c 40

PRICES
Too rare for accurate valuation

Veritas Dyna-Veritas

The 2-litre Veritas was preposterously expensive for its era, so Loof was keen to produce a more affordable model. He used the Dyna-Panhard as a basis: its air-cooled flat twin engine sat in a tubular chassis and drove the front wheels. The convertible bodywork was designed by Veritas and built by Baur of Stuttgart. The standard offering was a 2+2 which looked rather like a Gutbrod (not a good sign), but there were more attractive two-seater styles. The trouble was, the Dyna-Veritas still cost twice as much as a VW Beetle and sales were not high. Veritas disappeared in 1953.

Engine 744cc 79.5x75mm, 2 cyl OHV, F/F
Max power 32bhp at 5000rpm
Max speed 73mph
Body styles convertible
Prod years 1950-52
Prod numbers 176

PRICES
A 14,000 **B** 12,000 **C** 10,000

Vespa 400

The Italian firm Piaggio made its name with the Vespa motor scooter in the 1950s, but it also made a break for the car market in 1957. It could not be made in Italy thanks to a 'gentleman's agreement' with Fiat and so it was built in France. The Vespa 400 was a pretty 2+2 rolltop coupé with a rear-mounted air-cooled vertical twin engine, and advanced features such as unitary construction, all-independent suspension, hydraulic brakes and a four-speed synchromesh gearbox. In a different stratosphere to most microcars and a stylish alternative to a Fiat 500.

Engine 393cc 63x63mm, 2 cyl TS, R/R
Max power 14bhp at 4350rpm
Max speed 55mph
Body styles rolltop coupé
Prod years 1957-61
Prod numbers c 34,000

PRICES
A 4000 **B** 2500 **C** 1200

Vignale 850

Like Bertone, Alfredo Vignale had a design shop and factory in Grugliasco. Vignale's zenith was in the late 1950s, when it made bespoke bodies for many exotic sports car makers. In the mid-1960s, Vignale turned to making bodies on Fiat platforms. Its effort on the 850, one of many coachbuilt 850s but rare in that it actually went into production, was quite pretty. It was available as a Special coupé, a Spider and a Special Saloon. A handful were imported to Britain but, at £1225 (£250 more than an MGB), it had little chance.

Engine 843cc 65x63.5mm, 4 cyl OHV, R/R
Max power 47bhp at 6200rpm
Max speed 83mph
Body styles saloon, coupé, convertible
Prod years 1965-70
Prod numbers n/a

PRICES
Saloon **A** 1200 **B** 750 **C** 500
Coupé **A** 2000 **B** 1300 **C** 800
Spider **A** 3000 **B** 2000 **C** 1200

Vignale · Volga

Vignale Samantha

The prettiest of the Vignale-Fiats was the 125-based Samantha. Its extreme fastback, electric flip-up headlamps (ten years before the Porsche 928) and flowing lines were all exactly right. Actually heavier than the 125 saloon, so disappointing performance. Fiat never made a 125 coupé, so this is an interesting exercise. 125 floorpan is used, plus its front bulkhead, underbonnet panels, dashboard and rear lights. Bumpers are from a 124 Coupé, door skins shared with Vignale-designed Jensen Interceptor! Cost more than an E-Type in UK. Vignale build quality not up to much, rust a persistent enemy.

Engine 1608cc 80x80mm, 4 cyl DOHC, F/R
Max power 90bhp at 5600rpm to 100bhp at 6000rpm
Max speed 103-110mph
Body styles coupé
Prod years 1967-70
Prod numbers n/a

PRICES
A 3500 B 2000 C 1000

Vignale Eveline

Vignale's effort on the Fiat 124 was not entirely happy, having a rather dumpy, even Japanese, feel to it. It was based on the 124 saloon and shared not only its floorpan and mechanicals but most of its interior, windscreen and steel disc wheels, while the bumpers came from the 124 Coupé. As such, it is more practical to own as a classic than most coachbuilt Italian cars. Casino owner Frixos Demetriou imported a batch in RHD from Vignale, but it is believed all of these have now rusted into non-existence. Despite this, there are some LHD cars around.

Engine 1197cc 73x71.5mm/1438cc 80x71.5mm, 4 cyl OHV, F/R
Max power 60bhp at 5600rpm/ 70bhp at 5400rpm
Max speed 85-90mph
Body styles coupé
Prod years 1967-70
Prod numbers c 200

PRICES
A 2800 B 1700 C 1000

Vignale Gamine

Alfredo Vignale surely had no experience of Enid Blyton, so why does everyone call the Gamine a Noddy car? A piece of 'sixties frippery: a rebodied Fiat 500F with mock classic grille, cutaway doors, free-standing headlamps and sideways-opening bonnet. Novelty value exceeds its competence as a car: crudely built, very slow, lots of scuttle shake, no interior trim, rudimentary hood, lots of rust traps. At £700 (slightly less than an MG Midget), probably around 300 were sold in the UK, courtesy of Demetriou, so there are still plenty around. Vignale sold out to De Tomaso in 1969, and was killed the very same day.

Engine 499.5cc 67.4x70mm, 2 cyl OHV, R/R
Max power 18bhp at 4400rpm
Max speed 60mph
Body styles sports
Prod years 1967-70
Prod numbers n/a

PRICES
A 7500 B 4000 C 2000

Volga M21/M22

The GAZ works at Gorky made the Pobeda before it came up with the unitary-built Volga in 1955. This was built to be tough and was well equipped, having standard radiator blind, tool kit, radio, two-speed wipers, and seats which converted into a bed. Five-bearing 2.5-litre engine was standard, though Belgian importer Sobimpex fitted Perkins or Land-Rover diesel engines for export. Generally three speeds, though a few automatics made from '59. M21 was the saloon, M22 the estate. Limited UK imports 1959-67, including some RHD.

Engine 2445cc 92x92mm, 4 cyl OHV, F/R
Max power 70-97bhp at 4000rpm
Max speed 80-85mph
Body styles saloon/estate
Prod years 1955-71/1961-71
Prod numbers n/a

PRICES
A 1700 B 900 C 300

Volga M24

Shorter and more modern M24 arrived in 1968, boasting such improvements as 110bhp engine, servo brakes, all-synchromesh four speed 'box and floor-mounted gear lever. Estate version from '73 has three rows of seats. Just as solidly built as before (steel is 1mm thick), and ideal for surviving Siberian winters. Slotted in the Soviet hierarchy as a taxi and transport for middle-ranking officialdom. UK imports were slated in the early 1970s, but probably never happened. Peugeot 2.0- and 2.3-litre diesels listed as options. Revised in 1982 to become the 3102.

Engine 1948cc 88x80mm/2304cc 94x83mm/2445cc 92x92mm, 4 cyl diesel/diesel/OHV, F/R
Max power 50bhp at 4500rpm/70bhp at 4500rpm/85-110bhp at 4500rpm
Max speed 78-95mph
Body styles saloon, estate
Prod years 1968-82
Prod numbers n/a

PRICES
A 1200 B 800 C 300

Volkswagen

Volkswagen Type 51/Type 11

Ferry Porsche presented his first 'Kraft-durch-Freude' Wagen (KdF) in 1934. Most of the Wolfsburg works was destroyed during the war, and production of the Type 51 began in 1945, based on the tall Kübelwagen chassis. The Type 11 (or Standard) had the normal chassis and differed from the 1949-and-on Export (Type 11a) by its absence of any chromework and austere trim. Export gains hydraulic brakes in '50, new part-synchro 'box from '51, one-piece oval window from March '53. Optional sunroof. Split-windows are now highly collectable.

Engine 1131cc 75x64mm, 4 cyl OHV, R/R
Max power 25bhp at 3300rpm
Max speed 50-65mph
Body styles saloon
Prod years 1945-49/1945-53
Prod numbers 1285/527,508

PRICES
Type 51 Too rare for accurate valuation
Type 11 **A** 6500 **B** 5000 **C** 4000

Volkswagen 1200 '54-'65

The Type 11 evolved into what is now known as the 'oval window' Beetle. A bored-out 1200 engine boosted power up to 30bhp, there was better engine cooling and, from '55, twin exhaust tail pipes and bigger lights. Standard model still had no synchro and cable brakes until '62, and Export (known as De Luxe here) is the chrome-and-a-bit-of-luxury model. Bigger front and rear screens, new engine cover and bigger brakes in '57, all-synchro 'box, front anti-roll bar and 34bhp engine from '60, steel sunroof option from '61. Bigger side windows from '64. Lots around to choose from.

Engine 1192cc 77x64mm, 4 cyl OHV, R/R
Max power 30bhp at 3400rpm to 34bhp at 3600rpm
Max speed 70mph
Body styles saloon
Prod years 1954-65
Prod numbers 7,267,899

PRICES
'54-'57 oval window **A** 5000 **B** 3200 **C** 2000
'57-'65 **A** 4000 **B** 2500 **C** 1500

Volkswagen 1200/1300/1500 '65-'78

The Beetle gets into its stride: 1200 continues with 34bhp engine, drum brakes, 6V electrics and boring reliability. New 1300 has 40bhp and stronger suspension, while 1500 has 44bhp and dual circuit disc/drum brakes. Wider track for 1300 from '66, all 1500s have this. External fuel filler, 12V electrics, auto option and vertical headlamps from '67, bigger luggage compartment and more power for 1300/1500 from '70. '72-'73 1300S and '75-'77 1200L have 1584cc engine. 1300/1500 engines more thirsty and not as reliable as 1200. Rust between doors and rear arches is difficult to fix.

Engine 1192cc 77x64mm/1285cc 77x69mm/1493cc 83x69mm/1584cc 85.5x69mm, 4 cyl OHV, R/R
Max power 34bhp at 3600rpm/40bhp at 4000rpm to 44bhp at 4100rpm/44bhp at 4000rpm/50bhp at 4000rpm
Max speed 72-84mph
Body styles saloon
Prod years 1965-78/1965-75/1966-70
Prod numbers 585,817/2,726,154/1,888,282
PRICES
A 3500 **B** 2300 **C** 1000

Volkswagen 1302/1303

New top-of-the-range 1302 Beetle, with 1in longer wheelbase, extended and rounder front lid, MacPherson strut front suspension and front discs. Better interior and more boot space, too. In the UK, the 1302 always had the new 1600 engine, but other markets also got a 1300 version. Some had semi-automatic transmission. Replaced by 1303, with its new padded dashboard, wrap-around windscreen and larger rear light clusters. This time the UK got the 1285cc version as well as the 1584cc 1303S, though imports ceased in '75. Fuel injection for USA only. Too commonplace to be collectable – yet.

Engine 1285cc 77x69mm/1584cc 85.5x69mm, 4 cyl OHV, R/R
Max power 44bhp at 4000rpm/50bhp at 4000rpm
Max speed 75-84mph
Body styles saloon
Prod years 1970-72/1972-75
Prod numbers n/a/916,713

PRICES
1302S/1303 **A** 3500 **B** 2000 **C** 800
1303S **A** 3600 **B** 2200 **C** 800

Volkswagen Cabriolet Hebmüller

Inspired by Colonel Radclyffe's convertible Beetle, Karmann got the contract to build a four-seater rag-top Beetle, but Wuppertal-based Hebmüller did the 2+2, arguably a prettier car. Sills, bulkhead and rear were all strengthened, so it felt very rigid. The original windscreen was retained, though the side windows and engine lid were new, and the hood could be raised with one hand. After a serious factory fire in 1949, Hebmüller never really recovered, suffering financial problems, and production ground to a halt. The last 14 were made by Karmann. Not commonly seen in Britain.

Engine 1131cc 75x64mm, 4 cyl OHV, R/R
Max power 25bhp at 3300rpm
Max speed 65mph
Body styles convertible
Prod years 1949-53
Prod numbers 696

PRICES
A 13,000 **B** 9500 **C** 6000

Volkswagen

Volkswagen Cabriolet Karmann

Karmann's convertible Beetle became the definitive item. Flexing was solved by adding members under the sills and around the doors, which pushed the weight up by 90lb (40kg). Externally identical to a Beetle below the waist (except for repositioned semaphore indicators), but new winding front and rear windows and an elegant fabric hood with glass rear window, which grew larger as the years passed. Folded hood stacks up high behind rear seats. All mechanical and bodywork changes as per the saloons, though production ended at Osnabrück two years after Wolfsburg.

Engine 1131cc 75x64mm/1192cc 77x64mm/1285cc 77x69mm/1493cc 83x69mm/1584cc 85.5x69mm, 4 cyl OHV, R/R **Max power** 25bhp at 3300rpm to 50bhp at 4000rpm **Max speed** 65-84mph **Body styles** convertible **Prod years** 1949-80 **Prod numbers** 331,847
PRICES
1100 **A** 8500 **B** 6500 **C** 4500
1200/1300/1500 **A** 7500 **B** 5500 **C** 3500
1302/1303 **A** 8500 **B** 5500 **C** 3500

Volkswagen Karmann-Ghia Coupé/Cabriolet (Type 14)

Ghia designed its 2+2 coupé Beetle in secret (possibly Virgil Exner was the stylist) and asked Karmann to build it with VW's blessing. The Beetle floorpan was widened by 3in and an anti-roll bar added at the front; otherwise the 'Käfer' mechanicals were virtually unaltered. Early ones were hand-built, but too popular for that arrangement to last. Drives just like a Beetle. Cabriolet arrived in '57, 1300 engine in '65, 1500 engine and front discs in '66, dual-circuit brakes and semi-auto option in '67, 1600 engine in '70, bigger bumpers '71. Healthy following ensures prices remain high.

Engine 1192cc 77x64mm/1285cc 77x69mm/1493cc 83x69mm/1584cc 85.5x69mm, 4 cyl OHV, R/R **Max power** 30bhp at 3400rpm to 50bhp at 4000rpm **Max speed** 74-87mph **Body styles** coupé, convertible **Prod years** 1955-74/1957-74 **Prod numbers** 364,401/80,899

PRICES
Coupé **A** 7000 **B** 4000 **C** 1800
Cabriolet **A** 9000 **B** 6000 **C** 2800

Specialists in the restoration & repair of 50s and 60s VWs

- 10 years experience • Bodywork •
- Welding • Servicing • Resprays •
- Repairs • Collection & delivery
- available • New & Used parts •
- Cars bought & sold • Ring for details •

Why buy a VW with shiny paint that looks good, when you don't know what's hidden under it?

We will inspect any aircooled VW with yourselves prior to purchase
OR
We will inspect with you any air-cooled VW for restoration & give you an honest professional opinion, along with a quote for any work required.
Full photographic restoration.
Stage payments – All work fully guaranteed.

Tel. 01255 420796 Eves. 01255 420244

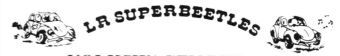

VOLKSWAGEN BEETLE SPECIALISTS

Beetles bought & sold
Repairs • renovations • spares • resprays etc
Telephone Colchester 01206 56433
WE ALSO UNDERTAKE CLASSIC CAR RESTORATIONS

Specialist in the restoration service & repair of the early Volkswagen
We offer an extremely competitive service.
- Accident repairs
- Welding
- Body strengthening
- Engine sales & fitting
- Stage payments

Resprays inside & out
Large stock of new parts
Large stock of SH parts
Top quality carpet sets

*Beetles Bought & Sold
Pick Up & Delivery
FREE vehicle inspections & valuation at our premises
Restored Vehicles to Order*

We also undertake classic car restorations (Special Orders only)
— please phone for a chat & helpful advice
UNIT 3, GOSBECKS FARM, GOSBECKS RD, COLCHESTER CO2 9XX

Volkswagen Karmann-Ghia Coupé (Type 34)

Supposedly up-market Karmann-Ghia based on the new VW 1500. It did offer more room inside, but its slab-sided body was awkward and heavy, for all its wrap-around rear screen, sculpted nose and standard foglamps. A convertible was shown at the 1961 Frankfurt Show alongside the coupé, but it never entered production (a mere 17 were built). Electric sunroof option from '62. 1600 engine and front discs from '65. Sold alongside the smaller Type 14 for only eight years, doing rather poorly in comparison. Technically identical to the 1500/1600 (see below).

Engine 1493cc 83x69mm/1584cc 85.5x69mm, 4 cyl OHV, R/R
Max power 45bhp at 3800rpm to 54bhp at 4200rpm/54bhp at 4000rpm
Max speed 85-94mph
Body styles coupé
Prod years 1961-69
Prod numbers 42,563

PRICES
A 4700 **B** 2800 **C** 1500

Volkswagen 1500/1600

The Type 3 VW was launched as a bigger, better Beetle. Whether it was or not is a matter of conjecture. The floorpan is basically the same, though there is space to fit a luggage box above the engine and five people inside. Variant is the three-door estate, 1500S has twin carbs and 54bhp. Cabriolet shown in 1961 never reached production (12 were made). Replaced by 1600 in 1965 – the 1493cc engine remained an option, though also called a 1600 – and a pretty fastback saloon (pictured) arrived at the same time. Semi-auto option from '67, fuel injection option from '68 (not in UK until '71 TE).

Engine 1493cc 83x69mm/1584cc 85.5x69mm, 4 cyl OHV, R/R
Max power 45bhp at 3800rpm/54bhp at 4000rpm
Max speed 85-94mph
Body styles saloon, fastback saloon, estate
Prod years 1961-65/1966-73
Prod numbers 2,542,059

PRICES
Saloon/Estate **A** 1300 **B** 700 **C** 400
Fastback **A** 2000 **B** 1200 **C** 600

Volkswagen 411/412

You could hardly call the 411 pretty but, as VW's first four-door production car, it was an important step up-market. The wheelbase was longer by 4in but VW kept its tried-and-trusted rear-engined layout. Dynamically this might have been a mistake, but things were helped by MacPherson strut front suspension, dual-circuit brakes and front discs. New 1700 engine, from 1969 (as the 411E) with fuel injection and double halogen headlamps. 412 has restyled front end and better interior. 1800 twin carb engine from '73 in low and high compression forms.

Engine 1679cc 90x66mm/1795cc 93x66mm, 4 cyl OHV, R/R
Max power 68bhp at 4500rpm to 80bhp at 4900rpm/75-85bhp at 5000rpm
Max speed 90-99mph
Body styles saloon, estate
Prod years 1968-72/1972-74
Prod numbers 355,200

PRICES
A 1000 **B** 500 **C** 200

Volkswagen 181

The Americans officially called this 'The Thing', which was less of a mouthful than the official label ('Mehrzweck-Fahrzeug/Kurierwagen', or multi-purpose/delivery vehicle). No four-wheel drive, though, just a Beetle chassis. Vinyl hood or optional hardtop. Initially 1500 engines, from '70 1600 units (with an extra 4bhp and coil springs from '73 and fuel injection for the USA from '74). Always very basic interiors, but durable. German production ended in '78, though Mexico made 181s 1972-79 (and Indonesia!). LHD only. Main source of supply is German army surplus dumps.

Engine 1493cc 83x69mm/1584cc 85.5x69mm, 4 cyl OHV, R/R
Max power 44bhp at 4000rpm/ 44bhp at 3800rpm to 48bhp at 4000rpm
Max speed 72-75mph
Body styles utility
Prod years 1969-79
Prod numbers 90,883

PRICES
A 3200 **B** 2000 **C** 1100

Volkswagen K70

This car was destined to become the NSU K70, as it had been developed in Neckarsulm, not Wolfsburg. VW bought NSU in 1969 and was faced with a problem. It took the plunge and launched its first front-engined car, which also had front-wheel drive, and an overhead cam engine with water cooling. All-independent suspension, servo front discs, all-synchro 'box. 1600 engine available with 75 or 90bhp, latter superseded by 100bhp 1800 K70S model in '73. No parts shared with any other VW and therefore highly marginal. Something of a motoring cuckoo: initiated VW's change of direction but not in itself a good car.

Engine 1605cc 82x76mm/1807cc 87x76mm, 4 cyl OHC, F/F
Max power 75-90bhp at 5200rpm/100bhp at 5300rpm
Max speed 92-101mph
Body styles saloon
Prod years 1970-74
Prod numbers 211,127

PRICES
A 700 **B** 500 **C** 200

Volkswagen Passat

Considering that VW gained a majority holding in Audi as early as 1964, it is surprising that it took so long for it to realise the technical superiority of its sister. The Passat was based on the Audi 80 platform, sharing its front-mounted water-cooled engines, front-drive and front discs. Launched as a fastback saloon (two- or four-door) or a five-door estate, but from January '75 also offered with three- and five-door hatchback bodies. 1.3 and 1.5-litre engines to start, latter supplanted by 1.6 in '75. 1272cc crossflow and 1.5 diesel from '78, 1.6 injection from '79.

Engine 1297cc 75x73.4mm/1272cc 75x72mm/1588cc 79.5x80mm, 4 cyl OHC, 1471cc 76.5x80mm 4 cyl OHC & diesel, F/F
Max power 50bhp at 5000rpm to 110bhp at 6100rpm **Max speed** 88-115mph **Body styles** saloon, hatchback, estate
Prod years 1973-80
Prod numbers over 2,000,000

PRICES
A 800 **B** 400 **C** 100

Volkswagen Golf

As with the Passat, VW turned to a newly-independent Giugiaro (Ital Design) for its new small car. The Golf was an incontrovertible masterstroke. It established the idea of the front-drive small hatchback, looked smart, handled well and was beautifully built. And the 1975 GTI was the world's first hot hatch: Bosch fuel injection, taut suspension, understated add-ons and amazing performance. LHD only until '79, when five-speed 'box becomes standard, 1800 engine from '82. Cabriolet launched in '79 (UK '83), and lasted in Mk1 bodyshell until 1992. GTIs are now true classics; early ones attract a premium.

Engine 1093cc/1272cc/1457cc/1781cc, 4 cyl OHC, 1471/1588cc OHC & diesel, F/F **Max power** 50bhp at 5600rpm to 112bhp at 5800rpm
Max speed 88-117mph
Body styles hatchback, convertible
Prod years 1974-83
Prod numbers 6,005,635
PRICES
Golf **A** 1200 **B** 700 **C** 250
GTI **A** 2500 **B** 1500 **C** 800
Convertible '79-'84 **A** 3000 **B** 1800 **C** 1200

Volkswagen Scirocco

The Scirocco was developed alongside the Golf (it actually arrived two months before it), and shared many common elements: platform, power trains, suspension. Giugiaro did another classic job according to the folded paper school of design, while Karmann built it. 1100 and 1300 versions not sold in UK, but we did get the 1500 and 1600. Golf GTI-engined versions best of all (badged GLI/GTI/Storm), though RHD edition not available until 1979. Automatic optional on all but base models and GLI. Build quality high, but rust can get hold in a severe way.

Engine 1093cc 69.5x72mm/1272cc 75x72mm/1457cc 79.5x73.4mm/1471cc 76.5x80mm/1588cc 76.5x86.4mm, 4 cyl OHC, F/F
Max power 50bhp at 6000rpm to 110bhp at 6100rpm **Max speed** 89-116mph **Body styles** hatchback coupé
Prod years 1974-81
Prod numbers 504,200
PRICES
Scirocco **A** 1600 **B** 1000 **C** 600
Scirocco GLI/GTI/Storm **A** 2000 **B** 1400 **C** 800

Volkswagen Polo

The Polo completed VW's comprehensive onslaught of launches over a two-year period. Like the Passat, Golf and Scirocco, it cleaned up in its market area: refinement, build quality and dynamics were all at new levels for a supermini – a new market for VW. All-new 895cc engine, plus Golf 1100 and 1300 choices, though latter never made it to UK, nor did the 771cc super-economy export engine (nor the 60bhp GT or Audi 50 sister model). Another very strong seller for VW, replaced by new generation Polo in '82, which then lasted 14 years. Significant, but no collector credentials.

Engine 771cc 64.5x59mm/895cc 69.5x59mm/1093cc 69.5x72mm/1272cc 75x72mm, 4 cyl OHC, F/F
Max power 34bhp at 6000rpm/40bhp at 5900rpm/50bhp at 5600rpm to 60bhp at 6000rpm/60bhp at 5600rpm
Max speed 77-95mph
Body styles hatchback
Prod years 1975-82
Prod numbers 768,200

PRICES
A 700 **B** 400 **C** 100

Volkswagen-Porsche 914 1.7/1.8/2.0

'Porsche for the people', or how to create controversy without reaping the reward. Porsche people were disdainful of the 914 because the engine sat in the middle and it was a mere VW 411 unit. On the other hand, that made it practical, it had five speeds (or optional Sportomatic), all-independent suspension, four-wheel discs and a targa roof panel – and it cost half as much as a Porsche 911E. Launched with 1700 engine, switched to 1.8 litres in '72, plus four-cylinder 2.0-litre fuel-injection option (though not for UK). Still under-rated.

Engine 1679cc 90x66mm/1795cc 93x66mm/1971cc 94x71mm, 4 cyl OHV, M/R
Max power 80bhp at 4900rpm/85bhp at 5000rpm/100bhp at 5000rpm
Max speed 106/112/119mph
Body styles targa coupé
Prod years 1969-72/1972-75/1972-75
Prod numbers 115,646 (65,351/17,773/32,522)

PRICES
A 5000 **B** 2800 **C** 1500

Volkswagen-Porsche 914/6

This was a half-way house between the four-cylinder 914 and the 911: the engine was taken from the 911T, the mildest current 911. That made its performance much better than the lack-lustre 914 'fours', but handling had to be watched, though wider alloy wheels improved grip. In America, all 914s were called Porsches, which partly explains why the majority were sold there. In Europe, the 914/6 was a sales disaster, and was withdrawn after just three years. Sportomatic is a rare option. LHD only and very rare in the UK; some parts problems and not cheap to restore.

Engine 1991cc 80x66mm, 6 cyl OHC, M/R
Max power 110bhp at 5800rpm
Max speed 125mph
Body styles targa coupé
Prod years 1969-72
Prod numbers 3332

PRICES
A 8000 **B** 5500 **C** 2500

Volvo PV444

Sweden's leading car maker decided to enter the small car market after the war. Designer Helmer Petterson was told to make it look 'not like a soapbox', and the result was certainly distinctive. PV means 'personvagn' and 444 means four seats, and 1944, though production actually started in '47. Split front and rear screens (though one-piece rear after '55). IFS by coils and wishbones, floor-mounted gear lever, 1583cc twin carb engine from '57. Built to last, but the PV444 was never marketed in Britain.

Engine 1414cc 75x80mm/1583cc 79.4x80mm, 4 cyl OHV, F/R
Max power 40bhp at 3800rpm to 70bhp at 5500rpm/60bhp at 6000rpm to 85bhp at 5500rpm
Max speed 75-94mph
Body styles saloon
Prod years 1947-58
Prod numbers 196,005

PRICES
A 3500 **B** 2700 **C** 2200

Volvo PV544

The 444's replacement had five seats, as the title suggests, but also a curved one-piece windscreen, larger rear window and a new dashboard similar to the Amazon. There was a choice of three-speed or four-speed gearboxes, which were revised in 1960, and single carb or twin carb (Sport) engines. From August '61, sold simply as '544' and boasting the larger B18A (75bhp) or B18D (90-95bhp) five-bearing engines developed for the Amazon, plus 12V electrics and improved steering gear. As reliable as they come and terrifically solidly built. Not especially glamorous, but a good one could outlast its owner.

Engine 1583cc 79.4x80mm/1778cc 84.1x80mm, 4 cyl OHV, F/R
Max power 60-85bhp at 4500rpm/75-95bhp at 5600rpm
Max speed 90-100mph
Body styles saloon
Prod years 1958-65
Prod numbers 246,995

PRICES
A 3500 **B** 2700 **C** 2000

Volvo PV445/P210 (Duett)

The separate-chassis estate version of the PV444 arrived in 1953. The official name was PV445, but many markets knew it as the Duett, so called because it could be used as a van or a passenger car. Five or seven seats could be fitted and most were duo-tone (the second colour wrapping around the windows). A roof rack was optional. From 1960 the car was called the P210. It received the larger B18 five-bearing engine in 1962, one year later than the saloon. The Duett acquired a reputation as a do-anything, hold-everything, go-anywhere no-nonsense machine, which seems fair. Outlived saloon by four years.

Engine 1414cc 75x80mm/1583cc 79.4x80mm/1780cc 84.1x80mm, 4 cyl OHV, F/R
Max power 40bhp at 3800rpm to 75bhp at 5600rpm
Max speed 75-92mph
Body styles estate
Prod years 1953-69
Prod numbers 73,944

PRICES
A 3200 **B** 2200 **C** 1600

Volvo P1900

Unusual project from Volvo: a sports car based on PV444 mechanicals using a glassfibre body. Though first shown in 1954, problems with the new plastic material delayed production until 1956. Mounted on a tubular chassis, the barrel-shaped two-seater bodywork featured a deeply recessed grille, was quite pretty, and was co-developed by American glassfibre pioneers Glasspar. Spec included a 70bhp twin carb version of the 1.4-litre engine, three-speed gearbox and 6V electrics. New Volvo chief Gunnar Engellau had no hesitation about axing it after just 67 were built.

Engine 1414cc 75x80mm, 4 cyl OHV, F/R
Max power 70bhp at 6000rpm
Max speed 96mph
Body styles 67
Prod years 1956-57
Prod numbers 67

PRICES
A 20,000 **B** 12,500 **C** 8000

Volvo 121/221

Everyone calls them Amazons, though only badged as such in Sweden. Everywhere else, it was the 120 series. PV444 mechanicals, handsome styling by Jan Wilsgaard. Initially four doors, B16A single carb 1.6-litre engine and only three speeds. A new cylinder head arrived in 1961, expanding the capacity to 1.8 litres (B18A), and a two-door model joined the range at this time. The estate was known as the 221 and arrived in '62 (confusingly, it was also known as the 121 Combi). Front discs from '64, optional auto '65, 85bhp '67, four-door version axed '67. Two-door gets B20 2-litre engine in '68.

Engine 1583cc 79.4x80mm/1778cc 84.1x80mm/1986cc 88.9x80mm, 4 cyl OHV, F/R
Max power 60bhp at 4500rpm/75-85bhp at 5000rpm/90bhp at 5000rpm
Max speed 87-93mph
Body styles saloon, estate
Prod years 1956-70/1962-69
Prod numbers 657,323 (all Amazons)

PRICES
A 3300 **B** 2200 **C** 1000

Volvo 122/222

These were the twin carb versions of the Amazon series. 122S saloon launched in 1958 as export model, with four speeds as standard. Expanded 1.8 B18D engine from '61 (when model becomes just '122'), and front discs were standardised. 222 estate from '62, two-door model from '63 (known internally as the 132 but never badged as such). 140 series replaces four-door in '67, two-door gains twin carb B20B engine and dual-circuit braking in '68. All Amazons are rust-beaters: high-quality steel means high survival rate. You can expect the ultimate in dependability, too. Deservedly popular.

Engine 1583cc 79.4x80mm/1778cc 84.1x80mm/1986cc 88.9x80mm, 4 cyl OHV, F/R
Max power 85bhp at 4500rpm/90-95bhp at 5000rpm/118bhp at 4700rpm
Max speed 90-105mph
Body styles saloon, estate
Prod years 1958-70/1962-69
Prod numbers see above

PRICES
A 3500 **B** 2400 **C** 1200

Volvo 123GT

The 120 series made quite an impact on the rally scene and there was a strong demand for a sporting member of the family. So in 1966 Volvo fitted the twin carb 1.8-litre engine from the 1800S in a two-door bodyshell, gave it standard overdrive, servo front disc brakes, bumper-mounted foglamps, a special steering wheel and a rev counter on top of the dashboard. Thus the 123GT was born. UK imports began in May 1967, but very few were brought in. Fine cruising abilities and very desirable in Amazon circles. In the UK, tuners Ruddspeed offered even more power (up to 132bhp).

Engine 1778cc 84.1x80mm, 4 cyl OHV, F/R
Max power 115bhp at 6000rpm
Max speed 108mph
Body styles saloon
Prod years 1966-68
Prod numbers c 2500

PRICES
A 4000 **B** 3000 **C** 2000

Volvo 142/144/145

Boxy replacement for the Amazon whose basic shape remained in production (as 240) until 1993. Safety plays a bigger part now: all-round servo discs, dual-circuit braking, rubber-faced bumpers. Limited slip diff and automatic are options. 142 is two-door, 144 is four-door, 145 is estate. All could be had in S specification, equating to twin carbs (115bhp) and the option of overdrive (but not automatic). 2-litre engines standard from '68, heated rear window and (on estate) rear wash/wipe from '69. Facelifted DL and fuel-injected leather-upholstered GL from '70. Enhanced Volvo's reputation for durability no end.

Engine 1778cc 84.1x80mm/1986cc 88.9x80mm, 4 cyl OHV, F/R
Max power 85bhp at 5000rpm to 115bhp at 6000rpm/90bhp at 4800rpm to 135bhp at 6000rpm
Max speed 90-112mph
Body styles saloon, estate
Prod years 1967-74/1966-74/1967-74
Prod numbers 1,205,111 (412,986/523,808/268,317)

PRICES
144/145 **A** 1300 **B** 700 **C** 400
142 **A** 1500 **B** 800 **C** 500

Volvo 164

Longer wheelbase and greater overall length for the six-cylinder evolution of the 140 series, but it was no Jaguar/Mercedes competitor, as Volvo hyped it up to be. Big 3-litre seven-bearing engine was a good lugger, and very powerful in post-'71 175bhp fuel-injected guise. If you have power steering, you'll get overdrive or automatic, and limited slip diff is optional. Top-of-the-range TE ('73-'75) has standard auto, leather, air con and PAS. All 164s have upright grilles and post-'69 ones have fitted foglamps. Thirsty but the most satisfying 1970s Volvo.

Engine 2978cc 88.9x80mm, 6 cyl OHV, F/R
Max power 145bhp at 5000rpm to 175bhp at 5500rpm
Max speed 112-120mph
Body styles saloon
Prod years 1968-75
Prod numbers 155,068

PRICES
A 2600 **B** 1500 **C** 900

Volvo P1800/1800S/1800E

Roger Moore as *The Saint* got Volvo's new sports car off to a good start. The body was designed by Swede Per Petterson, then employed by Frua, with mechanicals from the 120. Volvo didn't have the capacity to build it, so sub-contracted Pressed Steel to make the bodies and Jensen to assemble them. Quality concerns returned it to Sweden from '63. Standard twin carb B18 engine, servo front discs, optional overdrive. Post-'63 models are renamed 1800S (108bhp, later 115bhp), cowhorn bumpers deleted in '64, 2-litre engine from '68. 1800E gets Bosch fuel injection, four-wheel discs, alloy wheels, auto option from '70.

Engine 1778cc 84.1x80mm/1986cc 88.9x80mm, 4 cyl OHV, F/R
Max power 90bhp at 5500rpm to 115bhp at 6000rpm/118-130bhp at 6000rpm
Max speed 102-115mph
Body styles coupé
Prod years 1960-63/1963-69/1969-72
Prod numbers 6000/23,993/9414

PRICES
P1800/1800S **A** 5000 **B** 3000 **C** 1800
1800E **A** 4200 **B** 2300 **B** 1200

Volvo 1800ES

By the end of the 1960s, the 1800S was looking a little out-dated. Instead of rebodying it, Jan Wilsgaard took a leaf out of Reliant's book and made a GTE-style sports estate. The result was not unattractive, even if continentals nicknamed it Snow White's hearse. The rear hatch was a frameless glass affair, there was more room for rear passengers and practicality was enhanced. Same mechanicals as 1800E, which means heavy steering, unspectacular performance, but rugged reliability. Auto option from '72. Most were sold in America but you'll find plenty still around in Britain at reasonable prices.

Engine 1986cc 88.9x80mm, 4 cyl OHV, F/R
Max power 130bhp at 6000rpm
Max speed 115mph
Body styles estate
Prod years 1971-73
Prod numbers 8078

PRICES
A 4500 **B** 2600 **C** 1500

Volvo 240/260

Amazingly successful and long-running tanks. The front end is new, but the hull is the same as the old 144. Volvo updated its front suspension with MacPherson struts, but the old live rear axle stayed on. A brand new range of OHC engines was supplemented by a turbocharged 2.1 and a VW 2.4-litre diesel from '79. UK did not get either of these, nor the two-door 242: we had only the four-door 244 and 245 estate. 264/265 had the new PRV V6 (2664cc to '80, 2849cc thereafter). Bertone-built 264TE LWB limo (all 18ft 5in/560cm of it) from '77. Bertone also built (but did *not* style) the unusual 262C coupé (photo) of '77.

Engine 1986cc 4 cyl OHV, 2127cc/2315cc 4 cyl OHC, 2664cc/2849cc V6 cyl OHC, 2383cc 4 cyl diesel, F/R
Max power 82bhp at 4700rpm to 155bhp at 5500rpm **Max speed** 90-120mph **Body styles** saloon, estate, coupé **Prod years** 1974-93/1974-85
Prod numbers 2,442,395/179,072

PRICES
244/245 **A** 1700 **B** 800 **C** 400
264/265 **A** 2000 **B** 1000 **C** 400
262C **A** 3500 **B** 2300 **C** 1500

Volvo 66

Volvo took over Holland's only independent car maker in 1974/75 and inherited the DAF 66. It dropped the coupé but kept the saloon and estate going, adding its own variety of safety-related modifications: telescopic steering column, impact bumpers, head restraints, dual-circuit brakes, inertia reel seat belts front and rear, and laminated screen. Mechanically it was all but identical to the Marathon-spec DAF (the transmission was slightly modified) and it looked very much the same except for a new Volvo grille (on GL, incorporating foglamps). Slow and unromantic, but a doddle to drive.

Engine 1108cc 70x72mm/1289cc 73x77mm, 4 cyl OHV, F/R
Max power 47bhp at 5000rpm/57bhp at 5200rpm
Max speed 85/90mph
Body styles saloon, estate
Prod years 1975-80
Prod numbers 106,137

PRICES
A 1000 **B** 600 **C** 300

Volvo 343/345

DAF had been working on its 66 successor when Volvo took it over. Hence it received the Volvo treatment and was launched under the Swedish name, though it was built in Holland. Hatchback styling was by Trevor Fiore in three-door (343) and five-door (345) forms. Engine came from Renault, transmission was DAF's familiar Variomatic, but a manual option was offered from '78. Its peculiar charm appealed to British buyers, whose devotion to the model (it often featured in Top 10 best-sellers lists) kept its 1982 340 and 2-litre 360 successors in production until 1991 and made it a million-seller. A shopping car, not a classic.

Engine 1397cc 76x77mm, 4 cyl OHV, F/R
Max power 70bhp at 5500rpm
Max speed 90mph
Body styles hatchback
Prod years 1976-82
Prod numbers 1,086,405 (all 340/360s)

PRICES
A 800 **B** 500 **C** 200

Warszawa

This was the Polish version of the Russian Pobeda and is more likely to be found in the west, if only because the Warszawa was built for some 16 years after the end of the Pobeda. Initially, the body and engine were imported by FSO from the GAZ factory in Gorky, Russia but, from '55, the Warszawa became all-Polish. First ones (called 201) had a sidevalve engine, but the 202 had an OHV unit. The '65 203 had a notchback body, unitary construction and IFS. Miniature version sold as the Syrena. Like GAZ, FSO went on to make Fiats under licence.

Engine 2120cc 82x100mm, 4 cyl SV/OHV, F/R
Max power 50bhp at 3650rpm to 70bhp at 4000rpm
Max speed 72-80mph
Body styles saloon, estate
Prod years 1951-74
Prod numbers n/a

PRICES
A 1600 **B** 1000 **C** 400

Wartburg EMW 340

BMW was split apart by the war, its Eisenach plant lying in what became East Germany. That plant continued to build the BMW after the war (even calling it 'BMW' until legal proceedings switched the name to EMW – Eisenacher Motoren-Werke). The 340 was a BMW 326 with a different grille. That means twin carbs, hydraulic brakes, live rear axle with torsion bars, four speed 'box with synchro on third and top, plus freewheel on bottom two. SWB ex-BMW 320 also made post-war as the Awtovelo.

Engine 1971cc 66x96mm, 6 cyl OHV, F/R
Max power 55bhp at 3750rpm
Max speed 75mph
Body styles saloon
Prod years 1949-55
Prod numbers 21,249

PRICES
A 8000 **B** 5000 **C** 2500

Wartburg EMW 327

Pre-war BMW 327s are extremely rare and valuable. A good substitute for the impecunious fanatic would be an EMW 327, which is the East German post-war replica. It is so similar to the pre-war car that many are passed off as genuine BMWs by the simple expedient of substituting blue-and-white badging. Sold in many communist countries as the Eisenacher Sport. Chassis is a shortened version of the 340 with a semi-elliptic rear end but not the high-compression engine of pre-war BMWs. Very desirable nonetheless.

Engine 1971cc 66x96mm, 6 cyl OHV, F/R
Max power 55bhp at 3750rpm
Max speed 78mph
Body styles coupé, convertible
Prod years 1948-53
Prod numbers 505

PRICES
Too rare for accurate valuation

Wartburg 311/312

The Eisenach works were turned over to manufacture of the Wartburg from 1955. This made use of the pre-war DKW chassis (also seen in the Zwickau-built IFA F9), but had an all-new full-width four-door body. Mechanicals as per the DKW: 900cc two-stroke three-cylinder engine, beam rear axle, four-speed crash 'box with freewheel (synchro from '58 and optional Saxomat automatic clutch from '60). 312 has 1-litre engine, IRS from '65. Three-door Kombi estate from the start, but Glaser-built '57 Camping is fun, with its wrap-over rear windows and bed conversion (photo).

Engine 900cc 70x78mm/992cc 73.5x78mm, 3 cyl TS, F/F
Max power 37bhp at 4000rpm/45bhp at 4200rpm
Max speed 62-76mph
Body styles saloon, estate
Prod years 1956-62/1962-66
Prod numbers 292,723 (incl coupé/cabriolet)

PRICES
Saloon **A** 1000 **B** 600 **C** 300
Camping **A** 2500 **B** 1600 **C** 1200

Wartburg Coupé/Cabriolet

This pair was produced mainly for export, presumably because they were too decadent for communist citizens. They were actually quite good-looking: the Coupé had a wrap-around rear window and pillarless hardtop style, while the Cabriolet had a stacking hood (in the German fashion) and four winding windows. The drop-top's hood was of thick, high-quality material and its interior was upholstered in leather. Two-toning was obligatory. Coupés and Cabriolets have long since been 'discovered' in Germany, and prices reflect a growing interest, but they are almost unknown in Britain.

Engine 900cc 70x78mm/992cc 73.5x78mm, 3 cyl TS, F/F
Max power 37bhp at 4000rpm/ 45bhp at 4200rpm
Max speed 64/78mph
Body styles coupé, convertible
Prod years 1957-66/1956-66
Prod numbers see above

PRICES
Coupé **A** 3400 **B** 2400 **C** 1700
Cabriolet **A** 3800 **B** 2800 **C** 2000

Wartburg Sport

This was quite a racy departure for the East German firm. Though the chassis is identical to the rest of the range, the body is all-new: longer, lower and wider, with a notably extended bonnet. A strict two-seater, it came with a basic removable hood or an optional hardtop (pictured). The engine was given a higher compression and two carburettors, so performance went up (but you still wouldn't call 0-60mph in 27 secs sizzling). It cost more than a VW Karmann-Ghia Cabriolet in West Germany, so production was very limited. Less than 200 are known to remain.

Engine 900cc 70x78mm, 3 cyl TS, F/F
Max power 50bhp at 4500rpm
Max speed 87mph
Body styles sports
Prod years 1957-60
Prod numbers 469

PRICES
A 6000 B 3500 C 2000

Wartburg 353 (Knight)/1.3

This was state-of-the-art looks for 1966, and way ahead of the rest of Eastern Europe. Underneath, it was the same mildly depressing story, however: the latest incarnation of the pre-war DKW chassis, the same two-stroke three-cylinder engine and drum brakes. All-synchro 'box from '67, Tourist estate and sunroof option from '68, 50bhp from '69. Sold in UK at rock-bottom prices (cheapest car on sale during '70s), but effectively ousted from Western Europe by appalling emissions record (UK imports ceased '74). Dual-circuit brakes in '75 353W, VW Golf 1300 engines substituted in final three years.

Engine 992cc 73.5x78mm/1272cc 75x72mm, 3/4 cyl TS/OHC, F/F
Max power 45-50bhp at 4250rpm/58bhp at 5400rpm
Max speed 78-87mph
Body styles saloon, estate
Prod years 1966-88/1988-91
Prod numbers 1,225,190/152,775

PRICES
A 900 B 400 C 100

Willys Station Wagon

Willys made its mark with the wartime Jeep (of which 360,000 were made). It inevitably made civilian versions after the war, but the first specifically 'civvy street' one was the '46 Station Wagon, America's first steel-bodied estate. The body was recognisably Jeep, but was on a longer wheelbase, had a six-cylinder engine option and only two-wheel drive. It had non-structural wood applied to its sides and split rear doors. Official records class it as a truck, but it was mostly used as a five-seater passenger car.

Engine 2199cc 79.4x111.1mm/2433cc 76.2x88.9mm, 4/6 cyl SV & OHV, 2622cc 79.4x88.9mm, 6 cyl OHV, F/R
Max power 63-72bhp at 4000rpm
Max speed 57-65mph
Body styles estate
Prod years 1946-65
Prod numbers 148,880

PRICES
A 4500 B 2800 C 1500

Willys Jeepster

Like the Wagon, Brooks Stevens was responsible for this sporty Jeep, designed during the war. It wore bright colour schemes, whitewall tyres, luxurious fittings, a detachable hood and clip-on side screens. It shared the 104in long wheelbase and two-wheel drive of the Wagon and its side-valve engines (OHV from '50), while a larger six was added from '50. The Jeepster was necessarily a novelty product and demand tailed off rapidly after the first year. The Brooks Stevens connection makes the rare Jeepsters hot properties in America, and they are seldom seen in Europe.

Engine 2199cc 79.4x111.1mm/2433cc 76.2x88.9mm, 4/6 cyl SV & OHV, 2622cc 79.4x88.9mm, 6 cyl OHV, F/R
Max power 63-75bhp at 4000rpm
Max speed 65-70mph
Body styles convertible
Prod years 1948-51
Prod numbers 19,798

PRICES
A 8000 B 5000 C 2500

Wolseley 8

The only immediately post-war Wolseley not offered before the war was the 8. Aft of the A-pillars, it was a Morris 8, but it had a more powerful OHV engine, different bonnet and grille, separate headlamps and far superior leather-and-wood trim. Like the Morris, it had a semi-integral chassis, 6V electrics, four speeds and hydraulic brakes. No two-doors, though. Sold as a solidly-built, reasonably-priced, yet superior small car, which it certainly was, but 17cwt (864kg) and 33bhp spell extremely leisurely travel.

Engine 918cc 57x90mm, 4 cyl OHV, F/R
Max power 33bhp at 4400rpm
Max speed 64mph
Body styles saloon
Prod years 1946-48
Prod numbers 5344 (post-war)

PRICES
A 3800 B 2400 C 1300

Wolseley 10

This Wolseley appeared briefly from January 1939, but returned in 1946. Unlike the 8 and Morris 10, it had a tough separate chassis, which helps in restoration. The body – largely shared with the Morris 10 – was rather bulky for the slender track. The up-side was a comfortable ride and roomy interior; the down-side, rather wayward cornering behaviour. Specification includes beam axles, leaf springs all round, all-synchromesh four-speed 'box, horizontal SU carb, hydraulic brakes and built-in hydraulic jacking. The 10 also boasted a telescopically adjustable steering column.

Engine 1140cc 63.5x90mm, 4 cyl OHV, F/R
Max power 40bhp at 4600rpm
Max speed 70mph
Body styles saloon
Prod years 1939-48
Prod numbers 2715 (post-war)

PRICES
A 2700 **B** 1500 **C** 900

Wolseley 12/48 / 14/60 / 18/85

The big '37/'38 Wolseleys were all revived after the war. All shared the same construction of a separate cross-braced chassis and heavy semi-coachbuilt bodies of metal panels over a wood frame. The 12/48 had a 98in wheelbase and a 1½-litre engine, not really up to the job of lugging 26cwt (1321kg) around. The 14/60 and 18/85 shared the same hull, but had longer (104.5in) wheelbases, all taken up by the bonnets, under which sat six-cylinder twin carb engines of 1.8 and 2.3 litres respectively. Lots of leather and wood, reasonable cruising ability, sturdy construction.

Engine 1548cc 69.5x102mm/1818cc 61.5x102mm/2322cc 69.5x102mm, 4/6/6 cyl OHV, F/R
Max power 44bhp at 4000rpm/55bhp at 4200rpm/85bhp at 4000rpm
Max speed 62-80mph
Body styles saloon
Prod years 1937-48/1938-48/1938-48
Prod numbers 5602/5731/8213 (post-war)

PRICES
12/48 **A** 3500 **B** 2000 **C** 1000
14/60/ 18/85 **A** 4500 **B** 3000 **C** 1200

Wolseley 25

The largest of the so-called Series III Wolseleys was the Super-Six 25, first seen in 1938. It was the last to be reintroduced after the war, badged as the 25, from August 1947 and remaining in production for just over a year. Built to special order only, usually for government departments, the limousine had a huge 141in (358cm) wheelbase, measured 17ft 8in (538cm) long and weighed well over two tons. Huge grille, long bonnet, twin carb 3.5-litre 'six', cloth-covered rear bench seat plus two occasional seats, and a division with a telephone to the front compartment. Rare and imposing.

Engine 3485cc 82x110mm, 6 cyl OHV, F/R
Max power 105bhp at 3600rpm
Max speed 80mph
Body styles limousine
Prod years 1947-48
Prod numbers 75

PRICES
Too rare for accurate valuation

Wolseley 4/50 / 6/80

Rationalisation at Nuffield in 1948 caused all the Wolseley models to be replaced by one basic body style. Therefore the 4/50 was a Morris Oxford clone and the 6/80 a Morris Six, though both had traditional Wolseley grilles and contoured bumpers. Both Wolseleys shared the same bodywork and interiors aft of the scuttle, though the 6/80 (photo) had an 8in longer wheelbase to accommodate its six-cylinder engine. Both engines OHC with same bore and stroke. 6/80 was twin carbs against Morris Six's one. Lots of leather and wood, plus standard foglamps. 6/80 has nice engine, good performance.

Engine 1476cc 73.5x87mm/2215cc 73.5x87mm, 4/6 cyl OHC, F/R
Max power 51bhp at 4800rpm/72bhp at 4800rpm
Max speed 75/83mph
Body styles saloon
Prod years 1948-53/1948-54
Prod numbers 8925/25,281

PRICES
4/50 **A** 2200 **B** 1200 **C** 500
6/80 **A** 3500 **B** 2000 **C** 1000

Wolseley 4/44 / 15/50

No Morris equivalent of this one, though the shell was shared with the MG Magnette ZA. Unitary construction, single carb MG YB engine (the OHC 4/50 engine had not been an unqualified success), coil spring rather than transverse leaf IFS, imprecise column-changed four-speed 'box. Good handling, a fine standard of finish and refinement were its strengths. Wood facia from '53, twin foglamps on last 4/44s. 15/50 looked exactly the same from the outside, but underneath was the latest B-series engine, floor-mounted lever, higher axle ratio plus the option of Manumatic two-pedal transmission.

Engine 1250cc 66.5x90mm/1489cc 73.25x89mm, 4 cyl OHV, F/R
Max power 46bhp at 4800rpm/50-55bhp at 4400rpm
Max speed 75-80mph
Body styles saloon
Prod years 1952-56/1956-58
Prod numbers 29,845/12,353

PRICES
4/44 **A** 2800 **B** 1600 **C** 800
15/50 **A** 3000 **B** 1700 **C** 900

Wolseley

Wolseley 6/90 Series I/II/III

Nuffield's sister marque, Riley, was a smidgen higher in the pecking order. The bodywork of the 6/90 had been conceived a year earlier as the Riley Pathfinder, but its mechanicals differed: in place of the Riley 2.5 engine was the 2.6-litre C-series 'six' from the Austin Westminster, though tuned slightly to give 90bhp (soon 95bhp), and there was a column change until the Series II of '56. Optional overdrive from '55. Series II has semi-elliptic rear, wood dash, auto option. Series III has larger rear window, servo brakes and lower-geared steering. Separate chassis on all. Never very numerous and rare today.

Engine 2639cc 79.4x88.9mm, 6 cyl OHV, F/R
Max power 90-95bhp at 4500rpm
Max speed 90mph
Body styles saloon
Prod years 1954-56/1956-57/1957-59
Prod numbers 5776/1024/5052

PRICES
A 3200 **B** 2000 **C** 1000

Wolseley 1500

New Nuffield small saloon based on the Morris Minor floorpan and sharing its rack-and-pinion steering and torsion-bar IFS/semi-elliptic rear. Unlike the Minor, it had a close-ratio four-speed 'box and a BMC B-series engine (single carb). Internal bonnet/boot hinges and new camshaft on '60 Series II, side grilles incorporate sidelights on '61 Series III. New crank and optional higher compression engine from '62. A good seller for Wolseley, though not as endearing as its sister model, the twin carb Riley 1.5. Usual rust problems, but spares not too difficult, and a practical car to own.

Engine 1489cc 73x88.9mm, 4 cyl OHV, F/R
Max power 50bhp at 4200rpm
Max speed 80mph
Body styles saloon
Prod years 1957-65
Prod numbers 103,394

PRICES
A 2500 **B** 1500 **C** 800

Wolseley 15/60 / 16/60

Yet another incarnation of the BMC Farina, slotting in as the luxury model above Austin and Morris, yet without the performance add-ons of the Riley and MG. Therefore you got the single carb engine of the Austin Cambridge, Wolseley's famous grille and front bumper, plus leather and walnut inside. In line with the Austin/Morris, revised in September '61 to become the 16/60, with a larger engine, cut-down tail fins, anti-roll bars, and new exhaust system. Other minor visual changes were different side grilles and body flashes and repositioned overriders. 16/60 has optional automatic.

Engine 1489cc 73x88.9mm/1622cc 76.2x88.9mm, 4 cyl OHV, F/R
Max power 52bhp at 4350rpm/61bhp at 4500rpm
Max speed 81/84mph
Body styles saloon
Prod years 1958-61/1961-71
Prod numbers 24,579/63,082

PRICES
A 2200 **B** 1200 **C** 500

Wolseley 6/99 / 6/110

Mid-range choice in the larger six-cylinder Farina range, slotting in between the Austin Westminster and Vanden Plas 3-Litre. Exactly the same mechanically as the Austin A99, which means 2912cc C-series twin carb engine, all-synchro three-speed 'box with dual overdrive and servo-assisted front discs. As well as the upright grille and sculpted bonnet, you got foglamps, dog-leg chrome strip and luxury trim. 6/110 has twin exhausts, Panhard rod out back, plus PAS and air con options from '62. Series II ('64) has four speeds but no standard overdrive. The last Police Wolseleys.

Engine 2912cc 83.3x88.9mm, 6 cyl OHV, F/R
Max power 102bhp at 4500rpm/120bhp at 4750rpm
Max speed 98/102mph
Body styles saloon
Prod years 1959-61/1961-68
Prod numbers 13,108/24,101

PRICES
A 2500 **B** 1800 **C** 800

Wolseley Hornet

Mini-with-a-boot, and sister to the Riley Elf. The only external differences over the Elf were the Wolseley grille with its illuminated badge and side chrome grilles with vertical bars as well as horizontal ones. The interior was a little less well-appointed, having an oval three-instrument nacelle (with wood veneer), but normal Mini parcel shelves. The Hornet was the first of the Mini family to be fitted with the 998cc engine in its '63 MkII guise (when larger brakes with twin leading shoes were introduced). Hydrolastic from '64, winding windows and face-level ventilation on '66 MkIII.

Engine 848cc 62.9x68.3mm/998cc 64.6x76.2mm, 4 cyl OHV, F/F
Max power 34bhp at 5500rpm/38bhp at 5250rpm
Max speed 73/77mph
Body styles saloon
Prod years 1961-69
Prod numbers 28,455

PRICES
A 1800 **B** 1200 **C** 700

Wolseley 1100/1300

Up-market Morris 1100 with the same sort of kudos as the Riley and MG versions (it shared their twin carb engines). There was the inevitable illuminated-badge Wolseley chrome grille and walnut veneer facia. Many were two-tone, with a chrome flash separating the colours. As with all other ADO16 types, a 1300 version was launched in '67 with an all-synchro 'box (but not the MG/Riley close-ratio unit) and the option of automatic transmission, in which case only one carb was fitted. More power on '68 1300 MkII. 1100 MkII arrives '67 with auto option, but lasts only four months.

Engine 1098cc 64.6x83.7mm/1275cc 70.6x81.3mm, 4 cyl OHV, F/F
Max power 55bhp at 5500rpm to 70bhp at 6000rpm
Max speed 89-101mph
Body styles saloon
Prod years 1965-68/1967-73
Prod numbers 17,397/27,470

PRICES
A 2000 B 1200 C 600

Wolseley 18/85 / Six

This badge-engineered Landcrab boasts its customary chrome grille flanked by side grilles, walnut veneer dash, extra brightwork, rubber-faced overriders, some leather trim and standard power steering. More power and larger (14in) wheels from '68, MkII (1969) has more walnut, rocker switches, deeper front seats. Twin carb 18/85S arrives '69, identified by chrome waist strip; also has bigger brakes and an auto option. The Six supersedes 18/85 and has the Austin 2200's twin carb 2.2-litre 'six', plus black headlamp surrounds and Rostyle wheels, though PAS is now optional.

Engine 1798cc 80.3x88.9mm/2227cc 76.2x81.3mm, 4/6 cyl OHV/OHC, F/F
Max power 80bhp at 5000rpm to 96bhp at 5700rpm/110bhp at 5250rpm
Max speed 90-108mph
Body styles saloon
Prod years 1969-72/1972-75
Prod numbers 35,597/25,214

PRICES
A 1500 B 800 C 400

Wolseley Six

For devotees of the Harris Mann wedge, the Wolseley Six represents the ultimate. It is also noteworthy as the last ever Wolseley, but sadly that is virtually its only remarkable feature. Before it was pensioned off by the arrival of the Princess only six months after launch, it was the top dog in the '75 18-22 range. That meant the six-cylinder engine, Morris-type front end treatment (but with a tiny version of the illuminated Wolseley grille), vinyl roof, bumper inserts, wood-finished instrument display, standard PAS. Might just become obliquely historic.

Engine 2227cc 76.2x81.3mm, 6cyl OHC, F/F
Max power 110bhp at 5250rpm
Max speed 105mph
Body styles saloon
Prod years 1975
Prod numbers c 3800

PRICES
A 1500 B 900 C 400

Zagato Zele

For all the classic bodies it made for Italian exotics, it is curious that Zagato's first model made under its own name was this electric microcar. The Zele was a tall glassfibre two-seater only 6ft 5in (195cm) long. It had drum brakes, IFS and a solid rear axle to which was attached a Marelli electric motor powered by four 24V batteries. Zele 1000 claimed to do 25mph and travel for around 40 miles; Zele 2000 could reach 38mph. Quite a few were sold in the USA, mostly as open golf carts, but not in the UK, where the importer was Bristol, no less.

Engine electric, R/R
Max power 1kW/2kW
Max speed 25/38mph
Body styles saloon
Prod years 1974-91
Prod numbers n/a

PRICES
Too rare for accurate valuation

ZAZ 965

The Ukraine-based Zaporozhets (ZAZ) factory had made a prototype as early as 1954, but its first production car arrived in 1960. This was a fairly unpleasant two-door saloon powered by a rear-mounted air-cooled V4 engine. Suspension was all-independent by torsion bars up front and coils at the rear, and you got hydraulic brakes. There was even a separate petrol-burning heater. Larger 887cc engine had an extra 4bhp. Not easy to know why you'd want one, but look in Holland if you do: there it was marketed as the Yalta.

Engine 746cc 66x54.5mm/887cc 72x54.5mm, V4 cyl OHV, R/R
Max power 23-27bhp at 4000rpm
Max speed 58-62mph
Body styles saloon
Prod years 1960-67
Prod numbers n/a

PRICES
A 1300 B 700 C 300

ZAZ 966/968

If the 965 was inspired by the Fiat 600, the 966 was a blatant copy of the NSU Prinz. The rear-mounted air-cooled V4 format remained, but there was now a choice of 887cc or 1196cc engines; an export version called the Yalta 1000 was shown at Geneva in 1968 with a 956cc Renault engine, but it is doubtful that this entered production. 968 took over in 1970 with the option of slightly more power from 1.2-litre engine only. Amazingly, this model is still in production. Incidentally, all ZAZ cars have a removable panel in the floor for fishing on frozen lakes!

Engine 887cc 72x54.5mm/1196cc 76x66mm, V4 cyl OHV, R/R
Max power 27bhp at 4000rpm/41bhp at 4400rpm to 45bhp at 4600rpm
Max speed 62-74mph
Body styles saloon
Prod years 1967-70/1970-date
Prod numbers n/a

PRICES
A 1200 **B** 600 **C** 300

ZIL 111/111G

When the memory of Stalin became stale, the ZIS factory was renamed after Ivan Likhachev, the works director. The ZIL 111 therefore replaced the ZIS (see below), this time merely influenced by contemporary Packards, not a straight imitation. Power was now by a V8 engine and the automatic gearbox was operated by push-buttons. Kremlin chauffeurs were treated to standard power steering and servo-assisted brakes (plus their own 'bus lane' in Moscow!). '63 111G was restyled (photo) and rare 111V is the drop-top tourer.

Engine 5980cc 100x95mm, V8 cyl OHV, F/R
Max power 220-230bhp at 4200rpm
Max speed 105mph
Body styles limousine, convertible
Prod years 1956-63/1963-67
Prod numbers n/a

PRICES
A 3000 **B** 1800 **C** 1000

ZIL 114/117

The new razor-edge Soviet Kremlin-mobile arrived in 1967. Once again, it was overtly inspired by Detroit, this time looking rather like a Lincoln. A larger 7-litre V8 now sat under the imposing bonnet, driving through a two-speed push-button automatic gearbox. Air conditioning was standard. The 114's wheelbase was a massive 148in (376cm), and overall length stood at well over 20ft (628cm), though a shorter (130in wheelbase) 117 model was also available. Quite a few ZILs have now made their way to the west, but values are low unless you have cast-iron evidence it was once Brezhnev's.

Engine 6962cc 108x95mm, V8 cyl OHV, F/R
Max power 300bhp at 4400rpm
Max speed 118mph
Body styles saloon, limousine
Prod years 1967-87
Prod numbers n/a

PRICES
A 3000 **B** 1800 **C** 1000

ZIM

ZIM stands for Zavod Imieni Molotova, which simply means the Molotov Works, based at Gorky – in other words, the same factory responsible for the Pobeda and Volga. The ZIM (also known as the GAZ-12) was the predecessor of the Chaika (see page 53) and arrived in 1949. Styling was straight out of Buick's book and the engineering was very conservative: a sidevalve six-cylinder engine, fluid drive coupling and three-speed gearbox. Four-door convertibles very rare. Second-grade Party transport, though some sold in Finland.

Engine 3480cc 82x110mm, 6 cyl SV, F/R
Max power 94bhp at 3600rpm
Max speed 78mph
Body styles saloon, limousine, convertible
Prod years 1949-57
Prod numbers n/a

PRICES
A 2500 **B** 1300 **C** 800

ZIS 110

Zavod Imieni Stalina was founded in 1928 and built limousines for Stalin's Party officials. Its post-war product was the 110, a car whose resemblance to the 1941 Packard 180 should not be surprising, since a Russian delegation bought the body tooling while in America. This was an enormous straight-eight powered eight-seater, weighing some 5600lb (2540kg). Standard features included an all-synchro three-speed 'box, power windows and glass division. A handful of power-top four-door convertibles were also made.

Engine 6000cc 90x118mm, 8 cyl SV, F/R
Max power 140bhp at 3600rpm
Max speed 87mph
Body styles limousine, convertible
Prod years 1946-56
Prod numbers n/a

PRICES
A 3500 **B** 2000 **C** 1000

Zündapp Janus

This symmetrical device was designed by aircraft maker Dornier and named after the Greek god who faced two ways. Much more expensive than other micros, so sales were slow, and Zündapp returned to its core motorbike business. This could be the perfect car for divorcing couples: rear seat passengers sit with their backs to the driver, separated by the engine. You even have your own separate doors, which open up like the Heinkel bubble car. On the other hand, reconciliations are made possible by the fact that the seats fold to make a double bed.

Engine 248cc 67x70mm, 1 cyl TS, M/R
Max power 14bhp at 5000rpm
Max speed 50mph
Body styles saloon
Prod years 1957-58
Prod numbers 6902

PRICES
A 4000 **B** 2500 **C** 1200

DIRECTORY

A & R SHELDON
33 Bramhall Park Road, Bramhall, Stockport, Cheshire, SK7 3JN Tel/Fax: 0161 440 0821
Whit/BSF BA UNF and Metric. Every size for your classic vehicle. Spanners, sockets, greasers, taps and dies by Williams, King Dick, Osborn, traditional British makers. Modern range by Beta.

AH SPARES LTD
Unit 7/8, Westfield Road, Southam, Warwickshire, CV33 0JH Tel: 01926 817181 Fax: 01926 817868
The largest supplier totally committed to supplying parts for Austin Healey models. We offer quality parts at a reasonable price and a worldwide mail order service.

ANDREW BRODIE ENGINEERING LTD
50 Sapcote Estate, 374 High Road, London, NW10 2DJ Tel: 0181 459 3725 Fax: 0181 451 4379
All things Citroën and Maserati. Extensive parts stock for all models. Specialists in hydraulics and stainless exhausts. Full service and body repairs facilities. Free advice.

ARB VEHICLE RESTORATION COMPONENTS
Millvale House, Selsley Hill, Dudbridge, Stroud, GL5 3HF Tel: 01453 751731 Fax: 01453 759630
The company manufactures a wide range of suspension, engine and fastening products for classic cars and older vehicles which are distributed only through *bona fide* trade outlets.

THE ASTON WORKSHOP
THE ASTON WORKSHOP
Gosling Hill, Red Row, Beamish, County Durham, DH9 0RW Tel: 01207 233525 Fax: 01207 232202
Classic Aston Martin specialist. Car sales (own insured transport). Full parts range with express delivery. Full restoration facilities. Worldwide service.

AUTOTECH TRANSMISSIONS LIMITED
Unit 7, Balmoral Trading Estate, Stuart Road, Bredbury, Stockport, Cheshire, SK6 2SR Tel: 0161 406 8686 Fax: 0161 406 8305
Suppliers of transmission overhaul kits (Trans-Fix Kits®) for manual and automatic gearboxes, overdrives and axle units, gears, shafts, selector forks etc.; available for most UK based models to trade and retail.

TAYLOR MADE

AUTOTUNE RUSHTON LTD
Riverside Industrial Estate, Rushton, Blackburn, Lancashire, BB1 4NF Tel: 01254 886819 Fax: 01254 886 819
Company manufacturing low-cost replica cars with new spaceframe chassis and fibreglass bodyshells. Mechanics utilised from rusty 'donor' Jaguars (Aristocat), Ford (Gemini). Also Can Am McLaren M1 replicas built for the track and classic restoration work.

BMS LTD
92 Co-operative Street, Stafford, Staffordshire, ST16 3DA Tel: 01785 250850 Fax: 01785 250852
Specialists for engine mountings, gearbox and suspension mountings and bushes; track rod ends and ball joints, see advertisement on colour pages following page 72.

Bradleys™
BRADLEYS
Unit 6A, Eastlands Industrial Estate, Leiston, Suffolk, IP16 4LL Tel: 01449 616223
Complete repair and recolouring systems for hard and soft plastics, vinyl and leather.

BRIT BITS MG SPARES
Drinsey Nook Garage, Gainsborough Road, Saxilby, Lincoln, LN1 2JJ Tel: 01522 703774 Fax: 01522 704128
Massive stocks of MGB and Midget spares. Over 26,000 parts in stock. Fast worldwide mail order. 24hr telephone service.

British Sports Car Spares

BRITISH SPORTS CAR SPARES
303 Goldhawk Road, London, W12 8EU Tel: 0181 748 7823 Fax: 0181 563 0101
Suppliers of new and second-hand parts for Triumph Spitfire, GT6, Herald, Vitesse, MGB, MG Midget and Austin Healey Sprites. Try us for other sports cars.

BUGS CLASSIC CAR RESTORATION
3A St Osyth Road, Clacton on Sea, Essex, CO15 3BN Tel: 01255 420796 Fax: 01255 420796
Specialist in the restoration of 50s and 60s sports and classic vehicles. XJS Jaguars and VW air cooled. Two-pack paintwork to concours standard. We buy all air cooled VWs.

BURLEN FUEL SYSTEMS LIMITED
Spitfire House, Castle Road, Salisbury, Wiltshire, SP1 3SA Tel: 01722 412500 Fax: 01722 334221
The World's largest SU and Zenith carburetter, fuel pump and parts specialist for Jaguar, MG, Triumph, Lagonda and other classic vehicles. Full restoration and technical support. UK and international mail order service.

CALOREX
The Causeway, Maldon, Essex, CM9 5PU Tel: 01621 856611 Fax: 01621 850871
Calorex are specialists in heat pumps and dehumidifiers. Damp is the classic car killer. It can best be eliminated, not by expensive heat, but by dehumidification. Calorex has a worldwide reputation, throughout industry, as the expert in moisture removal.

CAMBRIDGE MOTORSPORT
Caxton Road, Great Gransden, Bedfordshire, SG19 3AH Tel: 01767 677969 Fax: 01767 677026
Specialists in 1950s and 1960s British sportscars. Competition engine building by Chris Conoley. Historic rally and race preparation. Servicing and restoration to the highest standards. Fast parts by mail order worldwide.

Directory

CARBANK LIMITED
London Road, Ashington, West Sussex, RH20 3AX
Tel: 01903 893000 Fax: 01903 893222
As the original car storage company, Carbank has been providing high security vehicle storage, transportation and shipping since 1980. Why take risks with your investment? Use the professionals – Carbank Ltd.

CENTRAL ENGLAND SPORTSCARS
Unit 4B, Enstone Airfield, Enstone, Chipping Norton, Oxfordshire, OX7 4NP Tel: 01608 677141 (evening 01608 642018) Fax: 01608 645411
The Frogeye specialists. Cars to drive away and restoration projects. Masses of new and secondhand spares: bonnets, hardtops, restored body shells, engines, gearboxes, etc. New catalogue now available. Beautiful restorations at competitive prices. Visitors welcome – please telephone first.

CHATER'S MOTORING BOOKSELLERS
8 South Street, Isleworth, Middlesex, TW7 7BG
Tel: 0181 568 9750 Fax: 0181 569 8273
Specialists of motoring books both new and out of print, plus an extensive range of videos. International mail order service available. Free catalogue upon request.

CHRIS ALFORD RACING & SPORTSCARS
Newland Cottage, Hassocks, West Sussex BN6 8NU Tel: 01273 845966 Fax: 01273 846736
The Lovejoy of historic racing. Cars for sale and for hire, agent for Merlyn sports and racing cars, also ABR Trailers. Noted historic racing driver with 30 years experience.

CLAREMONT CORVETTE
Snodland, Kent, ME6 5NA Tel: 01634 244444 Fax: 01634 244534
The original Corvette specialist since 1977. New and used Corvette parts, stainless exhausts, books, accessories. Europe's largest stockist. Fifteen warranted classic and modern Corvettes for sale in our showroom.

CLASSIC PANELS
Sunnyside, 10 Rhosnesni Lane, Wrexham, Clywd, LL12 7LY Tel/Fax 01978 357776 (mobile: 0378 795303)
Lancia specialists supplying original steel Lancia body panels, exhaust systems and spare parts. Over 30 years experience. Overseas orders welcome.

COVERCRAFT OF EUROPE
60 Maltings Place, Bagleys Lane, London, SW6 2BX
Tel: 0171 736 3214 Fax: 0171 384 2384
Covercraft, the world's largest manufacturer of tailor-made car covers, offer the finest outdoor and in-garage covers, with patterns for over 27,000 different models. Whatever car you have, Covercraft's got it covered!

DAVID MANNERS
991 Wolverhampton Road, Oldbury, West Midlands, B69 4RJ Tel: 0121 544 4040 Fax: 0121 544 5558
Parts for Jaguar and Daimler cars, with over 1200 lines always in stock. Free price lists for most models with next day delivery by TNT.

DSN CLASSICS
Unit 14, Bunns Bank Industrial Estate, Attleborough, Norfolk, NR17 1QD Tel: 01953 455551 Fax: 01953 455451
Mini and Minor parts specialists. Retail, trade and wholesale. UK and export.

FROST AUTO RESTORATION TECHNIQUES LTD
Crawford Street, Rochdale, OL16 5NU
Tel: 01706 58619 Fax: 01706 860338
Suppliers of specialist tools, equipment and paints for your restoration. Send for free catalogue.

GREENWOOD EXHIBITIONS
PO Box 49, Aylesbury, Buckinghamshire, HP22 5FF
Tel: 01296 631181 Fax: 01296 630394
Organisers of the London Classic Motor Show at Alexandra Palace on 16 & 17 March, plus other major classic motor shows and the London to Brighton Classic Car Run on Sunday 9 June.

HOLDEN VINTAGE & CLASSIC LTD
Linton Trading Estate, Bromyard, Herefordshire, HR7 4QT Tel: 01885 488000 Fax: 01885 488889
Worldwide mail order motoring company. Suppliers of Lucas electrical, current and obsolete. Hella, Wipac, SU, etc. Period accessories, Motolita, Smiths gauges. Historic race and rally parts, seat and harnesses, timing equipment. Period clothing, Barbour, Sparco, flying jackets. 72 page mail order catalogue available: £3.00 UK, £4.00 overseas.

JAYMIC LTD – The Classic BMW Specialists
Norwich Road, Cromer, Norfolk, NR27 0HF Tel: 01263 511710 (parts 01263 512883) Fax: 01263 514133
Anything and everything for classic BMWs. Full workshop facilities including AA/VBRA approved bodyshop. Trim and engine reconditioning no problem. Free parts catalogues for 02 & CS models. Worldwide mail order.

JME HEALEYS
4a Wise Terrace, Leamington Spa, Warwickshire, CU31 3AS Tel: 01926 425038 Fax: 01926 640031
A team of skilled craftsmen, led by ex-works employee Jonathan Everard, offers a comprehensive Austin Healey restoration and rally preparation service.

JOHN KIPPING
421 Aldermans Green Road, Coventry, CV2 1NP
Tel: 01203 645333 Fax: 01203 645030
Herald – Vitesse – Spitfire – GT6. For quality parts at reasonable prices. Worldwide mail order. Free illustrated catalogue.

KING STREET MOTOR SERVICES
40 King Street, West Malling, Kent, ME19 4QT
Tel: 01732 843135 Fax: 01732 871579
Morris Minor specialists.

L & M LANCIA LTD
Unit 2, The Ebor Works, Chapel Lane, High Wycombe, Buckinghamshire, HP12 4BS
Tel: 01494 538899 or 0860 614703
Specialists in new and used Lancia spares. Lancias bought and sold.

LANCASTER INSURANCE SERVICES LTD
8 Station Road, St Ives, Huntingdon, Cambridgeshire, PE17 4BH
Tel: 01480 484848 Fax: 01480 464104
Lancaster are one of the country's leaders in arranging insurance for cherished cars and vehicle enthusiasts clubs. Our agreed value policies and own claims department reflect our high standards of service.

LONGSTONE GARAGE
Main Street, Great Longstone, Bakewell, Derbyshire, Tel: 01629 640227 Fax: 01629 640533
Tyres and wheels for veteran, vintage and classic vehicles. Whatever your requirements call in and see us, we shall be pleased to help you. Worldwide mail order service available.

L R SUPER BEETLES
Unit 3, Gosbecks Farm, Gosbecks Road, Colchester, Essex, CO2 9JR Tel/Fax: 01206 563433
L R Super Beetles are specialists in restoration, service and repair of early Volkswagens. New and used spares. Beetles bought and sold. Classic car restoration also undertaken.

LUMENITION
Autocar Electrical Co Ltd, 49-51 Tiverton Street, London, SE1 6NZ
Tel: 0171 403 4334 Fax: 0171 378 1270
Manufacturers of ignition systems and performance related products. Electronic ignition conversions from contact points available for most 70s and 80s vehicles. Backed by over 25 years experience.

Directory

METEX (DARWEN) LTD
Wood Street Mill, Darwen, Lancashire, BB3 1LS
Tel: 01254 704625 Fax: 01254 776927
Metex are the manufacturers of the most popular dust cover in the world. We supply to over 30 countries and with prices from just £21.50 we are also the best value in the world.

MG OWNERS' CLUB
Octagon House, Swavesey, Cambridge, CB4 5QZ
Tel: 01954 231125 Fax: 01954 232106
Buying an MG? To get our booklet and MG Magazine with lists of MGs for sale send £2 to cover postage and packing and don't buy a rotten MG by mistake.

MORRIS MINOR CENTRE (BIRMINGHAM)
2 Camden Street, Parade, Birmingham, B1 3BN
Tel: 0121 236 1341 Fax: 0121 236 1342
Parts supplied and manufactured and car sales for Minors – 1948 to 1971.

MORRIS MINOR CENTRE LTD
Avon House, Lower Bristol Road, Bath, Avon, BA1 1ES
Tel: 01225 315449 Fax: 01225 444642
The specialist for all your needs. Restoration and repairs. Cars for sale. Spare parts. Our famous Catalogue £5 (including postage).

MORRIS MINOR COMPANY
Unit 3, Woodfield Road Industrial Centre, Balby, Doncaster, South Yorkshire, DN4 8EP
Tel: 01302 859331 Fax: 01302 850052
Manufacturers and suppliers of Morris Minor parts, including our famous 'Tourer Conversion Kits'. Our 1996 colour illustrated parts Catalogue is now available. Call us for your free copy now.

MORRIS MINOR MANIA
1-3 Hale Lane, Mill Hill, London, NW7 3NU
Tel: 0181 959 0818 Fax: 0181 959 0819
Established since 1981, specialising in the repair and restoration of all types of Morris Minor, as well as day to day running repairs, welding and MOT testing.

MORRIS MINOR SOUTH WEST
Unit 7, Willow Green Farm, Threemilestone, Truro, Cornwall, TR4 9AL Tel: 01872 70210 Fax: 01872 70210
Specialists in all aspects of Morris Minors. Fast and efficient mail order spares service (s.a.e. for Catalogue). Callers welcome. Workshop facilities. Free, friendly advice just a phone call away.

MORRISPARES SOUTH EAST
Unit 9, Bassett Business Centre, Hurricane Way, North Weald, Epping, Essex, CM16 6AA
Tel: 01992 524249 (out of hours 01799 526278) Fax: 01992 524542
We supply a large quantity of parts and accessories for the Morris Minor, ranging from series MM water pumps to assembled wood for the traveller. Prompt mail order and visitors welcome.

MOTORAPIDE

MOTORAPIDE
Hale Oaks Farm, Loxwood Road, Rudgwick, West Sussex, RH12 2BP Tel/Fax: 01403 753762 Mobile: 0468 112532
With fifteen years experience working on Lamborghini and Ferrari cars, from servicing to a complete restoration. Engine rebuilding specialist from standard to full race specifications. All at competitive rates.

MUNICH LEGENDS LTD
The Ashdown Garage, Lewes Road, Chelwood Gate, East Sussex, RH17 7DE
Tel: 01825 740456 Fax: 01825 740094
The leading classic and performance BMW specialists offering an exciting selection of carefully inspected cars, a comprehensive range of spares (new, used and rare), and engineering skills that are second to none – total commitment to the Munich marque.

THE NATIONAL TRAMWAY MUSUEM
Crich, Matlock, Derbyshire, DE4 5DP
Tel: 01773 852565 Fax: 01773 852326
This unique all-weather 'action stop' offers scenic journeys through a period street to open countryside and panoramic views. Special admission offer for classic vehicle drivers!

NIGEL COOPER COACHWORKS
38 Leysfield Road, London, W12 9JF
Tel: 0181 749 8282 Fax: 0181 743 9454
Twelve skilled craftsmen in our West London workshop undertake all aspects of coachwork restoration to one standard – the best.

NORMAN MOTORS LTD
100 Mill Lane, London, NW6 1NF
Tel: 0171 431 0940 Fax: 0171 794 5034
Jaguar spares specialists. Spares for all models.

ON TWO WHEELS DISTRIBUTION LTD
Unit 3, Castell Close, Swansea Enterprise Park, Llansamlet, Swansea, SA7 9FH
Tel: 01792 700396 Fax: 01792 700396
Crystal-Glo is a 'total' treatment product that works on all surfaces. It is acrylic based, containing no silicone, that leaves a non-abrasive, protective layer. It is easy to apply and use.

BEST CLASSIC

PETER BEST INSURANCE SERVICES LTD
Registered Insurance Brokers
The Farriers, Bull Lane, Maldon, Essex, CM9 4GB
Tel: 01621 840400 Fax: 01621 841401
Leading classic car insurance specialists. Limited mileage (2000, 3000, 4000, 5000, 6000 miles p.a.) and unlimited mileage with agreed value, normally for cars over 10 years old. Underwritten by leading insurers. Available Great Britain.

PRESCOTE MOTOR CARRIAGES
Mill House, Mill Road, Totton, Hampshire, SO4 3ZQ
Tel: 01703 666682 (out of hours 01590 623352) Fax: 01703 666682
Rolls-Royce and Bentley specialists. Restoration, servicing, mechanical, electrical, chassis and coachwork. We take care of your special car.

RIMMER BROS
Triumph House, Sleaford Road, Bracebridge Heath, Lincoln, LN4 2NA
Tel: 01522 568000 Fax: 01522 567600
British Motor Heritage Approved Dealers. One of the world's largest Triumph parts specialists supporting Triumph Stag, TR6, TR7, TR8, Spitfire, GT6, Dolomite (inc Sprint), 2000/2500/2.5Pi, Herald, Vitesse, and Rover SD1. Phone, fax or call for free Catalogue and price guide. Open Mon-Fri 8.30-5.30, Sat 8.30-1.30.

Directory

ROBSPORT INTERNATIONAL
Cokenach Esate, Barkway, Nr Royston, Hertfordshire
SG8 8DL Tel: 01763 848673 Fax: 01763 848167
Triumph TR7/8 specialist. Fast world wide mail order spares, all Triumphs 1952-1982. Free TR7/8 price list. Repair and restoration service economically priced. Always good stock of TRs for sale.

RR & B
Shaw Lane Industrial Estate, Shaw Lane, Stoke Prior, Bromsgrove, Worcestershire, B60 4DT
Tel: 01527 876513 Fax: 01527 876513
Specialists in the servicing and restoration of Rolls-Royce and Bentley motor cars, both pre and post war. All work completed to the highest standard and fully guaranteed. Brochure available upon request.

SNG BARRATT
The Heritage Building, Stourbridge Road, Bridgnorth, Shropshire,
WV15 6AP
Tel: 01746 765432 Fax: 01746 761144
As independent parts professionals specialising in Jaguar and Triumph Stag, we are always striving to provide quality parts for these classic marques, building our business on availability and customer service.

SPECIALISED CAR COVERS
Concours House, Main Street, Burley in Whafedale, West Yorkshire, LS29 7JP
Tel: 01943 864646 Fax: 01943 864365
Specialised car mats manufacture. Individually tailored over mats. Our unique coloured logos embroidered in silk embellish the exclusive pile carpet that is both hard wearing and easy to clean – leatherette edging with non slip rubber backing.

SPECIALISED CAR COVERS
Concours House, Main Street, Burley in Whafedale, West Yorkshire, LS29 7JP
Tel: 01943 864646 Fax: 01943 864365
Specialised car covers manufacture. Individually tailored car covers for all makes of cars for indoor and outdoor storage. High quality fabrics used with a wide choice of colours – indoor covers complete with logos.

SPITZBITZ
Jingles Farm, New Mill Road, Finchampstead, Berkshire, RG40 4QT
Tel/Fax: 01734 732648
We have the largest stock of Spitfire/GT6 used parts in the country, new panels, trim, soft-tops, boot racks, etc, etc. We take Access and Visa and post anywhere in the world. If you would prefer to call in, we are just 10 mins off Junction 11 of the M4 or 10 mins from Junction 4A of the M3.

S P TYRES UK LIMITED
Fort Dunlop, Birmingham, B24 9QT
Tel: 0121 384 4444 Fax: 0121 306 2359
S P Tyres manufacture and supply the widest available range of authentic Dunlop tyres for vintage, veteran and classic cars. Through sole distributor, Vintage Tyre Supplies, Tel: 01590 612261.

STEVE GILL SPORTS CARS
Stedmans Yard, Talywain, Pontypool, Gwent.
Tel: 01495 774963
New and used spares for all Triumph models. Cars for sale from restoration projects to full rebuilds. Full and part restorations, insurance and MOT work carried out. Mechanical overhauls and tuning. Free friendly advice given.

SUNBEAM SUPREME
Memory Lane, Leicester, LE1 3UL
Tel: 01162 536214 Fax: 01162 513231
The largest specialists of new and used parts for Sunbeam Alpines and Tigers. Free brochure and price list.

TECHNILOCK WELDING SERVICES LIMITED
Unit 7, Viking Business Centre, High Street, Woodville, Derbyshire, DE11 7EH
Tel: 01283 222202 Fax: 01283 222203
Cast iron and alloy welding specialists. Veteran and pre-1920 engines a speciality. Top-quality repairs for 20 years. All work guaranteed.

TRGB

TRGB
Unit 1, Sycamore Farm Industrial Estate, Long Drove, Somersham, Huntingdon, Cambridgeshire, PE17 3HJ Tel: 01487 842168 Fax: 01487 740274
Triumph specialists. Massive range of new and secondhand spares for: TR2-6, Spitfire, GT6, Herald and Vitesse. Full workshop and restoration services. Always have a range of TRs for sale. Triumphs purchased dead or alive!

TUBE TORQUE
Unit 10, Brook Street Mill, Brook Street, Macclesfield, Cheshire,
SK11 7AN
Tel/Fax: 01625 511153 & 01625 503866
Exhaust manifolds and systems in mild and stainless steel for race, rally, vintage, veteran, historics, classics, one-offs and conversions. Turbo specialists.

UROGLAS
17 Silver Birches Business Park, Aston Road, Aston Fields Industrial Estate, Bromsgrove, Worcestershire, B60 3EX
Tel: 01527 577477 Fax: 011527 576577
Uroglas are specialist automotive glass distributors. Windscreens, rearscreens and bodyglass are available for all types of vehicle, rare or classic. Complete remanufacturing service available subject to pattern. One-offs and low volume production available. American autoglass a speciality! Overnight carriage within UK, export service available.

UROGLAS
17 Silver Birches Business Park, Aston Road, Aston Fields Industrial Estate, Bromsgrove, Worcestershire, B60 3EX
Tel: 01527 577477 Fax: 011527 576577
Uroglas are specialist automotive glass distributors. Windscreens, rearscreens and bodyglass are available for all types of vehicle, rare or classic. Complete remanufacturing service available subject to pattern. One-offs and low volume production available. American autoglass a speciality! Overnight carriage within UK, export service available.

VERDI FOR FERRARI
9-10 Hayes Metro Centre, Springfield Road, Hayes, Middlesex, UB3 0LE
Tel/Fax: 0181 756 0066
Verdi for Ferrari are specialists in servicing, mechanical and body repairs, rebuilds, renovations, retrimming, spares and sales of Ferraris and other high quality performance cars.

VINTAGE TYRE SUPPLIES LTD
National Motor Museum, Beaulieu, Hampshire, SO42 7ZN
Tel: 01590 612261 Fax: 01590 612722
We stock over 16000 tyres to fit vehicles from the dawn of motoring to the classic 60s and 70s. All can be delivered within 24 hours both nationally and internationally.